MW00576555

Atlas of Endoscopy Imaging in Inflammatory Bowel Disease

Atlas of Endoscopy Imaging in Inflammatory Bowel Disease

Edited by

Bo Shen

The Edelman-Jarislowky Professor of Surgical Sciences, Professor of Medicine and Surgical Sciences (in Surgery), Director of IBD Section and Interventional IBD Center, Vice Chair for Innovation, Departments of Medicine and Surgery, Columbia University Irving Medical Center/New York Presbyterian Hospital, New York, NY, United States

Formerly the Ed and Joey Story Endowed Chair and Professor of Medicine, Cleveland Clinic, Cleveland, OH, United States

ELSEVIER

ACADEMIC PRESS
An imprint of Elsevier

Academic Press is an imprint of Elsevier
125 London Wall, London EC2Y 5AS, United Kingdom
525 B Street, Suite 1650, San Diego, CA 92101, United States
50 Hampshire Street, 5th Floor, Cambridge, MA 02139, United States
The Boulevard, Langford Lane, Kidlington, Oxford OX5 1GB, United Kingdom

Copyright © 2020 Elsevier Inc. All rights reserved.

No part of this publication may be reproduced or transmitted in any form or by any means, electronic or mechanical, including photocopying, recording, or any information storage and retrieval system, without permission in writing from the publisher. Details on how to seek permission, further information about the Publisher's permissions policies and our arrangements with organizations such as the Copyright Clearance Center and the Copyright Licensing Agency, can be found at our website: www. elsevier.com/permissions.

This book and the individual contributions contained in it are protected under copyright by the Publisher (other than as may be noted herein).

Notices
Knowledge and best practice in this field are constantly changing. As new research and experience broaden our understanding, changes in research methods, professional practices, or medical treatment may become necessary.

Practitioners and researchers must always rely on their own experience and knowledge in evaluating and using any information, methods, compounds, or experiments described herein. In using such information or methods they should be mindful of their own safety and the safety of others, including parties for whom they have a professional responsibility.

To the fullest extent of the law, neither the Publisher nor the authors, contributors, or editors, assume any liability for any injury and/or damage to persons or property as a matter of products liability, negligence or otherwise, or from any use or operation of any methods, products, instructions, or ideas contained in the material herein.

British Library Cataloguing-in-Publication Data
A catalogue record for this book is available from the British Library

Library of Congress Cataloging-in-Publication Data
A catalog record for this book is available from the Library of Congress

ISBN: 978-0-12-814811-2

For Information on all Academic Press publications
visit our website at https://www.elsevier.com/books-and-journals

Publisher: Stacy Masucci
Editorial Project Manager: Charlotte Kent
Production Project Manager: Sreejith Viswanathan
Cover Designer: Christian Bilbow

Typeset by MPS Limited, Chennai, India

Contents

List of contributors

Xinbo Ai Department of Gastroenterology, Zhuhai People's Hospital, Jinan University, Zhuhai, P.R. China

Raja Atreya The Ludwig Demling Endoscopy Center of Excellence, Erlangen University Hospital, Erlangen, Germany

Yan Chen Department of Gastroenterology, The Second Affiliated Hospital, School of Medicine, Zhejiang University, Hangzhou, P.R. China

Sara El Ouali Department of Gastroenterology, Hepatology and Nutrition, the Cleveland Clinic, Cleveland, OH, United States

Yu-Bei Gu Department of Gastroenterology, Ruijin Hospital of Shanghai Jiao Tong University, Shanghai, P.R. China

Badar Hasan Department of Gastroenterology and Hepatology, Cleveland Clinic Florida, Weston, FL, United States

Ying Huang Department of Gastroenterology, Children's Hospital of Fudan University, Shanghai, P.R. China

Marietta Iacucci Institute of Translational Medicine, Institute of Immunology and Immunotherapy, NIHR Birmingham Biomedical Research Centre, University Hospitals NHS Foundation Trust, University of Birmingham, Brimingham, United Kingdom

Jesse Kresak Department of Pathology, Immunology, and Laboratory Medicine, University of Florida School of Medicine, Gainesville, FL, United States

Geeta Kulkarni Center for Inflammatory Bowel Diseases, Digestive Disease and Surgery Institute, Cleveland Clinic, Cleveland, OH, United States

Danfeng Lan Department of Gastroenterology, The First Affiliated Hospital of Kunming Medical University, Yunnan Institute of Digestive Disease, Kunming, P.R. China

Ping Lan Department of Colorectal Surgery, The Sixth Affiliated Hospital of Sun Yat-Sen University, Guangzhou, P.R. China

Charles A. Lavender Division of Gastroenterology, University of Arkansas for Medical Sciences, Little Rock, AR, United States

Julia J. Liu Division of Gastroenterology, University of Arkansas for Medical Sciences, Little Rock, AR, United States

Side Liu Department of Gastroenterology, Nanfang Hospital, Southern Medical University, Guangzhou, P.R. China

Xiuli Liu Department of Pathology, Immunology, and Laboratory Medicine, University of Florida School of Medicine, Gainesville, FL, United States

Chanqing Ma Department of Pathology, University of Pittsburgh Medical Center, Pittsburgh, PA, United States

Tian Ma Department of Gastroenterology, Qilu Hospital, Shandong University, Jinan, P.R. China

Jose Melendez-Rosado Department of Gastroenterology, Cleveland Clinic Florida, Weston, FL, United States

Yinglei Miao Department of Gastroenterology, The First Affiliated Hospital of Kunming Medical University, Yunnan Institute of Digestive Disease, Kunming, P.R. China

Markus F. Neurath The Ludwig Demling Endoscopy Center of Excellence, Erlangen University Hospital, Erlangen, Germany

Francesca N. Raffa Inflammatory Bowel Disease Center, Division of Gastroenterology, Hepatology, and Nutrition, Department of Medicine, Vanderbilt University Medical Center, Nashville, TN, United States

Timo Rath The Ludwig Demling Endoscopy Center of Excellence, Erlangen University Hospital, Erlangen, Germany

David A. Schwartz Inflammatory Bowel Disease Center, Division of Gastroenterology, Hepatology, and Nutrition, Department of Medicine, Vanderbilt University Medical Center, Nashville, TN, United States

Peter A. Senada Department of Gastroenterology and Hepatology, Mayo Clinic Florida, Jacksonville, FL, United States

Bo Shen Center for Inflammatory Bowel Diseases, Columbia University Irving Medical Center-New York Presbyterian Hospital, New York, NY, United States

Zifei Tang Department of Gastroenterology, Children's Hospital of Fudan University, Shanghai, P.R. China

Michael B. Wallace Department of Gastroenterology and Hepatology, Mayo Clinic Florida, Jacksonville, FL, United States

Xiaoying Wang Department of Gastroenterology, The Second Affiliated Hospital, School of Medicine, Zhejiang University, Hangzhou, P.R. China

Xin-Ying Wang Department of Gastroenterology, Zhujiang Hospital, Southern Medical University, Guangzhou, P.R. China

Ying-Hong Wang Department of Gastroenterology, University of Texas MD Anderson Cancer Center, Houston, TX, United States

Yuhuan Wang Department of Gastroenterology, Children's Hospital of Fudan University, Shanghai, P.R. China

Xianrui Wu Department of Colorectal Surgery, The Sixth Affiliated Hospital of Sun Yat-Sen University, Guangzhou, P.R. China

Ziqing Ye Department of Gastroenterology, Children's Hospital of Fudan University, Shanghai, P.R. China

Xiuli Zuo Department of Gastroenterology, Qilu Hospital, Shandong University, Jinan, P.R. China

Preface

The spectrum of inflammatory bowel disease (IBD) has expanded beyond Crohn's disease and ulcerative colitis. In a broad sense, the disease entity should also include various forms of IBD-like conditions, after the exclusion of chronic inflammatory gastrointestinal disorders with identifiable etiologies. The list of differential diagnoses of IBD and IBD-like conditions is extensive. Endoscopy with tissue biopsy plays the key role in the diagnosis, differential diagnosis, disease monitoring, assessment of treatment response, and surveillance for dysplasia. Endoscopy provides accurate characterization of a variety of pathologies, ranging from mucosal inflammation, ulceration, strictures to polypoid lesions or masses. The main differential diagnoses of IBD in the industrialized country are nonsteroidal antiinflammatory drug—induced bowel injury, autoimmune-associated enterocolitis, and ischemic colitis. Worldwide, the list can extend to intestinal tuberculosis, chronic infectious enterocolitis, gastrointestinal lymphoma, and Behcet's syndrome. The past two decades have witnessed rapid advances in endoscopy and imaging technologies. While high-definition endoscopy has become a standard diagnostic modality, image-enhanced endoscopy is emerging as a valuable tool to further depict features of the mucosa and even submucosal vasculature. The detailed characterization of the mucosal features has made the grading and classification of inflammation, stricture, postoperative disease recurrence, and colitis-associated neoplasia possible. The histopathological and macroscopic examination also plays an important role in the diagnosis and differential diagnosis. The Editor is a strong advocate for teaching and training of gastrointestinal pathology, radiology, and surgery among endoscopists and IBD clinicians.

After a 20-year tenure at Cleveland Clinic, the Editor relocated his practice to another great institution, Columbia University Irving Medical Center—New York Presbyterian Hospital at the end of 2019. The collaboration among IBD clinicians, general gastroenterologists, pediatric gastroenterologists, colorectal surgeons, gastrointestinal pathologists, gastrointestinal radiologists, and other specialists has created the backbone for the IBD Centers in these two great institutions. The Editor feels so fortunate to have close interactions with wonderful colleagues. And above all, we have provided the best possible care for our patients. Most endoscopy images in this book are from the Editor's collection. Additional images are provided by a panel of national and international experts.

The Editor is grateful to the panel of national and international experts for their contribution to the book. The Editor is also deeply thankful to Mr. and Mrs. Story, Mr. and Mrs. Quint, Mr. and Mrs. Jarislowsky, Mr. and Mrs. Donaghy, Mr. Klise, Mr. Horing, Mr. and Mrs. Stuckey, and Mr. Spero and many other philanthropists for their generous support, and Mr. Joe Pangrace of Cleveland Clinic for his outstanding artwork. The Editor would like to pay special gratitude to Prof. Xian-Yong Meng, MD and Prof. Victor W. Fazio, MD, for mentoring, guidance, and being a role model. The Editor is also thankful for the grant support from National Institutes of Health, American Gastroenterological Association, American Society for Gastrointestinal Endoscopy, American College of Gastroenterology, American Society of Colorectal Surgery, Crohn's and Colitis Foundation, Broad Foundation, Columbia University, and Cleveland Clinic. Finally, the Editor deeply appreciates the help and support from our publisher Ms. Stacy Masucci and managing editors Ms. Charlotte Kent and Mr. Sreejith Viswanathan from Elsevier

I hope that our readers will enjoy the book and provide feedback for our future edition.

Bo Shen

Center for Inflammatory Bowel Disease, Columbia University Irving Medical Center-New York Prebyterian Hospital, New York, NY, United States

Chapter 1

Introduction and classification of inflammatory bowel diseases

Bo Shen

Center for Inflammatory Bowel Diseases, Columbia University Irving Medical Center-New York Presbyterian Hospital, New York, NY, United States

Chapter Outline

Abbreviations

AS	ankylosing spondylitis
CD	Crohn's disease
EIM	extraintestinal manifestations
GI	gastrointestinal
IBD	inflammatory bowel disease
IBD-V	inflammatory bowel disease variant
IC	indeterminate colitis
PG	pyoderma gangrenosum
TNF	tumor necrosis factor
UC	ulcerative colitis

Introduction

Generally speaking, inflammatory bowel disease (IBD) consists of two classic forms: Crohn's disease (CD) and ulcerative colitis (UC). CD and UC present with distinctive clinicopathological features. The diagnosis of IBD is made based on a combined assessment of clinical, endoscopic, imaging, and histologic features. Classic CD is characterized by the presence of transmural, granulomatous, skip lesions at any part of gastrointestinal (GI) tract, whereas classic UC is featured with diffuse mucosal inflammation starting from the rectum.

Spectrum of inflammatory bowel disease and classification

The scope of IBD is beyond the two classic disease entities. In fact, IBD represents a disease spectrum with ranging and overlapping, in etiology, pathogenesis, clinical phenotype, disease course, and prognosis. Even between CD and UC, there is a histopathologically defined disease category of indeterminate colitis (IC) or IBD-unclassified. A spectrum of disease phenotypes of IBD exists according to the location, ranged from ulcerative proctitis to Crohn's ileocolitis (Fig. 1.1).

In a broader sense, microscopic colitis, consisting of lymphocytic colitis and collagenous colitis, has been considered as variants of IBD. In addition, patients with IBD may have concurrent extraintestinal manifestations (EIM),

Atlas of Endoscopy Imaging in Inflammatory Bowel Disease. DOI: https://doi.org/10.1016/B978-0-12-814811-2.00001-3
© 2020 Elsevier Inc. All rights reserved.

involving joints [such as ankylosing spondylitis (AS)], liver (such as primary sclerosing cholangitis [PSC]), skin [such as pyoderma gangrenosum (PG)], eyes (such as uveitis), and other organs. On the other hand, the immune-mediated disorders, other than IBD, can also affect the gut. Autoimmune (such as autoimmune hepatitis, Hashimoto thyroiditis) and autoinflammatory diseases (such as Blau syndrome) can occur concomitantly with IBD. To confuse the picture even further, chronic infectious diseases of the gut, such as intestinal tuberculosis (ITB) and chronic *Salmonella* enteritis can mimic CD (Fig. 1.2). There is a spectrum of disease phenotypes based on the depth of the involved gut (Fig. 1.3).

Disease initiation and progression of IBD are complex. The current theory holds that IBD results from abnormal interactions of environmental and microbiological factors and immune systems in genetically susceptible hosts. Microbiota, host immune system, and genetic factors play key roles in the pathogenesis of IBD. However, the degree of contribution and interactions of these factors results in a wide spectrum of disease phenotypes in patients at different age of onset and different stage of the disease and different ethnic groups. For example, some adult-onset CD may be considered as a polygenetic disorder [1], while infant-onset CD may be a monogenic disorder [2,3]. In addition, the presence of IBD simply reflects an epiphenomenon of the involvement of one organ by immune-mediated systematic diseases.

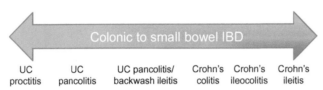

FIGURE 1.1 Range of "classic" IBD from the rectum to the distal ileum. *IBD*, Inflammatory bowel disease. *Modified from Chang S, Shen B. Classification and reclassification of inflammatory bowel diseases: from clinical perspective. In: Shen B, editor. Interventional inflammatory bowel: endoscopic management of complications. Cambridge, MA: Elsevier; 2018. p. 17–34.*

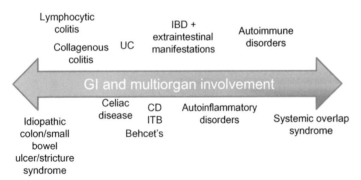

FIGURE 1.2 Range of overlap syndrome from isolated gut disorder to multiorgan involvement of immune-mediated disorders. *CD*, Crohn's disease; *IBD*, inflammatory bowel disease; *ITB*, intestinal tuberculosis; *UC*, ulcerative colitis. *Modified from Chang S, Shen B. Classification and reclassification of inflammatory bowel diseases: from clinical perspective. In: Shen B, editor. Interventional inflammatory bowel: endoscopic management of complications. Cambridge, MA: Elsevier; 2018. p. 17–34.*

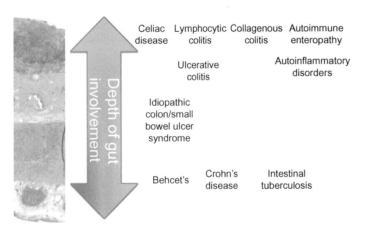

FIGURE 1.3 Depth of inflammation in various forms of immune-mediated chronic inflammatory bowel disease. *Modified from Chang S, Shen B. Classification and reclassification of inflammatory bowel diseases: from clinical perspective. In: Shen B, editor. Interventional inflammatory bowel: endoscopic management of complications. Cambridge, MA: Elsevier; 2018. p. 17–34.*

There are several classification systems for IBD, including molecular [4] and genetic [5]. However, the most commonly used instrument in clinical practice and clinical trials has been the Montreal classification (Tables 1.1 and 1.2) [6], with modifications, such as the one with further elaboration of age at onset [7]. This author believes that the current classification

TABLE 1.1 The Montreal classification of ulcerative colitis (UC) [6].

	Montreal classification	Definition
Disease extent		
Ulcerative proctitis	E1	Only involves rectum
Left sided UC	E2	Extending to the splenic flexure
Extensive UC	E3	Extending proximal to the splenic flexure
Disease severity		
Clinical remission	S0	Asymptomatic
Mild	S1	≤ 4 Stools per day (with or without blood), absence of systemic disease, normal inflammatory markers (ESR)
Moderate	S2	>4 Stools per day, but with minimal signs of systemic toxicity
Severe	S3	≥ 6 Bloody stools daily, pulse rate ≥ 90 beats/min, temperature > 37.5°C, hemoglobin < 10.5 g/dL, and ESR ≥ 30 mm/h

ESR, Erythrocyte sedimentation rate.

TABLE 1.2 The Montreal classification of Crohn's disease [6].

	Montreal classification
Age at diagnosis	
<16 years	A1
17–40 years	A2
>40 years	A3
Disease location	
Ileal disease	L1
Colonic disease	L2
Ileocolonic disease	L3
Upper isolated gastrointestinal disease	L4
Disease behavior	
Nonstricturing and nonpenetrating	B1
Stricturing	B2
Penetrating	B3
Perianal disease	P

systems do not reflect the complexity of the scope of IBD and IBD represents various etiopathogenesis pathways and phenotypes in a wide disease spectrum.

IBD is often associated with EIM. In addition, IBD can coexist with autoimmune or autoinflammatory disorders. From historical perspective, it is not clear why certain disease entities, such as rheumatoid arthritis and psoriasis in patients with IBD, have been considered as concurrent autoimmune disorders, while AS and PG are considered as EIM of IBD. The spectrum of immune-mediated disorders of the gut is also beyond CD, UC, and microscopic colitis.

Immune-mediated GI disorders represent a wide spectrum of phenotypes with histopathological characteristics, ranging from classic mucosal disease (such as lymphocytic colitis, collagenous colitis, and celiac disease), "pseudo" transmural disorders (such as IC) to true transmural disease (such as CD). The disease process of transmural disease can be the extrinsic or "outside-in," from the mucosa to deeper layers of the bowel wall, in which dysbiosis can be a triggering factor. In contrast the "inside-out" theory or the intrinsic pathway of IBD implies that the disease process starts from the mesentery or deep bowel wall to the mucosa, which may play an important role in the development of intrinsic IBD. In clinical practice, we have encountered patients with extensive enteroenteric fistulae wrapped with a thick mesentery but with minimum mucosal inflammation. In fact the diseased mesentery has been considered as a key factor for the disease process and progression of CD before and after surgery [8,9].

IBD has been traditionally considered as disease entities with unclear etiology or being idiopathic. However, there are secondary forms of IBD, with clearly identified etiological or triggering factors, such as infection, medications, and bowel surgery. For example, immunosuppressive medications used to treat autoimmune disorders can paradoxically trigger autoimmune conditions. Anti−tumor necrosis factor (TNF) α inhibitors are agents for the treatment of IBD and its EIMs. Interestingly, anti-TNF agents can cause drug-induced lupus [10], psoriasis [11], and multiple sclerosis [12]. The classic examples of drug-induced IBD or IBD-like conditions are mycophenolate mofetil [13], tacrolimus [14], Secukinumab (anti-IL-17A for treatment of psoriasis) [15], and checkpoint inhibitors (agents for the treatment of malignancies) [16].

Stem cell transplantations have been shown to be beneficial in IBD [17,18], presumably by resetting the immune thermostat. On the other hand, stem cell or organ transplantations can result in "IBD-like" conditions, which are exemplified by cord colitis syndrome [19]. De novo IBD occasionally occurs after solid organ transplantation [20−22].

De novo IBD can also be triggered by other abdominal surgeries, such as post−colectomy enteritis syndrome characterized by diffuse chronic active enteritis is triggered by colectomy for UC [23,24], CD-like bariatric surgery (Roux-en-Y gastric bypass) [25,26]. De novo CD may develop after restorative proctocolectomy with IPAA for the initial diagnosis of UC. It is estimated that 2.7%−13% after colectomy for UC or IC may develop de novo CD anytime from weeks to years after ileal pouch-anal anastomosis [27].

We have proposed a new classification to cover the wide spectrum of broad-sense IBD [28]. The disease spectrum is ideally classified based on known etiology, characteristic pathological features, or clinical phenotypes. However, it has been difficult to classify immune-mediated disorders, due to the complexity of disease process and interplay of genetic, environmental, microbiological, immunological, and vascular factors, at different age and different stages of diseases. In addition to the traditional classifications based on age at onset, disease location, extent, and phenotype, we proposed to classify a spectrum of IBD, based on disease location and the degree of shared etiopathogenetic pathways, causative factors, and disease processes. We speculate that genetic, environmental, immunological, and vascular factors play varied roles in the pathogenesis of infant- or very early−onset IBD versus early-onset versus elderly-onset IBD (Fig. 1.4). We propose to categorize classic IBD (UC and CD) and IBD variants (IBD-V), based on the degree of extraintestinal organ involvement and pathogenetic pathways. We further divide IBD-V into IBD$^+$ (IBD with classic EIM, e.g., IBD with PG); IBD^{++} (IBD with concurrent autoimmune or autoinflammatory disorders, with or without classic EIM, e.g., classic IBD with AS and psoriasis); and IBD$^{\pm}$, which shares clinical and histopathological features, and possible pathways with IBD, IBD$^+$, or IBD^{++}. The examples are microscopic colitis and autoimmune enteropathy (Fig. 1.5; Table 1.3).

In order to fit the diagnosis of IBD and IBD-V, at least two of the following histopathological criteria should be met: (1) infiltration of mononuclear cells in the lamina propria or epithelium; (2) crypt distortion; (3) basal lymphoplasmacytosis; (4) Paneth cell and/or pyloric gland metaplasia; and (5) mucin depletion or increase crypt apoptosis. Additional features, such as noncaseating granuloma, neuronal hyperplasia, and transmural infiltration of lymphocytes or microabscess, further support the diagnosis of IBD and IBD-V. Some degree of tissue eosinophilia, lymphangitis, and vasculopathy may be present in IBD and IBD-V.

We also propose the concept of secondary IBD, which differs from classic primary or idiopathic IBD. In secondary IBD, a triggering factor is readily identified (such as medications, surgery, or stem cell or organ transplantation).

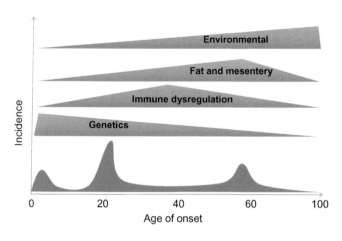

FIGURE 1.4 The three-peak incidences of IBD and relative contribution of etiopathogenic factors. *IBD*, Inflammatory bowel disease. *Modified from Chang S, Shen B. Classification and reclassification of inflammatory bowel diseases: from clinical perspective. In: Shen B, editor. Interventional inflammatory bowel: endoscopic management of complications. Cambridge, MA: Elsevier; 2018. p. 17—34.*

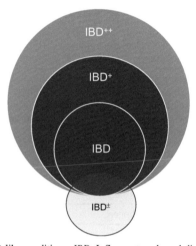

FIGURE 1.5 Range and overlaps of IBD and IBD-like conditions. *IBD*, Inflammatory bowel disease. *Modified Chang S, Shen B. Classification and reclassification of inflammatory bowel diseases: from clinical perspective. In: Shen B, editor. Interventional inflammatory bowel: endoscopic management of complications. Cambridge, MA: Elsevier; 2018. p. 17—34.*

TABLE 1.3 Proposed classification of inflammatory bowel diseases [28].

Criteria	Class	Description	Examples
Disease location, extent, and depth ± granulomas	Ulcerative colitis		Classic UC
	Crohn's disease		Classic CD
	Indeterminate colitis		
Age of onset	Very early onset	Age 0	IL-10/IL-10 R mutations
	Early onset	Age 0—10 years	
		Age 10—17 years	
	Regular onset	Age 17—40 years	
	Late onset	Age > 50 years	
Phenotype	Inflammatory		Inflammatory CD; classic UC
	Stricturing		Stricturing CD; UC with strictures
	Penetrating		Fistulizing CD

(Continued)

TABLE 1.3 (Continued)

Criteria	Class		Description	Examples
Locations	Oral			
	Upper GI			
	Jejunum			
	Ileum			
	Colon			
	Rectum			
	Perianal			
	Extraintestinal			Metastatic CD of the skin, lung, liver
Concurrent or immune-mediated disorders	IBD			Isolated UC or CD of the gut
	IBD-V	IBD[+]	IBD + classic extraintestinal manifestations	UC with concurrent PSC
		IBD[++]	IBD + autoimmune and/or autoinflammatory disorders ± classic extraintestinal manifestations	IBD with concurrent microscopic colitis, celiac disease, hidradenitis suppurativa
		IBD[±]	Diseases sharing clinical features and possible etiopathogenetic pathways with classic IBD ± classic extraintestinal manifestations of IBD, autoimmune disorders or autoinflammatory disorders	Lymphocytic colitis, collagenous colitis; Behcet disease, cryptogenic multifocal ulcerous stenosing enteritis, ulcerative jejunitis
Etiology of IBD	Primary or idiopathic		Monogenic	IL-10, IL-10Ra, IL-10Rb mutations
				Very early onset IBD
			Polygenic	Classic UC; classic CD
	Secondary		Identifiable pathogens	*Mycobacterial paratuberculosis*
			Medication-induced	Mycophenolate-associated colitis; Ipilimumab-associated colitis
			Organ transplantation-induced	Post–solid organ transplant IBD-like conditions, cord colitis syndrome;
			Surgery-induced	Pouchitis, Crohn's disease-like conditions of the pouch, post–colectomy enteritis, bariatric surgery-associated IBD
Genetic etiology	Monogenic			IL-10/IL-10R mutations, Familial Mediterranean fever
	Polygenic			Classic CD and classic UC
Disease spread process	Intrinsic ("inside-out")		Starting from the lymphatic system or mesentery, spreading to gut mucosa	Subset of obese CD patients; subset of sclerosing mesenteritis or lymphangitis
	Extrinsic ("outside-in")		External trigger (e.g., bacteria) leading to mucosal inflammation	Fulminant UC: from mucosal disease to transmural inflammation

CD, Crohn's disease; *GI*, gastrointestinal; *IBD*, inflammatory bowel disease; *IBD-V*, IBD-variant; *PSC*, primary sclerosing cholangitis; *UC*, ulcerative colitis.

Finally, IBD is classified into extrinsic versus intrinsic IBD, corresponding to the "outside-in" versus "inside-out" theory of disease mechanisms.

We recommend that our clinician readers think outside the box in clinical practice and realize the complexity of IBD. IBD represents a wide disease spectrum. Solid knowledge in etiology, pathogenesis, clinical presentations, and morphological feature in IBD is critical for clinicians to make correct diagnosis and differential diagnosis and delivery of proper therapy. Among them, accurate recognition of endoscopic features of various forms of IBD and IBD-like conditions is imperative. This is the main purpose of this book.

Diagnosis and differential diagnosis

Differential diagnosis of IBD should include infectious GI diseases, immune-mediated GI diseases, vascular disease—associated, medicine-associated GI disorders, organ transplant—associated IBD-like conditions, and GI malignancy. The endoscopic features of those IBD-like conditions or disorders are described in the following chapters of the book. In addition, histologic features of classic IBD and IBD-like conditions are detailed in the chapters of the book.

Summary and recommendations

In a broad sense, IBD represents a wide disease spectrum, with ranging etiopathogenic pathways, clinical phenotypes, and disease course. A comprehensive classification of IBD has been proposed. On the other hand, various infectious, autoimmune, autoinflammatory, drug-induced, and malignant disorders affect the GI tract, with similar clinical and radiographic presentations. Endoscopic and histologic evaluation plays a key role in the diagnosis and differential diagnosis.

Acknowledgment

Dr. Bo Shen is supported by the Ed and Joey Story Endowed Chair.

Disclosures

The author declares no financial conflicts of interest.

References

[1] Jostins L, Ripke S, Weersma RK, Duerr RH, McGovern DP, Hui KY, et al. Host-microbe interactions have shaped the genetic architecture of inflammatory bowel disease. Nature 2012;491:119—24.

[2] Shim JO, Seo JK. Very early-onset inflammatory bowel disease (IBD) in infancy is a different disease entity from adult-onset IBD; one form of interleukin-10 receptor mutations. J Hum Genet 2014;59:337—41.

[3] Pigneur B, Escher J, Elawad M, Lima R, Buderus S, Kierkus J, et al. Phenotypic characterization of very early-onset IBD due to mutations in the IL10, IL10 receptor alpha or beta gene: a survey of the Genius Working Group. Inflamm Bowel Dis 2013;19:2820—8.

[4] Vermeire S. Towards a novel molecular classification of IBD. Dig Dis 2012;30:425—7.

[5] Cleynen I, Boucher G, Jostins L, Schumm LP, Zeissig S, Ahmad T, et al. Inherited determinants of Crohn's disease and ulcerative colitis phenotypes: a genetic association study. Lancet 2016;387:156—67.

[6] Silverberg MS, Satsangi J, Ahmad T, Arnott ID, Bernstein CN, Brant SR, et al. Toward an integrated clinical, molecular and serological classification of inflammatory bowel disease: report of a Working Party of the 2005 Montreal World Congress of Gastroenterology. Can J Gastroenterol 2005;19(Suppl. A):5A—36A.

[7] Levine A, Griffiths A, Markowitz J, Wilson DC, Turner D, Russell RK, et al. Pediatric modification of the Montreal classification for inflammatory bowel disease: the Paris classification. Inflamm Bowel Dis 2011;17:1314—21.

[8] Coffey J, O'Leary P. The mesentery: structure, function, and role in disease. Lancet Gastroenterol Hepatol 2016;1:238—47.

[9] Li Y, Zhu W, Zuo L, Shen B. The role of the mesentery in Crohn's disease: the contributions of nerves, vessels, lymphatics, and fat to the pathogenesis and disease course. Inflamm Bowel Dis 2016;22:1483—95.

[10] Marques M, Magro F, Cardoso H, Carneiro F, Portugal R, Lopes J, et al. Infliximab-induced lupus-like syndrome associated with autoimmune hepatitis. Inflamm Bowel Dis 2008;14:723—5.

[11] Cleynen I, Vermeire S. Paradoxical inflammation induced by anti-TNF agents in patients with IBD. Nat Rev Gastroenterol Hepatol 2012;9:496—503.

[12] Robinson WH, Genovese MC, Moreland LW. Demyelinating and neurologic events reported in association with tumor necrosis factor alpha antagonism: by what mechanisms could tumor necrosis factor alpha antagonists improve rheumatoid arthritis but exacerbate multiple sclerosis? Arthritis Rheum 2001;44:1977—83.

[13] Star KV, Ho VT, Wang HH, Odze RD. Histologic features in colon biopsies can discriminate mycophenolate from GVHD-induced colitis. Am J Surg Pathol 2013;37:1319—28.

[14] Kurnatowska I, Banasiak M, Daniel P, Wagrowska-Danilewicz M, Nowicki M. Two cases of severe de novo colitis in kidney transplant recipients after conversion to prolonged-release tacrolimus. Transpl Int 2010;23:553—8.

[15] Hueber W, Sands BE, Lewitzky S, Vandemeulebroecke M, Reinisch W, Higgins PD, et al. Secukinumab, a human anti-IL-17A monoclonal antibody, for moderate to severe Crohn's disease: unexpected results of a randomised, double-blind placebo-controlled trial. Gut 2012;61:1693—700.

[16] Khan F, Funchain P, Bennett A, Hull TL, Shen B. How should we diagnose and manage checkpoint inhibitor-associated colitis? Cleve Clin J Med 2018;85:679—83.

[17] Jauregui-Amezaga A, Rovira M, Marin P, Salas A, Pino-Donnay S, Feu F, et al. Improving safety of autologous haematopoietic stem cell transplantation in patients with Crohn's disease. Gut 2016;65:1456—62.

[18] Panes J, Garcia-Olmo D, Van Assche G, Colombel JF, Reinisch W, Baumgart DC, et al. Expanded allogeneic adipose-derived mesenchymal stem cells (Cx601) for complex perianal fistulas in Crohn's disease: a phase 3 randomised, double-blind controlled trial. Lancet 2016;388:1281—90.

[19] Herrera AF, Soriano G, Bellizzi AM, Hornick JL, Ho VT, Ballen KK, et al. Cord colitis syndrome in cord-blood stem-cell transplantation. N Engl J Med 2011;365:815—24.

[20] Nepal S, Navaneethan U, Bennett AE, Shen B. De novo inflammatory bowel disease and its mimics after organ transplantation. Inflamm Bowel Dis 2013;19:1518—22.

[21] Hampton DD, Poleski MH, Onken JE. Inflammatory bowel disease following solid organ transplantation. Clin Immunol 2008;128:287—93.

[22] Kochhar G, Singh T, Dust H, Lopez R, McCullough AJ, Liu X, et al. Impact of de novo and preexisting inflammatory bowel disease on the outcome of orthotopic liver transplantation. Inflamm Bowel Dis 2016;22:1670—8.

[23] Rush B, Berger L, Rosenfeld G, Bressler B. Tacrolimus therapy for ulcerative colitis-associated post-colectomy enteritis. ACG Case Rep J 2014;2:33—5.

[24] Rubenstein J, Sherif A, Appelman H, Chey WD. Ulcerative colitis associated enteritis: is ulcerative colitis always confined to the colon? J Clin Gastroenterol 2004;38:46—51.

[25] Ahn LB, Huang CS, Forse RA, Hess DT, Andrews C, Farraye FA. Crohn's disease after gastric bypass surgery for morbid obesity: is there an association? Inflamm Bowel Dis 2005;11:622—4.

[26] Braga Neto MB, Gregory M, Ramos GP, Loftus Jr EV, Ciorba MA, Bruining DH, et al. De-novo inflammatory bowel disease after bariatric surgery: a large case series. J Crohns Colitis 2018;12:452—7.

[27] Shen B. Crohn's disease of the ileal pouch: reality, diagnosis, and management. Inflamm Bowel Dis 2009;15:284—94.

[28] Chang S, Shen B. Classification and reclassification of inflammatory bowel diseases: from clinical perspective. In: Shen B, editor. Interventional inflammatory bowel: endoscopic management of complications. Cambridge, MA: Elsevier; 2018. p. 17—34.

Chapter 2

Setup and principle of endoscopy in inflammatory bowel disease

Bo Shen

Center for Inflammatory Bowel Diseases, Columbia University Irving Medical Center-New York Presbyterian Hospital, New York, NY, United States

Chapter Outline

Abbreviations

CAN	colitis-associated neoplasia
CD	Crohn's disease
EGD	esophagogastroduodenoscopy
EUS	endoscopic ultrasound
IBD	inflammatory bowel disease
GI	gastrointestinal
ICV	ileocecal valve
PSC	primary sclerosing cholangitis
UC	ulcerative colitis
VCE	video capsule endoscopy

Introduction

Endoscopy with histology plays a major role in the diagnosis, disease monitoring, assessment of treatment response, and neoplasia surveillance in inflammatory bowel disease (IBD). Fortunately, any part of diseased gastrointestinal (GI) tract is accessible to various modalities of endoscopy. In addition, various therapeutic modalities can be delivered through the endoscopy. Colonoscopy with ileoscopy and esophagogastroduodenoscopy (EGD) are the two most commonly used diagnostic tools in IBD, along with capsule endoscopy, enteroscopy, and pouchoscopy. Endoscopy is accurate in grading the degree of mucosal inflammation, which has been considered the gold standard for the quantification of the severity of mucosal inflammation in IBD. Certain endoscopic features, such as the distribution of inflammation and characteristics of ulcer, may help differential diagnosis between different phenotypes of IBD and between IBD and non-IBD conditions.

Setup for inflammatory bowel disease endoscopy

Patients with IBD often have active disease, concurrent use of corticosteroids and other immunosuppressive medications, or altered bowel anatomy due to surgery. Endoscopy plays both diagnostic and therapeutic roles. It is recommended that preprocedure abdominal imaging is obtained to delineate intra- and extraluminal anatomy. Prior to

Atlas of Endoscopy Imaging in Inflammatory Bowel Disease. DOI: https://doi.org/10.1016/B978-0-12-814811-2.00002-5
© 2020 Elsevier Inc. All rights reserved.

endoscopy, we should avoid medicines that may cause bowel inflammation, including sodium phosphate—based bowel preparation and nonsteroidal antiinflammatory drugs. If patients are undergoing endoscopy and simultaneously taking those medications, the information should be documented in the endoscopy report and submission script to pathology. Insufflation of carbon dioxide, rather than room air, is also strongly recommended in endoscopy for patients with IBD. Although there is no published consensus on antibiotic prophylaxis in IBD endoscopy, this author recommends antibiotic use during therapeutic endoscopies, such as balloon dilation and endoscopic stricturotomy, or endoscopy in patients with a diverted colon or diverted pouch.

Principles

Colonoscopy with ileoscopy or ileocolonoscopy is considered a standard practice in the evaluation of IBD, as a majority of patients with Crohn's disease (CD) have disease in the distal or terminal ileum, that is, L1 or L3 in the Montreal classification [1], and all patients with ulcerative colitis (UC) have diseased bowel, including backwash ileitis, within the reach. For the diagnosis and differential diagnosis of IBD, the value of ileocolonoscopy is beyond quantification of severity and distribution of the bowel inflammation. Description of mucosal inflammation should include a spectrum of features ranging from edema, erythema, exudates, and erosions to friability, ulcers, and spontaneous bleeding. In addition, location and length of bowel involved and number, size, shape, and depth of ulcers should also be documented. Furthermore, abnormalities resulting from chronic inflammatory injury should be evaluated and documented, including stricture, fistula, polyps, mucosal scars or bridges, and stiffness or distensibility of bowel (Fig. 2.1). Finally, surgically altered bowel in IBD with or without active disease, such as ileostomy, ileal pouch, stricturoplasty, and ileocolonic anastomosis, should be reported in endoscopy.

UC is characterized by the presence of diffuse inflammation of the rectum and above, while CD (L2 or L3) is featured with a segmental disease with sparing of the rectum (Fig. 2.2). Mucosal patterns of the terminal ileum and proximal colon along with features of the ileocecal valve (ICV) and histology may help distinguishing CD from UC with backwash ileitis. Endoscopic characteristics favoring the diagnosis of CD ileitis include discrete ulcers and strictures of the terminal ileum or ICV. In contrast, diffuse colitis and ileitis with the continuous pattern across a widely patent ICV are the features of UC with backwash ileitis, which is common in patients with concurrent primary sclerosing cholangitis (PSC) (Fig. 2.3).

Conventional white-light colonoscopy and image-enhanced endoscopy, such as chromoendoscopy [2], are the standard care for surveillance of colitis-associated neoplasia (CAN). Adequate control of inflammation and bowel preparation are prerequisites for accurate detection of CAN.

Techniques of tissue biopsy

Endoscopic features of IBD and non-IBD colitides, such as infectious colitis, drug-induced colitis, ischemic colitis, and radiation colitis, overlap. Endoscopy alone provides only a limited value in distinction among those colitides. In contrast, tissue biopsy is critical for the diagnosis and differential diagnosis of CD and UC and other colitides. In addition, the importance of the index colonoscopy is never overemphasized in the evaluation of IBD. For example, one distinctive feature for the differential diagnosis between UC and Crohn's colitis is the distribution of disease (continuous pattern vs segmental disease with rectal sparing) (Fig. 2.2) before the initiation of medical therapy. At ileocolonoscopy for the evaluation of IBD or IBD-like conditions, at least four topographic locations should be biopsied, including the terminal ileum, right colon, left colon, and rectum and separately labeled (Fig. 2.4).

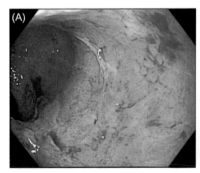 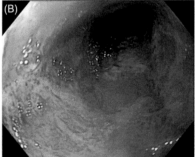

FIGURE 2.1 Description in IBD endoscopy is beyond the quantification of mucosal inflammation. It should also include other features. (A) Mucosal inflammation along with "lead piping" (or loss of haustra) of the lumen of the colon, from long-term refractory ulcerative colitis and (B) the same patient also had a stiff rectum with reduced distensibility. Those features are associated with poor outcomes of the disease. *IBD*, Inflammatory bowel disease.

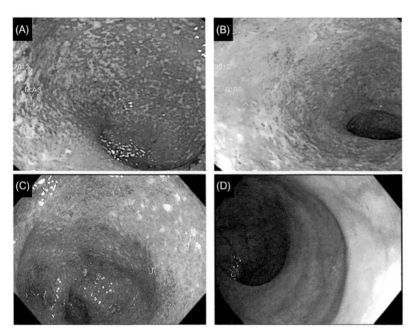

FIGURE 2.2 Distribution of mucosal inflammation in ulcerative colitis and Crohn's disease. Diffuse colitis with erythema, exudates, and friability in the colon (A) and rectum (B) in ulcerative colitis. Diffuse colitis with erythema, erosions, and exudates in the colon (C) with sparing of the rectum (D) in Crohn's colitis.

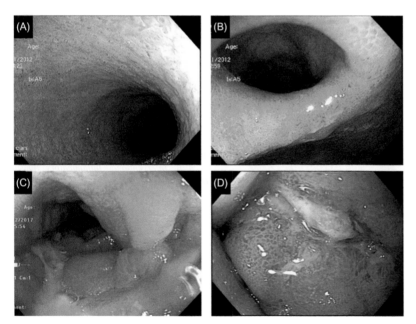

FIGURE 2.3 Distribution of ileitis in ulcerative colitis and Crohn's disease. Diffuse colitis with granular, flat mucosa, and loss of vascularity in the terminal ileum (i.e., backwash ileitis) (A) and widely patent or "fish mouth"–shaped configuration of the ileocecal valve (B) in ulcerative colitis. Linear ulcers and mucosal edema of the terminal ileum (C) and ulcerated, strictured, and deformed ileocecal valve (D) in Crohn's disease.

The endoscopist should exert extreme caution when patients have deep ulcers in the bowel or severe bowel inflammation resulting from IBD or IBD conditions, as the risk of bleeding and perforation is high. The endoscopist should minimize air insufflation and avoid the biopsy of the ulcer base.

The endoscopist should be familiar with the orientation of bowel anatomy, which is particularly important for tissue biopsy and endoscopic therapy during distal bowel endoscopy (Fig. 2.5). Location of biopsy at endoscopy is determined by various factors, including the purpose and pretest probabilities. For example, the accurate location of biopsy for rectal prolapse is the anterior wall of the distal rectum (Fig. 2.6), and biopsy for the evaluation of IBD should be in the lateral or posterior wall of the rectum. The cellularity and its pattern of colon epithelia and lamina propria in the healthy or diseased are different. For example, the number of lymphocytes in the epithelia and mononuclear cell infiltration in the right colon are greater than that in the left colon (Fig. 2.7). Paneth cell metaplasia has been considered a marker for

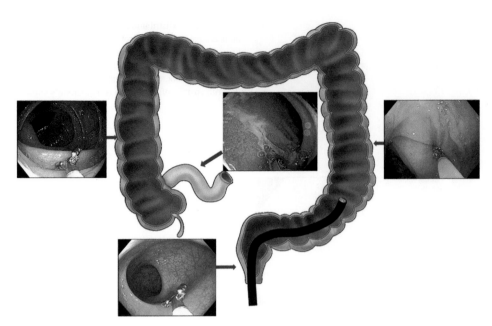

FIGURE 2.4 The topographic location of tissue biopsy in colonoscopy. For the diagnosis and differential diagnosis of inflammatory bowel disease, at least four segments of the lower GI tract during colonoscopy, particularly index colonoscopy, that is, terminal ileum, right colon, left colon, and rectum (*blue arrows*). *GI,* Gastrointestinal.

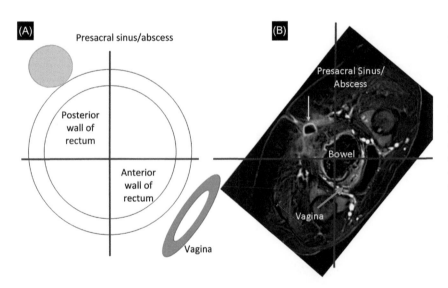

FIGURE 2.5 The orientation of distal colon/rectum on colonoscopy when the patient is on a left lateral decubitus position. (A) Artist's sketch of the location of the bowel, presacral abscess or sinus, and vagina. The anterior wall of the rectum is at right lower quadrant, and posterior wall is at the left upper quadrant. (B) MRI of pelvis showed the location of the vagina at 4−5 o'clock (*green arrow*) and presacral sinus at 10−11 o'clock (*yellow arrow*). *Reproduced with permission from Shen B. Interventional Inflammatory Bowel Disease. Elsevier 2018.*

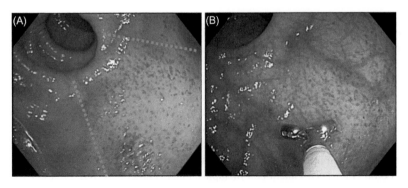

FIGURE 2.6 Biopsy of the rectum. (A) The anterior wall of the rectum is at 4−5 o'clock when the patient is on left lateral decubitus position, highlighted by green dot lines. Isolated inflammation in the area is more common in rectal prolapse, rectocele, and topical radiation injury for prostate cancer, and rare in Crohn's disease or ulcerative colitis. (B) Biopsy in the area was not for ruling out inflammatory bowel disease but for the evaluation of prolapse (this patient had a symptom of dyschezia and was labeled having ulcerative colitis).

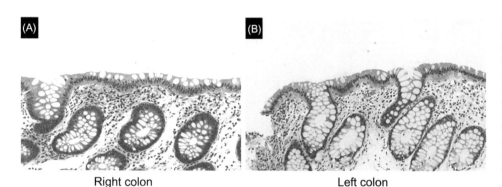

Right colon   Left colon

FIGURE 2.7 Different cellularity in the right and left colon in healthy individuals. Right colon biopsy (A) reveals slightly more mononuclear inflammatory cells in the lamina propria and a few more surface intraepithelial lymphocytes than the left colon (B). Therefore biopsy specimens from the right and left colon should be placed in separate jars. *Courtesy Xiuli Liu, MD, PhD.*

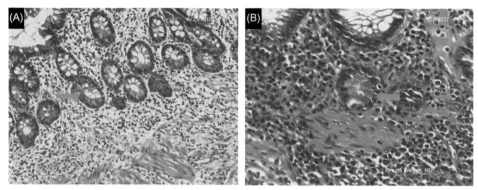

FIGURE 2.8 Paneth cell metaplasia of colon in inflammation bowel disease. Paneth cells are epithelial cells of the small intestine, which are characterized by the intracellular eosinophilic granules. The presence of Paneth cell metaplasia in the colon (A and B) is indicative of chronic mucosal injury from inflammatory bowel disease and other chronic colitides. However, few Paneth cells may be present in the right colon in healthy individuals. Their presence in the left colon or rectum is always abnormal.

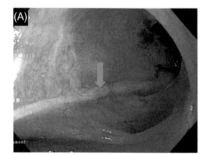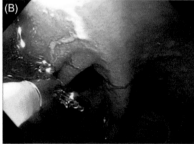

FIGURE 2.9 Tissue biopsy in patients with surgery. (A) Staple line (*green arrow*) in an ileal pouch and (B) avoidance of biopsy at the staple line.

chronic mucosal injury, which is occasionally present in UC or Crohn's colitis. However, Paneth cell metaplasia in the right colon can be "physiological," and its presence in the left colon or rectum is always abnormal (Fig. 2.8). Therefore the biopsy specimens from the right and left colon should be separately labeled and submitted.

A higher yield can be obtained by biopsy from the ulcer base, rather than from the edge of the ulcer, for the identification of cytomegalovirus colitis. However, the risk of bleeding would be higher. Biopsy of the suture line or staple line in patients with bowel surgery should be avoided to minimize the confusion of foreign body–associated granuloma versus true disease–associated (e.g., CD or tuberculosis) granulomas (Fig. 2.9). A higher detection rate of granulomas may be achieved on biopsy from the edge of the ulcer [3]. Disease extent on endoscopy and histology in IBD does not entirely overlap. In fact, colonoscopy may underestimate the extent of disease as compared to histology. Therefore an accurate estimation of the extent of colitis should be based on histology rather than colonoscopy. In addition, endoscopically normal-appearing mucosa in patients with chronic diarrhea should also be biopsied to rule out microscopic colitis.

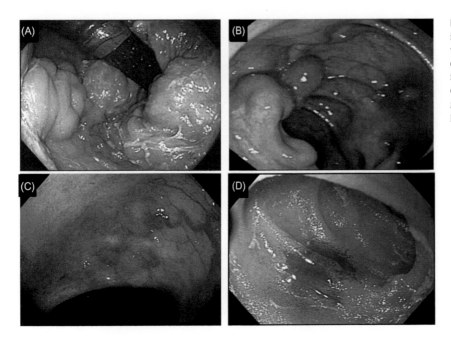

FIGURE 2.10 Bowel lesion prone to bleeding during endoscopy and biopsy. (A) Rectal varices; (B) cecal varices; (C) rectal cuff varices in the ileal pouch, from underlying primary sclerosing cholangitis and ulcerative colitis; (D) Bleeding of the ileum even from a gentle biopsy in a patient small bowel transplant for Crohn's disease.

FIGURE 2.11 Biopsy forceps used in endoscopy in inflammatory bowel disease. Superficial biopsy with nonspiked forceps with shallow caps (blue, right) is used in those diseases or lesions prone to bleeding, such as portal hypertension and small bowel transplantation. Deep biopsy with spiked forceps with longer cups (orange, left) is used for the evaluation of strictures and cancer surveillance.

Systemic evaluation prior to endoscopy is important for the achievement of accurate diagnosis and minimization of procedure-associated adverse events. For example, IBD patients with concurrent PSC or portal hypertension or thrombocytopenia carry a higher risk for endoscopy (especially tissue biopsy)−associated bleeding (Fig. 2.10). Patients with CD who underwent small bowel transplantation are prone to biopsy-associated bleeding (Fig. 2.10). Therefore superficial biopsy with small forceps should be taken from nonvessel-rich area. In contrast, deep biopsy, even tunnel biopsy, may be taken in IBD patients who are at risk for the development of laterally spreading neoplasia, such as area with CAN undergoing endoscopic ablation (Fig. 2.11).

Special endoscopy

At upper and lower endoscopies the endoscopist should also pay attention to the anal and perianal area and oral cavities, which is particularly important in the evaluation of CD. The endoscopist should also evaluate the peristomal area in patients with ostomies during ileoscopy or colonoscopy via stoma (Fig. 2.12).

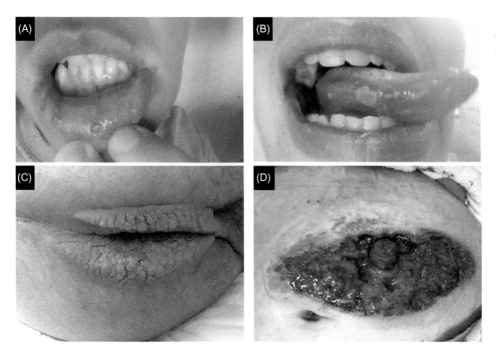

FIGURE 2.12 A bonus of endoscopy in inflammatory bowel disease. Oral (A and B), perianal (C), and peristome lesions (D) are evaluated during upper and lower endoscopy, and ileoscopy via a stoma. Photo documentation of the area and lesions, if present, is recommended for all patients undergoing endoscopy.

Flexible sigmoidoscopy is mainly indicated for disease monitoring, particularly the assessment of treatment response and exclusion of cytomegalovirus infection, in patients with UC.

EGD and push enteroscopy are the main tools for the evaluation of upper GI involvement of CD. In patients with indeterminate colitis an identification of the upper GI tract involvement helps establish a diagnosis of CD. However, patients with UC may also have upper GI inflammation, such as duodenitis [4]. Other applications of EGD and push enteroscopy with small bowel biopsy in IBD patients include the evaluation of celiac disease [5,6], eosinophilic enteritis [7], and common variable immune deficiency [8].

Ileoscopy via stoma [9,10] and pouchoscopy [11] are routinely used in patients with CD or UC who have ileostomies or ileal pouches for the diagnosis, differential diagnosis, monitoring of postoperative complication and disease recurrence, and delivery of therapy.

Proctoscopy, flexible sigmoidoscopy, pouchoscopy, or colonoscopy in the diverted rectum, ileal pouch, or colon should be performed with extreme caution. Due to the lack of nutrients (such as short-chain fatty acid) to the epithelia, the bowel mucosa can be extremely friable and the risk for bleeding, perforation, and bacterial translation is high. The endoscopist should minimize air insufflation and deep tissue biopsy.

The role of video capsule endoscopy (VCE) in IBD with its high sensitivity has been controversial, due to its inherited inability of tissue biopsy and delivery of endoscopic therapy, specificity, and concern on capsule retention (especially in CD). Besides, its utility in preoperative exclusion of small bowel CD in patients with UC undergoing restorative proctocolectomy and ileal pouches has not been established [12]. To overcome the shortcomings of VCE comes with retrograde and antegrade balloon−assisted endoscopy [13].

Endoscopic ultrasound (EUS) has been used to transmural disease [14], fistulae [15], abscess, perianal disease [16,17], and even regional lymphadenopathy. The EUS detection of transmural disease may help distinguish CD from UC [18−20].

Summary and recommendations

The diagnosis of CD and UC is based on the quality of colonoscopy−ileoscopy with tissue biopsy. The pretreatment index colonoscopy is critical for the assessment of disease distribution, which provides the important clue for the differential diagnosis of CD and UC. At least four topographic locations should be biopsied and separately labeled: terminal ileum, right colon, left colon, and rectum. Special endoscopy modalities, such as ileoscopy, pouchoscopy, enteroscopy, VCE, and EUS, can be selected for various special indications.

References

[1] Silverberg MS, Satsangi J, Ahmad T, Arnott ID, Bernstein CN, Brant SR, et al. Toward an integrated clinical, molecular and serological classification of inflammatory bowel disease: report of a Working Party of the 2005 Montreal World Congress of Gastroenterology. Can J Gastroenterol 2005;19(Suppl. A):5A−36A.

[2] Laine L, Kaltenbach T, Barkun A, McQuaid KR, Subramanian V, Soetikno R, et al. SCENIC international consensus statement on surveillance and management of dysplasia in inflammatory bowel disease. Gastroenterology 2015;148:639−51.

[3] Potzi R, Walgram M, Lochs H, Holzner H, Ganl A. Diagnostic significance of endoscopic biopsy in Crohn's disease. Endoscopy 1989;21:60−2.

[4] Valdez R, Appelman HD, Bronner MP, Greenson JK. Diffuse duodenitis associated with ulcerative colitis. Am J Surg Pathol 2000;24:1407−13.

[5] Curtis WD, Schuman BM, Griffin JW. Association of gluten-sensitive enteropathy and Crohn's colitis. Am J Gastroenterol 1992;87:1634−7.

[6] Gillberg R, Dotevall G, Ahren C. Chronic inflammatory bowel disease in patients with coeliac disease. Scand J Gastroenterol 1982;17:491−6.

[7] Talley NJ, Shorter RG, Phillips SF, Zinsmeister AR. Eosinophilic gastroenteritis: a clinicopathological study of patients with disease of the mucosa, muscle layer, and subserosal tissues. Gut 1990;31:5−8.

[8] Washington K, Stenzel T, Buckley RH, Gottfried MR. Gastrointestinal pathology in patients with common variable immunodeficiency and X-linked agammaglobulinemia. Am J Surg Pathol 1996;20:1240−52.

[9] Vadlamudi N, Alkhouri N, Mahajan L, Lopez R, Shen B. Ileoscopy via stoma after diverting ileostomy: a safe and effective tool to evaluate for Crohn's recurrence of neoterminal ileum. Dig Dis Sci 2011;56:866−70.

[10] Chongthammakun V, Fialho A, Fialho A, Lopez R, Shen B. Correlation of the Rutgeerts score and recurrence of Crohn's disease in patients with end ileostomy. Gastroenterol Rep (Oxf) 2017;5:271−6.

[11] Shen B, Achkar JP, Lashner BA, Ormsby AH, Remzi FH, Bevins CL, et al. Endoscopic and histologic evaluation together with symptom assessment are required to diagnose pouchitis. Gastroenterology 2001;121:261−7.

[12] Murrell Z, Vasiliauskas E, Melmed G, Lo S, Targan S, Fleshner P. Preoperative wireless capsule endoscopy does not predict outcome after ileal pouch-anal anastomosis. Dis Colon Rectum 2010;53:293−300.

[13] Yen HH, Chang CW, Chou JW, Wei SC. Balloon-assisted enteroscopy and capsule endoscopy in suspected small bowel Crohn's disease. Clin Endosc 2017;50:417−23.

[14] Dagli U, Over H, Tezel A, Ulker A, Temucin G. Transrectal ultrasound in the diagnosis and management of inflammatory bowel disease. Endoscopy 1999;31:152−7.

[15] Orsoni P, Barthet M, Portier F, Desjeux A, Grimaud JC. Prospective comparison of endosonography, magnetic resonance and surgical findings in anorectal fistula and abscess complicating Crohn's disease. Br J Surg 1999;86:360−4.

[16] Schwartz DA, Harewood GC, Wiersema MJ. EUS in rectal disease. Gastrointest Endosc 2002;56:100−9.

[17] Schwartz DA, Wiersema MJ, Dudiak KM, Fletcher JG, Clain JE, Tremaine WJ, et al. A comparison of endoscopic ultrasound, magnetic resonance imaging, and exam under anesthesia for evaluation of Crohn's perianal fistulas. Gastroenterology 2001;121:1064−72.

[18] Hildebrandt U, Kraus J, Ecker KW, Schmit T, Schüder G, Feifel G. Endosonographic differentiation of mucosal and transmural nonspecific inflammatory bowel disease. Endoscopy 1992;24(Suppl. 1):359−63.

[19] Shimizu S, Tada M, Kawai K. Value of endoscopic ultrasonography in the assessment of inflammatory bowel disease. Endoscopy 1992;24 (Suppl. 1):354−8.

[20] Shimizu S, Tada M, Kawai K. Endoscopic ultrasonography in the assessment of inflammatory bowel diseases. Gastrointest Endosc Clin N Am 1995;5:851−9.

Chapter 3

Normal gastrointestinal tract and variations on endoscopy

Bo Shen

Center for Inflammatory Bowel Diseases, Columbia University Irving Medical Center-New York Presbyterian Hospital, New York, NY, United States

Chapter Outline

Abbreviations

CD	Crohn's disease
GI	gastrointestinal
IBD	inflammatory bowel disease
ICV	ileocecal valve
PSC	primary sclerosing cholangitis
UC	ulcerative colitis

Introduction

Crohn's disease (CD) can affect any part of the gastrointestinal (GI) tract, while classic ulcerative colitis (UC) only involves colon and rectum. Differential diagnosis between inflammatory bowel disease (IBD) and other inflammatory, neoplastic, medication-related, and organ transplantation−related disease conditions is discussed in separate chapters. On the other hand, identification of normal endoscopic features of the esophagus, stomach, small and large intestine, and anorectum is critical for the diagnosis and differential diagnosis. In addition, there are "normal variations" on endoscopy of the GI tract. Some of these variations can mimic IBD, IBD-associated complications, and other inflammatory or neoplastic conditions. The normal anatomy and their common alterations in IBD- or non-IBD conditions are demonstrated.

Esophagus

The inner layer of the esophagus is covered with whitish squamous cells, in contrast to the salmon-color columnar epithelia in the stomach. The longitudinal vascular pattern is visible, especially in the distal esophagus, close to the esophagogastric junction (Fig. 3.1A). Congenital cervical inlet patch is occasionally found in the proximal esophagus (Fig. 3.1B). The cervical inlet patch consists of the ectopic gastric mucosa. Due to acid secretion from ectopic gastric mucosa may lead to esophagitis, ulcer, web, or stricture, mimicking CD of the esophagus.

Atlas of Endoscopy Imaging in Inflammatory Bowel Disease. DOI: https://doi.org/10.1016/B978-0-12-814811-2.00003-7
© 2020 Elsevier Inc. All rights reserved.

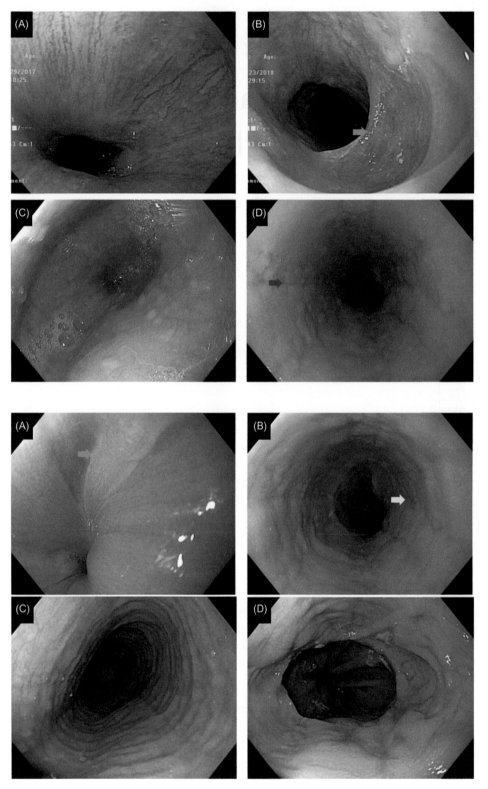

FIGURE 3.1 Normal esophagus: (A) normal distal esophagus and esophagogastric junction. Small linear vasculature is clearly seen; (B) cervical inlet patch in the proximal esophagus, which is covered with pinkish gastric columnar mucosa (*green arrow*); and (C and D) glycogenic acanthosis with small white epithelial nodules throughout the esophagus (*blue arrow*).

FIGURE 3.2 Normal esophagus and eosinophilic esophagitis: (A) linear white specks (*green arrow*), which has no clinical significance; (B) pseudolinear furrows (*yellow arrow*) in a patient with Crohn's disease, resembling eosinophilic esophagitis; and (C and D) true eosinophilic esophagitis with rings and linear furrows (*blue arrow*) and erosive distal esophagus erosions and Schatzki ring.

Glycogenic acanthosis is characterized by multifocal plaques of hyperplastic squamous epithelium with abundant intracellular glycogen deposits in the entire esophagus. Glycogenic acanthosis needs to be differentiated from Candida esophagitis (Fig. 3.1C and D).

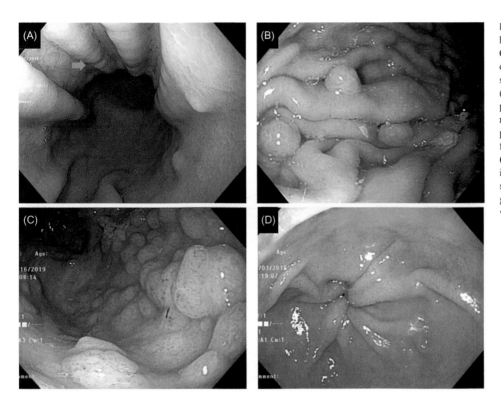

FIGURE 3.3 Stomach in the healthy and in patients with Crohn's disease: (A) bamboo joint of the stomach, resembling that seen in gastric Crohn's disease (*green arrow*); (B) fundic gland polyps in the body of stomach; resulting from long-term use of proton pump inhibitors in a noninflammatory bowel disease patient; (C) extensive fundic gland polyps in a Crohn's disease patient with short-gut syndrome; and (D) antral gastritis in ulcerative colitis patient with *Helicobacter pylori* infection.

Linear white specks are occasionally seen in the normal esophagus (Fig. 3.2A), which should be differentiated from white plaques in those with Candida esophagitis. Patients with esophageal spasms, a functional disorder of the esophagus, may present with linear furrows and esophageal rings (Fig. 3.2B), which resemble to that seen in eosinophilic esophagitis (Fig. 3.2C and D). The distinction may only be made by histologic evaluation.

Stomach

IBD rarely involves the stomach. One of the endoscopic features of CD of the stomach is the presence of "bamboo joints," which was initially described by Japanese investigators, as shown in Fig. 4.4 in Chapter 4, Crohn's disease: inflammatory type [1]. Patients with normal stomach may have a similar endoscopic appearance (Fig. 3.3A). Therefore histologic evaluation is critical.

With the extensive use of proton pump inhibitors in the general population, fundic gland polyps are commonly seen in the stomach (Fig. 3.3B). However, the presence of fundic gland polyps may reflect the body's compensatory reaction to various stimuli, such as short-gut syndrome in CD (Fig. 3.3C). Antral gastritis in IBD is also common, which is not necessarily a part of the disease (Fig. 3.3D), with etiologies such as *Helicobacter pylori* infection and the use of nonsteroidal antiinflammatory drugs [2].

Duodenum

The duodenum is a common location of CD of the upper GI tract. CD of the duodenum may present with nonulcerative inflammation, ulcers, or strictures. The distinction between the inflammatory phenotype of CD and "normal" variations of the duodenum sometimes is difficult on endoscopy. The duodenum bulb in healthy individuals can have various appearances. The bulbar mucosa can be granular or nodular in a non-disease setting. Granularity and nodularity of the bulbar mucosa mainly consist of Bruner's glands or Brunner's gland hyperplasia (Figs. 3.4 and 3.5). Large nodules may result from gastric heterotopia which is assumed to be congenital in origin, which also may be associated with fundic gland polyps (Fig. 3.6A). Those features in the healthy individuals should be differentiated from IBD involvement of the duodenum (Fig. 3.6B and D) or neoplasia there (Fig. 3.6C).

The duodenum cap is located in the junction between the bulb and second portion of the duodenum, which is a common place for CD strictures in the upper GI tract (Fig. 3.7). The second to fourth portions of the normal duodenum are

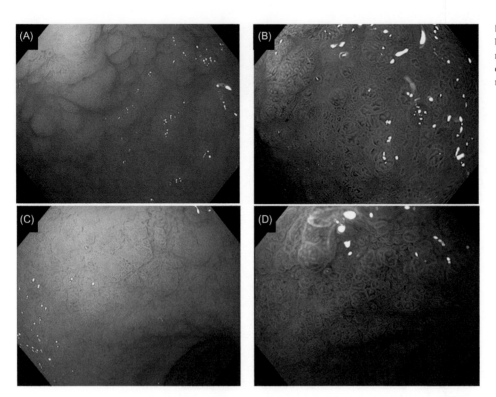

FIGURE 3.4 Normal duodenum bulb on white-light (left panel) and narrow-band imaging (right panel) endoscopy: (A—D) granular and nodular mucosa with short villi.

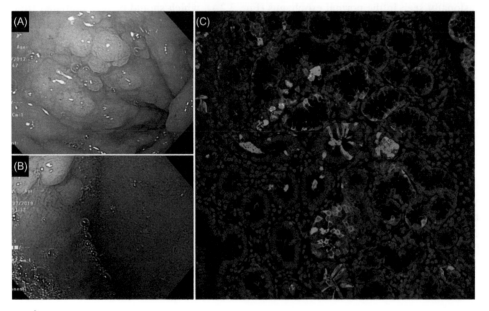

FIGURE 3.5 Duodenum bulb with nodular mucosa from Brunner's gland hyperplasia: (A and B) localized Brunner's gland hyperplasia in the anterior wall of duodenum bulb; (C) Brunner's gland hyperplasia highlighted with lysozyme (*green*) on immunofluorescence. Paneth cells in the intestinal epithelia are highlighted with immunostain of human-defensing-5 (*red*).

characterized by the folds, villi, and occasionally lacteals, which are different from the bulbar mucosa (Fig. 3.8). Lacteals may present with punctate white spots which mostly represent some degree of intestinal lymphangiectasia (Figs. 3.8C and D, 3.9A and B, and 3.10A and B) [3,4]. Diffuse white duodenum mucosa, however, may be associated with nonspecific inflammation, infections (such as Whipple's disease), or lymphoma [1]. In between the white mucosa, there may be erythema mucosa with short villi, which can be part of IBD in the upper GI tract (Fig. 3.10C and D).

Duodenum papilla may present with different shapes and sizes (Fig. 3.11A and B). The papilla and surrounding areas are common locations for familial adenomatous polyposis (Fig. 3.6C). Patients with primary sclerosing cholangitis (PSC) or IgG4-associated systemic disorders may present with diffuse duodenitis, predominantly at the second portion,

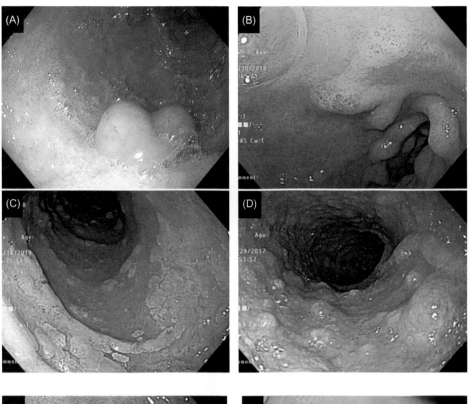

FIGURE 3.6 Nodular lesions in the healthy and diseased: (A) gastric (fundic) heterotopia with nodules found in the dilated lumen of the second part of the duodenum; (B) duodenitis with erythema and nodules in a patient with Crohn's disease; (C) diffuse flat polyps in the duodenum in a patient with familial adenomatous polyposis; and (D) diffuse duodenitis with mucosal nodularity in a patient with ulcerative colitis and ileal pouch—anal anastomosis.

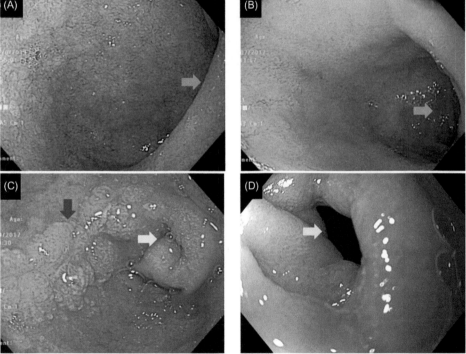

FIGURE 3.7 Normal and abnormal duodenum bulb and cap. (A and B) Normal granular mucosa in the duodenum bulb. Duodenum cap is presented between the bulb and second portion (*green arrows*). (C and D) Strictured duodenum cap resulting from Crohn's disease (*yellow arrows*). In addition, there are linear erythematous nodules of Crohn's disease (*blue arrow*) (C).

on endoscopy and histology. Also, the duodenum involvement, particularly the lesions adjacent to the papilla may explain an etiology of CD-associated pancreatitis, cholangitis, or cholangiopathy (Fig. 3.11C and D) [5—7].

It is important to document the size of the lumen of the duodenum. The narrowed lumen can be seen in CD, while dilated lumen may be seen in diseases such as small bowel bacterial overgrowth or autoimmune enteropathy or duodenopathy (Fig. 3.9C and D).

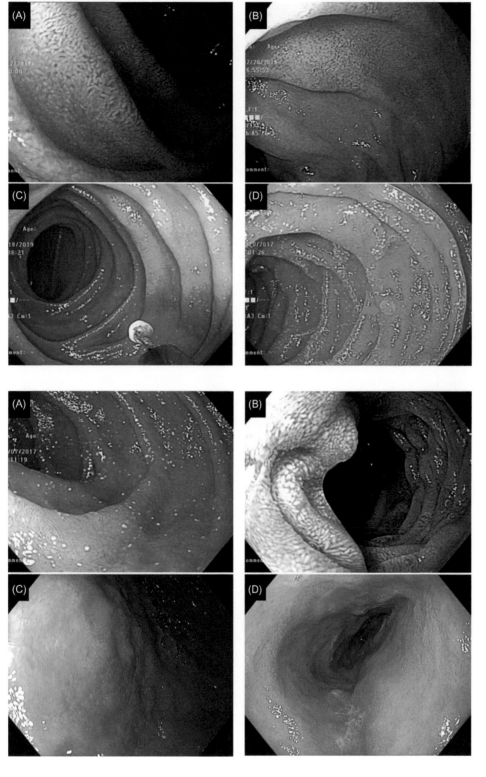

FIGURE 3.8 Normal mucosa in the second to fourth portions of the duodenum: (A and B) prominent folds and villi and (C and D) tiny white spots from lacteals. Lymphangiectasia with a white nodule is highlighted (*green arrow*).

FIGURE 3.9 The lumen and mucosa of the duodenum in the healthy and diseased: (A) normal duodenum with white spots from lacteals with normal-sized lumen and number and normal size and number of the folds; (B) diffuse white duodenum with prominent lymphangiectasia and normal-sized lumen and normal size and number of the folds; and (C and D) dilated lumen of the second to fourth portions of the duodenum with flatten villi and flatten and dilated lumen, resulting from autoimmune enteropathy.

Distal ileum

Folds and villi of the distal ileum are not as prominent as that in the duodenum (Fig. 3.12A and B). Lymphoid hyperplasia (or lymphoid hypertrophy) with small submucosal nodules is a normal endoscopic finding (Fig. 3.13A−C). However, more diffuse and prominent nodules that have been described as nodular lymphoid hyperplasia may be

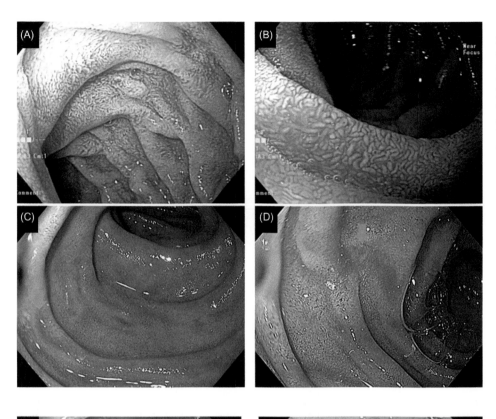

FIGURE 3.10 White duodenum mucosa in the healthy and diseased: (A and B) diffuse white duodenum with prominent lacteals with a velvet appearance in a healthy individual; and (C and D) diffuse white villi with erythema and depletion of villi in a patient with Crohn's disease (*green arrows*).

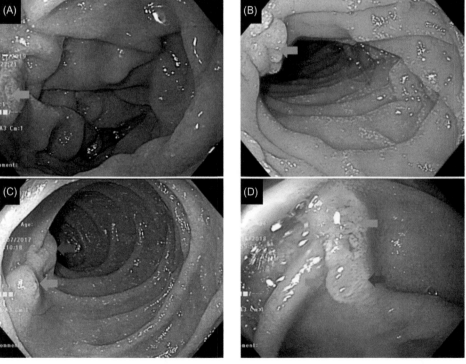

FIGURE 3.11 Duodenum papilla in the healthy and Crohn's disease. (A–D) Normal duodenum papilla (*green arrows*). Two patients with ileocolitis with additional nodules close to the papilla and otherwise normal duodenum mucosa (*blue arrows*). Biopsy showed chronic active enteritis (C and D). The presence of the ampulla lesions may explain Crohn's disease–associated pancreatitis.

associated with diarrhea, abdominal pain, hematochezia and hypoproteinemia [8], infectious enterocolitis, GI-mediated allergy [9], or common variable immune deficiency (Fig. 3.13D) [10]. Lymphoid hyperplasia may be seen in patients ileal pouches (Fig. 3.14). Differential diagnosis of granular or nodular lesions of the terminal ileum also includes nonspecific ileitis, CD (Figs. 3.15 and 3.16), intestinal tuberculosis, and lymphoma [11].

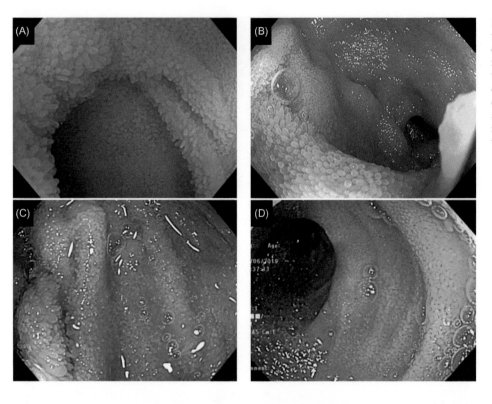

FIGURE 3.12 Terminal ileum in the healthy and quiescent Crohn's disease: (A and B) normal terminal ileum with shorter villi than that in the second to fourth portions of the duodenum and (C and D) Crohn's disease with a short disease course achieved remission after effective medical therapy. The height of the villi was completely restored.

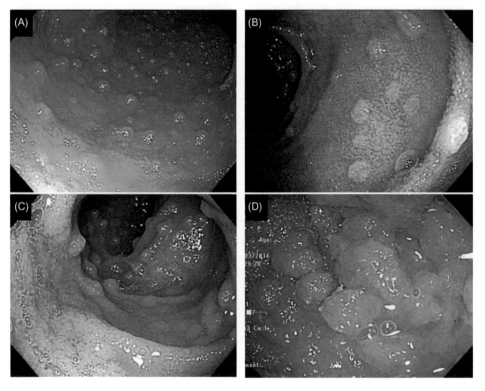

FIGURE 3.13 Lymphoid hyperplasia and nodular lymphoid hyperplasia in the healthy and diseased: (A–C) lymphoid hyperplasia in three healthy asymptomatic patients and (D) nodular lymphoid hyperplasia in a patient with chronic diarrhea.

Mild CD of the ileum or CD in remission may present with depleted or shorten villi (Figs. 3.12C and D and 3.17). To fully assess the disease status of the terminal ileum, bile and mucus should be washed out. Folds, villi, and the size of the lumen of the small bowel should be assessed and photo-documented (Fig. 3.18). The dilated lumen of the distal ileum can be seen in those with downstream stricture, backwash ileitis, and small intestinal bacterial overgrowth (Fig. 3.18C and D).

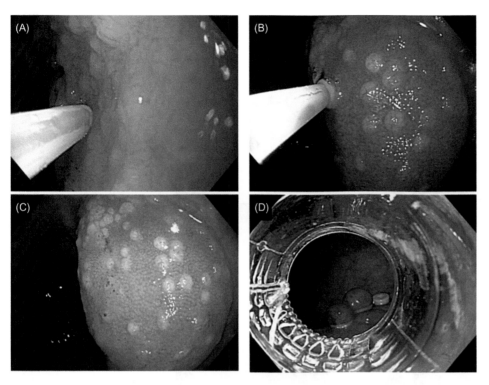

FIGURE 3.14 Lymphoid hyperplasia of prolapse distal pouch mucosa in an ulcerative patient with ileal pouch−anal anastomosis: (A−C) nodular mucosa with lymphoid hyperplasia highlighted with submucosal injection of hypertonic glucose and (D) endoscopic band ligation was performed to treat the prolapse which had caused the patient's dyschezia.

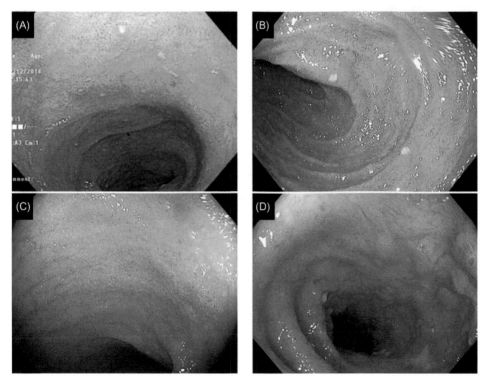

FIGURE 3.15 Erosions, granularity, and nodularity of the terminal ileum in healthy and Crohn's disease: (A and B) small nonspecific erosions in tow healthy patients undergoing screening colonoscopy; (C) patchy granularity of the mucosa in a patient with Crohn's disease; and (D) nodular mucosa in a patient with Crohn's disease.

Ileocecal valve, cecum, and appendix

The shape and size of the ileocecal valve (ICV) vary greatly in healthy individuals, which should be described in the endoscopic report. Sometimes, ICV can only be assessed by retroflex view during colonoscopy (Fig. 3.19). Strictured

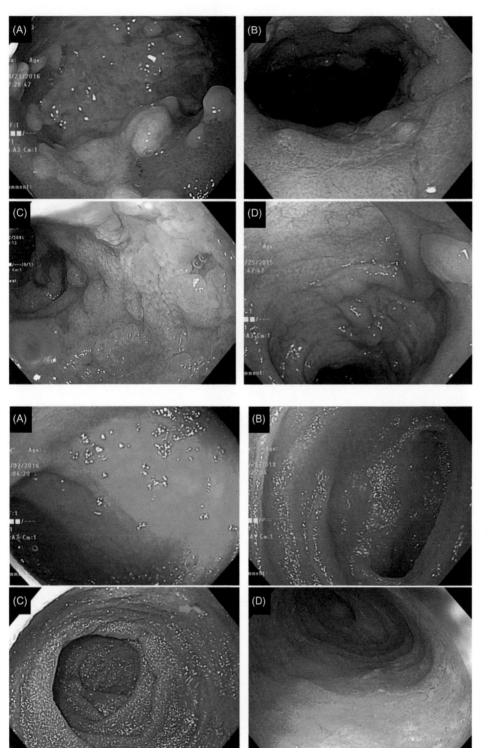

FIGURE 3.16 Nodular lesions of the terminal ileum in Crohn's disease: (A) ulceration and mucopurulent nodules; (B and C) nodular mucosa; and (D) elongated pseudopolyps.

FIGURE 3.17 Terminal ileum in the healthy and diseased: (A) normal terminal ileum mucosa covered with bile; (B) normal terminal ileum mucosa with light reflection; (C) spastic terminal ileum with small Crohn's disease—associated erosion (*green arrow*); and (D) Crohn's disease of the terminal ileum in remission.

or ulcerated ICVs are common in patients with ileal or ileocolonic CD (Fig. 3.20A and B), while patulous ICV is found in patients with UC, particularly in those with concurrent UC and PSC (Fig. 3.20C and D).

Patchy erythema at the cecal base may be found in patients with IBD or with the use of sodium phosphate—based bowel preparation (Fig. 3.21). The appendiceal orifice is normally identified without difficulty. However, it is important to differentiate the appendiceal orifice from strictured or deformed ICV (Fig. 3.22).

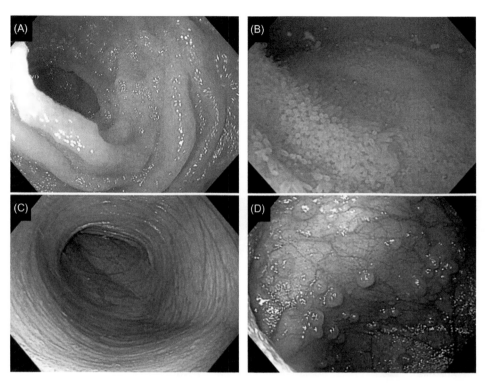

FIGURE 3.18 Size of the lumen of the terminal ileum in the healthy and diseased: (A and B) normal size of the lumen of the terminal ileum with intact villi in a healthy individual; (C) dilated terminal ileum with normal mucosa in a Crohn's disease patient who had ileocecal a valve stricture; and (D) dilated lumen of the terminal with nodular lymphoid hyperplasia in a Crohn's disease patient.

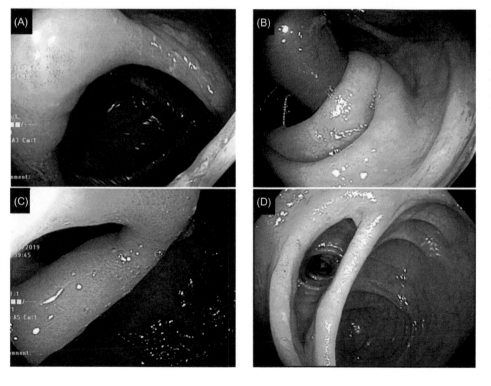

FIGURE 3.19 Various forms of the ileocecal valves in healthy individuals: (A) The prominent fold is the ileocecal valve; (B) occasionally the ileocecal valve is best viewed with endoscopic retroflex; and (C and D) normal appearance of the ileocecal valves.

Colon and rectum

The landmarks for the identification of geographic locations of the large bowel include splenic and hepatic flexure, triangular-shaped lumen of the traverse colon (Fig. 3.23), and rectal sigmoid junction. Those landmarks are used for the disease extent and distribution of IBD.

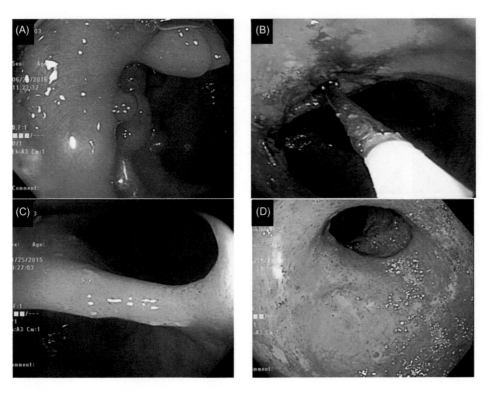

FIGURE 3.20 Ileocecal valves in inflammatory bowel disease: (A) Strictured and deformed ileocecal valve in Crohn's disease; (B) endoscopic balloon dilation of the ileocecal valve in a patient with Crohn's disease; (C) patulous ileocecal valve in a patient with quiescent ulcerative colitis; and (D) patulous ileocecal valve in a separate patient with active ulcerative colitis.

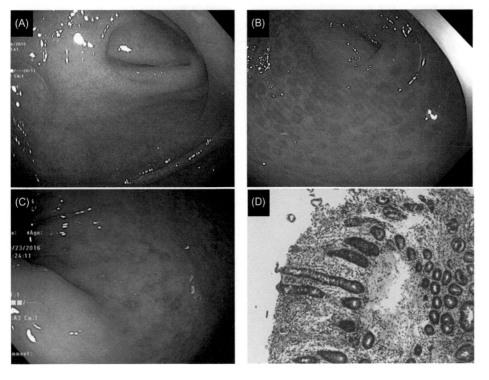

FIGURE 3.21 Normal cecum: (A) Normal appendiceal orifice and cecum; (B and C) patchy erythema with (C) or without (B) erosions in the cecum; and (D) histology of biopsy of the erythematous area showed nonspecific active colitis, likely resulting from sodium phosphate−based bowel preparation agent.

Healthy colon and rectum should have light reflection and transparent view of mucosal vasculature (Fig. 3.23). Subtle pathological changes such as excessive mucus in the spastic colon in irritable bowel syndrome and mucosal edema and erythema can easily be missed (Fig. 3.24). Attention should also be paid to the lumen and luminal contents of the colon, which sometimes yield diagnostic clues for disease entities, for example, prolapse, intussusception, malabsorption, and mucosal scars from chronic bowel injury (Figs. 3.25 and 3.26).

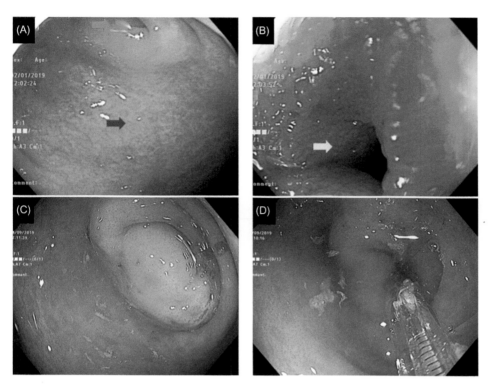

FIGURE 3.22 Appendix in the healthy and diseased: (A) appendiceal orifice (*green arrow*) with surrounding mosaic-appearing cecal mucosa (*blue arrow*) and (B–D) nonulcerated stricture at the ileal cecal valve (*yellow arrow*) was supposed to undergo endoscopic balloon dilation. The mistook endoscopist the appendiceal orifice as a strictured ileocecal valve and attempted to perform endoscopic dilation of the orifice. Fortunately, the nonintended procedure was not eventful.

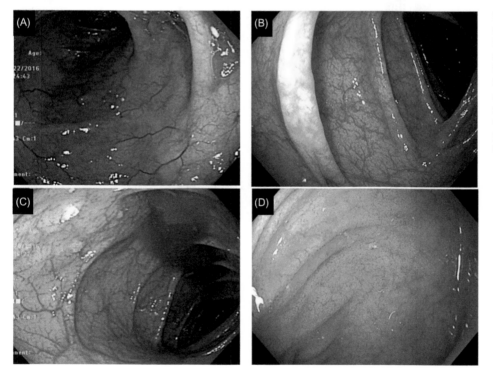

FIGURE 3.23 Landmarks of the colon in the healthy. Colon mucosa is characterized by light reflection and mucosal vasculature: (A) hepatic flexure of the colon with bluish liver seen thorough the colon wall; (B and C) triangular-shaped lumen indicates the transverse colon; and (D) left-colon proximal to the triangular-shaped transverse colon.

The anterior distal rectum is prone to inflammatory and mechanical injury. Isolated inflammation in this particular location often results from mechanical and functional etiologies, such as rectal prolapse, rectocele, and dyssynergic defecation (Fig. 3.27). Differential diagnosis includes rectal involvement of CD and UC.

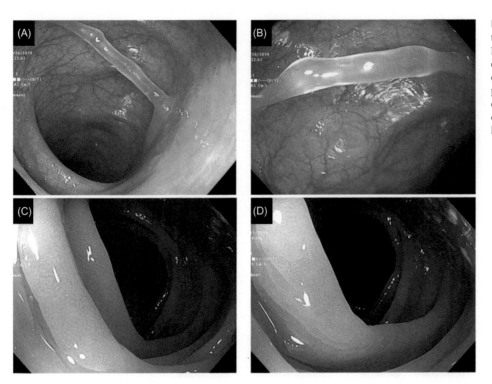

FIGURE 3.24 Endoscopic features of the colon suggesting underlying disorders: (A and B) excessive mucus in the spastic colon in a patient with diarrhea-predominant irritable bowel syndrome and (C and D) edema of the colon mucosa in a patient with hypoalbuminemia.

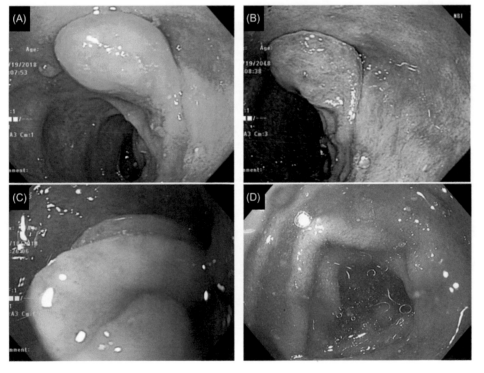

FIGURE 3.25 Endoscopic features of the colon suggesting underlying disorders: (A and B) lipoma of the colon with overlying normal mucosa on white-light and narrow-band imaging; (C) prolapsed mucosa of the sigmoid colon in a patient with chronic constipation and dyschezia; and (D) floating oil droplets on the luminal content in a patient with malabsorption.

Anus and anal canal

The anatomy of the anorectum is complex. Various IBD and non-IBD disease entities can involve the area. Retroflex evaluation during endoscopy is encouraged in patients without proctitis. Retroflex of the endoscopy provides a better

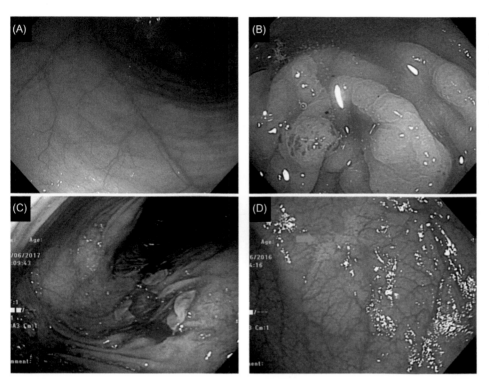

FIGURE 3.26 Colon mucosa in the healthy and diseased: (A) normal colon mucosa; (B) intussusception of colon mucosa in a patient with endometriosis; (C) endoscopic trauma to colon mucosa in a patient on systemic corticosteroids; and (D) mucosal scars (*green arrow*) from ischemic colitis, which would easily be missed.

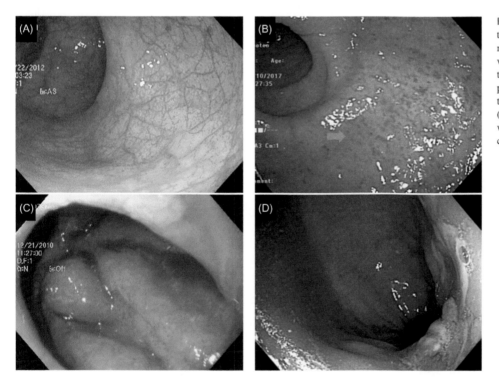

FIGURE 3.27 Anterior rectum in the healthy and diseased: (A) normal rectum with a clear mucosal vascular pattern; (B) mucosal erythema at the anterior wall from prolapse; (C) mucosal prolapse at the anterior wall of the rectum; and (D) nodular mucosa at the anterior wall of the rectum from long-term constipation.

view of the distal rectum and anal canal (Fig. 3.28). Polypoid or nodular lesions in the areas could be from hemorrhoids, hypertrophic anal papillae, skin tags, and warts (Figs. 3.29–3.31). Besides, IBD- or non-IBD-related anal fissures or erosions (Figs. 3.28B and C, 3.29D, and 3.30C and D) and perianal or vaginal fistulae (Fig. 3.31D) can be identified with a proper endoscopy.

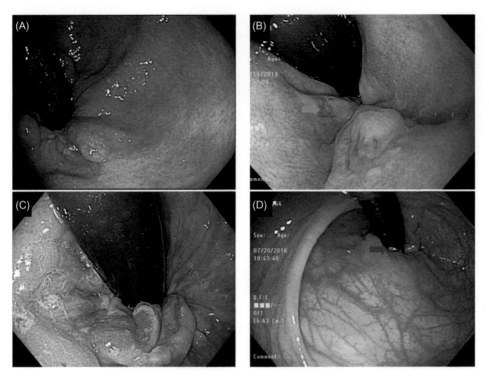

FIGURE 3.28 Distal rectum under the retroflex view of endoscopy: (A) normal distal rectum and proximal anal canal; (B) anal fissures at the dentate line (*green arrow*); (C) linear erosions in the distal rectum and anal transitional zone in a patient with Crohn's ileitis; and (D) nodular anal canal in a patient with perianal Crohn's disease (*blue arrow*).

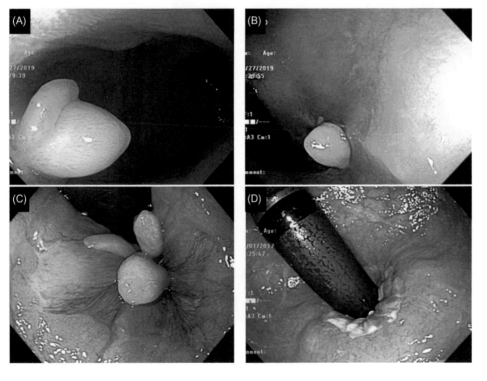

FIGURE 3.29 Hypertrophic anal papillae and inflammation: (A–C) anal papillae with overlying squamous cells originated from the dentate line in forward and retroflex views and (D) erosions of the anal canal (*green arrow*).

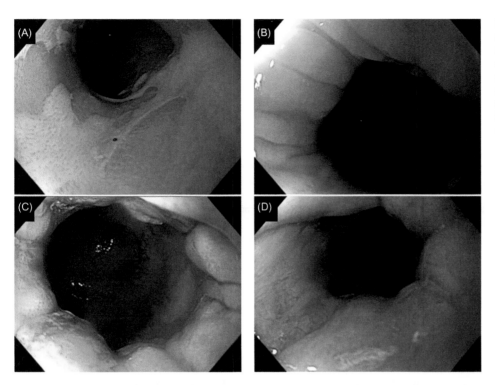

FIGURE 3.30 Anal transition zone and anal canal: (A) normal anal transition zone between columnar epithelia (*salmon colored*) and squamous epithelia (*white colored*); (B) normal anal canal; and (C and D) anal fissures in two patients with Crohn's disease.

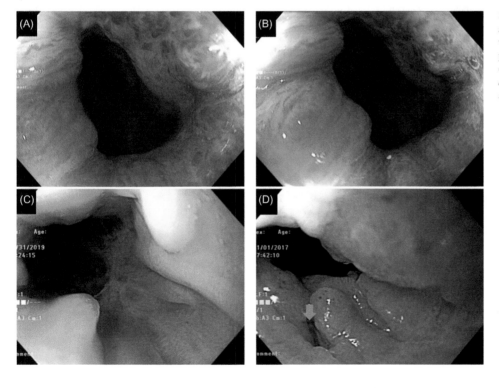

FIGURE 3.31 Anal lesions: (A and B) hemorrhoids; (C) anal stricture with a nodular anal canal in a patient with Crohn's disease; and (D) anal vaginal fistula (*green arrow*) with a nodular anal canal in a patient with Crohn's disease.

Summary and recommendations

Endoscopic presentations of IBD are nonspecific. The list for the differential diagnosis is long. Accurate recognition of the normal anatomy and landmarks of the GI tract is the first step. Some physiological changes, such as granular or nodular duodenum bulb mucosa and lymphoid hyperplasia, may mimic IBD. Certain anatomic locations of the GI tract are prone to injury with endoscopic features resembling that in IBD, such as distal rectal inflammation from prolapse and hypertrophic anal papillae from chronic constipation. In addition to mucosal features, endoscopic evaluation should include the shape and size of the lumen of the GI tract and characteristics of luminal contents. Histologic evaluation often provides useful information.

References

[1] Fujiya M, Sakatani A, Dokoshi T, Tanaka K, Ando K, Ueno N, et al. A bamboo joint-like appearance is a characteristic finding in the upper gastrointestinal tract of Crohn's disease patients: a case-control study. Medicine (Baltimore) 2015;94:e1500.

[2] Long MD, Kappelman MD, Martin CF, Chen W, Anton K, Sandler RS. Role of nonsteroidal anti-inflammatory drugs in exacerbations of inflammatory bowel disease. J Clin Gastroenterol 2016;50:152−6.

[3] Taş A, Koklu S, Beyazit Y, Akbal E, Kocak E, Celik H, et al. The endoscopic course of scattered white spots in the descending duodenum: a prospective study. Gastroenterol Hepatol 2012;35:57−64.

[4] Kim JH, Bak YT, Kim JS, Seol SY, Shin BK, Kim HK. Clinical significance of duodenal lymphangiectasia incidentally found during routine upper gastrointestinal endoscopy. Endoscopy 2009;41:510−15.

[5] Navaneethan U, Liu X, Bennett AE, Walsh RM, Venkatesh PG, Shen B. IgG4-associated ampullitis and cholangiopathy in Crohn's disease. J Crohns Colitis 2011;5:451−6.

[6] Spiess SE, Braun M, Vogelzang RL, Craig RM. Crohn's disease of the duodenum complicated by pancreatitis and common bile duct obstruction. Am J Gastroenterol 1992;87:1033−6.

[7] Newman LH, Wellinger JR, Present DH, Aufses Jr. AH. Crohn's disease of the duodenum associated with pancreatitis: a case report and review of the literature. Mt Sinai J Med 1987;54:429−32.

[8] Lin R, Lu H, Zhou G, Wei Q, Liu Z. Clinicopathological and ileocolonoscopic characteristics in patients with nodular lymphoid hyperplasia in the terminal ileum. Int J Med Sci 2017;14:750−7.

[9] Krauss E, Konturek P, Maiss J, Kressel J, Schulz U, Hahn EG, et al. Clinical significance of lymphoid hyperplasia of the lower gastrointestinal tract. Endoscopy 2010;42:334−7.

[10] Van den Brande P, Geboes K, Vantrappen G, Van den Eeckhout A, Vertessen S, Stevens EA, et al. Intestinal nodular lymphoid hyperplasia in patients with common variable immunodeficiency: local accumulation of B and CD8(+) lymphocytes. J Clin Immunol 1988;8:296−306.

[11] Ueno N, Fujiya M, Moriichi K, Ikuta K, Nata T, Konno Y, et al. Endosopic autofluorescence imaging is useful for the differential diagnosis of intestinal lymphomas resembling lymphoid hyperplasia. J Clin Gastroenterol 2011;45:507−13.

Chapter 4

Crohn's disease: inflammatory type

Bo Shen

Center for Inflammatory Bowel Diseases, Columbia University Irving Medical Center-New York Presbyterian Hospital, New York, NY, United States

Chapter Outline

Abbreviations

CD   Crohn's disease
GI   gastrointestinal
IBD   inflammatory bowel disease
ICV   ileocecal valve
ITB   intestinal tuberculosis
PSC   primary sclerosing cholangitis
UC   ulcerative colitis

Introduction

Crohn's disease (CD), by location, can involve any part of gastrointestinal (GI) tract, from the mouth to anus. CD can also involve the GI tract with different depths. Therefore several classifications have been proposed to categorize CD. The Montreal classification was proposed to categorize CD, on the basis of the age of onset, disease location, and disease phenotype [1] (Table 4.1), which was modified from earlier Vienna classification [2]. The classifications were recently further expanded by the incorporation of age groups, disease locations, and etiopathogenetic factors [3] (Table 4.2).

This chapter is focused on inflammatory phenotype of CD. Fibrostenotic (Chapter 5: Crohn's disease: fibrostenotic type), fistulizing (Chapter 6: Crohn's disease: penetrating type), and perianal (Chapter 7: Crohn's disease-perianal) phenotypes are discussed in separate chapters.

Esophagogastroduodenoscopy and ileocolonoscopy play a key role in the evaluation, diagnosis, differential diagnosis, disease monitoring, and assessment of treatment response of CD (Table 4.2).

Oral, esophageal, and gastric Crohn's disease of inflammatory type

Upper GI CD is defined as the disease involvement in segment(s) of the tract proximal to the ligament of Treitz. It is estimated that the frequency of upper GI CD ranges from 0.5% to 16%, as summarized by Laube et al. [4]. Oral CD is common in both pediatric and adult patients. The presentation of oral CD is variable. The lip, buccal gingiva, and tongue are common locations of disease involvement. The disease presentations can be swelling, induration, and various forms of aphthae, erosions and ulcers, plaques, and even "cobblestoning" (Fig. 4.1). The oral lesions are often painful. In fact, aphthae, by strict definition, are painful, small, shallow ulcer lesions.

Atlas of Endoscopy Imaging in Inflammatory Bowel Disease. DOI: https://doi.org/10.1016/B978-0-12-814811-2.00004-9
© 2020 Elsevier Inc. All rights reserved.

TABLE 4.1 The Montreal classification of Crohn's disease [1].

Age at diagnosis (A)		
A1 16 years or younger		
A2 17–40 years		
A3 over 40 years		
Location (L)	*Upper GI modifier (L4)*	
L1 terminal ileum	L1 + L4	Terminal ileum + upper GI
L2 colon	L2 + L4	Colon + upper GI
L3 ileocolon	L3 + L4	Ileocolon + upper GI
L4 upper GI	–	–
Behavior (B)	*Perianal disease modifier (p)*	
B1 nonstricturing, nonpenetrating	B1p	Nonstricturing, nonpenetrating + perianal
B2 stricturing	B2p	Stricturing + perianal
B3 penetrating	B3p	Penetrating + perianal

A, Age; *B*, behavior; *GI*, gastrointestinal; *L*, location.

TABLE 4.2 Proposed classification of inflammatory bowel diseases.

Criteria	Class		Description	Examples
Disease location, extent and depth ± granulomas	Ulcerative colitis			Classic UC
	Crohn's disease			Classic CD
	Indeterminate colitis			
Age of onset	Very early onset		Age 0	IL-10/ILR mutations
	Early onset		Age 0–10 years	
			Age 10–17 years	
	Regular onset		Age 17–40 years	
	Late onset		Age >50 years	
Phenotype	Inflammatory			Inflammatory CD; classic UC
	Stricturing			Stricturing CD; UC with stricture
	Penetrating			Fistulizing CD
Locations	Oral			
	Upper GI			
	Jejunum			
	Ileum			
	Colon			
	Rectum			
	Perianal			
	Extraintestinal			Metastatic CD of the skin, lung, and liver
Concurrent or immune-mediated disorders	IBD			Isolated UC or CD of the gut
	IBD-V	IBD⁺	IBD + classic extraintestinal manifestations	UC with concurrent PSC
		IBD⁺⁺	IBD + autoimmune and/or autoinflammatory disorders ± classic extraintestinal manifestations	IBD with concurrent microscopic colitis, celiac disease, hidradenitis suppurativa
		IBD⁺/⁻	Diseases sharing clinical features and possible etiopathogenetic pathways with classic IBD ± classic extraintestinal manifestations of IBD, autoimmune disorders or autoinflammatory disorders	Lymphocytic colitis, collagenous colitis; Behcet disease, cryptogenic multifocal ulcerous stenosing enteritis, ulcerative jejunitis

(Continued)

TABLE 4.2 (Continued)

Criteria	Class	Description	Examples
Etiology of IBD	Primary or idiopathic	Monogenic	IL-10, IL-10Ra, IL-10Rb mutations
			Very early—onset IBD
		Polygenic	Classic UC; classic CD
	Secondary	Identifiable pathogens	*Mycobacterial paratuberculosis*
		Medication-induced	Mycophenolate-associated colitis; ipilimumab-associated colitis
		Organ transplantation-induced	Post—solid organ transplant IBD-like conditions, cord colitis syndrome
		Surgery induced	Pouchitis, Crohn's disease—like conditions of the pouch, post—colectomy enteritis, bariatric surgery—associated IBD
Genetic etiology	Monogenic		IL-10/IL-R mutations, familial Mediterranean fever
	Polygenic		Classic CD and classic UC
Disease spread process	Intrinsic (inside-out)	Starting from the lymphatic system or mesentery, spreading to gut mucosa	Subset of obese CD patients; subset of sclerosing mesenteritis or lymphangitis
	Extrinsic (outside-in)	External trigger (e.g., bacteria) leading to mucosal inflammation	Fulminant UC: from mucosal disease to transmural inflammation

CD, Crohn's disease; *IBD*, inflammatory bowel disease; *IBD-V*, IBD-variant; *PSC*, primary sclerosing cholangitis; *UC*, ulcerative colitis.

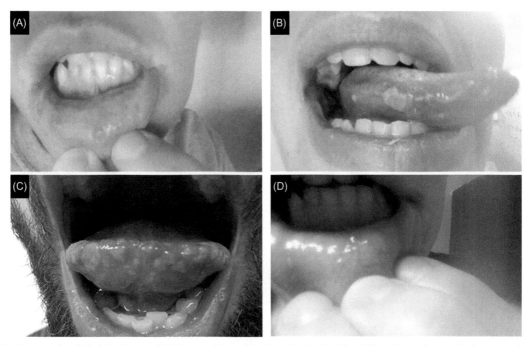

FIGURE 4.1 Patterns of oral Crohn's disease. (A) Linear aphthous-like ulcers in the lip; (B and C) multiple ulcers with plaques on the tongue; and (D) "cobblestoning" of the lower lip.

Esophageal CD is rare, which is often presented with mucosal inflammation, erosions, and ulcers, leading to strictures in some. It appears that the distal esophagus is the most common location of the disease. Endoscopic features of esophageal CD range from hyperemic mucosa, erosions, ulcers, granularity (Fig. 4.2), and strictures [5]. Patients with esophageal CD are considered as having an aggressive form of disease often having poor nutrition in protein and micronutrients and current use of immunosuppressive medications. The endoscopic features can overlap with esophagitis from nutrition deficiencies [such as iron (Fig. 4.2D) and zinc] or superimposed bacterial, fungal, or viral infection (such as herpes simplex).

Gastric CD appears more commonly in patients in Asia than those in the United States or Europe. Gastric CD may present with erosions, ulcers, nodularity, and verrucous gastritis. The erosions or ulcers are usually discrete and can be single or multiple or small or large with regular or irregular border. In the majority of patients the ulcer is clean-based, with a raised edge or inflammation or nodularity in the surrounding area (Fig. 4.3). A "bamboo joint"−like appearance has been described in gastric and duodenal CD, characterized by swollen longitudinal folds traversed by erosive fissures or linear furrows (Fig. 4.4). This endoscopic feature is hardly observed in patients with ulcerative colitis (UC) or non−inflammatory bowel disease (non-IBD) conditions. In fact, it has been considered a specific marker for gastroduodenal CD [5,6]. The rate of yield of granulomas on histology with biopsy of the lesion was reported to be 45.5% [5]. It appears that this particular endoscopic feature is often seen in CD patients in Asia. Confounding factors, such as *Helicobacter pylori* infection, and the use of nonsteroidal antiinflammatory drugs should be evaluated and treated accordingly.

Isolated or concurrent involvement of the duodenum can occur in patients with CD. Duodenum bulb and cap are the most locations. Endoscopic features of duodenal CD include edematous or erythematous mucosa, aphthous-like lesions, various forms or shapes of ulcers, erosions, patchy erythematous mucosa, thickened folds, fissures, granularity or nodularity, and "bamboo joint"−like appearances (Figs. 4.5 and 4.6). Strictures are also common, while fistulae are rare. IgG4-associated duodenitis characterized by diffuse inflammation and ulcers, predominantly at the second portion, can occur in patients with CD (which may be associated with concurrent hepatobiliary involvement of IBD) (Fig. 4.7) [7].

Jejunum and ileum Crohn's disease of inflammatory type

Three phenotypes of CD in the jejunum and ileum are discussed in Chapter 18, Enteroscopy in inflammatory bowel disease and inflammatory bowel disease−like conditions.

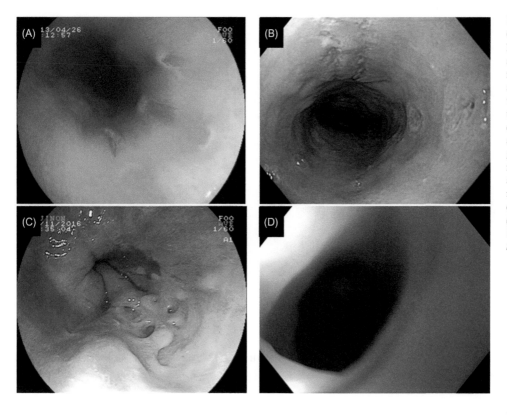

FIGURE 4.2 Esophageal Crohn's disease. (A) Longitudinal superficial ulcers at the distal esophagus; (B) round, medium-sized ulcers arranged with a linear pattern with surrounding mucosal erythema and edema in the whole esophagus; (C) mucosal scars and bridges in the distal esophagus, after therapy with infliximab; and (D) web-like stricture at the proximal esophagus from malnutrition and severe iron deficiency in Crohn's disease. *(A and C) Courtesy Dr. Gu Yubei of Ruijin Hospital of Shanghai Jiaotong University.*

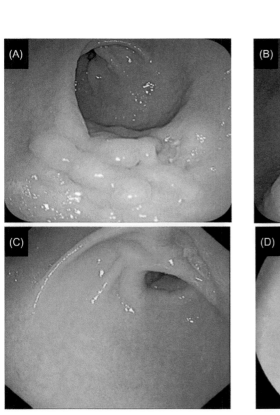
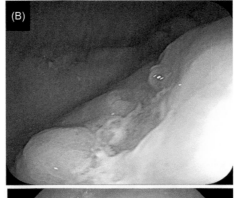
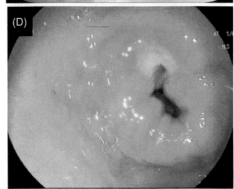

FIGURE 4.3 Gastric Crohn's disease pattern of erosion and ulcer. (A) Irregular erosion with adjacent nodular mucosa at the antrum; (B) deep ulcer with exudates at the gastric body; (C) large, deep, clean-based ulcer with clear border at the prepyloric region; and (D) asymmetric ulcer at the pylorus. *(A and B) Courtesy Dr. Mei Wang of Affiliated Hospital of Yangzhou University; (C and D) courtesy Dr. Yubei Gu of Ruijin Hospital of Shanghai Jiaotong University.*

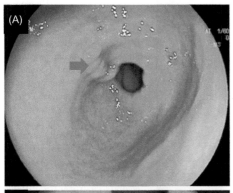
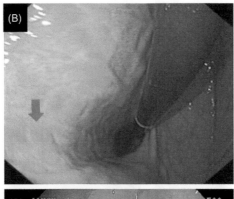
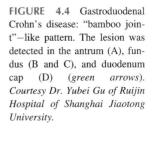
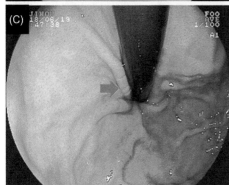
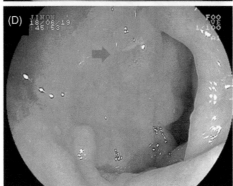

FIGURE 4.4 Gastroduodenal Crohn's disease: "bamboo joint"—like pattern. The lesion was detected in the antrum (A), fundus (B and C), and duodenum cap (D) *(green arrows). Courtesy Dr. Yubei Gu of Ruijin Hospital of Shanghai Jiaotong University.*

    CD in patients with ileostomy or strictureplasty is discussed in Chapter 8, Crohn's disease: postsurgical. The "normal" adaptive changes of the distal small bowel in patients with ileostomies or jejunostomies, such as dilated lacteals, should be distinguished from true recurrent CD in the bowel segment (Fig. 4.8).

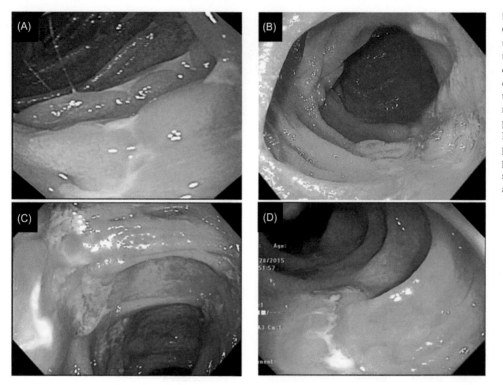

FIGURE 4.5 Duodenum Crohn's disease patterns of ulcers. (A) Discrete erosions and ulcers along the folds in a circumferential pattern; (B) discrete clean-based large ulcer at the second part of the duodenum; (C) longitudinal ulcers, perpendicular to the folds; (D) longitudinal ulcer at the second portion of the duodenum in a patient with concurrent primary sclerosing cholangitis and biliary stent in place.

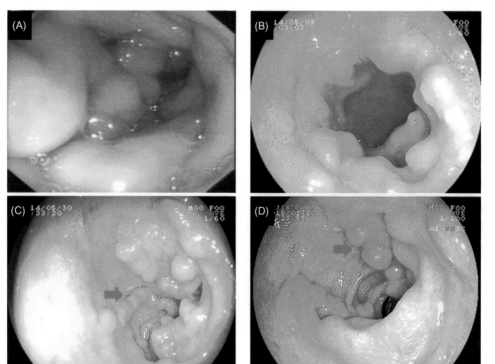

FIGURE 4.6 Duodenum Crohn's disease—nodular and "bamboo joint"—like patterns. (A and B) Nodular mucosa at the bulb and second portion of the duodenum and (C and D) "bamboo joint"—like appearance of duodenum mucosa (*green arrow*). (A) Courtesy Dr. Xinying Wang of Zhujiang Hospital Southern Medical University; (B) courtesy Dr. Mei Wang of Affiliated Hospital of Yangzhou University; (C and D) courtesy Dr. Yubei Gu of Ruijin Hospital of Shanghai Jiaotong University.

Distal ileum inflammatory Crohn's disease

The distal or terminal ileum is the most common location of CD. The disease distribution can be segmental in the area. It is in this location that CD has a wide range of endoscopic presentations, ranging from villous depletion to serpiginous

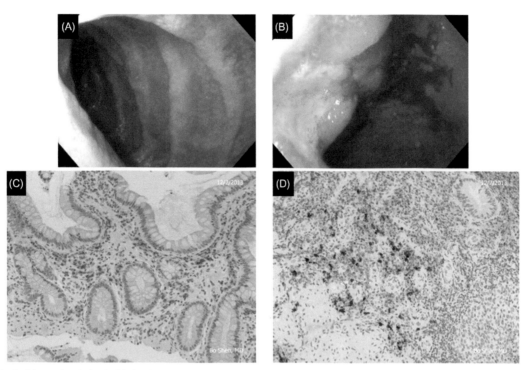

FIGURE 4.7 IgG4-associated duodenitis in Crohn's disease. (A and B) Diffuse granular mucosa with ulcers and loss of vascular pattern and (C and D) infiltration of plasma cells (pink in CD138 stain) with expression of IgG4 (brown with 3,3'-diaminobenzidine). *CD*, Crohn's disease.

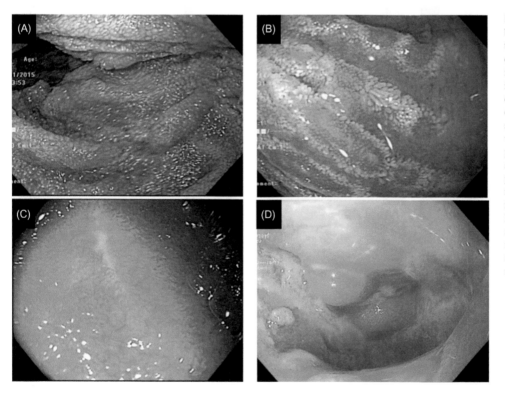

FIGURE 4.8 Neo-distal small bowel in patients with ileostomies for CD. (A) Diffusely distributed lacteals with "white" ileal mucosa in a patient with ileostomy and short-gut syndrome. The mucosal pattern is considered a "normal," compensatory changes; (B) depleted villi in the proximal ileum in a patient with high ileostomy and short-gut syndrome, with the pattern not consistent with recurrent CD; (C) single circumferential erosion at the fascia level of ileostomy, which is not active CD; and (D) true recurrent CD in the ileum with deep, stellate ulcers and nodular mucosa. *CD*, Crohn's disease.

ulcers. In order to obtain accurate characterization of mucosal disease, adequate wash of mucus, exudates, and plaques is required (Fig. 4.9). Terminal ileitis may present with villous depletion, erythema, or edema, which are considered mild or partially treated forms of disease (Fig. 4.10). Erosions and ulcers, however, are more common endoscopic

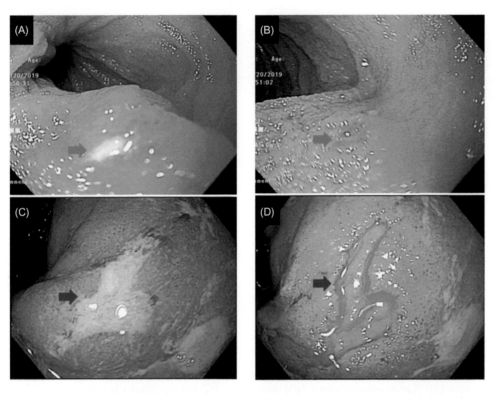

FIGURE 4.9 Importance of mucosal washing in the characterization of ulcers in inflammatory bowel disease. (A and B) Aphthous-like erosion covered with exudate in the terminal ileum before and after washing (*green arrows*); and (C and D) irregular ulcer covered with exudates before and after washing (*blue arrows*).

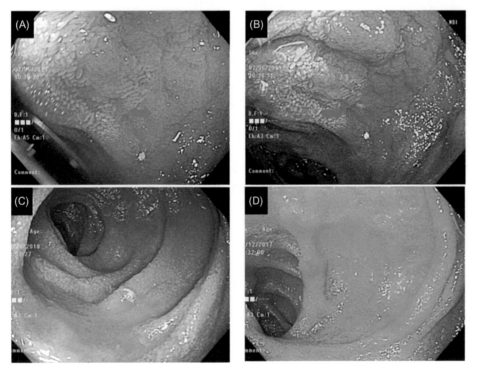

FIGURE 4.10 Villous patterns of Crohn's disease in the distal ileum. (A and B) Patchy villous depletion before therapy on white-light colonoscopy and narrow banding imaging (*green arrows*); (C) area of erythema with villous depletion surrounded by normal appearing mucosa; and (D) villous depletion with mucosal scars and mild luminal stenosis after anti-tumor necrosis factor therapy.

features, reflecting active CD. They can be aphthous-like, small or large, shallow or deep, regular or irregular in border, clean-based or exudate- or plaque-covered, isolated or surrounded by mucosal edema, erythema or granularity (Fig. 4.11). Longitudinal ulcers, especially single ones along the mesentery border, are considered classic features for Crohn's ileitis and Crohn's colitis (Fig. 4.12). However, variants of these classic ulcer patterns can be found in the small bowel and colon in CD. The ulcers can be circumferentially distributed, especially in the neo-terminal ileum after

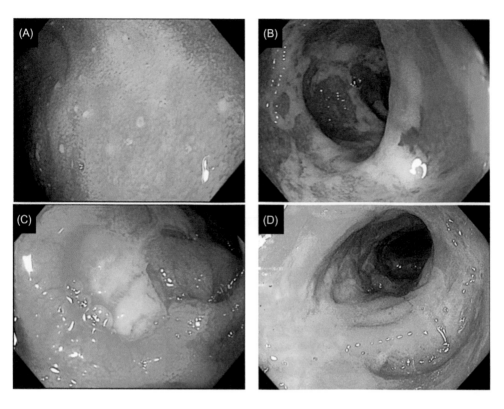

FIGURE 4.11 Patterns of erosions and ulcers in Crohn's disease involving the terminal ileum. (A) Multiple aphtha-like erosions; (B) diffuse ulcers covered with exudates on the background of edematous mucosa; (C) irregular large deep ulcers with adjacent edema; and (D) multiple longitudinal and serpiginous ulcers.

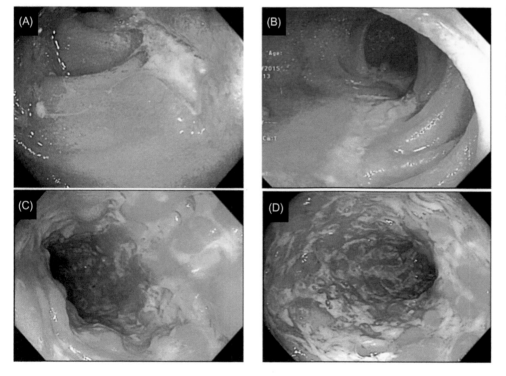

FIGURE 4.12 Patterns of serpiginous ulcers in Crohn's disease in the terminal ileum and colon. (A and B) Short and long serpiginous ulcers along the mesentery edge in the terminal ileum and (C and D) multiple columns of serpiginous ulcers in the colon.

ileocolonic resection and ileocolonic or ileorectal anastomosis (Fig. 4.13). Therefore the presence of circumferentially distributed or transverse ulcers does not necessarily exclude the diagnosis of CD, although the pattern is more commonly seen in intestinal tuberculosis (ITB) [8,9] and also seen in ischemic colitis [10]. The "cobblestoning"-like

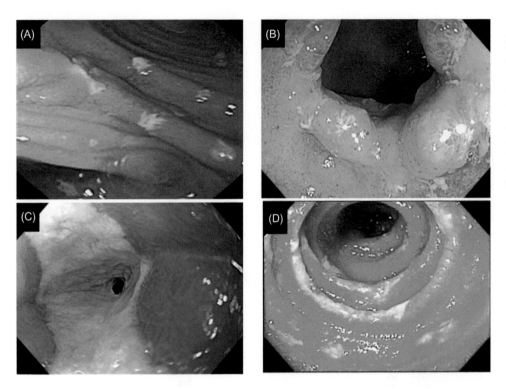

FIGURE 4.13 Variant ulcer patterns in Crohn's disease in the distal ileum. (A) Discrete ulcers arranged along the folds; (B) circumferential ulcer and edema, about to form a short stricture; (C) longitudinal kissing ulcers; and (D) superficial ulcers with exudates, circumferentially or transversely distributed along the neo-terminal ileum fold after ileocolonic resection and ileocolonic anastomosis.

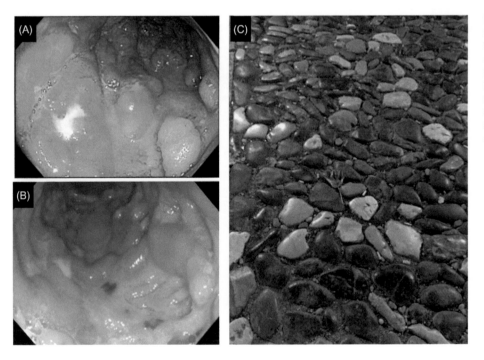

FIGURE 4.14 Cobblestoning-like mucosa of the small and large bowels in Crohn's disease. The endoscopic feature results from deep longitudinal ulcers or fissures in between islet of edematous mucosa. (A) Cobblestoning mucosa of the terminal ileum; (B) cobblestoning mucosa of the colon; and (C) the cobblestones.

appearance of the ileum and less so of colon mucosa has been considered another classic endoscopic and radiographic feature in CD (Fig. 4.14). The formation of "cobblestoning" appearance results from deep fissures or ulcers with nodular mucosal inflammation in between. The nodular or "bamboo joint"—like pattern in gastroduodenal CD (Figs. 4.4—4.6) may also stem from the same disease process.

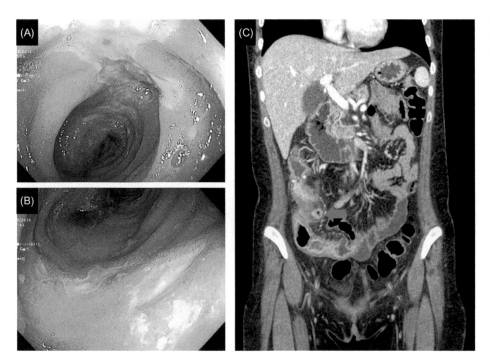

FIGURE 4.15 Correlation between mucosal inflammation and transmural inflammation in Crohn's disease. (A and B) Longitudinally scattered ulcers in the terminal ileum and (C) terminal ileitis with mucosal hyperenhancement and mesenteric fistula on computed tomography enterography (*green arrow*).

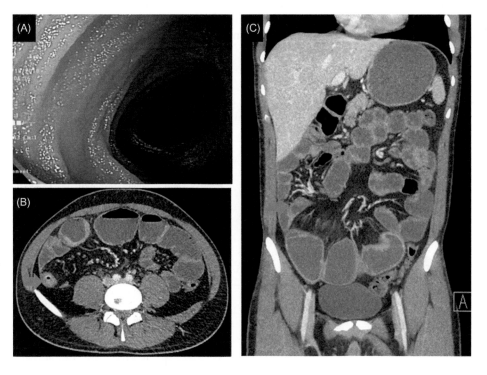

FIGURE 4.16 Poor correlation between mucosal inflammation and transmural inflammation in Crohn's disease. (A) Normal distal ileum and (B and C) multiple strictures with thickened bowel wall in the mid and distal ileum (*green arrows*).

In most cases, pattern and severity of mucosal disease predict transmural disease process, which follows the traditional "outside-in" theory of CD, that is, the disease from the mucosa to serosa, then mesentery (Fig. 4.15). Patients with severe mucosal disease may be associated with transmural disease and formation of stricture, fistula, and even abscess. However, the disease process may start from the serosa or mesentery to the mucosa, resulting in severe transmural inflammation or fistula in the presence of minimum or no mucosal inflammation (Fig. 4.16). The latter pattern of the disease process reflects the "inside-out" theory in CD [3]. Therefore the absence of mucosal inflammation does not exclude active CD, and in patients clinically suspected of active CD, transmural imaging is needed.

The presence of diffuse colitis and terminal ileitis in IBD was designated as confluent ileitis—colitis. Backwash ileitis can occur in patients with diffuse UC [11], particularly in those with primary sclerosing cholangitis (PSC) [12,13]. Backwash ileitis can also occur in patients with diffuse right-sided Crohn's colitis [13]. Histologic features, such as crypt distortion, acute lamina propria inflammation, and lamina propria expansion by mononuclear cells, have been reported to distinguish Crohn's colitis—associated backwash ileitis from UC-associated backwash ileitis [13]. Endoscopic distinction between Crohn's colitis—associated backwash ileitis and UC or UC + PSC—associated backwash ileitis can be subtle. There is diffuse inflammation with edema, erythema, loss of vascularity, granularity, exudates, and erosions or small ulcers of the bowel, which are shared by the terminal ileum and right colon. Strictures and fistulae are rare in Crohn's colitis—associated backwash ileitis. Both Crohn's colitis—associated and UC + PSC—associated backwash ileitis can have rectal sparing [12,14]. Patulous or "fish mouth"—like ileocecal valve (ICV) is considered a classic feature for both Crohn's colitis—associated and UC-associated backwash ileitis. However, ulceration on the patulous ICV may be more common in Crohn's colitis—associated backwash ileitis than UC-associated entity (Fig. 4.17). Endoscopic features of UC- and UC + PSC—associated backwash ileitis are also discussed in Chapter 9, Ulcerative colitis.

There is no consensus on whether Crohn's colitis—associated backwash ileitis should be classified into L1 (ileitis), L2 (colitis), or L3 (ileocolitis). It may be reasonable to put Crohn's colitis—associated backwash ileitis into L2 category, as the backwash ileitis may be considered being secondary to diffuse colitis. The endoscopic features of CD ileitis (L1) and Crohn's colitis—associated backwash ileitis appear to be different. Isolated, larger, irregular, serpiginous or deep ulcers, and strictures are common in CD ileitis (Figs. 4.8D, 4.9, 4.11—4.15). In addition, CD ileitis is often associated with ulcerated, strictured, or deformed ICV (Fig. 4.18). However, isolated involvement of ICV by CD can also present, sparing other parts of GI tract. This author has speculated that CD in ICV may share features of etiopathogenetic factors with achalasia of the esophagus, which may be "cured" by endoscopic valvectomy, equivalent to endoscopic or surgical myotomy [15]. Patulous and nodular ICV is one of distinct endoscopic features of ITB [9].

Inflammatory CD often evolves into fibrostenotic or fistulizing CD. Some patients may have strictures and fistulas as initial presentations (see Chapter 5: Crohn's disease: fibrostenotic type and Chapter 6: Crohn's disease: penetrating type).

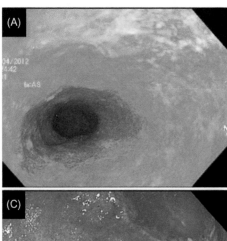
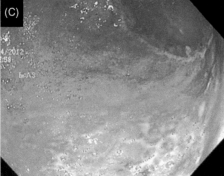
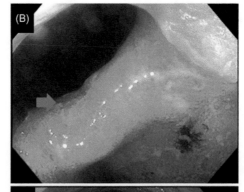
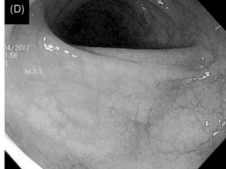

FIGURE 4.17 Backwash ileitis from diffuse Crohn's colitis and primary sclerosing cholangitis with rectal sparing. Backwash ileitis can occur in patients with diffuse Crohn's colitis, as well as UC. (A) Diffuse ileitis with loss of vascularity; (B) patulous ICV with ulcers (*green arrow*). In contrast, ICV in UC and backwash ileitis typically does not present with ulcers; (C) diffuse active colitis, similar pattern to that in the distal ileum. That is why this pattern of bowel inflammation has also been termed confluent colitis and ileitis; (D) disease-spared rectum. *ICV,* Ileocecal valve; *UC,* ulcerative colitis.

Colonic Crohn's disease/Crohn's colitis

It is estimated that 10%−15% of patients with CD have disease limited to the large bowel only. The pattern of Crohn's colitis pattern varies greatly in the distribution, extent, depth, and severity. Crohn's colitis can be presented as segmental or diffuse inflammation with edema, erythema, exudates, granularity, nodularity, loss of vascularity, friability, and various forms of ulcers (Fig. 4.19). Although segmental distribution and rectal sparing are common in patients with

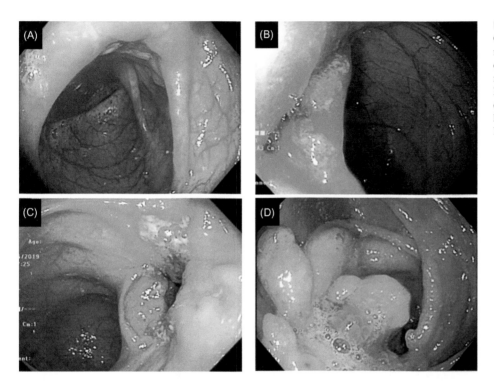

FIGURE 4.18 Abnormal ICV in Crohn's disease. (A) Strictured ICV, making endoscopic intubation difficult; (B) inflamed ICV with granular mucosa; (C) ulcerated and deformed ICV; and (D) nodular, strictured, and deformed ICV, with stricture and hidden ileo-ileal fistula. *ICV*, Ileocecal valve.

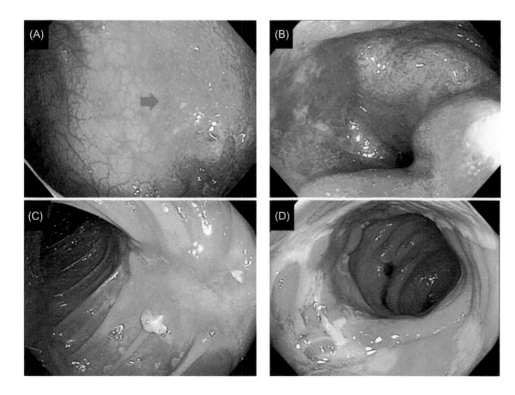

FIGURE 4.19 Patterns of Crohn's colitis. (A) Inflamed (erythema) and noninflamed parts of the colon, with a sharp demarcation (*green arrow*), similar to that in left-side UC or UC proctitis; (B) edematous and erythematous colon mucosa with exudates; (C) partially healed longitudinal ulcer with scar along the mesentery border in the transverse colon; and (D) discrete and confluent ulcers in the rectum in a patient with subtotal colectomy with ileorectal anastomosis. Rectal involvement can occur before or after this type of surgery. *UC*, Ulcerative colitis.

Crohn's colitis (L2) or Crohn's ileocolitis (L3), rectal involvement can occur, typically presented with discrete ulcers and inflammation. Patients with ileocolitis with rectal sparing undergoing subtotal colectomy with ileorectal anastomosis may develop de novo Crohn's proctitis after surgery (Fig. 4.19D). Isolated diffuse Crohn's proctitis is rare; and if present, it is often presented with distal rectal or anal inflammation with perianal disease. Sometimes, various degrees and patterns of colitis can be found in a given session of colonoscopy in an individual patient (Fig. 4.20). Long-standing Crohn's colitis can develop chronic inflammatory changes, such as pseudopolyps and mucosal bridges (Fig. 4.21).

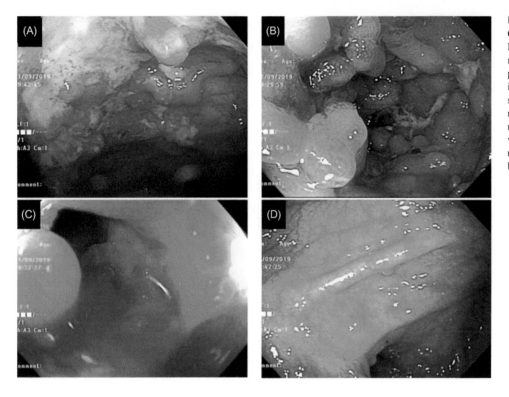

FIGURE 4.20 Pattern of Crohn's colitis. (A) Large stellate ulcer with surrounding mucosal edema, nodularity, and pseudopolyps; (B) multiple inflammatory polyps with fissures in between causing luminal narrowing; (C) ulcerated and nodular stricture in the transverse colon; and (D) sparing of nearby segments of the large bowel.

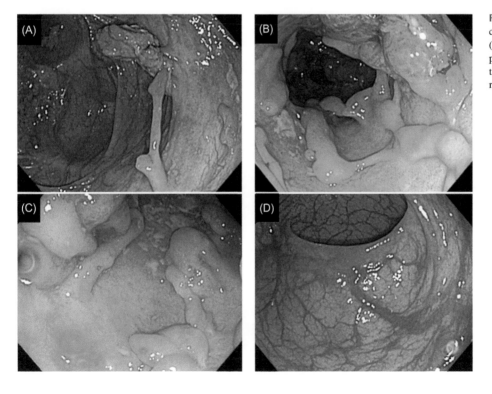

FIGURE 4.21 Pattern of Crohn's colitis. (A−C) Mucosal bridge (*green arrow*) along with elongated pseudopolyps and mucosal ulceration in the proximal colon and (D) rectal sparing with normal mucosa.

Crohn's disease in surgically altered small bowel

Patients with refractory CD may require fecal diversion with ileostomy, jejunostomy, or colostomy. Strictureplasty CD of the small bowel in patients with ostomies or strictureplasty is discussed in Chapter 8, Crohn's disease: postsurgical. CD of the ileal pouch can also occur in UC patients undergoing restorative proctocolectomy. Endoscopic features of inflammatory, fibrostenotic, or fistulizing CD of the pouch is discussed in Chapter 11, Ulcerative colitis postsurgical.

Small bowel transplantation has been performed in patients with severe, refractory CD, or CD with short-gut syndrome [16]. Endoscopic features of recurrent CD in patients with small bowel transplant are discussed in Chapter 27, Inflammatory bowel disease after organ transplant.

Summary and recommendations

Inflammatory phenotype of CD can occur in any parts of the GI tract. Natural history of CD dictates that patients with poorly controlled inflammatory CD are likely to develop further tissue structural damage and complications, such as stricture and fistula. The patterns of inflammation vary depending on the disease location and severity, prior history surgery with degree of altered bowel anatomy. Recognition and appreciation of the common endoscopic features are a key for the diagnosis, differential diagnosis, disease monitoring, and therapy.

References

[1] Silverberg MS, Satsangi J, Ahmad T, Arnott ID, Bernstein CN, Brant SR, et al. Toward an integrated clinical, molecular and serological classification of inflammatory bowel disease: report of a Working Party of the 2005 Montreal World Congress of Gastroenterology. Can J Gastroenterol 2005;19(Suppl. A):5A–36A.

[2] Gasche C, Scholmerich J, Brynskov J, D'Haens G, Hanauer SB, Irvine EJ, et al. A simple classification of Crohn's disease: report of the Working Party for the World Congresses of Gastroenterology, Vienna 1998. Inflamm Bowel Dis 2007;6:8–15.

[3] Chang S, Shen B. Classification and reclassification of inflammatory bowel diseases: from clinical perspective. In: Shen B, editor. Interventional inflammatory bowel disease. Cambridge, MA: Elsevier; 2018. p. 17–34.

[4] Laube R, Liu K, Schifter M, Yang JL, Suen MK, Leong RW. Oral and upper gastrointestinal Crohn's disease. J Gastroenterol Hepatol 2018;33:355–64.

[5] Nomura Y, Moriichi K, Fujiya M, Okumura T. The endoscopic findings of the upper gastrointestinal tract in patients with Crohn's disease. Clin J Gastroenterol 2017;10(4):289–96.

[6] Yokota K, Saito Y, Einami K, Ayabe T, Shibata Y, Tanabe H, et al. A bamboo joint-like appearance of the gastric body and cardia: possible association with Crohn's disease. Gastrointest Endosc 1997;46:268–72.

[7] Navaneethan U, Liu X, Bennett AE, Walsh RM, Venkatesh PG, Shen B. IgG4-associated ampullitis and cholangiopathy in Crohn's disease. J Crohns Colitis 2011;5:451–6.

[8] He Y, Zhu Z, Chen Y, Chen F, Wang Y, Ouyang C, et al. Development and validation of a novel diagnostic nomogram to differentiate between intestinal tuberculosis and Crohn's disease: a 6-year prospective multicenter study. Am J Gastroenterol 2019;114:490–9.

[9] Kedia S, Das P, Madhusudhan KS, Dattagupta S, Sharma R, Sahni P, et al. Differentiating Crohn's disease from intestinal tuberculosis. World J Gastroenterol 2019;25:418–32.

[10] Beppu K, Osada T, Nagahara A, Matsumoto K, Shibuya T, Sakamoto N, et al. Relationship between endoscopic findings and clinical severity in ischemic colitis. Intern Med 2011;50:2263–7.

[11] Hamilton MJ, Makrauer FM, Golden K, Wang H, Friedman S, Burakoff RB, et al. Prospective evaluation of terminal ileitis in a surveillance population of patients with ulcerative colitis. Inflamm Bowel Dis 2016;22:2448–55.

[12] de Vries AB, Janse M, Blokzijl H, Weersma RK. Distinctive inflammatory bowel disease phenotype in primary sclerosing cholangitis. World J Gastroenterol 2015;21:1956–71.

[13] Sahn B, De Matos V, Stein R, Ruchelli E, Masur S, Klink AJ, et al. Histological features of ileitis differentiating pediatric Crohn disease from ulcerative colitis with backwash ileitis. Dig Liver Dis 2018;50:147–53.

[14] Loftus Jr EV, Harewood GC, Loftus CG, Tremaine WJ, Harmsen WS, Zinsmeister AR, et al. PSC-IBD: a unique form of inflammatory bowel disease associated with primary sclerosing cholangitis. Gut 2005;54:91–6.

[15] Yang Y, Lyu W, Shen B. Endoscopic valvectomy of ileocecal valve stricture resulting in resolution of ileitis in Crohn's disease. Gastrointest Endosc 2018;88:195–6.

[16] Nyabanga C, Kochhar G, Costa G, Soliman B, Shen B, Abu-Elmagd K. Management of Crohn's disease in the new era of gut rehabilitation and intestinal transplantation. Inflamm Bowel Dis 2016;22:1763–76.

Chapter 5

Strictures in Crohn's disease

Bo Shen
Center for Inflammatory Bowel Diseases, Columbia University Irving Medical Center-New York Presbyterian Hospital, New York, NY, United States

Chapter Outline

Abbreviations

CD	Crohn's disease
EBD	endoscopic balloon dilation
ESt	endoscopic stricturotomy
IBD	inflammatory bowel disease
ICV	ileocecal valve
NSAID	nonsteroidal antiinflammatory drugs
UC	ulcerative colitis

Introduction

Intestinal stricture, defined as abnormal narrowing of bowel lumen, has also been termed stenosis. A stricture can lead to a spectrum of narrowing from subtle to complete obstruction. In patients with inflammatory bowel disease (IBD), either Crohn's disease (CD) or ulcerative colitis (UC), intrinsic stricture may result from disease process ranging from inflammation and fibrosis to malignancy in the mucosa, muscularis mucosae, submucosa, or muscularis propria, or combination. Extraintestinal or extraluminal disease process can also cause bowel stricture or obstruction (extrinsic stricture), such as abscess, adhesion, and compression from benign or malignant mass.

In the Montreal classification system, CD was divided into nonstricturing/nonpenetrating (B1), stricturing (B2), and penetrating (B3), based on clinical behavior, as shown in Chapter 4, Crohn's disease: inflammatory type (Table 4.1) [1]. Stricturing disease is believed to result from persistent inflammation (B1) and fibrosis. On the other hand, stricture is a major contributing factor for the formation of fistula (B3) and abscess. Strictures often lead to significant morbidities, particularly bowel obstruction. The main treatment strategies for stricture are the control of inflammation with medical therapy and relief of obstruction with mechanical force, such as endoscopic balloon dilation (EBD), endoscopic stricturotomy (ESt), bowel resection, and stricturoplasty [2,3].

Stricture in CD represents a spectrum of clinical phenotypes, underlying disease process, and prognosis. A classification of IBD-related stricture has been proposed (Table 5.1) [4]. In the classification system, IBD-related strictures are categorized based on the source, clinical presentation, underlying disease, the presence of prior surgery, malignant potentials, degree, location and length, and associated disease conditions. Proper diagnosis and classification of stricture will guide monitoring of disease progression and medical, endoscopic, and surgical therapy and improve quality of life

Atlas of Endoscopy Imaging in Inflammatory Bowel Disease. DOI: https://doi.org/10.1016/B978-0-12-814811-2.00005-0
© 2020 Elsevier Inc. All rights reserved.

TABLE 5.1 Classification of strictures in inflammatory bowel disease [4].

	Category	Description	Examples
Source	Intrinsic	Inflammation, fibrosis, or malignancy in any layers of bowel wall	Terminal ileum stricture of Crohn's disease
	Extrinsic	Extraintestinal compression, pushing, and pulling	Adhesion, abscess compression
Clinical presentation	Symptomatic		
	Asymptomatic		
Underlying disease and surgery	Crohn's disease		Ileocecal valve stricture, anal stricture, terminal ileum strictures
	Ulcerative colitis		
	Postsurgical	Bowel resection and anastomosis	Ileocolonic stricture, ileal rectal stricture
		Ileal pouch	Inlet and anastomosis strictures, loop ileostomy site stricture, afferent limb site strictures
		Stricturoplasty	Inlet and outlet strictures
		Bypass	Gastrojejunostomy stricture
		Ileostomy/Jejunostomy/Colostomy	Skin, stoma, and bowel stricture
	Primary (disease, drug, ischemia)	Disease-associated	
		Drug associated	NSAID, pancreas enzyme
	Secondary	Anastomotic	
		Near suture or staple lines	Pouch inlet, stricturoplasty outlet/inlet
Malignant potential	Benign		Ileocolonic anastomotic stricture
	Malignant	Adenocarcinoma, lymphoma, squamous cell cancer (in anal canal)	Colon cancer from colitis-associated dysplasia
Inflammation and fibrosis component	Inflammatory		
	Fibrotic		
	Mixed		
Length	Short	<4 cm	
	Long	≥4 cm	
Characteristic of stricture	Ulcerated		
	Web like		Concurrent NSAID use
	Spindle shaped		
	Angulated		
	Symmetry	Circumferentially asymmetric	Some ileocecal valve stricture
		Longitudinally asymmetric	Ileocolonic or ileorectal anastomotic strictures
Location in nonsurgical patients	Esophagus		
	Pylorus		
	Small bowel		
	Ileocecal valve		
	Colon		
	Rectum		
	Anus		

(Continued)

TABLE 5.1 (Continued)

	Category	Description	Examples
Degree	No stricture	No stricture	
	Mild	Passage of scope with mild resistance	
	Moderate	Passage of scope with moderate resistance	
	Severe	Pinhole stricture, not traversable to endoscope	
Number	Single		
	Multiple		
Complexity	Simple		
	Complex with associated conditions	Fistula and/or abscess	
		Prestenotic luminal dilation	

NSAID, Nonsteroidal antiinflammatory drug.

and long-term outcome. For example, the treatment approach and outcome for the primary (or disease related) and secondary (surgery or medicine related) strictures are different.

Stricture is evaluated and diagnosed based on a combined assessment of clinical presentation, imaging, and endoscopy. The presence or absence, and degree and length of stricture on abdominal imaging and on endoscopy are not necessarily correlated. Endoscopy plays a key role in the assessment of degree, number, length, and associated mucosal inflammation as well as delivery of therapy (such as balloon dilation and ESt) in patients with strictures [5].

Esophageal and gastroduodenal stricturing Crohn's disease

Isolated involvement of the esophagus CD is rare. Often esophageal strictures in patients with CD result from the use of medications (such as nonsteroidal antiinflammatory drugs and potassium tablets) or metabolic complications [such as iron deficiency (Fig. 5.1)]. Superimposed viral, bacterial, or fungal infection in the esophagus may manifest as esophageal ulcers but rarely as strictures.

The most common location of stricturing CD of the stomach is the pylorus or prepyloric area, which can occur in isolation or concurrently with gastroduodenal CD. Pyloric narrowing can manifest as nonulcerated tight stenosis or ulcerated stricture, ranging from 1 to 3 cm in length. Concurrent pre- or postpyloric nodularity or ulcers may be present (Fig. 5.2). Severe pyloric or prepyloric stenosis is often associated with gastric outlet obstruction. Extreme precaution should be taken to avoid aspiration during endoscopy. Since medical and surgical treatment options of CD-associated pyloric strictures are limited, endoscopic therapy has been explored. EBD, ESt with needle knife or insulated-tip knife, and topical injection of botulin toxin have been used by this author's team (Figs. 5.2D and 5.3). For patients with symptomatic persistent gastric outlet obstruction, ventilation gastrostomy tube may be helpful.

Duodenum CD can present with inflammation, ulcers, or stricture on endoscopy. The most common location of stricturing CD in the duodenum is at the cap, that is, the junction between the first and second portions. Concurrent duodenum inflammation may be present or absent. Duodenum strictures may be subtle or severe, with normal, edematous, or ulcerated overlying mucosa (Fig. 5.4). While concurrent mucosal or transmural inflammation may respond to aggressive medical therapy, duodenum strictures, particularly fibrotic stricture, poorly respond to medications. On the other hand, surgical strictureplasty or bypass with gastrojejunostomy has been performed for long, medically refractory duodenum strictures (Fig. 8.21). EBD [6,7] or ESt [8] may be attempted in short duodenum strictures (<3 cm) (Fig. 5.5).

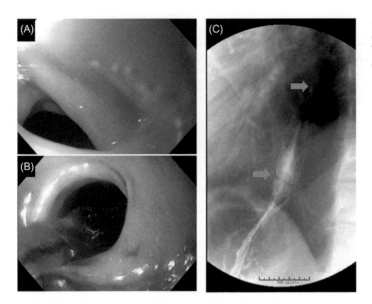

FIGURE 5.1 Esophageal stricture from iron deficiency (Plummer–Vinson syndrome) in Crohn's disease. (A and B) Web-like tight stricture undergoing endoscopic balloon dilation and (C) multiple esophageal strictures in barium esophagram.

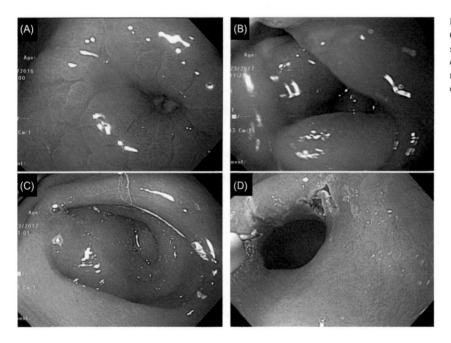

FIGURE 5.2 Pyloric stenosis in gastric Crohn's disease. (A) Severe ulcerated pyloric stricture; (B) nodular and strictured pyloric channel; (C) inflammatory, nonulcerated pyloric stricture; and (D) endoscopic stricturotomy of pyloric stricture with needle knife.

Deep small bowel stricturing Crohn's disease

Endoscopic and histologic evaluation of jejunal and ileal CD require deep enteroscopy mainly with device-assisted endoscopy, which is described in a separate chapter (see Chapter 18: Enteroscopy in inflammatory bowel disease and inflammatory bowel disease–like conditions).

Ileal stricturing Crohn's disease

The distal or terminal ileum is the most common location of CD with L1 and L3 disease. A segmental distribution of diseases in the small bowel is common (Fig. 5.6). The stricture can be short or long, single or multiple. There are various other features of primary, disease-associated strictures, including ulcerated and nonulcerated strictures,

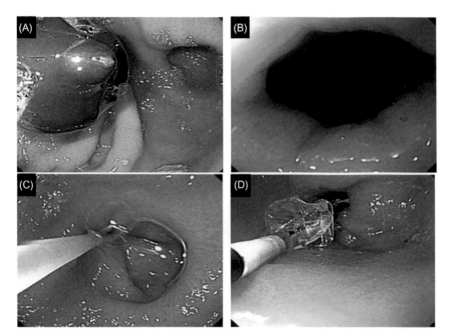

FIGURE 5.3 Pyloric stricture in gastric CD. (A) Ventilation gastrostomy button tube in a patient with CD-associated pyloric stenosis; (B) a close view of nonulcerated pyloric stenosis; (C) botulin toxin injection into pyloric stenosis; and (D) endoscopic balloon dilation of pyloric stenosis. *CD*, Crohn's disease.

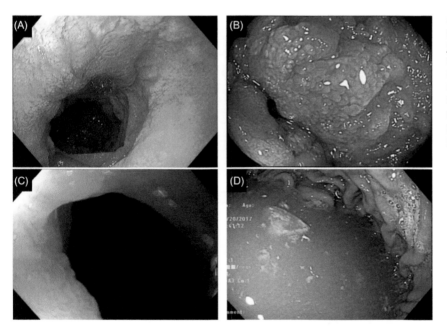

FIGURE 5.4 Duodenal stricture resulting from Crohn's disease. (A) Mild stricture in the junction between the first and second portions of duodenum, that is, the duodenum cap; (B) asymmetric inflammatory stricture at the duodenum cap with surrounding nodular mucosa; (C) a close view of a web-like stricture at the second portion of duodenum; and (D) retained gastric from the duodenum stricture-associated partial outlet obstruction.

inflammatory and fibrostenotic strictures (Figs. 5.7–5.10). This author has noticed a growing number of patients with nonulcerated, spindle-shaped strictures, especially in those with "mucosal healing" by a long-term biological therapy (Figs. 5.10 and 5.11). It is speculated that this type of stricture results from tissue healing and remodeling from the medical therapy. It is also important to document the presence or absence of bowel inflammation in pre- and poststenotic segments, and the presence or absence of prestenotic luminal dilation (Figs. 5.11–5.13). Distinction between inflammatory and fibrotic stricture can be difficult, even with radiographic features, such as mucosal hyperenhancement. Sometimes, the distinction has to be made by the treating endoscopist on site. Inflammatory strictures typically are friable, even with gentle endoscope contact. Local trauma resulting from endoscopic therapy is more severe in

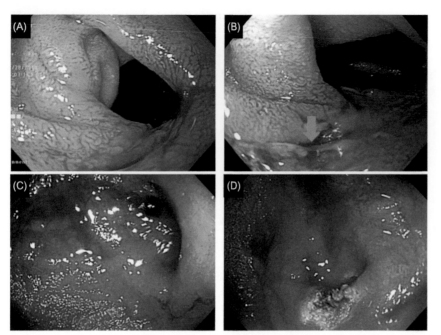

FIGURE 5.5 Duodenum strictures and endoscopic therapy. (A and B) A 1-cm-long, asymmetric, nonulcerated stricture at the proximal second portion of the duodenum, treated with 18-mm through-the-scope balloon dilation. Notice the mucosal tear after the treatment [*green arrow* in (B)]. (C and D) A 2-cm-long nonulcerated fibrotic stricture at the duodenum cap, treated with endoscopic stricturotomy.

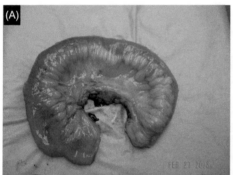
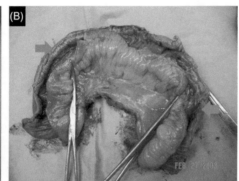

FIGURE 5.6 Classic small bowel strictures in Crohn's disease in surgically resected specimen. (A and B) Skip lesions with long strictures [*green arrows* in (B)] and fat creeping at the mesentery border.

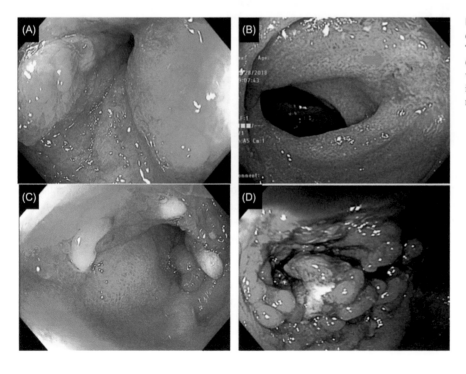

FIGURE 5.7 Inflammatory strictures in distal small bowel Crohn's disease. (A) Tight, edematous, nonulcerated stricture; (B) web-like, asymmetric, mild stricture, with ulcer (*green arrow*); and (C and D) inflammatory polyps leading to luminal narrowing.

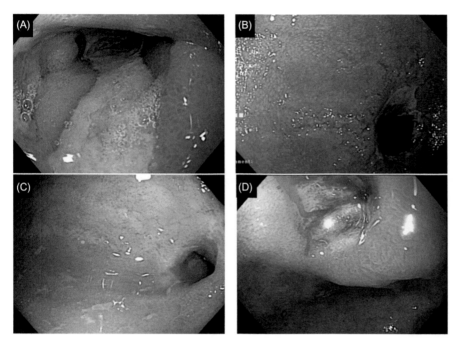

FIGURE 5.8 Patterns of ulcerated strictures in distal small bowel Crohn's disease. (A) Longitudinal ulcers across the stricture; (B and C) superficial, circumferential (B) or semicircumferential (C) ulcers on the stricture; and (D) edematous and ulcerated stricture.

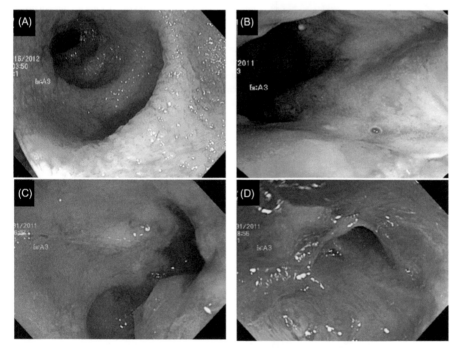

FIGURE 5.9 Patterns of primary strictures in Crohn's disease. (A) Multiple ring-shaped, nonulcerated strictures along the folds, from tissue healing after biological therapy; (B) mild stricture with a large ulcers; (C) inflammatory asymmetric stricture with edema and exudates; and (D) severe inflammatory stricture with friable tissue.

inflammatory strictures than fibrotic strictures (Figs. 5.14 and 5.15). Our previous studies showed that ESt was more effective and carried a lower risk for perforation, but a higher risk for bleeding, than EBD [9].

There can be hidden enteric fistulae which originate from diseased bowel segment proximal to the stricture. The bowel lumen around the fistula, proximal to the stricture may or may not be dilated (Fig. 5.13). Therefore cross-sectional imaging, such as computed tomography and magnetic resonance imaging, should be obtained prior to diagnostic and therapeutic endoscopy.

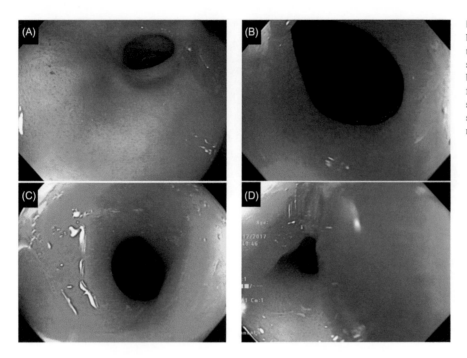

FIGURE 5.10 Nonulcerated distal small bowel strictures in Crohn's disease. Those types of strictures typically result from tissue injury, remodeling, and healing with biological therapy. (A) Nonulcerated short fibrotic stricture; (B and C) close view of short, nonulcerated, severe, spindle-shaped, fibrotic strictures; and (D) severe, nonulcerated, edematous stricture.

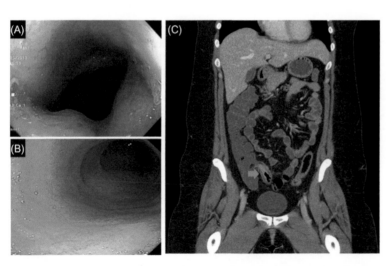

FIGURE 5.11 Mixed fibrotic and inflammatory strictures in distal small bowel Crohn's disease. (A) Nonulcerated stricture with overlying normal mucosa with a history of use of antitumor necrosis factor; (B) normal diameter of bowel lumen proximal to the stricture. The mucosa was also normal; and (C) narrowed lumen (3-cm stricture) at the distal ileum with mucosal hyperenhancement. Prestenotic bowel lumen was dilated (*green arrow*).

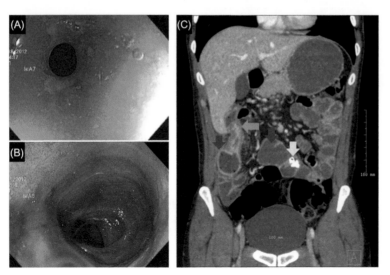

FIGURE 5.12 Fibrotic stricture at the terminal ileum in Crohn's disease with prestenotic luminal dilation. (A) Tight, nonulcerated fibrotic stricture resulting from long-term use of antitumor necrosis factor antagonists; (B) dilated bowel lumen proximal to the stricture; (C) dilated loop of bowel (*blue arrows*) proximal to the stricture (*green arrow*), with a fecal bezoar trapped (*yellow arrow*).

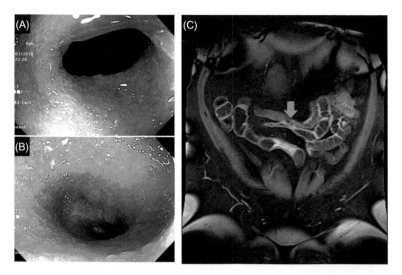

FIGURE 5.13 Stricture and hidden fistula at the distal ileum in Crohn's disease. (A) Web-like stricture, 3 cm proximal to the ICV; (B) another 5-cm-long stricture, 15 cm from ICV; (C) proximal to the two strictures, there was an enteroenteric fistula (ileoileal fistula on CT enterography) (*green arrow*). *ICV*, ileocecal valve.

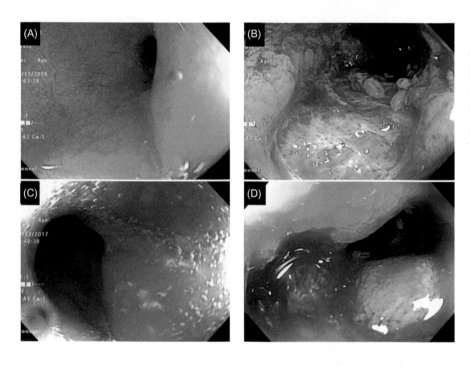

FIGURE 5.14 Fibrotic and inflammatory strictures in distal small bowel Crohn's disease. (A and B) 2-cm, nonulcerated fibrotic stricture, which was dilated with balloon. Notice large, superficial tear after the therapy; (C and D) 2-cm mixed inflammatory (with edema) and fibrotic stricture, which was dilated with the same size balloon, resulting in a deeper tear with more bleeding.

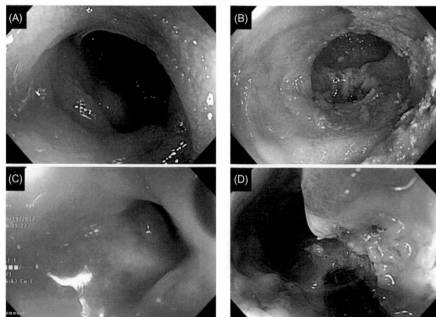

FIGURE 5.15 Endoscopic stricturotomy of distal ileum strictures in Crohn's disease. (A and B) Tattooed 2-cm, nonulcerated inflammatory stricture, treated with stricturotomy in a circumferential fashion; (C and D) mixed inflammatory and fibrotic, ulcerated stricture, which was treated with stricturotomy in a radial fashion.

Stricturing ilececal valve in Crohn's disease

Ileocecal valve (ICV) is a commonly affected location of CD. Patients with distal ileum CD often have coexisting inflammatory or strictured ICV. CD may affect ICV only. Endoscopic presentations of ICV stricturing CD vary, ranging from inflammatory stricture or fibrotic stricture at the valve to a nodular deformed valve (Figs. 5.16−5.22). Patients with ICV stricturing CD have various degrees of inflammation, fibrosis, or deformity at a certain degree. ICV valve strictures can be long or short. Endoscopic therapy with balloon dilation or stricturotomy may be attempted in the majority of patients (Figs. 5.16−5.20 and 5.22).

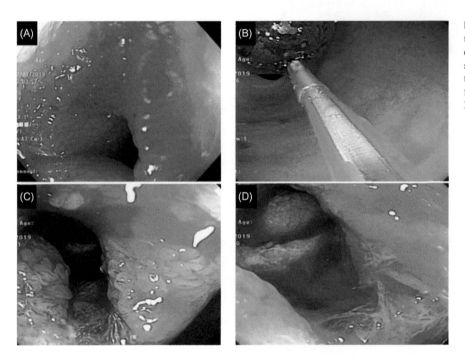

FIGURE 5.16 Inflammatory stricture at the ICV in Crohn's disease. (A) Tight, edematous ICV; (B) balloon dilation of the strictured ICV; and (C and D) tearing to the level of muscularis propria resulting from the endoscopic therapy. *ICV*, Ileocecal valve.

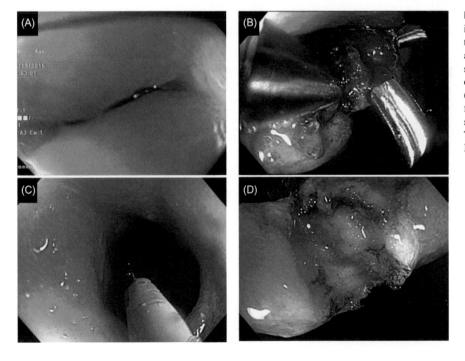

FIGURE 5.17 Fibrotic stricture at the ileocecal valve (ICV) in Crohn's disease in two patients. (A and B) Fibrotic stricture at ICV, which was treated with endoscopic stricturotomy, followed by the placement of endoclips to keep the luminal patency; (C and D) A separate patient had tight, fibrotic ICV stricture undergoing endoscopic stricturotomy with needle knife. (C) The whole wall of ICV was fibrotic. *ICV*, Ileocecal valve.

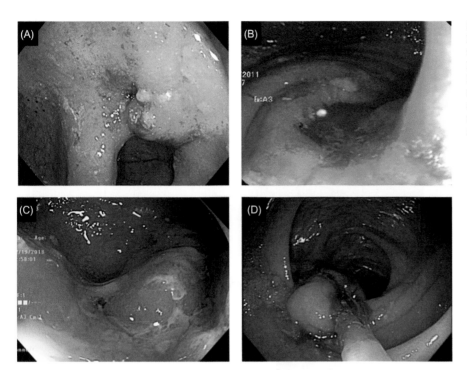

FIGURE 5.18 Deformed and strictured ICV in Crohn's disease. (A) Pinhole stricture at ICV; (B) near complete sealed ICV; (C) ulcerated and nodular stricture at ICV; (D) attempt to the dilation of an ICV stricture. These patterns of ICV strictures are different from backwash ileitis in ulcerative colitis or intestinal tuberculosis. *ICV*, Ileocecal valve.

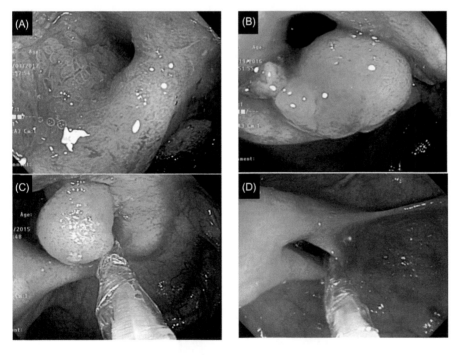

FIGURE 5.19 Deformed and strictured ICV in Crohn's disease. (A) Mixed inflammatory and fibrotic, nonulcerated ICV stricture with a long-term use of antitumor necrosis factor agent; (B) narrowed ICV from an inflammatory polyp; (C) balloon dilation of stricture ICV from a polypoid lesion; (D) deformed, nonulcerated, and narrowed ICV undergoing balloon dilation. *ICV*, Ileocecal valve.

    Isolated stricturing CD in ICV may be present in patients with or without prior medical therapy. Pure inflammatory (as opposed to mixed or fibrotic) strictures in ICV are rare (Fig. 5.17). Therefore the role of medical therapy for ICV strictures resulting from CD is limited. However, this author believes that some isolated fibrotic, stricturing ICV may represent a form of "achalasia" of the location, which may share etiopathogenetic process with achalasia in the esophagus. Similar to myotomy for achalasia of the esophagus, endoscopic valvectomy may offer a cure for selected patients with isolated ICV stricturing CD (Fig. 5.22) [10].

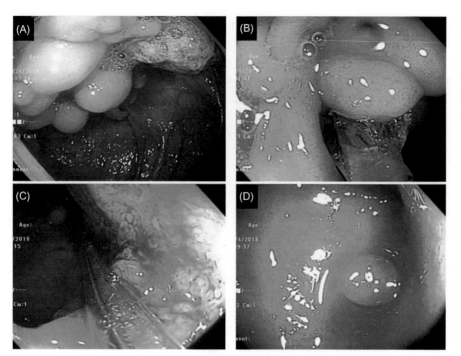

FIGURE 5.20 Deformed ICV in isolated Crohn's disease. (A) Nodular and narrowed ICV; (B and C) Endoscopic balloon dilation of the deformed ICV; and (D) normal terminal ileum proximal to ICV. This pattern of ICV disease should be differentiated from intestinal tuberculosis and intestinal lymphoma. *ICV*, Ileocecal valve.

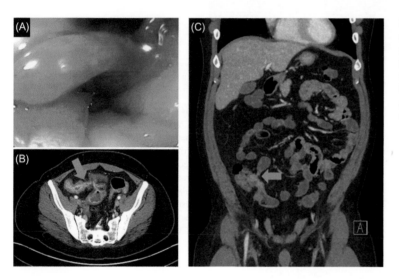

FIGURE 5.21 Enteric fistula hidden beyond deformed and strictured ICV in Crohn's disease. (A) Deformed, strictured, and edematous ICV; (B and C) enteroenteric fistula was demonstrated near ICV on CT enterography (*green arrows*). Preprocedural cross-sectional imaging is important when endoscopic intervention, such as balloon dilation or stricturotomy, is planned. *ICV*, Ileocecal valve.

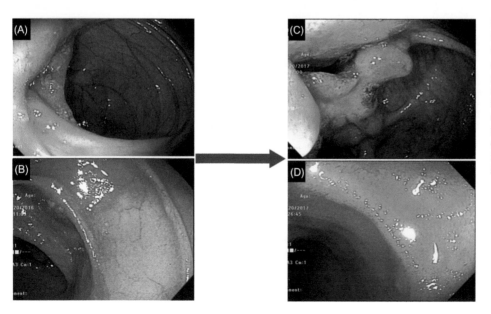

FIGURE 5.22 Fibrostenotic ICV in Crohn's disease. (A) Tight ICV with granular overlying mucosa, a similar pattern to achalasia at the gastroesophageal junction; (B) few aphthous erosions of the terminal ileum; (C) treatment with endoscopic stricturotomy, an equivalent to myotomy for achalasia; (D) resolution of erosions of the terminal ileum after the endoscopic therapy. *ICV*, Ileocecal valve.

Large bowel and anal stricures in Crohn's disease

Approximately 10%−15% of the patients with CD have colon disease only (L2 in the Montreal classification). In patients with Crohn's colitis, strictures are common, which can affect any parts of the large bowel. Like small bowel stricturing CD, colonic strictures can be mild or severe, short or long, ulcerated or nonulcerated, inflammatory or fibrotic, and benign or malignant (Figs. 5.23 and 5.24). EBD or ESt may be attempted in patients with single, short, and fibrotic colonic strictures (Figs. 5.23B and 5.24).

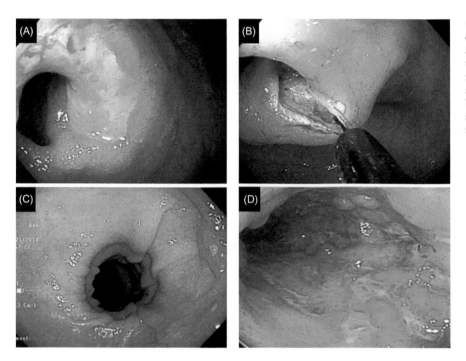

FIGURE 5.23 Patterns of primary colonic strictures in Crohn's disease. (A and B) Ulcerated primary colonic stricture treated with endoscopic stricturotomy; (C) mild, nonulcerated colon stricture with small erosions in the distal segment of the bowel; and (D) mild, long segment of stricture with serpiginous ulcers across the narrowed bowel lumen.

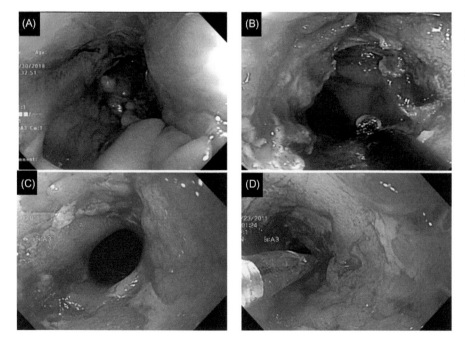

FIGURE 5.24 Distal rectal strictures in CD. These strictures can occur isolated or be a part of perianal CD. Patients may also have concurrent dyssynergic defecation, prolapse, or intussusception. (A and B) Distal rectal inflammatory stricture undergoing endoscopic stricturotomy with insulated-tip knife and (C and D) a similar distal rectal stricture in a separate patient undergoing endoscopic balloon dilation therapy. *CD*, Crohn's disease.

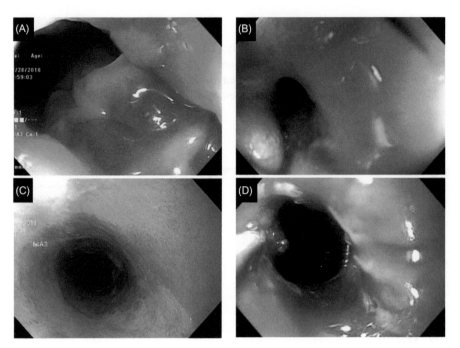

FIGURE 5.25 Anal strictures in CD. The stricture is commonly located at the anal transition zone or anal canal. Notice white squamous epithelia in the distal portion of the stricture. (A and B) Tight, mixed inflammatory and fibrotic strictures and (C and D) long, severe, smooth, nonulcerated stricture undergoing endoscopic stricturotomy with a needle knife. *CD*, Crohn's disease.

Anal ulcers and strictures may or may not be correlated with luminal CD. The presence of anal ulcers or strictures normally indicates the presence of aggressive form of CD [11,12]. Anal strictures can be covered with columnar or squamous epithelia, or both. The strictures are often of ulcerated type and can be mild or severe and short or long. Bougie dilation, EBD or ESt can be performed in patients with anal strictures (Fig. 5.25). ESt may be safer and more effective than bougie or balloon dilation, with a lower risk for iatrogenic trauma to nearby organs, such as the vagina [8]. CD-associated malignancy should be excluded in patients with refractory anal strictures.

Anal strictures can be a part of perianal CD, which is described further in a separate chapter (see Chapter 7: Endoscopic evaluation of perianal Crohn's disease).

Summary and recommendation

Stricturing in CD can occur anywhere from the esophagus to anus. A careful endoscopic examination may help to distinguish inflammatory strictures from fibrotic strictures. The distinction is important as the inflammatory stricture may have room for medical therapy, and fibrotic stricture needs to be treated with endoscopic or surgical intervention. Cross-sectional imaging should be periodically performed before diagnostic and therapeutic endoscopy to provide a road map for the stricturing disease and its possible complications. With an extensive use of biological agents in CD, a growing number of patients develop nonulcerated, spindle-shaped strictures in the small and large bowels. It is important to endoscopically document number, degree, type, location, distribution, and length of strictures.

Disclosure

The author declared no financial conflict of interest.

References

[1] Silverberg MS, Satsangi J, Ahmad T, Arnott ID, Bernstein CN, Brant SR, et al. Toward an integrated clinical, molecular and serological classification of inflammatory bowel disease: report of a Working Party of the 2005 Montreal World Congress of Gastroenterology. Can J Gastroenterol 2005;19(Suppl. A):5−36.

[2] Lian L, Stocchi L, Remzi FH, Shen B. Comparison of endoscopic dilation vs surgery for anastomotic stricture in patients with Crohn's disease following ileocolonic resection. Clin Gastroenterol Hepatol 2017;15:1226−31.

[3] Lan N, Shen B. Endoscopic stricturotomy with needle knife in the treatment of strictures from inflammatory bowel disease. Inflamm Bowel Dis 2017;23:502−13.

[4] Shen B. Classification of strictures from Crohn's disease, ulcerative colitis, and IBD-related surgery. In: Shen B, editor. Interventional inflammatory bowel disease: endoscopic management and treatment of complications. Cambridge, MA: Elsevier; 2018. p. 17–34.

[5] Chen M, Shen B. Endoscopic therapy in Crohn's disease: principle, preparation, and technique. Inflamm Bowel Dis 2015;21:2222–40.

[6] Singh A, Agrawal N, Kurada S, Lopez R, Kessler H, Philpott J, et al. Efficacy, safety, and long-term outcome of serial endoscopic balloon dilation for upper gastrointestinal Crohn's disease-associated strictures—a cohort study. J Crohns Colitis 2017;11:1044–105.

[7] Bettenworth D, Mücke MM, Lopez R, Singh A, Zhu W, Guo F, et al. Efficacy of endoscopic dilation of gastroduodenal Crohn's disease strictures: a systematic review and meta-analysis of individual patient data. Clin Gastroenterol Hepatol 2018; [Epub ahead of print].

[8] Shen B. Interventional IBD: the role of endoscopist in the multidisciplinary team management of IBD. Inflamm Bowel Dis 2018;24:298–309.

[9] Lan N, Shen B. Endoscopic stricturotomy versus balloon dilation in the treatment of anastomotic Strictures in Crohn's disease. Inflamm Bowel Dis 2018;24:897–907.

[10] Yang Y, Lyu W, Shen B. Endoscopic valvectomy of ileocecal valve stricture resulting in resolution of ileitis in Crohn's disease. Gastrointest Endosc 2018;88:195–6.

[11] Wallenhorst T, Brochard C, Bretagne JF, Bouguen G, Siproudhis L. Crohn's disease: is there any link between anal and luminal phenotypes? Int J Colorectal Dis 2016;31:307–11.

[12] Brochard C, Siproudhis L, Wallenhorst T, Cuen D, d'Halluin PN, Garros A, et al. Anorectal stricture in 102 patients with Crohn's disease: natural history in the era of biologics. Aliment Pharmacol Ther 2014;40:796–803.

Chapter 6

Crohn's disease: penetrating type

Bo Shen

Center for Inflammatory Bowel Diseases, Columbia University Irving Medical Center-New York Presbyterian Hospital, New York, NY, United States

Chapter Outline

Abbreviations

CD	Crohn's disease
CT	computed tomography
EBD	endoscopic balloon dilation
ECF	enterocutaneous fistula
EEF	enteroenteric fistula
ESt	endoscopic stricturotomy
EVF	enterovesical fistula
IBD	inflammatory bowel disease
ISF	ileosigmoid fistula
MRI	magnetic resonance imaging
PVF	pouch vaginal fistula
RVF	rectal vaginal fistula

Introduction

A fistula is defined as a pathologic channel connecting two or more epithelialized surfaces. While abdominal imaging, such as computed tomography (CT), magnetic resonance imaging (MRI) and small bowel series, and enterclysis, is the main stay for the diagnosis of fistula, endoscopy can provide important information on the primary and secondary openings in the bowel and status of inflammation around the orifices. Endoscopy may be combined with examination under anesthesia for the diagnosis of fistulas, especially perianal fistulae. Approximately 14%–50% of the patients with Crohn's disease (CD) present or eventually present with fistulae [1–3]. Perianal fistula or abscess can also occur in patients with ulcerative colitis [4]. Furthermore, fistula can develop after surgery for inflammatory bowel disease (IBD), with main cause being anastomotic leaks. Therefore fistulas in IBD can be primary (i.e., disease-associated) or secondary (e.g., anastomotic or suture-line leak associated). Fistulizing CD with the formation of abscess has been considered the ultimate adverse complication of the disease. While fistula can be the initial presentation of patients, the formation is preceded by transmural inflammation and stricture formation in the downstream bowel segment. We believe, "no inflammation, no stricture; and no stricture, no fistula." For example, ileosigmoid fistula (ISF) is often associated with terminal ileum or ileocecal valve (ICV) strictures (Fig. 6.1).

CD-associated perianal fistula is discussed in a separate chapter (Chapter 7: Endoscopic evaluation of perianal Crohn's disease). The classification of CD-associated fistula is proposed (Table 6.1).

Atlas of Endoscopy Imaging in Inflammatory Bowel Disease. DOI: https://doi.org/10.1016/B978-0-12-814811-2.00006-2
© 2020 Elsevier Inc. All rights reserved.

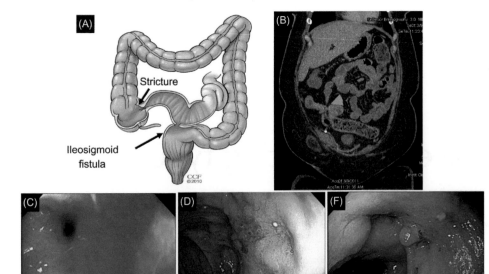

FIGURE 6.1 Ileosigmoid fistula associated with ICV stricture in Crohn's disease: (A) illustration of relationship between ICV stricture and ileosigmoid fistula; (B) ileosigmoid fistula (*green arrow*); (C) primary orifice of the fistula at the distal ileum with chronic inflammatory changes of adjacent mucosa; (D) strictured ICV; and (F) secondary orifice of the fistula with nodularity. *ICV*, Ileocecal valve.

TABLE 6.1 Classification of inflammatory bowel disease—associated fistula.

Category	Subcategory	Examples
Etiology	Primary or disease associated	Crohn's disease—associated enterocutaneous fistula
	Secondary or anastomotic	Enterocutaneous fistula from ileocolonic anastomosis leak, parastomal enterocutaneous fistula
Underlying diseases	Crohn's disease	Crohn's disease—associated jejuno-colonic fistula
	Ulcerative colitis	Mucus fistula from Hartmann pouch after subtotal colectomy
	Ileal pouch	Enterocutaneous fistula from the tip of the "J" of the pouch to skin
Symptomatology	Dry	
	Draining	
	Abscess ± systemic symptoms	
Organ involved	Gut-to-gut	Gastro-colonic fistula, ileosigmoid fistula, duodeno-colonic fistula, pouch—pouch fistula
	Gut-adjacent hollow organs	Rectal vaginal fistula, ileal pouch—bladder fistula, esophagobroncheal fistula
	Gut-to-skin	Enterocutaneous fistula
Length	Short	<3 cm
	Long	≥3 cm
Depth (from lumen of fistula track to bowel lumen)	Shallow	<2 cm
	Deep	≥2 cm
Concurrent inflammation adjacent to the primary orifice of fistula	Absent	
	Present	
Concurrent stricture	Absent	
	Present	
Complexity	Simple	Single track
	Complex	Multiple, branched, multiexit, associated abscess
Malignant potential	Benign	
	Malignant	Adenocarcinoma, squamous cell carcinoma

Enteroenteric fistula

It is estimated that one-third of patients with CD may eventually developed enterocutaneous fistula (ECF). Common enteroenteric fistula (EEF) include ileoileal, ileocolonic, jejunoileal or ileojejunal, duodeno-colonic, and pouch—pouch fistulae. The origin of the first can be small bowel or large bowel. However, EEF from the small bowel to large bowel often presents with the former being the origin and the latter being the target organ, such as ileocolonic or duodeno-colonic fistula. EEF often coexists with bowel inflammation around the primary orifice of the fistula and with stricture(s) of intestine distal to the primary fistula opening (Figs. 6.1 and 6.2). The stricture often prevents the passage of endoscope. Therefore therapy with endoscopic balloon dilation (EBD) or endoscopic stricturotomy (ESt) may be needed to get access to the bowel segment at which the orifice of EEF is located. The primary opening of the originating bowel in EEF is typically small, which is often accompanied by adjacent mucosal inflammation (Figs. 6.1 and 6.2). However, the primary orifice can be insidious and difficult to identify (Fig. 6.3). Therefore cross-sectional abdominal imaging, such as CT and MRI, is routinely performed in those with a clinical suspicion of penetrating CD with or without complications.

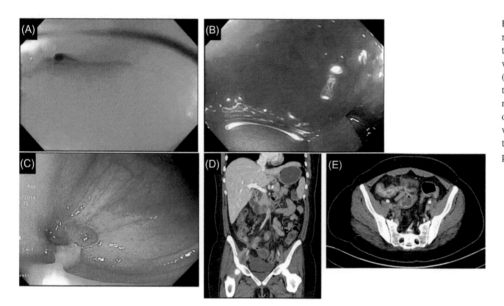

FIGURE 6.2 Another ileosigmoid fistula: (A) primary orifice of the fistula at the terminal ileum with adjacent edematous mucosa; (B) ileocecal valve stricture, distal to the fistula opening at the terminal ileum; (C) secondary or exit orifice of the fistula at the sigmoid with pus drainage; and (D and E) the fistula on computed tomography enterography (*green arrows*).

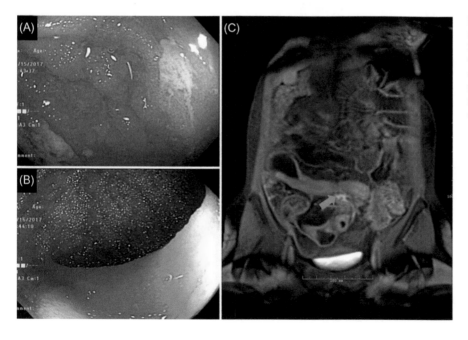

FIGURE 6.3 Hidden enteric fistula in Crohn's disease: (A and B) no visible lesions on ileocolonoscopy in a patient with ileal Crohn's disease who has been on long-term therapy with mesalamine and (C) ileo-mesenteric fistula on magnetic resonance enterography (*green arrow*).

The secondary or exit orifice in the target organs, such as the sigmoid colon in ISF or vagina in rectal vaginal fistula (RVF) or pouch vaginal fistula (PVF), typically has minimal or no inflammation surrounding mucosa (Fig. 6.4). However, the exit orifice or may have nodularity or even have small inflammatory polyps (Figs. 6.1 and 6.2).

Due to the nature of underlying chronic transmural disease process, free perforation with peritonitis in CD is not common. The chronic penetrating disease may present with fistula, sinus, or abscess to nonhallow organs, such as psoas muscle (Fig. 6.5) and mesentery (Fig. 6.3). The origin of fistula could be CD or anastomotic leak. The orifice of the primary opening of fistula can be obvious (Figs. 6.1 and 6.2) or insidious (Fig. 6.3) on endoscopy.

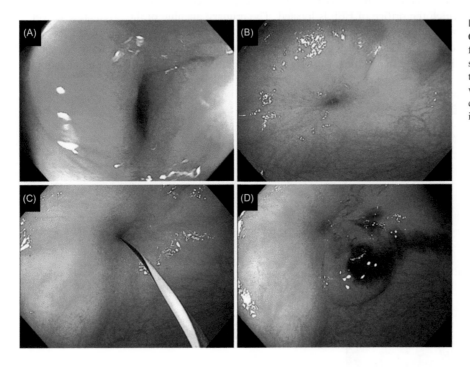

FIGURE 6.4 Ileosigmoid fistula in Crohn's disease: (A) primary orifice of the fistula at the terminal ileum and (B–D) secondary or exit orifice of the fistula at the sigmoid, detected with a soft-tip guide wire and tattooed. Notice that the sigmoid colon is a target organ with minimum inflammation.

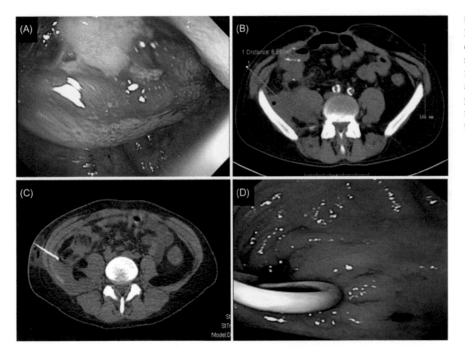

FIGURE 6.5 Ileocolonic anastomosis leak leading to psoas abscess in Crohn's disease: (A) bulging mass lesion at the right lower quadrant with draining pus from the anastomotic leak; (B and C) attempted CT-guided drainage of the psoas abscess; and (D) endoscopic placement of pigtail stent for the drainage of the abscess. *CT*, Computed tomography.

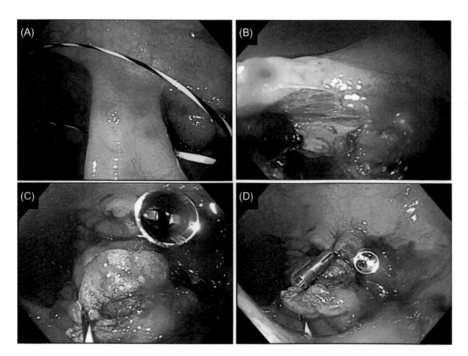

FIGURE 6.6 Ileocolonic fistula with endoscopic fistulotomy: (A) ileocolonic fistula across the ileocecal valve, detected by a soft-tip guide wire; (B) endoscopic fistulotomy with insulated-tip knife; and (C and D) complete excised fistula track, with placement of endoclips to keep the fistula from reformation and to prevent bleeding and perforation.

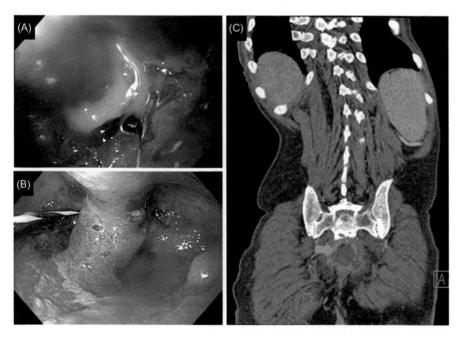

FIGURE 6.7 Peripouch fistula and abscess in Crohn's disease: (A and B) pouch−pouch fistula with draining pus, resulting from chronic anastomosis leak, which was detected with a soft-tip guide wire and (C) the presacral abscess on CT enterography (*green arrow*). *CT*, Computed tomography.

Since EEF is often associated with intestinal stricture distal to the originating bowel segment, such as stricture at the ICV with ISF, endoscopic therapy with EBD or ESt of the stricture may help reduce the drainage of ISF or the risk of fistula-associated abscess. In addition, shallow, short EEF, such as ileo-cecal fistula, may be treated with endoscopic fistulotomy (Fig. 6.6) [5]. Patients with pouch-to-pouch fistula with or without concurrent abscess can be managed endoscopically with fistulotomy (Fig. 6.7).

Enterocutaneous fistula

ECF can result from underlying CD (primary ECF) or anastomotic leak or ischemia (secondary). The diseased bowel of the origin of ECF can be the duodenum, jejunum, ileum, colon, or rectum. In patients with CD, ECF commonly

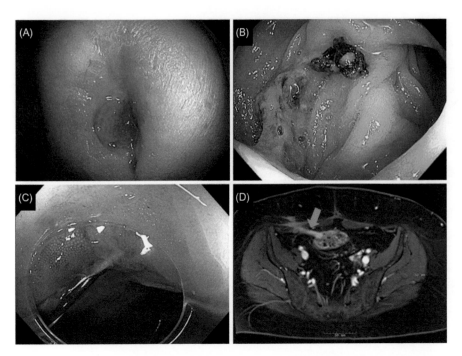

FIGURE 6.8 ECF from ileocolonic anastomosis leak: (A) exit orifice of ECF at the skin side; (B) ileocolonic anastomosis with sutures; (C) detection of the primary orifice of ECF from the anastomosis, by a soft-tip guide wire; and (D) the fistula track on computed tomography enterography (*green arrow*). *ECF*, Enterocutaneous fistula.

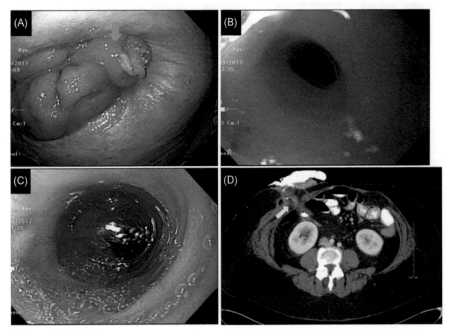

FIGURE 6.9 Paracolostomy enterocutaneous fistula in Crohn's disease: (A) colostomy was created for refractory perianal Crohn's disease. The external opening of the fistula along the side of the stoma (*green arrow*); (B) the internal opening of the fistula 7 cm from the stoma, at the fascia level; (C) normal proximal colon and ileum; and (D) the enterocutaneous fistula on CT enteroscopy (*yellow arrow*). *CT*, Computed tomography.

responds to medical therapy, particularly to antitumor necrosis factor agents, more favorably than EEF and other entero-hollow organ fistulae [6]. In addition, disease-associated ECF responds better to the medical therapy than anastomotic leak—associated ECF (Fig. 6.8) or mechanical factor—associated ECF (such as those at the ileostomy or colostomy site) (Fig. 6.9). The prognosis is worse in those with ECF originating from the duodenum or jejunum than those with ECF from the distal ileum or colon [7]. ECF can occur in patients with ileostomy or colostomy. The internal orifice of ECF in patients with ostomies is often located at the fascia level (Fig. 6.9). ECF can also occur in patients with colectomy and Hartmann pouch (i.e., diverted rectum). The origin of fistula is the leak at the sutured or stapled rectal stump (Figs. 6.10 and 6.11). Often ECF is intentionally recreated as a mucus fistula to prevent pelvic abscess, in the setting of colectomy and Hartmann pouch.

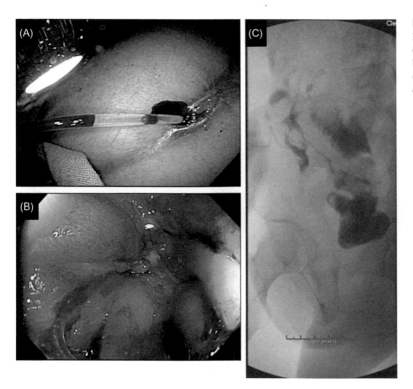

FIGURE 6.10 Enterocutaneous fistula from stump leak of the Hartmann pouch, that is, diverted rectum: (A) exit orifice at the skin level with administration of betadine; (B) leak at the Hartmann pouch on endoscopy (*blue arrow*); and (C) fistula track on fistulogram with contrast instilled from the orifice at the skin.

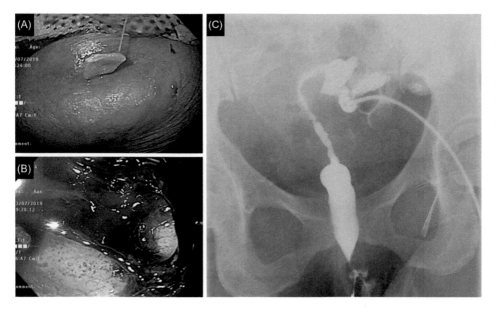

FIGURE 6.11 Enterocutaneous fistula from the rectum stump in Crohn's disease: (A) exit orifice of the fistula in the mid abdomen, with a drainage tab; (B) diverted rectum or Hartmann pouch with spontaneous bleeding; and (C) antegrade fistulogram showed a long fistula track from skin to the top of the rectum (*green arrow*).

The recognition of the internal opening of ECF can be challenging. The administration of betadine, methylene blue, or hydrogen peroxide (Fig. 6.10), or probing with of a soft-tip guide wire (Fig. 6.8) via cutaneous orifice of ECF at the time of endoscopy, often helps identify the internal orifice.

The purpose of endoscopic identification of the internal opening of ECF is to potentially delivery therapy, such as clipping [8,9]. Otherwise, for the diagnosis purpose only, cross-sectional imaging or fistulogram would provide better characterization of ECF (Figs. 6.8–6.11).

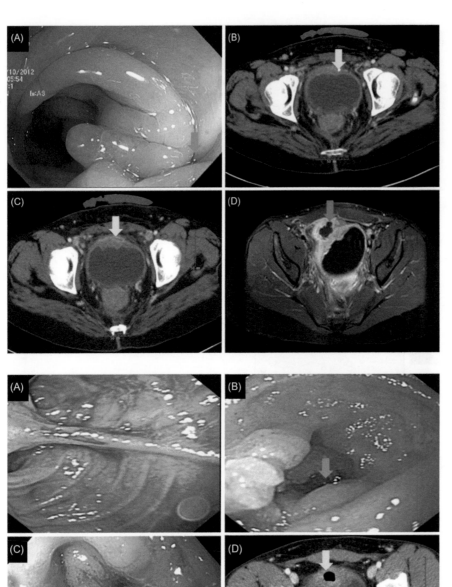

FIGURE 6.12 Sigmoid colon −bladder fistulae: (A) fistula opening at the sigmoid colon, surrounded by a converged mucosa folds (*green arrow*); (B and C) thickened bladder wall with the fistula track inside (*yellow arrows*); and (D) inflammatory mass with fluid collection at the right aspect of the bladder resulting from Crohn's disease in the sigmoid colon (*red arrow*).

FIGURE 6.13 Ileal pouch−bladder fistula: (A) normal configuration of proximal ileal J pouch with two openings, with one leading to the afferent limb and the other heading to the tip of the "J"; (B and C) a leak at the tip of "J" (*green arrow*), which was treated with endoclips; and (D) thickened bladder wall with air pocket inside (*yellow arrow*).

Enterovesical and entero-vaginal fistulae

The anatomy of the pelvis dictates the close relation between pelvic organs in the healthy and diseased. CD in the distal bowel or anus may involve adjacent organs, such as urinary bladder and vagina.

Enterovesical fistulae (EVF) in CD are rare. The diagnosis of EVF is mainly based on clinical presentation (e.g., pneumaturia, fecaluria, or recurrent urinary tract infection), laboratory testing (e.g., urine culture), and abdominal imaging. On cross-sectional imaging, there can be thickened bladder wall, mass-like lesion in the bladder, fistula track across bladder wall, or intra- or peribladder abscess. The primary opening of EVF in the distal large bowel may be identified through colonoscopy (Fig. 6.12). Vesical fistula can also result from surgical leak, such as the leak the tip of the "J" in patients with ileal pouch−anal anastomosis. Small fistula may be treated with endoscopic clipping (Fig. 6.13).

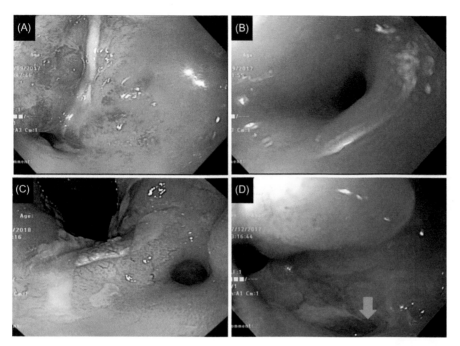

FIGURE 6.14 Patterns of vaginal fistula in Crohn's disease. (A and B) Rectal vaginal fistula. The primary opening of the fistula is surrounded with mucosal inflammation (A). The exit opening at the proximal vagina on vaginoscopy (B). (C) Rectal vaginal fistula with the primary orifice at the rectum on endoscopic retroflex view. There is no concurrent inflammation in the distal rectum. (D) Ileal pouch vaginal fistula in a patient with Crohn's disease of the pouch. The primary opening at the anal transition zone of the pouch, with adjacent cuffitis (*green arrow*).

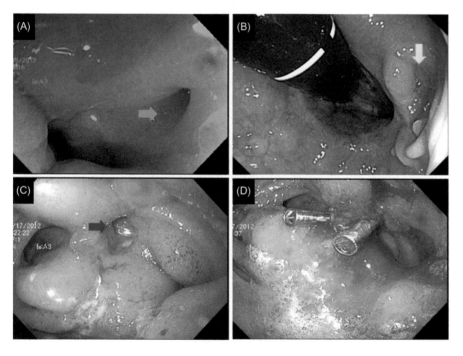

FIGURE 6.15 Patterns of vaginal fistula in Crohn's disease: (A) primary opening of the fistula at the dentate line (*green arrow*); (B) primary opening of the fistula at the distal rectum on endoscopic retroflex view, surrounded by nodular mucosa (*yellow arrow*); and (C and D) primary opening of the fistula at the distal rectum with adjacent prolapsed mucosa (*blue arrow*). Endoclips were placed for temporary symptom relief.

CD-associated vaginal fistulae have posed one of greatest challenges for medical, endoscopic, or surgical therapy. The origin of fistulae can be rectum (i.e., RVF), anal canal (i.e., anal rectal fistula), or ileal pouch (i.e., PVF) (Figs. 6.14–6.16). Endoscopic evaluation of the anorectum, along with imaging and examination under anesthesia, plays a key role in the identification of the internal opening of the fistula. Anovaginal fistula should be distinguished from RVF, as their treatment modalities are different.

The underlying etiology could be disease process of CD, iatrogenic trauma from endoscopic or surgical procedures, or cryptoglandular source. An accurate assessment of the anatomotic location, shape of fistula opening, and adjacent inflammation may help the distinction. Colonoscopy, flexible sigmoidoscopy, proctoscopy, or pouchoscopy may be performed in combination with vaginoscopy (Fig. 6.14B). Endoscopic features of the size, shape, and location of the

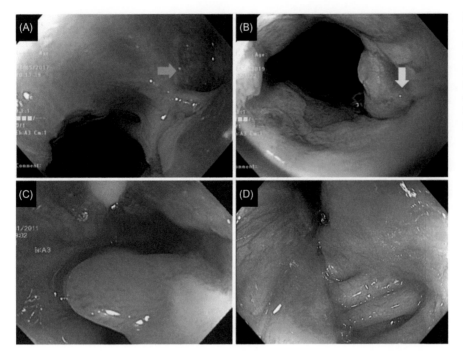

FIGURE 6.16 Patterns of vaginal fistula in Crohn's disease: (A) fistula opening at the anal canal (*blue arrow*); (B) fistula opening at the anal canal covered with an inflammatory nodule (*yellow arrow*); (C) seton in place as a part of treatment for anal vaginal fistula (*blue arrow*); and (D) recurrent fistula after multiple attempts of surgical repair (*red arrow*).

internal opening of the fistula, and adjacent mucosa, should be carefully documented. Endoscopic clipping (Fig. 6.15D) or seton placement (Fig. 6.16C) can be performed along with diagnostic evaluation.

Summary and recommendations

Penetrating CD with or without formation of abscess is considered an end of the phenotype spectrum. Although abdominal imaging is the main diagnostic modality, endoscopy plays an important role in the diagnosis, differential diagnosis, and prognosis, and, in selected patients, treatment. Accurate classification of CD fistula is also important for the management. Endoscopy may help identify the internal opening of fistulae and disease conditions around the fistula. The characteristics of fistula opening and adjacent bowel may also help distinguish the etiology of fistula.

References

[1] Schwartz DA, Loftus Jr EV, Tremaine WJ, Panaccione R, Harmsen WS, Zinsmeister AR, et al. The natural history of fistulizing Crohn's disease in Olmsted County, Minnesota. Gastroenterology 2002;122:875–80.

[2] Tang LY, Rawsthorne P, Bernstein CN. Are perineal and luminal fistulas associated in Crohn's disease? A population-based study. Clin Gastroenterol Hepatol 2006;4:1130–4.

[3] Ingle SB, Loftus Jr. EV. The natural history of perianal Crohn's disease. Dig Liver Dis 2007;39:963–9.

[4] Hamzaoglu I, Hodin RA. Perianal problems in patients with ulcerative colitis. Inflamm Bowel Dis 2005;11:856–9.

[5] Kochhar G, Shen B. Endoscopic fistulotomy in inflammatory bowel disease (with video). Gastrointest Endosc 2018;88:87–94.

[6] Present DH, Rutgeerts P, Targan S, Hanauer SB, Mayer L, van Hogezand RA, et al. Infliximab for the treatment of fistulas in patients with Crohn's disease. N Engl J Med 1999;340:1398–405.

[7] Quinn M, Falconer S, McKee RF. Management of enterocutaneous fistula: outcomes in 276 patients. World J Surg 2017;41:2502–11.

[8] Kochhar GS, Shen B. Use of over-the-scope-clip system to treat ileocolonic transverse staple line leak in patients with Crohn's disease. Inflamm Bowel Dis 2018;24:666–7.

[9] Shen B, Kochhar G, Navaneethan U, Liu X, Farraye FA, Gonzalez-Lama Y, et al. Role of interventional inflammatory bowel disease in the era of biologic therapy: a position statement from the Global Interventional IBD Group. Gastrointest Endosc 2019;89:215–37.

Chapter 7

Endoscopic evaluation of perianal Crohn's disease

Bo Shen

Center for Inflammatory Bowel Diseases, Columbia University Irving Medical Center-New York Presbyterian Hospital, New York, NY, United States

Chapter Outline

Abbreviations

AGA the American Gastroenterological Association
CD Crohn's disease
CT computed tomography
pCD perianal Crohn's disease
EUA examination under anesthesia
IPAA ileal pouch–anal anastomosis
MRI magnetic resonance imaging
PVF pouch-vaginal fistula
RVF rectovaginal fistula

Introduction

Perianal fistulas are the most common phenotype of penetrating Crohn's disease (CD), with a prevalence of 20%–24% in the patients with CD [1,2]. The cumulative incidence was estimated to be 12% and the incidence to be 0.7 per 100 patient-years [2]. The terms of perianal fistulizing CD and perianal CD (pCD) have been used interchangeably. pCD can be associated with strictures, abscesses, perianal skin lesions, and even malignant changes, Diagnostic evaluation of pCD has been largely relied on cross-sectional imaging [such as computed tomography (CT), magnetic resonance imaging (MRI), transanal ultrasound] or contrasted fistulography. For both diagnostic and therapeutic purposes, examination under anesthesia (EUA) is routinely performed. Based on clinical, radiographic, or operative features, various classifications have been proposed [3–13]. The most commonly used is the Parks' classification (Table 7.1) [2], St. James Hospital classification (Table 7.2) [5], and the American Gastroenterological Association (AGA) classification (Table 7.3) [6], which mainly delineate characteristics of perianal fistulas (e.g., route and number of openings) in relationship to the internal and external anal sphincters. To assess patients' quality of life (i.e., pain or restructure of physical and sexual activities) and perianal disease activity (i.e., fistula discharge, type of perianal disease, and perianal induration), investigators have proposed the Perianal Crohn's Disease Activity Index in a graded scale [14]. In addition,

Atlas of Endoscopy Imaging in Inflammatory Bowel Disease. DOI: https://doi.org/10.1016/B978-0-12-814811-2.00007-4
© 2020 Elsevier Inc. All rights reserved.

TABLE 7.1 Parks classification [2].

Superficial	Superficial fistula without crossing any sphincter or muscular structure
Intersphincteric	Fistula tract between the internal and external anal sphincters in the intersphincteric space
Transsphincteric	Fistula tract crosses the external anal sphincter
Suprasphincteric	Fistula tract penetrates the intersphincteric space and continues over the top of the puborectalis and penetrates the levator muscle before reaching the skin
Extrasphincteric	Fistula tract outside the external anal sphincter and penetrating the levator muscle

TABLE 7.2 St. James' Hospital classification [5].

Grade 1	Simple linear intersphincteric fistula
Grade 2	Intersphincteric fistula with intersphincteric abscess or secondary fistulous tract
Grade 3	Transsphincteric fistula
Grade 4	Transsphincteric fistula with abscess or secondary tract within the ischioanal or ischiorectal fossa
Grade 5	Supralevator or translevator disease

TABLE 7.3 The American Gastroenterological Association classification [6].

Simple fistula	• Low (superficial or low intersphincteric or low transsphincteric origin of the fistula tract) • Single external opening • No pain or fluctuation to suggest perianal abscess • No evidence of a rectovaginal fistula • No evidence of anorectal stricture
Complex fistula	• High (high intersphincteric or high transsphincteric or extrasphincteric or suprasphincteric origin of the fistula tract) • Multiple external openings • Presence of pain or fluctuation to suggest a perianal abscess • Rectovaginal fistula • Anorectal stricture

imaging-based classification of pCD has been proposed (Table 7.4) [11]. The pros and cons of the various classifications are appraised (Table 7.5) [15].

The role of conventional endoscopy in the evaluation of pCD has been peripheral, mainly for the identification of the primary or internal opening of fistula or concurrent proctitis, distal bowel or anal strictures, or pouchitis or cuffitis (in patients with ileal pouches). An endoscopic evaluation may also help to identify pCD-associated adenocarcinoma or squamous cell cancer [16]. However, the role of endoscopy has been evolving from a diagnostic modality to a tool of delivery of therapy. Endoscopic documentation of perianal or perineal skin lesions would provide additional "bonus." Therefore a new Columbia Classification of Perianal Crohn's Disease is here proposed (Table 7.6). The contents of this chapter will follow the framework of this classification.

Etiologies of perianal Crohn's disease

The pathogenesis of pCD is not entirely clear. It is believed that increased production of proinflammatory factors, such as transforming growth factor β, tumor necrosis factor, as well as interleukin-13 induce epithelial-to-mesenchymal transition and upregulation of matrix metalloproteinases, resulting in tissue remodeling and fistula formation [15]. Although the perianal disease is a form of presentation of CD, not all pCD results solely from the underlying disease process. Perianal fistula and perianal abscess can result from surgical staple line or suture line leaks, or cryptoglandular source. The characterization of the internal opening of the fistula and the status of inflammation in the distal large bowel or anal canal are important for the differential diagnosis.

TABLE 7.4 The 22-point magnetic resonance imaging—based Van Assche index for perianal disease activity [12].

Descriptor	Categories	Scoring
Number of fistula tracts	None	0
	Single, unbranched	1
	Single, branched	2
	Multiple	3
Location	Extrasphincteric or intersphincteric	1
	Transsphincteric	2
	Suprasphincteric	3
Extension	Infralevatoric	1
	Supralevatoric	2
Hyperintensity on T2-weighted images	Absent	0
	Mild	4
	Pronounced	8
Collections (cavities >3 mm in diameter)	Absent	0
	Present	4
Rectal wall involvement	Normal	0
	Thickened	2

TABLE 7.5 Pros and cons of different clinical and imaging indices for measuring fistula activity.

Index	Pros	Cons
PCDAI	• Simple to apply in clinical practice • Validated against physicians' and patients' global assessment	• Limited to clinical assessment, no objective measurement of healing
Anal disease activity index	• Includes parameters to assess disease activity	• Incomplete evaluation of manifestations of perianal disease • Not validated
Fistula drainage assessment	• Simple to use • Accepted by regulatory agencies as an endpoint	• Limited to clinical assessment, no objective measure of healing • Fistula compression is investigator dependent
MRI score (Van Assche)	• Partially validated (against PCDAI) • Combined assessment of activity and complexity of fistulas • Simple to calculate	• Limited to clinical assessment, no objective measure of healing • Fistula compression is investigator dependent

MRI, Magnetic resonance imaging; *PCDAI*, perianal Crohn's Disease Activity Index.
Source: Modified from Irvine EJ. Usual therapy improves perianal Crohn's disease as measured by a new disease activity index. McMaster IBD Study Group.
J Clin Gastroenterol 1995;20:27—32.

CD-associated perianal disease typically presents with inflammation of the distal rectum, ileal pouch, or anal canal. The internal opening may not be obvious to be identified, as it is often covered with granular tissue and surrounded by inflammation at the adjacent bowel (Fig. 7.1). Cryptoglandular fistulae or abscesses can also occur in patients with CD, which is often short and superficial; respond favorably to medical and surgical therapy; and sometimes, maybe self-limited. A careful endoscopic examination or EUA may identify an internal opening of the fistula at the dentate line. The patients may or may not have concurrent proctitis or pouchitis or cuffitis [in patients with ileal pouch—anal anastomosis (IPAA)] (Figs. 7.2 and 7.3). Cryptoglandular abscesses are common causes of anal vaginal fistulas (Fig. 7.4C and D).

TABLE 7.6 Proposed Columbia Classification of perianal Crohn's disease.

Category	Description
Etiology	Crohn's disease
	Cryptoglandular
	Anastomotic or suture/staple line leaks
Associated conditions	Strictures
	Abscesses
	Fistula to adjacent organ(s)
	Inflammation around internal opening of fistula
	Skin lesions around external opening of fistula
	Skin lesions around the anus
	Perineal Crohn's diseases
	Malignant transformation
Characteristics of fistula in relation to sphincters	Superficial
	Intersphincter
	Transsphincter
	Suprasphincter
	Extrasphincter
Length	Short (<3 cm)
	Long (≥3 cm)
Complexity	Simple
	Complex (presence of stricture, abscess, branches of fistula with multiple internal and/external openings, high-positioned internal openings, malignancy)

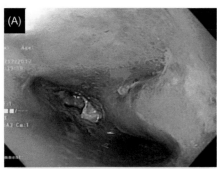
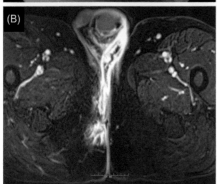
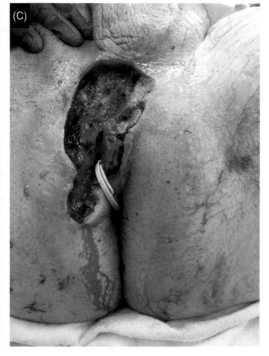

FIGURE 7.1 Perianal Crohn's disease. (A) Crohn's disease involving distal rectum and anal canal, with a fistula opening (*green arrow*); (B) complex branched fistula with openings to the scrotum and perianal skin; and (C) fistulotomy and seton placement.

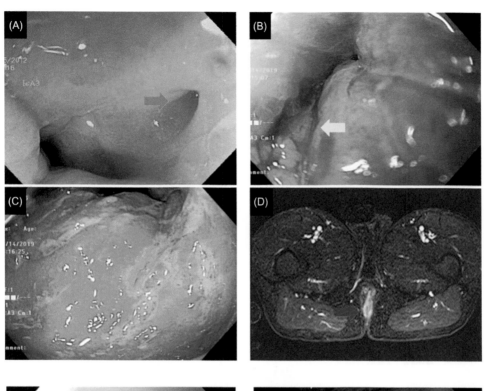

FIGURE 7.2 Cryptoglandular fistula in a patient with Crohn's proctitis. (A) The fistula opening was identified at the dentate line (*green arrow*); (B) concurrent anal fissures (*yellow arrow*); (C) distal proctitis with linear ulcers and exudates on endoscopic retroflex view; and (D) MRI showed simple, superficial fistula (*blue arrow*). *MRI*, Magnetic resonance imaging.

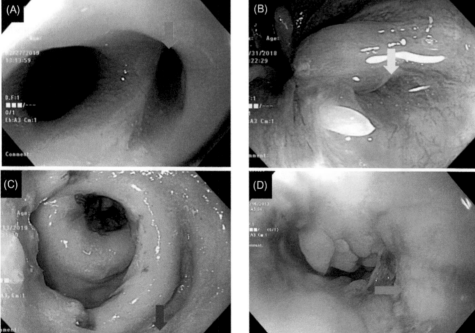

FIGURE 7.3 Fistula openings at the dentate line in patients with perianal Crohn's disease. Almost all cryptoglandular fistulas originate at the dentate line. (A) The fistula failed to surgical repair, which made the original fistula opening larger (*green arrow*); (B) a hidden fistula opening on retroflex view (*yellow arrow*); (C) a dentate line fistula in a patient with Crohn's disease of an "S" pouch (*blue arrow*); and (D) a seton placed in the fistula (*red arrow*).

Surgical bowel resection is an effective treatment modality for refractory CD. Currently performed bowel surgery includes resection followed by colorectal anastomosis, rectal anal anastomosis, colo-anal anastomosis, and IPAA. The lower anastomoses may be complicated with leaks, fistulas, or abscesses. The internal opening of the leak or fistula can more readily be identified than that from CD. The opening often shows a clear orifice with normal surrounding mucosa. Anastomosis leak-associated fistula or abscesses can be difficult to differentiate from that resulting from underlying CD (Fig. 7.5).

Fecal diversion with an ileostomy or colostomy, which may be performed along with partial colectomy, is an effective treatment modality for refractory pCD or CD in the distal bowel. Diverted bowels, such as diverted colon, diverted

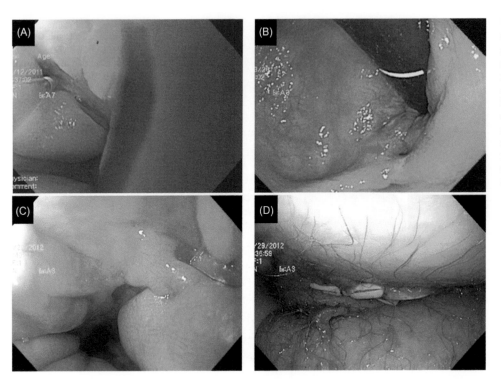

FIGURE 7.4 Vaginal fistulas in Crohn's disease. (A) A hidden internal opening of rectovaginal fistula originated from the inflamed anal canal, which was detected by the administration of hydrogen peroxide via an endoscopic catheter; (B) a large vaginal fistula opening at the distal rectum under endoscopic retroflex view (green arrow); and (C and D) superficial anal vaginal fistula treated with seton.

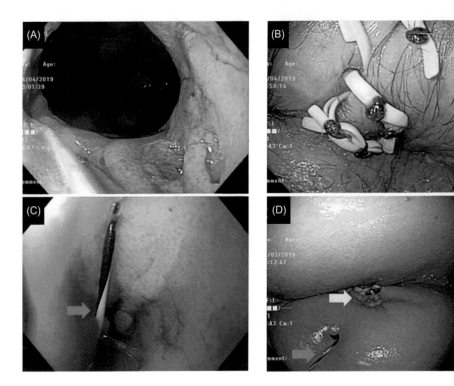

FIGURE 7.5 Perianal disease resulting from surgical anastomotic leaks. The anastomosis leak-associated fistula is difficult to differentiate from the cause of underlying inflammatory bowel disease. (A and B) The internal fistula opening originated from the IPAA with a seton. The leak also caused branched fistulae with multiple external openings; (C and D) an internal opening was detected at IPAA with a guidewire (green arrows) in a separate patient. The same internal opening also led to another external opening around the anus (yellow arrow). IPAA, Ileal pouch–anal anastomosis.

rectum (i.e., Hartmann pouch), or diverted pelvic ileal pouch, often develop diversion–association inflammation or stricture. Perianal skin lesions, fistulae, or abscesses may be present before or de novo after fecal diversion (Fig. 7.6).

Another form of pCD is the persistent perineal sinus or fistula after total proctocolectomy or completion proctectomy following colectomy. Patients with preproctectomy are at risk (Fig. 7.7).

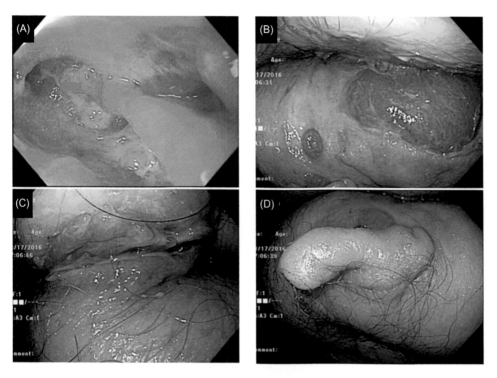

FIGURE 7.6 Perianal disease in a patient with diverted rectum for refractory Crohn's disease. (A) Diverted rectum with a distal rectum stricture (*green arrow*); (B and C) perineal ulcerated and nodular lesions; and (D) a pedunculated skin tag at the external opening of perianal fistula.

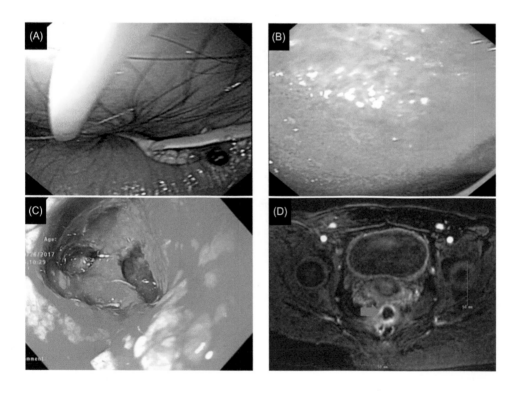

FIGURE 7.7 Persistent perineal sinus after proctectomy in a patient with refractory Crohn's disease. (A and B) Perianal fistulas with setons (A) and the presence of active proctitis (B) before proctectomy; (C) persistent perineal sinus on endoscopy after the surgery; (D) the sinus on pelvic MRI (*green arrow*). *MRI,* Magnetic resonance imaging.

Associated conditions

pCD with the presence of stricture, abscess, and fistula to the adjacent organs has been classified into complex fistula, according to the AGA classification [1].

Strictures

Strictures can be present in the anal canal, distal or mid rectum, or ileal pouch in patients with pCD. Strictures can be proximal or distal to the internal opening of perianal fistula. Strictures can be short or long, nonulcerated or ulcerated, with various degrees of severity. Although there are scant data, it is believed that proper treatment of the stricture, such

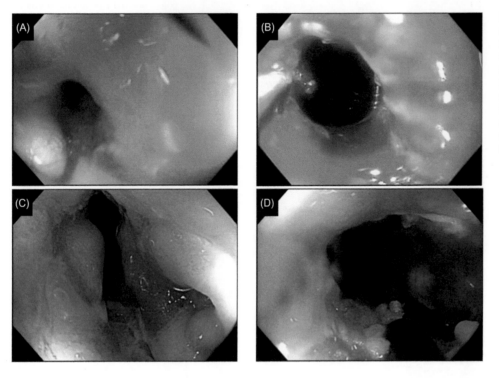

FIGURE 7.8 Anal strictures and treatment in Crohn's disease. (A and B) A tight stricture at the anal canal, which was not traversable to a gastroscope. Endoscopic stricturotomy with a needle knife was performed; and (C and D) an anal stricture in a separate patient, which was treated by endoscopic stricturotomy with an insulated-tip knife.

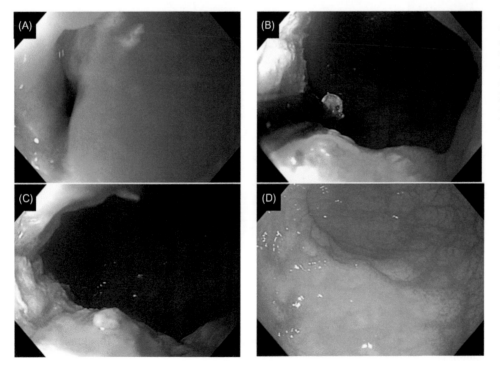

FIGURE 7.9 Anal stricture of Crohn's disease treated with endoscopic stricturotomy. (A) A tight anal stricture, not traversable to a gastroscope; (B) endoscopic stricturotomy was in action; (C) posttreatment appearance of the stricture; and (D) the rectum in this patient was normal.

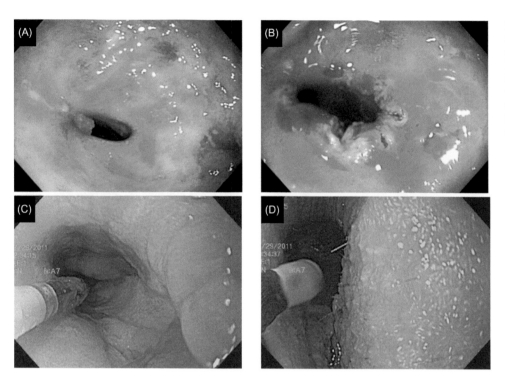

FIGURE 7.10 Distal bowel strictures in patients with perianal Crohn's disease. (A and B) Distal rectal stricture with concurrent proctitis. The stricture was treated with endoscopic stricturotomy; (C and D) distal ileal pouch stricture undergoing endoscopic balloon dilation.

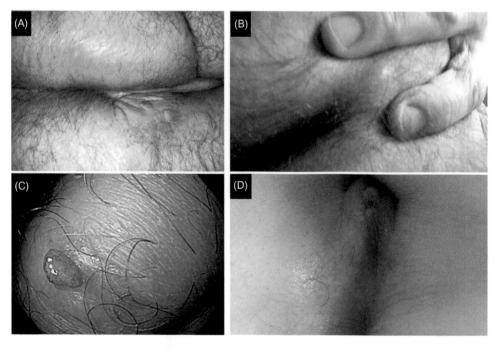

FIGURE 7.11 Perianal anal abscesses in Crohn's disease. (A) An abscess presented with a large red, bulging mass; (B) a small abscess in the perineum; (C) abscess with granular tissue on top; and (D) abscess may only present with topical erythema.

as bougie or balloon dilation and stricturotomy, may help the healing of perianal fistula, rectovaginal fistula (RVF), or pouch-vaginal fistula (PVF) (Figs. 7.8 and 7.9). Strictures at the distal rectum, if present, can also be detected and treated endoscopically (Fig. 7.10).

Abscesses

Perianal abscesses are the most common adverse sequelae of perianal fistulae, although the abscesses may occasionally develop in the absence of fistula. The latter is often of cryptoglandular source (Figs. 7.11−7.17). Examination of

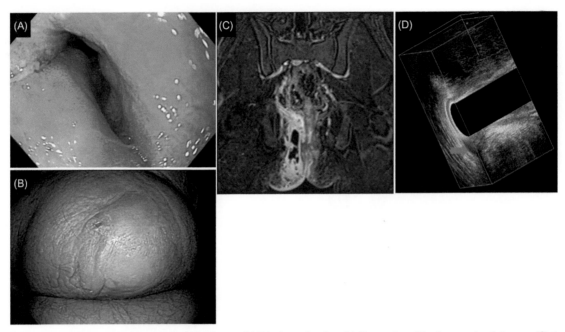

FIGURE 7.12 Perianal fistula with abscess in Crohn's disease. (A) Distal rectal and anal inflammation; (B) a large perianal abscess; (C) the abscess shown on MRI; and (D) 3D transanal ultrasound may also be used to detect perianal abscess (*green arrow*). *MRI*, Magnetic resonance imaging.

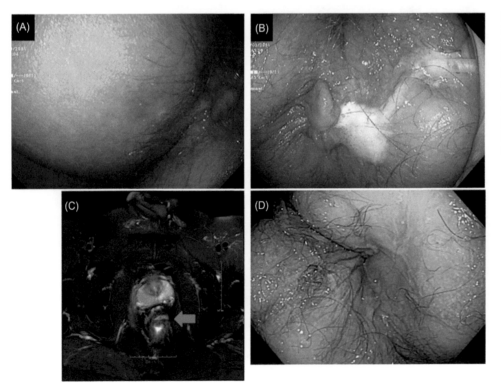

FIGURE 7.13 Simple perianal fistula in Crohn's disease. (A and B) Detection of simple fistula to injection of hydrogen peroxide; (C) the fistula tract viewed on MRI (*green arrow*); and (D) healed simple fistula after a combination therapy with anti-tumor necrosis factor and hyperbaric oxygen. *MRI*, Magnetic resonance imaging.

perianal abscesses can be performed in the endoscopy suite or during EUA. The abscesses can be small or large. It is recommended that photo documentation is performed as a routine clinical practice. Often cord-like structures, that is, fistulas, may be palpated around or underneath abscesses. To enhance the accuracy, hydrogen peroxide, methylene

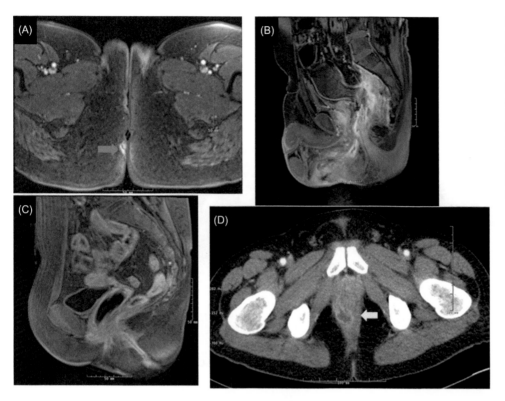

FIGURE 7.14 Short, superficial, and simple versus long, complex fistulas in four patients with Crohn's disease on MRI or CT. (A) Superficial short (<3 cm) anal fistula (*green arrow*); (B and C) long (>3 cm), complex fistulae leading to external openings at the perianal skin; and (D) fistula-associated perianal fistula on CT (*yellow arrow*). *CT*, Computed tomography; *MRI*, magnetic resonance imaging.

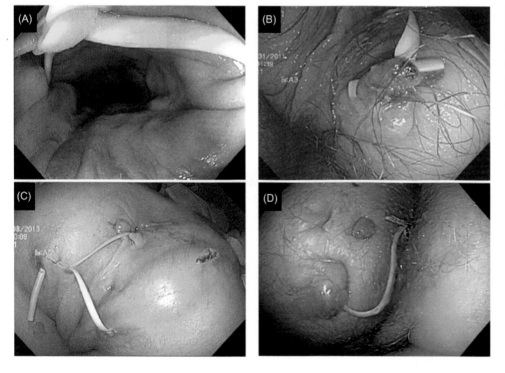

FIGURE 7.15 Complex fistula with multiple external openings in Crohn's disease. (A) Horseshoe perianal fistula with two internal openings linked by a single seton; and (B–D) multiple external openings with some being treated with setons.

blue, or betadine can be administered through an endoscopic catheter (Fig. 7.13). Perianal abscesses may be managed by incision and drainage with needle knife under conscious sedation and topical anesthesia during endoscopy (Figs. 7.16 and 7.17).

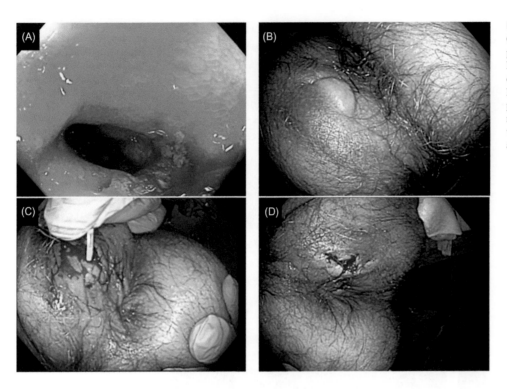

FIGURE 7.16 Perianal Crohn's disease along with the presence of ileitis. (A) Edematous and nodular mucosa of the terminal ileum with mild luminal narrowing and (B–D) perianal abscess from underneath fistula, which was treated with needle knife electroincision and drainage.

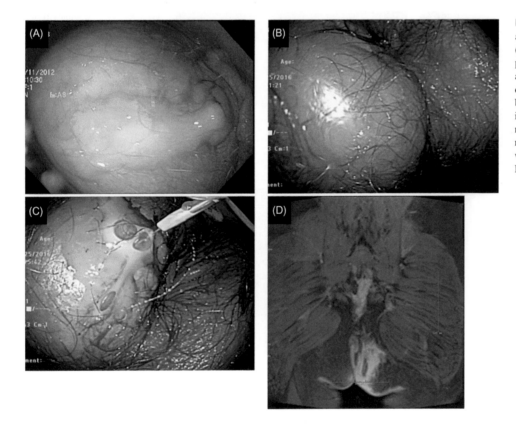

FIGURE 7.17 Perianal anal abscess in Crohn's disease. (A and B) Two separate patients presented with perianal abscesses, one with active pus drainage (A) and the other with bulging mass (B); (C and D) incision and drainage of a perianal abscess with endoscopic needle knife (C). The abscess was shown on MRI (D). *MRI*, Magnetic resonance imaging.

Fistula to adjacent organs

The presence of nonsurgery-associated RVFs, recto-vesicle fistulas, and PVFs, pouch vesicle fistulas, is considered having a complex perianal fistula (Fig. 7.4B−D). The fistula can originate from the distal rectum, distal ileal pouch, anal canal, or dentate line.

Inflammation around internal opening of fistula

It is important to document the status of inflammation in the distal bowel (i.e., rectum, pouch, or cuff) or anal canal, especially at or around the internal opening of fistulas (Figs. 7.18−7.21). The presence of proctitis in pCD was shown to be a predictive factor in the persistent nonhealed fistula and a higher frequency of proctectomy [17].

Perianal skin lesions in the presence or absence of external opening of fistula

The term "skin" here may be termed for structures covered with squamous cells, which including a part of the anal transition zone, anal canal, anus, and perianal or perineal areas. The skin lesions can be related to pCD skin tags (Fig. 7.22), fissures (Fig. 7.23), and ulcers (Fig. 7.24). Those lesions can easily be documented during endoscopy.

Perineal Crohn's disease

It is believed that perineal CD may belong to a separate disease category, different from classic pCD. Perineal CD may present with ulceration in the perineum or anal canal, nonhealing, painless fissures, or waxy perineal, perianal, or natal cleft edema. The perianal area involved includes the anal verge, the squamous part of the anal canal, and the perineum (Fig. 7.24) [18].

Perineal CD may be considered as a form of metastatic CD with a frequent finding of granulomas on biopsy. Perineal CD was found more often in female, to be presented at a younger age, less small bowel, and more colonic CD than classic pCD. The presence of perineal CD is found to be an indicator of poor healing of CD, requiring proctectomy [17].

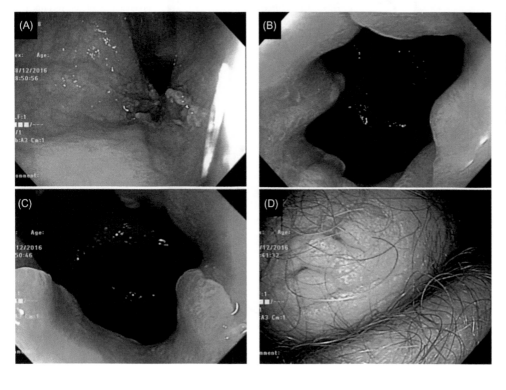

FIGURE 7.18 Inflammation at the distal rectum and anal canal around the internal opening of perianal fistula in Crohn's disease. (A) Erythema, nodularity, and ulcers at the distal rectum; (B and C) nodular anal canal; and (D) perianal skin induration with a hidden fistula underneath.

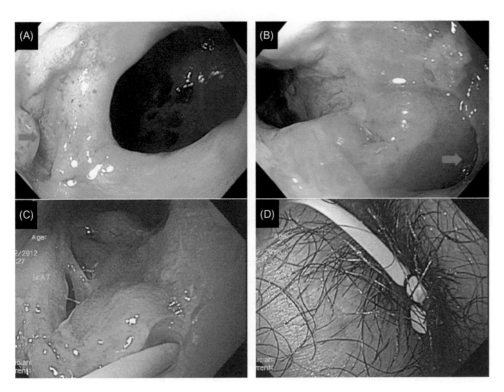

FIGURE 7.19 Inflammation at the anal canal around the internal opening of perianal fistula in CD. (A and B) Nodular and edematous anal canal with fistula openings (*green arrows*) in a patient with complex perianal CD; and (C and D) nodular anal canal with a seton placed across the short transsphincter fistula. *CD*, Crohn's disease.

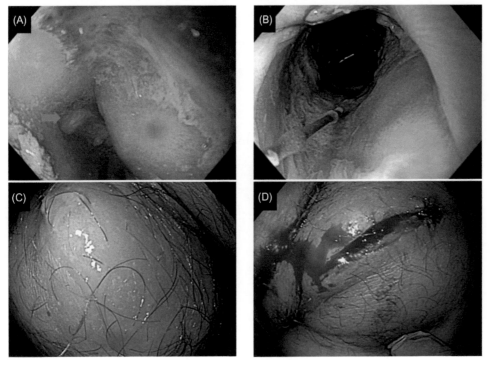

FIGURE 7.20 Perianal Crohn's disease with the involvement of the rectum. (A) Ulcers at the distal rectum with a suspected fistula opening (*green arrow*); (B) a soft-tip guidewire can be used for the detection of the opening of fistula; and (C and D) perianal abscess that was treated with needle knife incision and drainage.

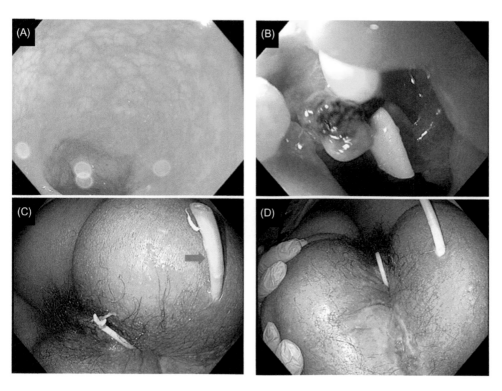

FIGURE 7.21 Perianal and perineal CD with sparing of the rectum. (A) Normal rectum; (B and C) nodular anal canal with a seton placed across a short fistula. There is also a nearby abscess with a mush-room catheter in place (*green arrow*); and (D) the same patient also had perineal CD with large skin ulceration at 6 o'clock. *CD*, Crohn's disease.

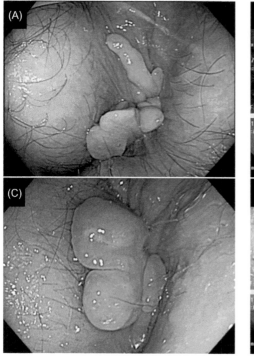
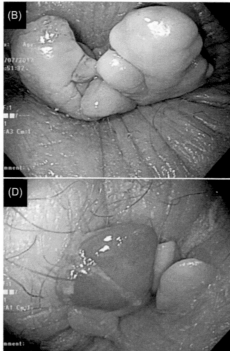

FIGURE 7.22 Various forms of skin tags, mostly with an "elephant ear" appearance. (A–D) The skin tags can be single or multiple, thin and thick, flat or cylindrical. The surgical excision of Crohn's disease-associated skin tags is not recommended.

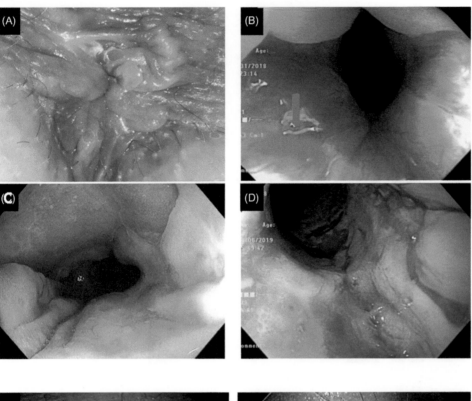

FIGURE 7.23 Anal fissures in Crohn's disease. (A) Skin break down on hemorrhoid lesions; (B) linear friable skin break down at the anal canal (*green arrow*); (C) superficial linear fissures in the anal canal; and (D) deep linear fissures along the anal verge in a patient with Crohn's disease of the ileal pouch.

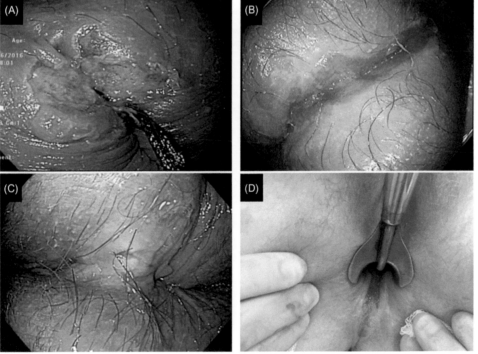

FIGURE 7.24 Perineal CD. Perineal CD has been considered a form of metastatic CD in the perineum area. The biopsy may show noncaseating granulomas. (A) Large deep ulcer with a nearby string seton in the perineum; (B) a long linear ulcer in the midline of the perineum area; (C) ulcerated induration at the perineum; and (D) perineal ulcer shown on examination under anesthesia. *CD*, Crohn's disease.

Other benign or malignant lesion in the distal bowel or perianal areas

Chronic perianal dermatitis or dermatopathy in various forms are common with chronic diarrhea or fistula drainage being major contributing factors (Figs. 7.25–7.27). Long-term use of immunosuppressive agents may cause bacterial, fungal, or viral infection of the perianal skin (Fig. 7.28A and B). Patients with pCD are at risk for the development of

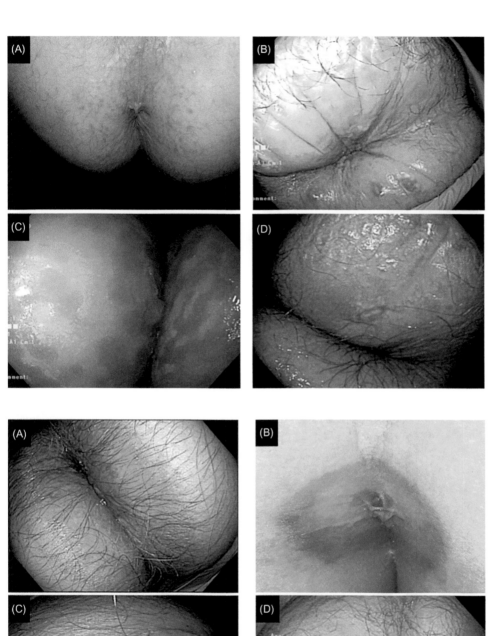

FIGURE 7.25 Perianal skin lesions in Crohn's disease. (A–C) Perianal dermatitis and (D) dermatitis around a healed fistula.

FIGURE 7.26 Perianal skin lesions in Crohn's disease. (A) Diffuse perianal dermatitis from chronic diarrhea; (B) severe perianal dermatitis with ulcer posterior to the anus; (C) verrucous skin lesions around the anus; and (D) perianal folliculitis, which it was not associated with fistula.

28 adenocarcinoma or squamous cell carcinoma in the distal rectum, rectal cuff (in IPAA), anal canal, or perianal skin (Fig. 7.28C and D).

Characteristics of fistula to sphincters

Conventional endoscopy plays a minimum role in the evaluation fistula tracks outside of the lumen of the bowel or perianal skin. In contrast, endoscopic ultrasonography can be used (please see Chapter 35: Transluminal imaging in

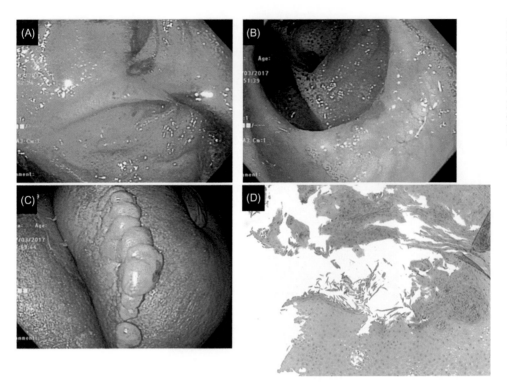

FIGURE 7.27 Perianal skin lesion in a patient with irritable pouch syndrome. (A and B) Despite symptoms, the patient had normal pouch and cuff mucosa; (C) verrucous perianal skin lesion from chronic diarrhea; and (D) skin biopsy showed hyperparakeratosis without malignancy.

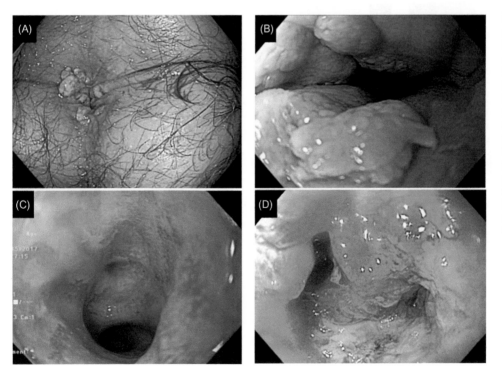

FIGURE 7.28 Infectious and malignant complications of long-standing CD. (A and B) Anal and perianal wart after long-term use of antitumor necrosis factor agent for refractory CD; (C) ulcerated anal canal with histology show squamous cell cancer in an 80-year-old female patient with a long-standing CD; (D) nodular, ulcerated cuff mucosa in a 43-year-old male patient with restorative proctocolectomy and ileal pouch–anal anastomosis, which turned out to be adenocarcinoma. *CD*, Crohn's disease.

inflammatory bowel disease), in combination with cross-sectional imaging, such as CT or MRI. CT and MRI are routinely used to assess the characteristics of perianal fistulae, including the length, number, branching, associated abscess, and internal and external openings. Also, CT and MRI can also provide information on the bowel wall, mesentery, and lymphadenopathy (Figs. 7.1, 7.7, 7.12–7.14). However, conventional endoscopy can provide adjacent information on the complexity of perianal fistula.

Summary and recommendations

It is important to characterize disease status of the rectum, ileal pouch, or cuff (such as inflammation and stricture) and to delineate location, size, and shape of the internal and external fistula openings and diseased conditions of surrounding tissues. Photo documentation of anal and perianal lesions should be incorporated into the routine practice of diagnostic and therapeutic endoscopy in patients with CD. Conventional endoscopy may be performed in combination with endoscopic ultrasonography, cross-sectional imaging, or contrasted fistulogram. Endoscopic evaluation can be a part of EUA.

References

[1] Schwartz DA, Loftus Jr EV, Tremaine WJ, Panaccione R, Harmsen WS, Zinsmeister AR, et al. The natural history of fistulizing Crohn's disease in Olmsted County, Minnesota. Gastroenterology 2002;122:875−80.

[2] Panes J, Reinisch W, Rupniewska E, Khan S, Forns J, Khalid JM, et al. Burden and outcomes for complex perianal fistulas in Crohn's disease: systematic review. World J Gastroenterol 2018;24:4821−34.

[3] Parks AG, Gordon PH, Hardcastle JD. A classification of fistula-in-ano. Br J Surg 1976;63:1−12.

[4] Buchmann P, Alexander-Williams J. Classification of perianal Crohn's disease. Clin Gastroenterol 1980;9:323−30.

[5] Hughes LE. Clinical classification of perianal Crohn's disease. Dis Colon Rectum 1992;35:928−32.

[6] Morris J, Spencer JA, Ambrose NS. MR imaging classification of perianal fistulas and its implications for patient management. Radiographics 2000;20:623−35 discussion 35−37.

[7] Sandborn WJ, Fazio VW, Feagan BG, Hanauer SB. American Gastroenterological Association Clinical Practice C. AGA technical review on perianal Crohn's disease. Gastroenterology 2003;125:1508−30.

[8] Present DH, Korelitz BI, Wisch N, Glass JL, Sachar DB, Pasternack BS. Treatment of Crohn's disease with 6-mercaptopurine. A long-term, randomized, double-blind study. N Engl J Med 1980;302:981−7.

[9] Allan A, Linares L, Spooner HA, Alexander-Williams J. Clinical index to quantitate symptoms of perianal Crohn's disease. Dis Colon Rectum 1992;35:656−61.

[10] Present DH, Rutgeerts P, Targan S, Hanauer SB, Mayer L, van Hogezand RA, et al. Infliximab for the treatment of fistulas in patients with Crohn's disease. N Engl J Med 1999;340:1398−405.

[11] Pikarsky AJ, Gervaz P, Wexner SD. Perianal Crohn disease: a new scoring system to evaluate and predict outcome of surgical intervention. Arch Surg 2002;137:774−7 discussion 8.

[12] Van Assche G, Vanbeckevoort D, Bielen D, Coremans G, Aerden I, Noman M, et al. Magnetic resonance imaging of the effects of infliximab on perianal fistulizing Crohn's disease. Am J Gastroenterol 2003;98:332−9.

[13] Gecse KB, Bemelman W, Kamm MA, Stoker J, Khanna R, Ng SC, et al. A global consensus on the classification, diagnosis, and multidisciplinary treatment of perianal fistulising Crohn's disease. Gut 2014;63:1381−92.

[14] Irvine EJ. Usual therapy improves perianal Crohn's disease as measured by a new disease activity index. McMaster IBD Study Group. J Clin Gastroenterol 1995;20:27−32.

[15] Panés J, Rimola J. Perianal fistulizing Crohn's disease: pathogenesis, diagnosis and therapy. Nat Rev Gastroenterol Hepatol 2017;14:652−64.

[16] Beaugerie L, Carrat F, Nahon S, Zeitoun JD, Sabaté JM, Peyrin-Biroulet L, et al. High risk of anal and rectal cancer in patients with anal and/or perianal Crohn's disease. Clin Gastroenterol Hepatol 2018;16:892−9.

[17] Bell SJ, Williams AB, Wiesel P, Wilkinson K, Cohen RC, Kamm MA. The clinical course of fistulating Crohn's disease. Aliment Pharmacol Ther 2003;17:1145−51.

[18] Figg RE, Church JM. Perineal Crohn's disease: an indicator of poor prognosis and potential proctectomy. Dis Colon Rectum 2009;52:646−50.

Chapter 8

Crohn's disease: postsurgical

Bo Shen

Center for Inflammatory Bowel Diseases, Columbia University Irving Medical Center-New York Presbyterian Hospital, New York, NY, United States

Chapter Outline

Abbreviations

CD Crohn's disease
EI end ileostomy
IBD inflammatory bowel disease
ICA ileocolonic anastomosis
ICR ileocolonic resection
IRA ileorectal anastomosis
LI loop ileostomy
RS the Rutgeerts score
RVF rectovaginal fistula
STX strictureplasty

Introduction

Despite advances in medical therapy, approximately 70% of patients with Crohn's disease (CD) would eventually require surgery [1,2]. Various surgical modalities have been designed and performed for patients with medically or endoscopically refractory CD or inflammatory bowel disease (IBD)—associated neoplasia (Table 8.1). Multiple factors contribute to the selection of proper surgical treatment modalities. Those factors include degree and severity of underlying CD, disease phenotype and location, general health conditions (especially nutrition, anemia, and immunosuppression), concurrent medical therapy, history of prior surgery, and local expertise.

In this chapter the author discusses common surgical modalities, postsurgical anatomy, and monitoring disease recurrence in patients with CD. CD surgery-associated complications are discussed in a separate chapter (Chapter 16: Postoperative complications in Crohn's disease).

Bowel resection and anastomosis

Ileocolonic resection (ICR) followed by ileocolonic anastomosis (ICA) is the most commonly performed surgical procedure for patients with B1−B3 and L1−L3 lesions, based on the Montreal classification (Figs. 8.1 and 8.2) [3]. ICR and ICA may be performed in staged with diverting end ileostomy (EI) (Fig. 8.3) or loop ileostomy (LI) (Figs. 8.2 and 8.4). Variants of ICR and ileorectal anastomosis (IRA) include subtotal colectomy or partial colectomy with IRA, colo-colonic anastomosis, rectal anastomosis, and small bowel resection with entero-enteric anastomosis. The anastomosis

Atlas of Endoscopy Imaging in Inflammatory Bowel Disease. DOI: https://doi.org/10.1016/B978-0-12-814811-2.00008-6
© 2020 Elsevier Inc. All rights reserved.

TABLE 8.1 Surgical treatment modalities for Crohn's disease.

	Examples
Resection and anastomosis	Ileocolonic resection with ileocolonic anastomosis
	Ileocolonic resection with ileorectal anastomosis
	Jejunal resection with jejunoileal anastomosis
Strictureplasty	Heineke–Mikulicz
	Finney
	Michelassi
Stoma and fecal diversion	Loop or end ileostomy
	End colostomy
	Jejunostomy
	Hartmann procedure with later completion proctectomy
	Mucus fistula
Bypass	Duodenum disease with gastrojejunostomy
Surgical treatment of perianal disease	Incision and drainage
	Seton and mushroom

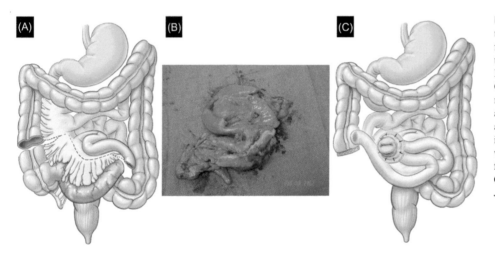

FIGURE 8.1 Ileocolonic resection with ileocolonic anastomosis for refractory inflammatory, stricturing, or penetrating CD. (A) Ileocolonic resection for refractory CD; (B) surgically resected specimen with terminal ileum, cecum, appendix, cecum, and proximal ascending colon; (C) end-to-side ileocolonic handsewn anastomosis with diverting loop ileostomy, following the resection. *CD*, Crohn's disease. *Courtesy of Mr. Joe Pangrace of Cleveland Clinic.*

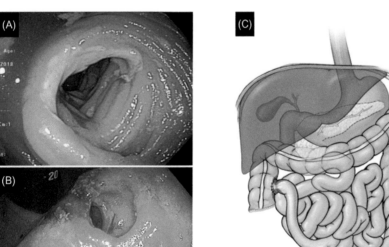

FIGURE 8.2 Ileocolonic resection with end-to-side ileocolonic anastomosis for refractory CD. (A) A circumferential ulcer along the anastomosis, possibly resulting from surgery-associated ischemia; (B) occasionally, the anastomosis is better viewed with a retroflexed endoscope; (C) end-to-side handsewn ileocolonic anastomosis is typically located at the right or right upper abdomen. Strictures at the anastomosis are common. *CD*, Crohn's disease. *Courtesy Mr. Joel Pangrace of Cleveland Clinic.*

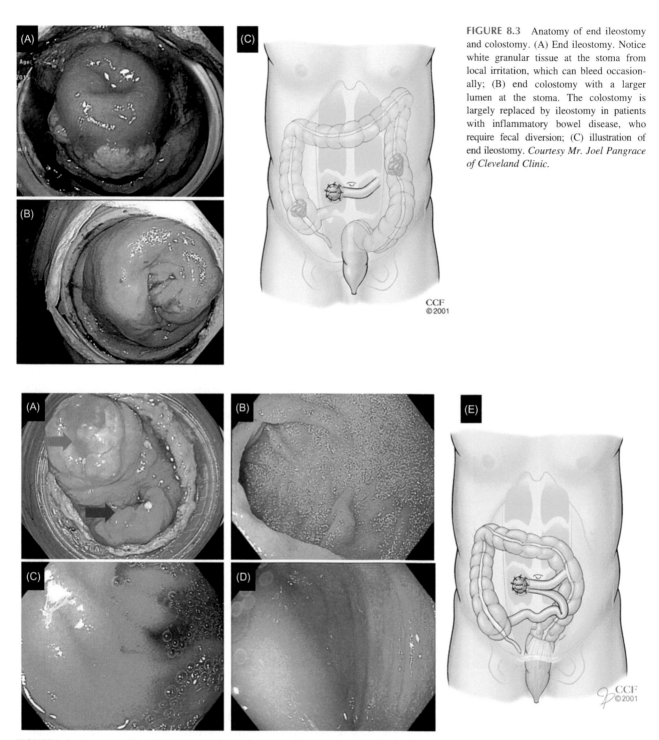

FIGURE 8.3 Anatomy of end ileostomy and colostomy. (A) End ileostomy. Notice white granular tissue at the stoma from local irritation, which can bleed occasionally; (B) end colostomy with a larger lumen at the stoma. The colostomy is largely replaced by ileostomy in patients with inflammatory bowel disease, who require fecal diversion; (C) illustration of end ileostomy. *Courtesy Mr. Joel Pangrace of Cleveland Clinic.*

FIGURE 8.4 Anatomy of loop ileostomy. (A) Loop ileostomy with a proximal opening leading to the afferent limb (*green arrow*) and a distal opening leading to the efferent limb (*blue arrow*). Notice that distal opening is often hidden underneath the proximal opening; (B and C) afferent limb mucosa slightly bile tinged; (D) efferent limb with copious secretion in the lumen from mucosa due to fecal diversion; (E) illustration of loop ileostomy. *Courtesy Mr. Joel Pangrace of Cleveland Clinic.*

can be fashioned with handsewn or staples and with end-to-side or side-to-side (Figs. 8.1, 8.2, and 8.5). Sometimes, submucosal staples can be visual in those with stapled anastomosis. In contrast, bioabsorbable suture threads are not seen in the handsewn anastomosis. Anastomosic ulcer may be present in those with or without IBD (Fig. 8.6). ICA following ICR is commonly located at the right abdomen or right upper quadrant (Fig. 8.7). The endoscopist should keep

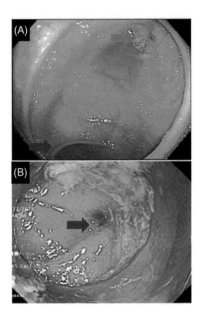

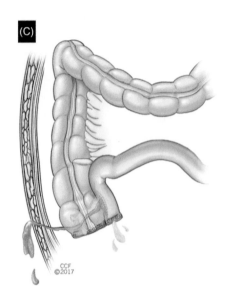

FIGURE 8.5 Ileocolonic resection with side-to-side stapled ileocolonic anastomosis for refractory CD. (A) The anastomosis with transverse staple line (*green arrow*) and lumen to the neo-terminal ileum (*blue arrow*); (B) a leak at the transverse staple line (*blue arrow*); (C) leak at the anastomosis leading to enterocutaneous fistula and leak at the transverse staple line leading to intraabdominal abscess or sepsis. *CD*, Crohn's disease. *Courtesy Mr. Joel Pangrace of Cleveland Clinic.*

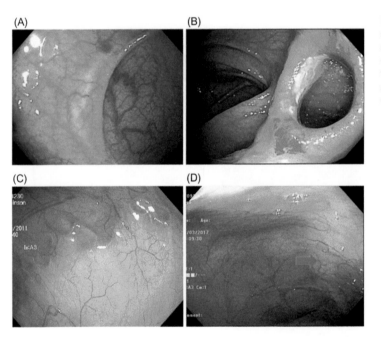

FIGURE 8.6 Patterns of anastomosis in Crohn's disease surgery. (A) Handsewn colorectal anastomosis; (B) end-to-side handsewn ileocolonic anastomosis with superficial ulcer; (C) stapled colorectal anastomosis with occasional visualization of staples underneath the mucosa; (D) stapled side-to-side ileocolonic anastomosis with submucosal lines of staples (*green arrow*) dividing neo-terminal ileum and colon.

this in mind when performing endoscopic treatments (such as balloon dilation of a stricture), as procedure-associated complications (such as perforation) in the location may result in a detrimental outcome. Sometimes, the identification of the orifice of an end-to-side anastomosis can be challenging from a tangential view (Fig. 8.7).

Surgical therapy with bowel resection and anastomosis does not offer a cure for the majority of patients with CD. Within 1 year after ICR and ICA, subclinical endoscopic recurrence occurs at the anastomosis was reported in 90%, symptomatic clinical recurrence in 30%, and surgery reintervention in 5% of patients with CD [4–6]. Ileocolonoscopy plays a key role in the monitoring of disease activity and prediction of poor disease outcomes and the requirement for additional surgery. Rutgeerts et al. developed an endoscopic score system (range: i,0–i,4) in CD patients with ICR and IRA, based on the number and characteristics of ulcers and inflammation in the neo-terminal ileum and anastomosis (Table 8.2) [5]. The representative endoscopic lesions listed in the Rutgeerts score (RS) are illustrated in Fig. 8.8. Multiple studies have shown that advanced RS was associated with clinical recurrence and need for further surgery. RS

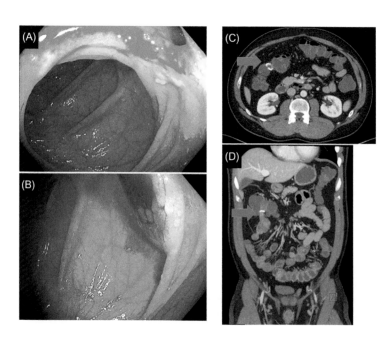

FIGURE 8.7 Ileocolonic resection with end-to-side ileocolonic anastomosis for refractory CD. The anastomosis is located at the right upper quadrant. (A) Tangential view of the strictured anastomosis, which was difficult to identify; (B) the anastomosis was tattooed for future diagnostic and therapeutic colonoscopy; (C and D) the stapled anastomosis viewed on CT enterography (*green arrows*). *CD*, Crohn's disease.

TABLE 8.2 The Rutgeerts endoscopic scoring system [1].

Classification	Endoscopic description
i,0	No lesions
i,1	<5 aphthous lesions
i,2	>5 aphthous lesions with normal mucosa in between the lesions, or skip area of larger lesions, lesions confined to the ileocolonic anastomosis (i.e., <1 cm in length)
i,3	Diffuse aphthous ileitis with diffusely inflamed mucosa
i,4	Diffuse inflammation with already large ulcers, nodules, and/or narrowing

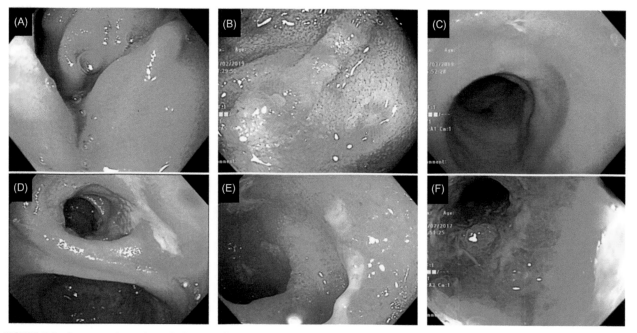

FIGURE 8.8 Representative endoscopic features of the Rutgeerts endoscopic score of postoperative Crohn's disease in the neo-terminal ileum after ileocolonic resection and anastomosis. (A) i,0—normal neo-terminal ileum; (B) i,1— <5 aphthous lesions; (C) i,2— >5 aphthous lesions with normal mucosa in between the lesions; (D) i,2—lesions confined to the ileocolonic anastomosis (i.e., <1 cm in length); (E) diffuse aphthous ileitis with diffusely inflamed mucosa; (F) diffuse inflammation with large ulcers and narrowing of lumen of the neo-terminal ileum.

has routinely been used in clinical practice and clinical trials. There is heterogeneity in scoring and implication of i,2 lesions, largely in the consideration or discard of anastomotic ulcers. Some investigators believed that postsurgical ischemia may contribute to anastomotic ulcers and distal neo-terminal ileum lesions, leading to a reduced predictive value for progressive disease. Therefore modified RS was proposed: i,2a, lesions confined to the anastomosis with or without <5 isolated aphthous ulcers in the ileum; i,2b, >5 aphthous ulcers in the ileum with normal mucosa in between, with or without anastomotic lesions (Table 8.3) [7,8]. The clinical implication of modified RS needs to be validated [9,10]. Common patterns of ICA ulcers are demonstrated in Fig. 8.9. It is believed that the presence of anastomotic ulcer may likely cause stricture later on.

Strictureplasty

Many patients with CD may require multiple bowel section surgery during their lifelong course of the disease. As a result, bowel-saving procedures, particularly strictureplasty (STX) for small bowel stricture, and less commonly colonic strictures are performed (Fig. 8.10). STX is indicated in the treatment of multiple and/or long-segment fibrotic strictures, especially in those at the risk for short-gut syndrome with multiple bowel resection surgery. Therefore STX is

TABLE 8.3 Modified Rutgeerts score [4].

Classification	Endoscopic description
i,0	No lesions
i,1	<5 aphthous ulcers
i,2a	Lesions confined to the anastomosis ± <5 isolated aphthous ulcers in the ileum
i,2b	>5 aphthous ulcers in the ileum with normal mucosa in between, with or without anastomotic lesions
i,3	Diffuse aphthous ileitis with diffusely inflamed mucosa
i,4	Diffuse inflammation with large ulcers, nodules, and/or narrowing

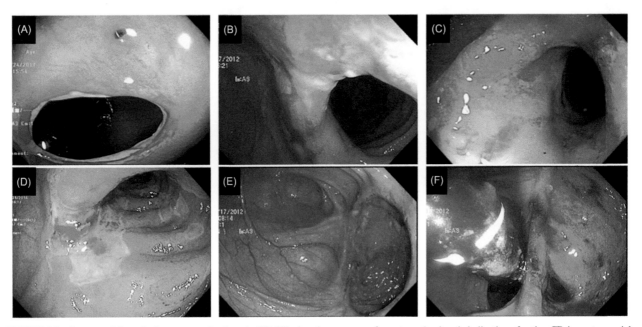

FIGURE 8.9 Patterns of ileocolonic anastomosic ulcers in CD. Whether the presence of anastomotic ulcer is indicative of active CD is controversial. (A) Circumferential ulcer along a thin side-to-side anastomosis; (B) semicircumferential ulcer along the anastomosis; (C) anastomotic ulcer extending more than 2.5 cm; (D) anastomotic ulcer along with multiple neo-terminal ileum ulcers; (E and F) side-to-side anastomosis with forward and retroflex views. The anastomosis was friable. *CD*, Crohn's disease.

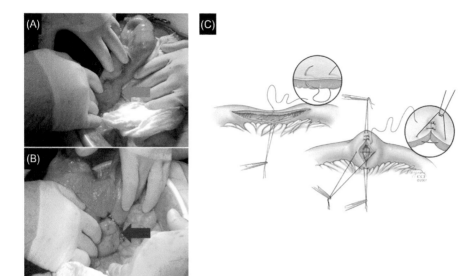

FIGURE 8.10 Heineke—Mikulicz stric-tureplasty. (A) A 7-cm long tight duode-num stricture (*green arrow*); (B) status post strictureplasty (*blue arrow*); (C) illus-tration of strictureplasty. *Courtesy Mr. Joel Pangrace of Cleveland Clinic.*

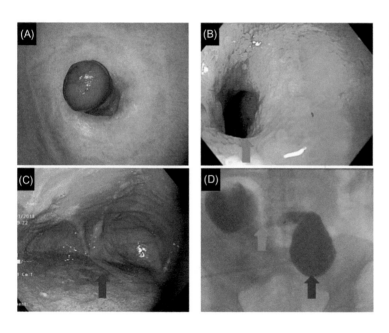

FIGURE 8.11 Ileostomy and strictureplasty. Patients with ileostomies may also have concurrent strictureplasty surgery to save the small intestine. (A) End ileostomy; (B) luminal narrowing 5 cm from the stoma, resulting from a pinch of fascia (*green arrow*); (C) strictureplasty close to the stoma (*blue arrow*); (D) gastrografin enema via stoma demonstrat-ing the pinch (*green arrow*) and strictureplasty (*blue arrow*).

often seen in patients with ileostomies to save valuable gut from being resected (Fig. 8.11). Three most commonly con-structed STXs are Heineke—Mikulicz (for strictures <10 cm in length) (Fig. 8.12), Finney (for strictures 10—20 cm in length) (Fig. 8.13), and Michelassi (for strictures >20 cm in length).

The anatomy of STX normally consists of inlet, outlet, and lumen (Figs. 8.12—8.14). There may be strictures at the inlet and outlet. The lumen of the body of STX site is usually enlarged, and mucosal inflammation in the body is rare. It is believed that patients with STX have some degree of small intestinal bacterial overgrowth. However, strictures and ulcers at the pouch inlet and outlet sites are common. Whether those ulcers and strictures should be considered recurrent CD is controversial. In this author's experience, aggressive medical therapy, even with biological agents, has a mini-mum impact on those lesions, suggesting an ischemic etiology. Endoscopic therapy of strictures at the STX site has been challenging, due to the altered anatomy and accessibility to an endoscope. Balloon-assisted enteroscopy is nor-mally not recommended for the evaluation of small bowel lesions in patients with multiple STXs, for the concern of barotrauma from the endoscope.

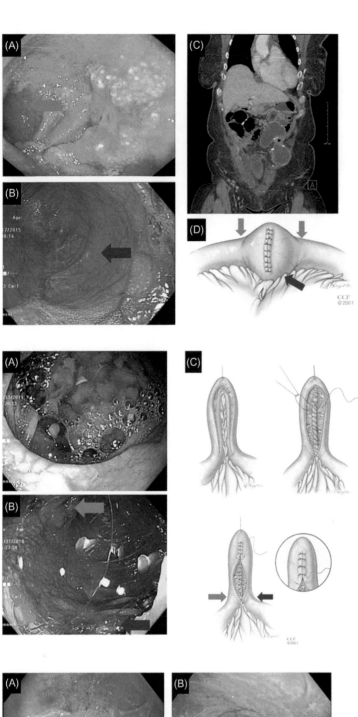

FIGURE 8.12 Heineke–Mikulicz strictureplasty. (A) Mildly narrowed outlet of strictureplasty site (*green arrow*); (B) enlarged lumen of strictureplasty site (*blue arrow*); (C) multiple dilated loop of the small bowel from prior bowel resection and strictureplasties; (D) illustration of Heineke–Mikulicz strictureplasty with inlet and outlet being highlighted in red arrows and lumen being highlighted in blue arrow. *Courtesy Mr. Joel Pangrace of Cleveland Clinic.*

FIGURE 8.13 Finney strictureplasty. (A) The lumen of the strictureplasty filled with food residues; (B) inlet and outlet of the strictureplasty (*green and blue arrows*); (C) illustration of Finney strictureplasty procedure with inlet and outlet highlighted in smaller green and blue arrows. *Courtesy Mr. Joel Pangrace of Cleveland Clinic.*

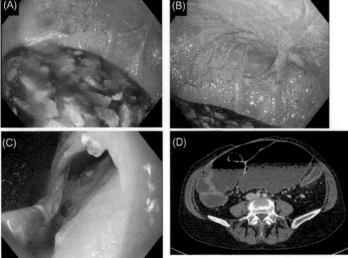

FIGURE 8.14 "Normal" anatomy of strictureplasty. (A and B) Enlarged lumen of strictureplasty with food residues. Small intestinal bacterial overgrowth is considered as a "norm" in those with strictureplasty; (C) outlet of the strictureplasty detected by a through-the-scope balloon; (D) dilated lumens of strictureplasties.

Fecal diversion with ostomy

Fecal diversion is an effective way to treat refractory CD in the downstream bowel. An ileostomy is commonly performed, followed by colostomy and jejunostomy. An ostomy has been performed as a temporary or permanent treatment modality for CD, as well as ulcerative colitis. An ileostomy is typically constructed as loop and end fashion, depending on the long treatment goal and technical factors. A temporary ileostomy is often performed in the form of LI, while a permanent ileostomy is usually in the form of EI. To help maintain the integrity of ileostomy, the distal opening with the efferent limb is commonly placed interiorly to the proximal opening. The luminal content of the afferent limb is usually bile tinged, while the content in the efferent limb is commonly mucus (Fig. 8.4). LI has the advantage of easier surgical closure while having the disadvantage of being harder to maintain than EI (Fig. 8.3A). Occasionally, end colostomy is constructed in patients with CD (Fig. 8.3B). Granular tissue at the stoma may be considered being "normal." In addition, there might be a stricture or ulceration from extrinsic pinch by fascia, which is also regarded as "normal" (Fig. 8.15). Long-term ileostomy and colostomy can result in downstream diversion-related colitis or proctitis (Fig. 8.16). A cutaneous fistula near ileostomy, namely, mucous fistula may also be intentionally created in patients with a risk for stump leak in colectomy, which should not be misdiagnosed as CD-related fistula (Fig. 8.16).

Recurrent inflammatory, stricturing, or fistulizing CD can occur in the distal ileum in CD patients with an ileostomy, which can reliably be assessed with ileostomy via the stoma. Ulcer, stricture, or fistula of the ileum at distal-to-fascia level likely result from technical factors, such as ischemia (Fig. 8.16). Inflammation, strictures, or fistulae at any segment of the ileum proximal to fascia level, in the absence of the use of nonsteroidal antiinflammatory drugs, are considered forms of recurrent CD. RS, which was designed for the monitoring disease activity at the neo-terminal ileum in those with ICR and ICA, has also been adopted to assess recurrence of CD in neo-distal ileum in patients with ileostomies (Fig. 8.17) [11]. Advanced RS in CD patients with ileostomies was found to be associated with poor disease outcomes [11]. Ileoscopy via stoma provides a valuable tool for the diagnosis and treatment of strictures in the distal ileum (Fig. 8.18).

Bypass surgery

Bypass surgery as a treatment modality for CD is largely abandoned (Fig. 8.19). The only exception is gastrojejunostomy bypass surgery for refractory duodenum CD (Fig. 8.20). The bypassed duodenum segment often remains to have active disease.

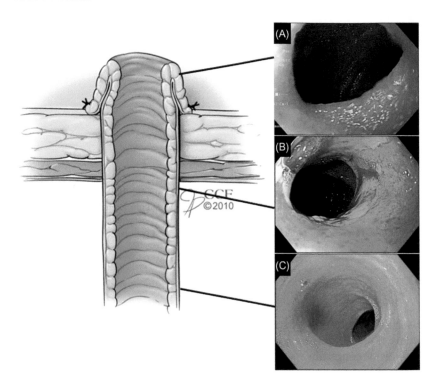

FIGURE 8.15 Ileostomy and strictures. (A) Mild stenosis at the stoma; (B) ulcerated, isolated stricture at the fascia level due to mechanical and ischemic factors; (C) normal distal ileum. These strictures are not considered as active Crohn's disease. *Courtesy Mr. Joel Pangrace of Cleveland Clinic.*

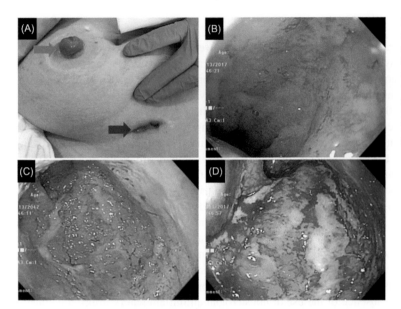

FIGURE 8.16 Diverting ileostomy combined with mucous fistula. The procedure is performed in the management of large bowel stump in patients with severe ulcerative colitis or Crohn's colitis undergoing colectomy, to prevent stump leak—associated abdominal or pelvic abscess. (A) Ileostomy (*green arrow*) and mucous fistula (*blue arrow*); (B) internal opening of mucous fistula at the sigmoid colon stump; (C and D) diverted distal large bowel with mucous exudates and severe bleeding from air insufflation.

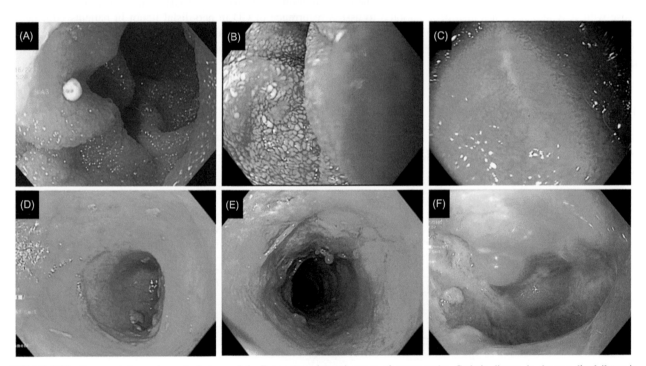

FIGURE 8.17 Representative endoscopic features of the Rutgeerts endoscopic score of postoperative Crohn's disease in the neo-distal ileum in patients with ileostomies. (A and B) i,0—normal neo-distal ileum before and after additional small bowel resection and reconstruction of an ileostomy. There are dilated lacteals (B); (C) i,2a—lesions confined to the fascia level (i.e., <1 cm in length); (D) i,2b− > 5 aphthous lesion with normal mucosa in between the lesions;(E) diffuse aphthous ileitis with diffusely inflamed mucosa; (F) diffuse inflammation with large ulcers, and narrowed lumen of the neo-terminal ileum.

Other surgeries

Refractory duodenum CD has been treated with Whipple's procedure, the one traditionally for pancreas case, that is, severe bile gastritis proximal Gastrojejunostomy (Fig. 8.21). Patients with Hartmann pouch or diverted rectum following colectomy carry the risk for diversion proctitis or colitis, diversion-associated stricture, or colitis-associated neoplasia. Completion colectomy may be required (Fig. 8.22).

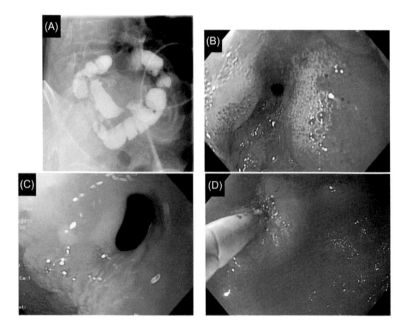

FIGURE 8.18   Recurrent CD with strictures in patients with ileostomies. (A) Multiple CD-associated strictures in the neo-distal ileum on gastrografin enema via stoma; (B) a pinhole stricture with bowel edema at neo-distal ileum; (C) a short, ulcerated stricture at the neo-distal ileum; (D) stricture at the neo-distal ileum undergoing endoscopic stricturotomy with needle knife. *CD*, Crohn's disease.

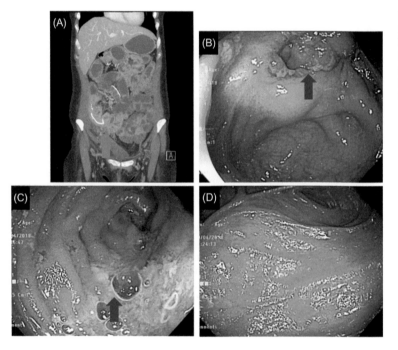

FIGURE 8.19   Bypass surgery for severe ileal Crohn's disease. The surgical procedure in the area has been largely replaced by bowel resection or stricturoplasty. (A) Bypassed diseased distal neo-terminal ileum (*green arrow*) on computed tomography; (B and C) bypassed diseased distal neo-terminal ileum; (D) previous bowel resection with ileocolonic anastomosis.

Surgical therapy for perianal disease is discussed in a separate chapter (Chapter 7: Endoscopic evaluation of perianal Crohn's disease).

Summary and recommendations

Various surgical modalities have been designed and performed in patients with medically and endoscopically refractory CD. Altered bowel anatomy by the surgery has posed challenges for clinicians and endoscopists. Accurate recognition of landmarks and anatomy is critical for disease monitoring and management. The distinction between "normal" postsurgical anatomy and recurrent CD is important, as surgery-associated factors can manifest any phenotypes of CD, including mucosal ulcer, bowel stricture, fistula, and abscess. For example, ulcer along the anastomosis

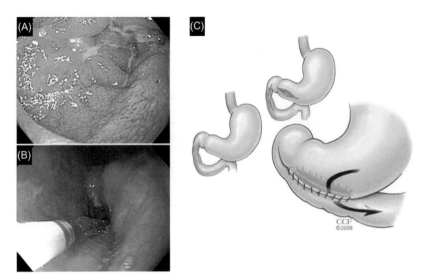

FIGURE 8.20 Bypass surgery for duodenum CD. (A) Severe duodenum CD with inflammatory stricture; (B) the duodenum stricture was surgically bypassed with a gastrojejunostomy, which was strictured. Balloon dilation of the gastrojejunostomy stricture was performed; (C) illustration of gastrojejunostomy for duodenum CD. *CD*, Crohn's disease. *Courtesy Mr. Joel Pangrace of Cleveland Clinic.*

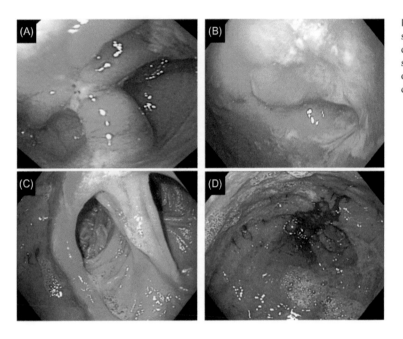

FIGURE 8.21 Whipple procedure for the treatment of severe refractory duodenum CD. (A and B) Severe duodenum CD with ulcers, edema, and stricture; (C and D) status of post—Whipple procedure with gastrojejunostomy. There was severe bile gastritis. *CD*, Crohn's disease.

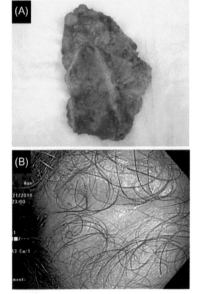
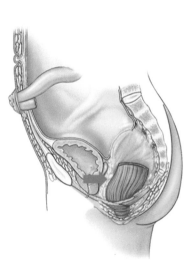

FIGURE 8.22 Completion proctectomy in Crohn's disease. (A) Surgically resected Hartmann pouch or rectum stump, which was contracted due to long-standing fecal diversion; (B) healed perineal detect after proctectomy; (C) illustration of Hartmann pouch or diverted rectum (*green arrow*) in a patient with loop-end ileostomy. *Courtesy Mr. Joel Pangrace of Cleveland Clinic.*

may be related to ischemia or recurrent CD. The distinction can often be reliably made by a careful evaluation of disease extent, distribution, and location to the surgical site.

References

[1] Bernell O, Lapidus A, Hellers G. Risk factors for surgery and postoperative recurrence in Crohn's disease. Ann Surg 2000;231:38−45.

[2] Peyrin-Biroulet L, Harmsen WS, Tremaine WJ, Zinsmeister AR, Sandborn WJ, Loftus Jr. EV. Surgery in a population-based cohort of Crohn's disease from Olmsted County, Minnesota (1970−2004). Am J Gastroenterol 2012;107:1693−701.

[3] Silverberg MS, Satsangi J, Ahmad T, Arnott ID, Bernstein CN, Brant SR, et al. Toward an integrated clinical, molecular and serological classification of inflammatory bowel disease: report of a Working Party of the 2005 Montreal World Congress of Gastroenterology. Can J Gastroenterol 2005;19(Suppl. A):5A−36A.

[4] Olaison G, Smedh K, Sjodahl R. Natural course of Crohn's disease after ileocolic resection: endoscopically visualised ileal ulcers preceding symptoms. Gut 1992;33:331−5.

[5] Rutgeerts P, Geboes K, Vantrappen G, Beyls J, Kerremans R, Hiele M. Predictability of the postoperative course of Crohn's disease. Gastroenterology 1990;99:956−63.

[6] Rutgeerts P, Geboes K, Vantrappen G, Kerremans R, Coenegrachts JL, Coremans G. Natural history of recurrent Crohn's disease at the ileocolonic anastomosis after curative surgery. Gut 1984;25:665−72.

[7] Domènech E, Mañosa M, Bernal I, Garcia-Planella E, Cabré E, Piñol M, et al. Impact of azathioprine on the prevention of postoperative Crohn's disease recurrence: results of a prospective, observational, long-term follow-up study. Inflamm Bowel Dis 2008;14:508−13.

[8] Gecse K, Lowenberg M, Bossuyt P, Rutgeerts PJ, Vermeire S, Stitt L, et al. Agreement among experts in the endoscopic evaluation of postoperative recurrence in Crohn's disease using the Rutgeerts score. Gastroenterology 2014;146:S227.

[9] Bayart P, Duveau N, Nachury M, Zerbib P, Gerard R, Branche J, et al. Ileal or anastomotic location of lesions does not impact rate of postoperative recurrence in Crohn's disease patients classified i2 on the Rutgeerts score. Dig Dis Sci 2016;61:2986−92.

[10] Rivière P, Vermeire S, Irles-Depe M, van Assche G, Rutgeerts P, de Buck van Overstraeten A, et al. No change in determining Crohn's disease recurrence or need for endoscopic or surgical intervention with modification of the Rutgeerts scoring system. Clin Gastroenterol Hepatol. 2019;17:1643−5.

[11] Chongthammakun V, Fialho A, Fialho A, Lopez R, Shen B. Correlation of the Rutgeerts score and recurrence of Crohn's disease in patients with end ileostomy. Gastroenterol Rep (Oxf) 2017;5:271−6.

Chapter 9

Ulcerative colitis

Bo Shen

Center for Inflammatory Bowel Diseases, Columbia University Irving Medical Center-New York Presbyterian Hospital, New York, NY, United States

Chapter Outline

Abbreviations

BWI	backwash ileitis
CAN	colitis-associated neoplasia
CD	Crohn's disease
CP	cap polyposis
GI	gastrointestinal
IBD	inflammatory bowel disease
ICV	ileocecal valve
PSC	primary sclerosing cholangitis
UC	ulcerative colitis
UP	ulcerative proctitis

Introduction

In clinical practice the Montreal classification of ulcerative colitis (UC) has routinely been used to guide diagnosis and treatment (Table 1.1) [1]. In addition to the clinical assessment, the identification of endoscopic and histologic features is the key to the diagnosis and differential diagnosis of UC as well as Crohn's disease (CD). Colonoscopy also plays a critical role in the disease monitoring, assessment of treatment response, surveillance of colitis-associated neoplasia (CAN), and endoscopic therapy [2]. The matrix of measuring a quality colonoscopy in UC includes (1) adequate level of intubation, especially the intubation of the terminal ileum; (2) assessment of disease extent and distribution pattern; (3) grading of degree and features of mucosal inflammation; and (4) proper biopsy. Detailed information on grading the severity of mucosal inflammation in UC as well as CD is discussed in Chapter 11, Ulcerative colitis: postsurgical and Chapter 14, Endoscopic scores in inflammatory bowel disease. Colonoscopic features of superimposed infections are discussed in Chapter 23, Superimposed infections in inflammatory bowel diseases.

Adequate bowel preparation is needed for a quality colonoscopy in UC. Active mucosal inflammation often presents with mucopurulent or fibrin exudates or plaques, which should be washed to evaluate true pattern and severity of mucosal inflammation (Fig. 9.1).

Atlas of Endoscopy Imaging in Inflammatory Bowel Disease. DOI: https://doi.org/10.1016/B978-0-12-814811-2.00009-8
© 2020 Elsevier Inc. All rights reserved.

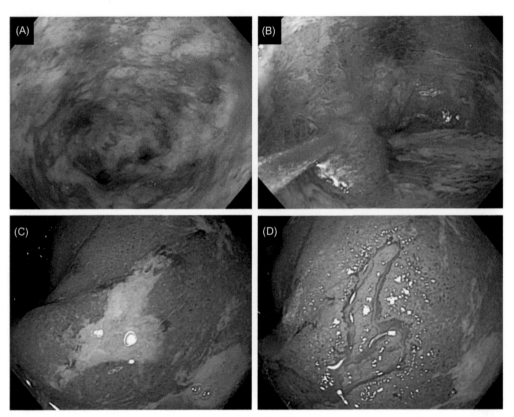

FIGURE 9.1 Washing and accurate grading inflammation in UC. (A) and (B) Diffuse mucosal inflammation with edema, erythema, and granularity of erosions was revealed by washing out mucopurulent exudates; (C) and (D) washing out mucous exudates uncovered edematous mucosa with erosion and friability. *UC*, Ulcerative colitis.

Disease extent and distribution

Patients with chronic diarrhea (i.e., the symptom lasting more than 4 months) should have colonoscopy evaluation for noninfectious etiologies, including inflammatory bowel disease (IBD) and microscopic colitis. The assessment on index colonoscopyprior to medical therapy more accurately reflects true disease extent [3]. The classic UC at initial presentation demonstrates continuous inflammation extending proximally from the anal verge and rectum.

Terminal ileum intubation

Evaluation of the terminal ileum and ileocecal valve (ICV) is important for the differential diagnosis of CD and UC. Features of the terminal ileum and ICV may also distinguish UC with backwash ileitis (BWI) from CD ileitis (see Chapter 10: Indeterminate colitis and inflammatory bowel disease unclassified). This requires intubation of the terminal ileum for all patients undergoing colonoscopy for suspected of IBD. However, this standard practice has not been satisfactorily achieved even in the tertiary care hospital [4].

Cecal patch, rectal sparing, and patchy distribution

In the majority of patients, the assessment of disease extent of UC is straightforward, as the diseased or nondiseased segments of UC often have a sharp demarcation or abrupt transition (Fig. 9.2). Histologic measurement of the disease extent of UC may be more extensive than endoscopic measurement [5]. Nonetheless, the endoscopic measurement of disease extent is considered as a gold standard. The disease extent should be measured by the segment of the bowel involved as well as the length from the anal verge. This practice is critical for subsequent endoscopic monitoring of disease extent. UC may progress proximally over time [6]. One quarter of patients with limited UC extend proximately with most extension occurring during the first 10 years after diagnosis [7]. The extension of the disease length is considered as a sign of poor prognosis for UC [8].

Patients with ulcerative proctitis (UP) or left-sided UC, especially pediatric patients or young adults, may have coexisting periappendiceal or cecal patch and distal UC (Figs. 9.3–9.5 and 10.8). The finding of the cecal patch or

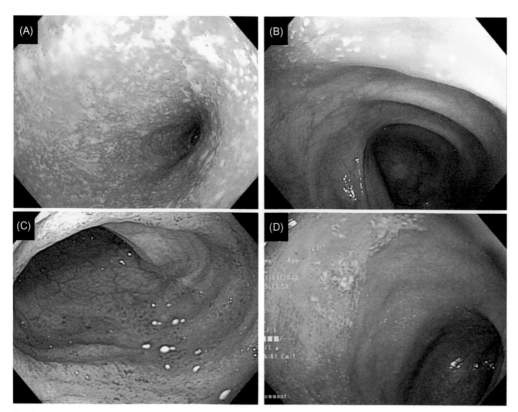

FIGURE 9.2 Classic patterns of the distribution of mucosal inflammation in UC. (A) Diffuse continuous mucosal inflammation with erythema, mucopurulent exudates, and loss of vascular pattern from the anal verge to the proximal segment of the colon; (B)–(D) abruption transition or sharp demarcation between the inflamed distal colon and noninflamed proximal parts of the colon. *UC*, Ulcerative colitis.

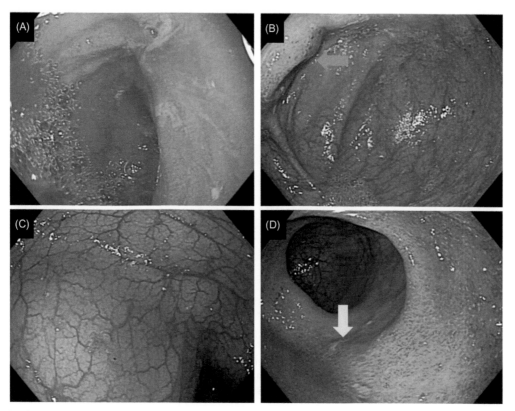

FIGURE 9.3 Cecal or periappendiceal patch in ulcerative proctitis. (A) Normal terminal ileum; (B) periappendiceal mucosal inflammation with erythema (*green arrow*); (C) normal colon; (D) active proctitis with erythema and granularityalong with a normal proximal colon showing a sharp demarcation between inflamed and noninflamed bowel segments (*yellow arrow*).

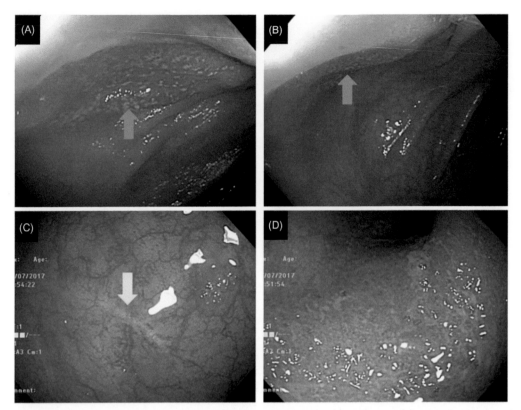

FIGURE 9.4 Periappendiceal patch in left-side UC. (A) and (B) Periappendiceal patch with erythema and mucopurulent exudates (*green arrows*); (C) mucosal scars (*yellow arrow*)in the sigmoid colon; (D) active proctitis with erythema and granularity. *UC*, Ulcerative colitis.

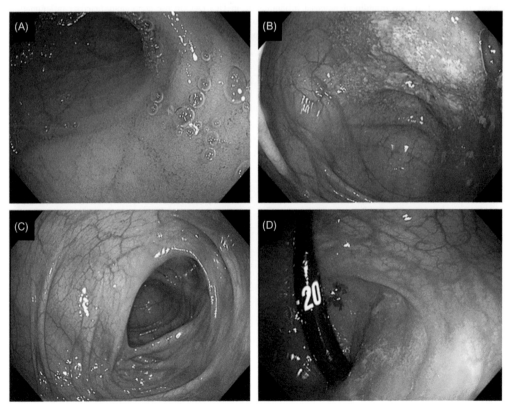

FIGURE 9.5 Cecal patch with distal ulcerative proctitis. Even though periappendiceal inflammation is more severe than distal proctitis, the patient should be treated with topical mesalamine therapy as the first-line regimen. (A) Normal terminal ileum; (B) moderate inflammation at the cecal base with erythema, erosions, and mucopurulent exudates; (C) normal ascending colon; (D) distal 10 cm of the rectum with mild erythema. Biopsy showed chronic active colitis.

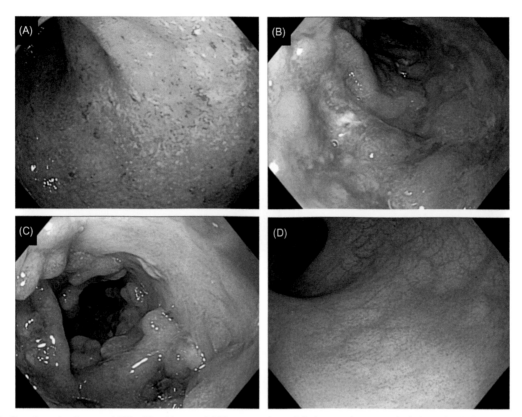

FIGURE 9.6   Rectal sparing in treated UC with a history of diffuse extensive colitis from anal verge at the index colon before medical therapy. (A) Diffuse inflammation with erythema and absent vascular pattern in the ascending colon; (B) and (C) severe colitis with ulcers, edema, nodularity, and pseudopolyps in the transverse and descending colon; (D) normal rectum. *UC*, Ulcerative colitis.

periappendiceal patch in the setting of UC with an otherwise normal right colon should not be confused with CD [9,10]. The cecal disease in UC patients should not be considered as a part of extensive UC. The clinical implication of cecal patch is not clear, although a recent controlled study revealed that UC patients with cecal patch had a similar rate of remission, relapse, and proximal extension to UC patients with no cecal patch [9]. The presence of cecal patch does not seem to impact the outcome of left-sided UC or UP [11]. Therefore an accurate distinction between extensive UC and left-sided UC or UP with cecal patch is important for clinical management and prognosis. For example, different mesalamine (oral vs topical) and corticosteroid (oral vs topical; and pH-dependent, delayed release vs extended release) regimens for treating the two disease phenotypes are different [12].

In an adult population with UC, the index colonoscopy typically demonstrates rectal involvement. However, pediatric or young adult UC patients may initially present with segmental colitis and/or rectal sparing (Fig. 9.6) [13]. True rectal sparing in UC, either before or after medical therapy, may leave a room for the future candidate of total colectomy with ileorectal anastomosis if surgical intervention is needed for medically refractory disease or for CAN [14]. In patients with active colitis, retroflex maneuvering is not recommended, to reduce the risk for iatrogenic trauma.

Patients with extensive UC may have concurrent BWI, particularly in those with coexisting primary sclerosing cholangitis (PSC) (Fig. 9.7). BWI in UC is discussed in a separate chapter (Chapter 10: Indeterminate colitis and inflammatory bowel disease unclassified) in details.

Patterns of mucosal inflammation

Common endoscopic features of UC are highlighted with a diagram, ranging from erosions and ulceration to pseudopolyps and strictures (Fig. 9.8). These features can overlap and coexist in a synchronous pattern. An individual patient may present one or several lesions in the same or different segments of the large bowel.

Erosions and ulcers

Gastric or intestinal erosion is defined as inflamed mucous membrane with damage limited to the epithelium, basement membrane, and lamina propria. Gastric or intestinal ulcer is defined as the damage to the gastrointestinal (GI) tract wall

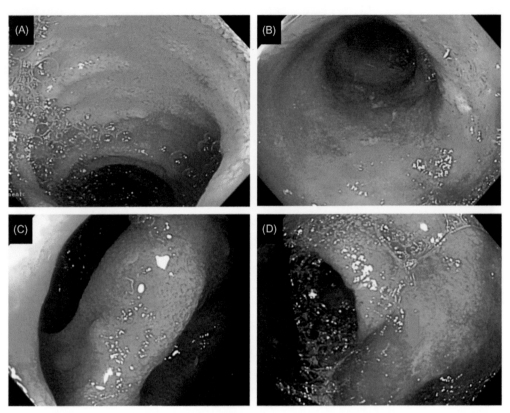

FIGURE 9.7 UC with primary sclerosing cholangitis after liver transplantation. (A) Normal distal ileum; (B) mild inflammation with erythema and erosions at the distal 3 cm of the terminal ileum; (C) patulous ileocecal valve covered with normal mucosa; (D) colitis in remission with mucosa scars (*green arrow*). Liver transplantation may help keep the patient's colitis in remission. *UC*, Ulcerative colitis.

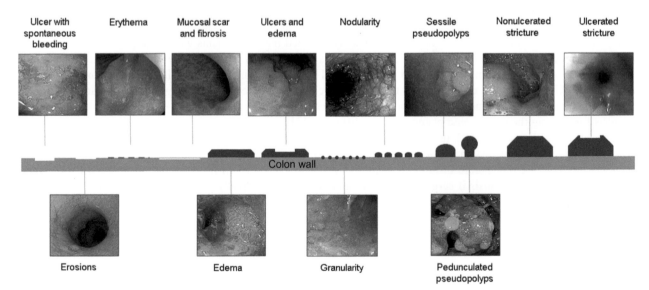

FIGURE 9.8 Patterns of inflammation in UC. From left to right: ulcers, erosions, mucosal scars/fibrosis, edema, nodularity, sessile and pedunculated pseudopolyps, and strictures. *UC*, Ulcerative colitis.

that extends deeper through the wall than an erosion. An ulcer can extend anywhere from beyond the lamina propria to right through the wall. On histology, there is acute loss of surface epithelium in a response to fibrin deposition and neutrophilic inflammation. Ulceration means that surface and crypt injury is accompanied by granulation tissue formation (Fig. 9.9).

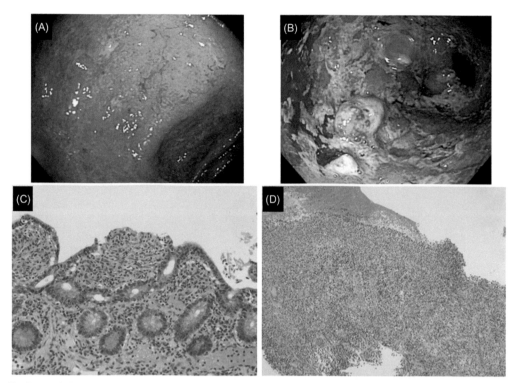

FIGURE 9.9 Erosions and ulcers on colonoscopy and histology. (A) Erosions on edematous erythematous mucosa; (B) diffuse ulcers with pseudopolyps, nodular mucosa, and spontaneous bleeding; (C) erosions: the small bowel mucosa shows villous blunting, detachment of surface epithelium from the basement membrane, with minimal inflammation; (D) ulceration: the small bowel mucosa shows complete loss of surface and crypt epithelium, accompanied by granulation and inflammation. *Histology photo courtesy Xiuli Liu, MD, PhD, of University of Florida.*

Aphthous or isolated erosions of the mucosa are not common in UC. Classic aphthous erosions are tiny (2−3 mm), raised or flat red lesion with a white center. Erosions often coexist with other forms of mucosal inflammation, such as erythema, edema, and granularity (Figs. 9.10 and 9.11).

Ulcers in the UC can be small or large, diffused or localized, shallow or deep, clean-based or exudate-covered, raised or flat border, or irregular, serpiginous, or confluent (Fig. 9.12). Round, oval, well-demarcated ulcers or circumferential or longitudinal ulcers are not common. The distinguishing endoscopic features of UC and CD are discussed in Chapter 10, Indeterminate colitis and inflammatory bowel disease unclassified, and listed in Table 10.1.

Erosions and ulcers in UC or CD can be triggered or exacerbated by medications, such as nonsteroidal antiinflammatory drugs, sodium phosphates, or even corticosteroids (Fig. 9.13).

Erythema, edema, loss of vascularity, friability, and spontaneous bleeding

Erythema is defined as redness of mucous membranes resulting from hyperemia (increased blood flow) in superficial capillaries. However, the terms of erythema and hyperemia have been used interchangeably (Fig. 9.14).

Edema or bowel swelling with abnormal accumulation of fluid in UC results from active inflammation and concurrent protein malnutrition. Edema often coexists with other forms of mucosal inflammation. In non-IBD patients with severe protein malnutrition may present bowel wall edema in the absence of endoscopic inflammatory features (Fig. 32.13).

Normal colon lining has a smooth and glistening pinkish color with a transparent surface mucosa and a visible network of underneath branching vessels (Fig. 9.15A). Absent vascular pattern or vascularity can occur in mild, moderate, and severe active colitis. It may also be seen in patients with endoscopic signs of disease in remission (Fig. 9.15B−D).

Friability of the mucosa is defined as break up of the surface and easy bleeding with contact. Spontaneous bleeding is a condition in which active bleeding of the surface of gut mucosa without contact. The latter condition reflects more severe inflammation than the former (Fig. 9.16).

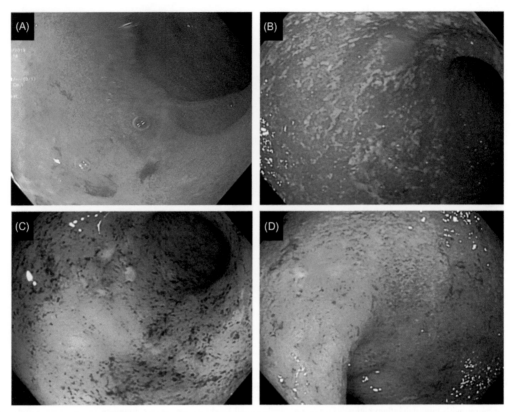

FIGURE 9.10 Erosions in UC commonly associated with other patterns of mucosal inflammation. (A) Erosions with friable mucosa and absent vascularity; (B) erosions covered with mucopurulent exudates and absent vascularity; (C) erosions with mucopurulent exudates, punctate erythema, and absent vascularity; (D) erosions with erythema and absent vascularity. *UC*, Ulcerative colitis.

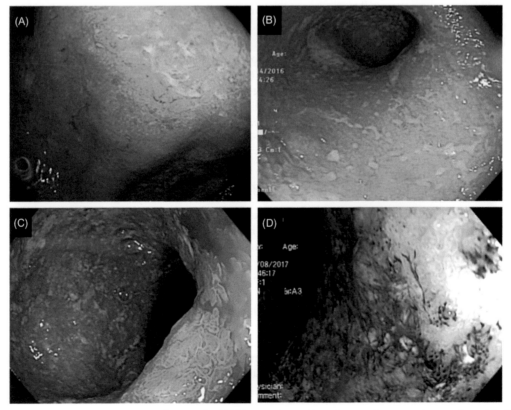

FIGURE 9.11 Additional patterns of erosions in UC. (A) and (B) Diffuse erosions with erythema and mucopurulent exudates; (C) diffuse erosions, erythema, edema, and mucopurulent exudates; (D) erosions covered with particles of oral iron supplements. *UC*, Ulcerative colitis.

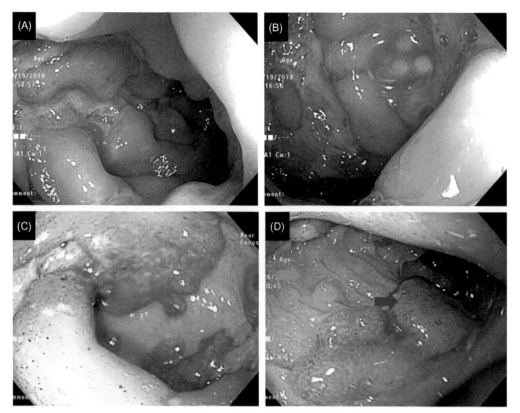

FIGURE 9.12 Distinction between ulcer-associated denuded mucosa and pseudopolyps. (A)–(C) Deep ulcers with large areas of denuded mucosa surrounded by "surviving" edematous and nodular mucosa; (D) pseudopolyps (*blue arrow*) near normal-appearing mucosa (*green arrow*), which can be mistaken as deep ulcerated mucosa.

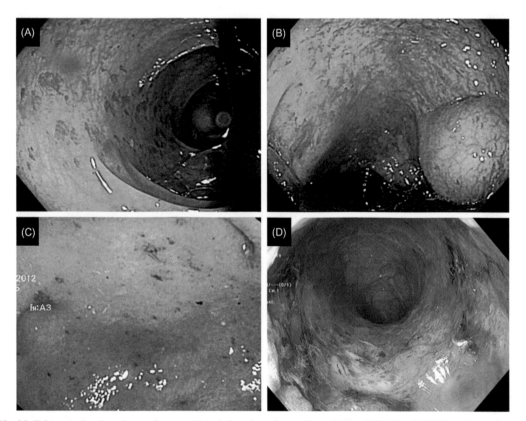

FIGURE 9.13 Medicine-associated erosions and mucosal hemorrhage in patients with underlying UC. (A) and (B) The patient treated with oral prednisone 60 mg/day; (C) another patient with daily use of nonsteroidal antiinflammatory drugs; (D) a UC patient undergoing flexible sigmoidoscopy who had sodium phosphate−based enema preparation. *UC*, Ulcerative colitis.

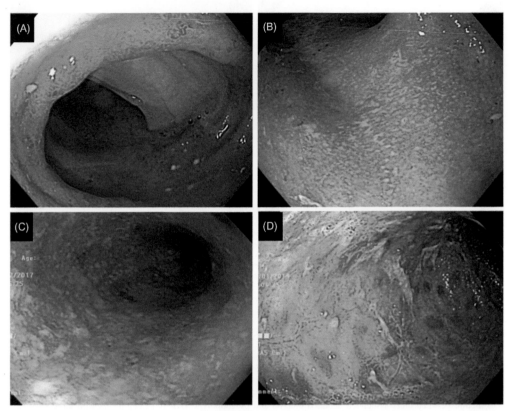

FIGURE 9.14 Diffuse and patchy erythema in UC. (A)–(C) Diffuse erythema and loss of vascular pattern of the colon; (D) patchy erythema with intervening mucosal scars. *UC*, Ulcerative colitis.

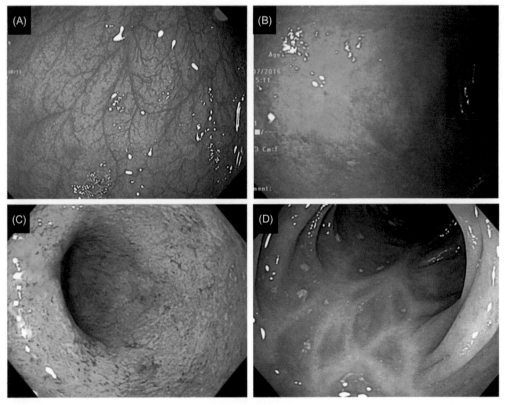

FIGURE 9.15 Loss of vascular patterns in UC. (A) Normal vascular pattern underneath the colon epithelia; (B) mild colitis with erythema and loss of vascular pattern; (C) moderate colitis with vascular pattern as well as erythema, hemorrhage, and lead piping of the lumen; (D) loss of vascular pattern with mucosal scars, a sign of disease in remission. *UC*, Ulcerative colitis.

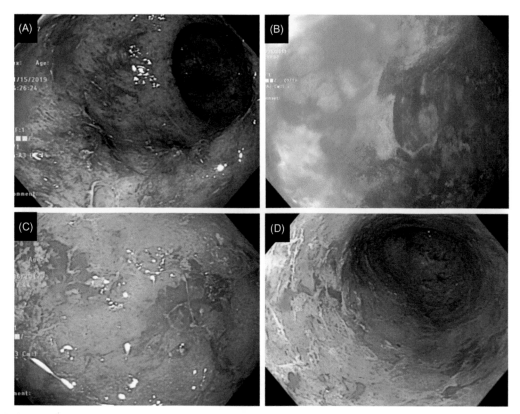

FIGURE 9.16 Severe UC with spontaneous bleeding in the absence of deep ulcers. (A)−(D) Diffuse spontaneous bleeding with erythema, edema, and large exfoliative erosions. *UC*, Ulcerative colitis.

Granularity, nodularity, and pseudopolyps

The term of granularity is used loosely to describe roughened mucosa. The distinction between granularity and nodularity may be their sizes with shared underlying pathological processes (Fig. 9.17).

Nodularity is typically observed in active UC, which can be coarse, consisting of small or large nodules. Nodules become elongated, becoming pseudopolyps. Mucosal nodularity in UC should be distinguished from cobblestoning in CD. Cobblestoning of the distal ileum or colon on ileocolonoscopy, a feature of CD, is characterized by linear, longitudinal, or serpiginous ulcers along the longitudinal axis of the colon, and intervening areas of normal or nodular, inflamed tissues (Figs. 9.18 and 9.19).

Inflammatory or pseudopolyps are common in patients with long-standing colitis. The polyps can be sessile, pedunculated, or elongated tubular. The polyps can be small or large in number and size and some of them may have white caps. Those polyps can be in patients with active or quiescent UC (Figs. 9.20−9.22). UC with extensive deep ulceration and denuded mucosa or superficial submucosa can present endoscopic an appearance-resembling pseudopolyps (Fig. 9.23). The presence of inflammatory polyps may contribute to patients' diarrhea, urgency, or bleeding. Whether UC patients with inflammatory polyps carry an increased risk for CAN is debatable [15−17].

Inflammatory pseudopolyps should be differentiated from cap polyposis (CP). The latter is characterized by erythematous, inflammatory colonic polyps with a mucofibrinopurulent cap. The etiology and pathogenesis of CP are not clear. Current theories hold infection, mucosal ischemia, T cell−mediated inflammation, mucosal prolapse, bowel dysmotility, and trauma from excessive straining. It may share some etiopathogenetic pathways with diverticular colitis, as both disorders are commonly located at the rectosigmoid colon (Fig. 32.17). The polyps may be sessile or pedunculated, often located at the apices of prolapsed mucosal folds. However, the intervening mucosa is normal on endoscopy and histology, different from IBD-associated pseudopolyps. In addition, histology of CP is featured with acute and chronic inflammatory changes and elongated, hyperplastic-appearing glands, fibromuscular obliteration of the lamina propria, and inflammatory exudates composed of mucus, fibrin, and leukocytes.

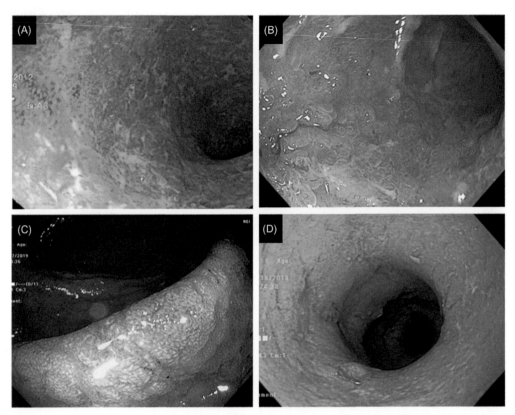

FIGURE 9.17 Patterns of granularity in UC with loss of normal smooth shining mucosa. (A) "Sandpaper"-like granularity, erythema, and mucopurulent exudates of the mucosa; (B) the more coarse granularity of the mucosa; (C) mucosal granularity can be highlighted with narrowband imaging; (D) smoother granularity with possible mucosal healing with adequate medical management. *UC*, Ulcerative colitis.

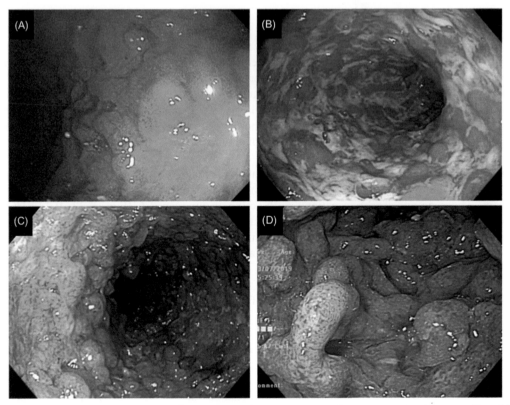

FIGURE 9.18 The transition from granularity to nodularity and pseudopolyps in UC. (A) Coarse granularity of mucosa; (B) nodularity resulting from adjacent extensive ulceration; (C) nodular mucosa; (D) nodular mucosa on the way to form pseudopolyps. Notice that nodularity patterns in UC lack longitudinal ulcers, in contrast to cobblestoning in Crohn's disease. *UC*, Ulcerative colitis.

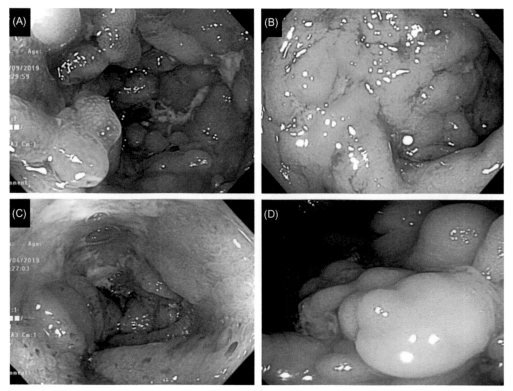

FIGURE 9.19   Nodularity and cobblestoning in colonic Crohn's disease for the differential diagnosis. The patterns result from long linear or serpiginous ulcers course along the longitudinal axis of the colon, with intervening areas of normal or inflamed tissues. (A) Diffuse nodularity of mucosa; (B) and (C) cobblestoning mucosa almost formed; and (D) large-sized cobblestoning.

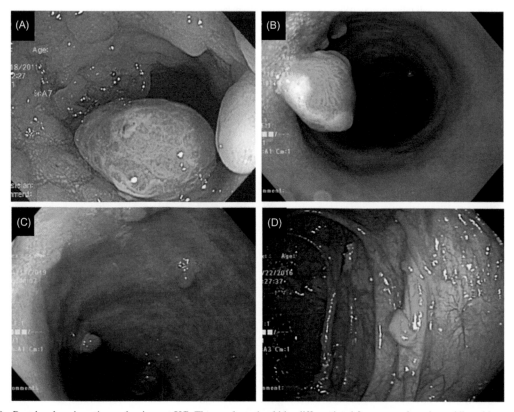

FIGURE 9.20   Pseudopolyps in active and quiescent UC. These polyps should be differentiated from cap polyposis, an idiopathic, noninflammatory bowel disease condition. (A) Large pedunculated inflammatory polyps with surrounding nodular mucosa; (B) isolated pedunculated polyp with white fibrinomucopurulent exudate on the top; (C) and (D) small sessile polyps in quiescent UC. *UC*, Ulcerative colitis.

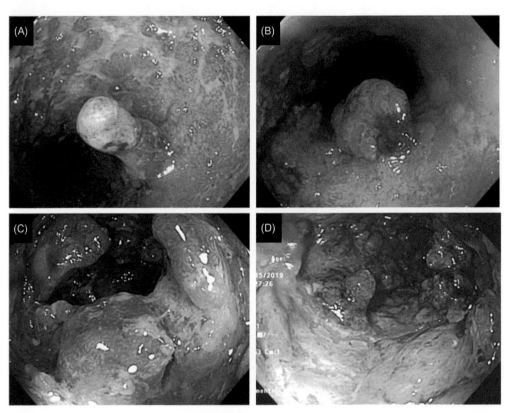

FIGURE 9.21 Inflammatory and vascular pseudopolyps in the setting of severe active UC. (A)−(C) Inflammatory pedunculated polyps in erythematous and ulcerated mucosa; (D) small inflammatory polyps in the middle of deep, large ulcers. *UC*, Ulcerative colitis.

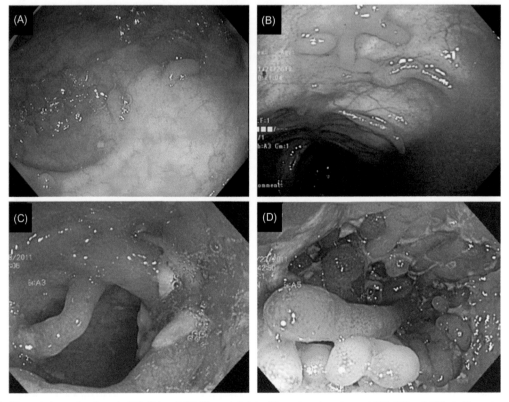

FIGURE 9.22 Elongated pseudopolyps in UC. (A) and (B) Elongated, tube-like polyps in quiescent UC; (C) and (D) similar polyps in active UC. *UC*, Ulcerative colitis.

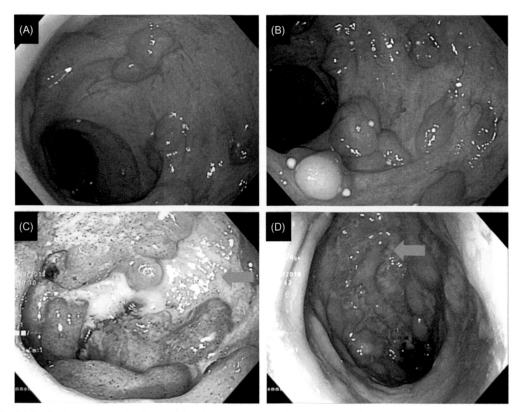

FIGURE 9.23   Distinction between pseudopolyps and denuded mucosa. (A) and (B) Multiple pedunculated pseudopolyps with surrounding quiescent mucosa; (C) and (D) large deep ulcers resulting in denuded mucosa, which makes adjacent survival mucosa resemble pseudopolyps.

Fibrosis and muscular hyperplasia

Intramucosal and submucosal fibrosis and hyperplasia of muscularis mucosae can occur in patients with UC, resulting in mucosal scars, mucosal bridges (Fig. 9.24), "lead pipe" appearance of the lumen (Fig. 9.25), and benign strictures (Fig. 9.26A–C). Chronic mucosal inflammation can result in CAN or even malignant strictures (Fig. 9.26D).

Fibrosis is considered as a part of the healing process, which can be present in quiescent or active UC (Fig. 9.25). Diffuse large bowel fibrosis, especially fibrosis in the rectum, may reduce the extensibility and compliance of bowel wall, contributing to rectal urgency, increased bowel frequency, and decreased bowel consistency. Furthermore, the fibrosis process can make endoscopic ablation procedures (such as endoscopic mucosal resection and endoscopic submucosal dissection) for CAN difficult [18].

Histologic evaluation

Adjunct to colonoscopy, histologic evaluation is an integral part of the diagnosis, differential diagnosis, disease monitoring, assessment of superimposed infections from cytomegalovirus, Epstein–Barr virus, or *Clostridium difficile* (by trapping luminal contents during colonoscopy) and surveillance for CAN. Histologic evaluation of IBD is discussed in a separate chapter (Chapter 15: Histologic evaluation of disease activity in inflammatory bowel disease).

Summary and recommendations

Ulcerative colitis can present with an array of acute, chronic, or quiescent endoscopic features. Accurate identification and documentation of these endoscopic characteristics along with disease location and tissue biopsy are important for the diagnosis, differential disease, disease activity monitoring, assessment of treatment response, and dysplasia surveillance. These features constitute components of various endoscopic disease activity instruments. To assess the disease extent and later on disease extension, the pretreatment index colonoscopy with intubation of the terminal ileum is a key.

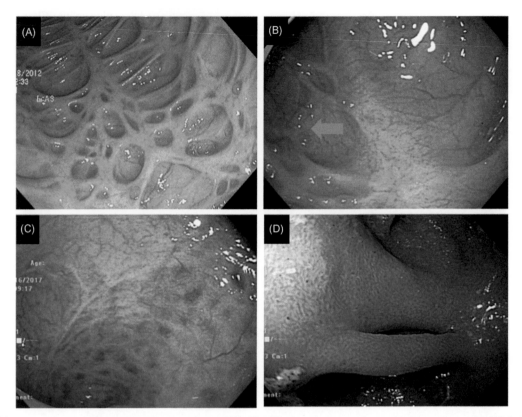

FIGURE 9.24 Mucosal scars and mucosal bridges in quiescent UC. (A)–(C) Extensive mucosal scars in three patients with UC in long-standing remission; (D) mucosal bridge in the right colon. *UC*, Ulcerative colitis.

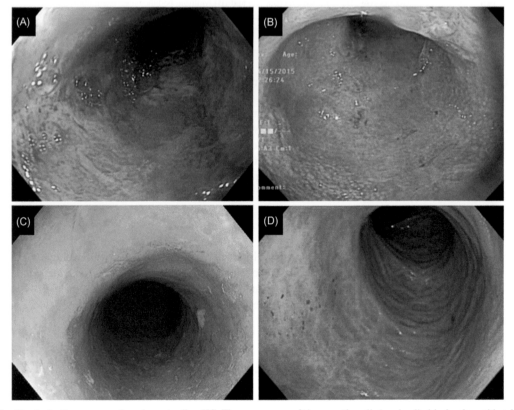

FIGURE 9.25 "Lead pipe" appearance from long-standing UC. The appearance of the smooth-walled and cylindrical colon with a loss of normal haustral marks is best illustrated by barium enema. It may also be seen with colonoscopy. (A) and (B) The "lead pipe" appearance with active colitis; (C) and (D) the "lead pipe sign" with quiescent disease, manifesting a tubular lumen and loss of vascularity and scars of the mucosa. *UC*, Ulcerative colitis.

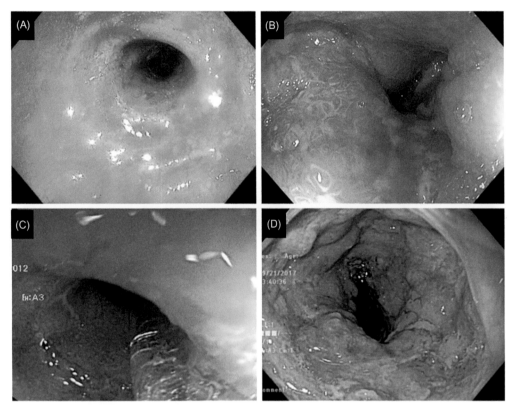

FIGURE 9.26 Benign and malignant strictures in UC. (A) Distal rectal nonulcerated fibrotic benign stricture; (B) inflammatory strictures in the proximal rectum; (C) tight nonulcerated benign stricture at the distal rectum undergoing endoscopic balloon dilation; (D) ulcerated malignant stricture at the mid-rectum from long-standing UC. *UC*, Ulcerative colitis.

References

[1] Silverberg MS, Satsangi J, Ahmad T, Arnott ID, Bernstein CN, Brant SR, et al. Toward an integrated clinical, molecular and serological classi-fication of inflammatory bowel disease: report of a Working Party of the 2005 Montreal World Congress of Gastroenterology. Can J Gastroenterol 2005;19(Suppl. A):5A–36A.

[2] Fausel RA, Kornbluth A, Dubinsky MC. The first endoscopy in suspected inflammatory bowel disease. Gastrointest Endosc Clin N Am 2016;26:593–610.

[3] ASGE Standards of Practice Committee, Shen B, Khan K, Ikenberry SO, Anderson MA, Banerjee S, et al. The role of endoscopy in the man-agement of patients with diarrhea. Gastrointest Endosc 2010;71:887–92.

[4] Makkar R, Lopez R, Shen B. Clinical utility of retrograde terminal ileum intubation in the evaluation of chronic non-bloody diarrhea. J Dig Dis 2013;14:536–42.

[5] Moum B, Ekbom A, Vatn MH, Elgjo K. Change in the extent of colonoscopic and histological involvement in ulcerative colitis over time. Am J Gastroenterol 1999;94:1564–9.

[6] Geboes K, Ectors N, D'Haens G, Rutgeerts P. Is ileoscopy with biopsy worthwhile in patients presenting with symptoms of inflammatory bowel disease? Am J Gastroenterol 1998;93:201–6.

[7] Roda G, Narula N, Pinotti R, Skamnelos A, Katsanos KH, Ungaro R, et al. Systematic review with meta-analysis: proximal disease extension in limited ulcerative colitis. Aliment Pharmacol Ther 2017;45:1481–92.

[8] Torres J, Billioud V, Sachar DB, Peyrin-Biroulet L, Colombel JF. Ulcerative colitis as a progressive disease: the forgotten evidence. Inflamm Bowel Dis 2012;18:1356–63.

[9] Okawa K, Aoki T, Sano K, Harihara S, Kitano A, Kuroki T. Ulcerative colitis with skip lesions at the mouth of the appendix: a clinical study. Am J Gastroenterol 1998;93:2405–10.

[10] Byeon J-S, Yan S-K, Myung S-J, Pyo SL, Park HJ, Kim YM, et al. Clinical course of distal ulcerative colitis in relation to appendiceal orifice inflammation status. Inflamm Bowel Dis 2005;11:366–71.

[11] Byeon JS, Yang SK, Myung S, Pyo SI, Park HJ, Kim YM, et al. Clinical course of distal ulcerative colitis in relation to appendiceal orifice inflammation status. Inflamm Bowel Dis 2005;11:366–71.

[12] Waterman M, Knight J, Dinani A, Xu W, Stempak JM, Croitoru K, et al. Predictors of outcome in ulcerative colitis. Inflamm Bowel Dis 2015;21:2097–105.

[13] Joo M, Odze RD. Rectal sparing and skip lesions in ulcerative colitis: a comparative study of endoscopic and histologic findings in patients who underwent proctocolectomy. Am J Surg Pathol 2010;34:689—96.

[14] da Luz Moreira A, Kiran RP, Lavery I. Clinical outcomes of ileorectal anastomosis for ulcerative colitis. Br J Surg 2010;97:65—9.

[15] Velayos FS, Loftus EVJ, Jess T, Harmsen WS, Bida J, Zinsmeister AR, et al. Predictive and protective factors associated with colorectal cancer in ulcerative colitis: a case—control study. Gastroenterology 2006;130:1941—9.

[16] Baars JE, Looman CWN, Steyerberg EW, Beukers R, Tan AC, Weusten BL, et al. The risk of inflammatory bowel disease-related colorectal carcinoma is limited: results from a nationwide nested case—control study. Am J Gastroenterol 2011;106:319—28.

[17] Mahmoud R, Shah SC, Ten Hove JR, Torres J, Mooiweer E, Castaneda D, et al. Dutch Initiative on Crohn and Colitis. No association between pseudopolyps and colorectal neoplasia in patients with inflammatory bowel diseases. Gastroenterology 2019;156:1333—44.

[18] Shen B, Kochhar G, Navaneethan U, Liu X, Farraye FA, Gonzalez-Lama Y, et al. Global Interventional Inflammatory Bowel Disease Group. Role of interventional inflammatory bowel disease in the era of biologic therapy: a position statement from the Global Interventional IBD Group. Gastrointest Endosc 2019;89:215—37.

Chapter 10

Indeterminate colitis and inflammatory bowel disease unclassified

Bo Shen

Center for Inflammatory Bowel Diseases, Columbia University Irving Medicine Center-New York Presbyterian Hospital, New York, NY, United States

Chapter Outline

Abbreviations

BWI	backwash ileitis
CD	Crohn's disease
IBD	inflammatory bowel disease
IBD-U	inflammatory bowel disease unclassified
ICV	ileocecal valve
GI	gastrointestinal
IC	indeterminate colitis
IPAA	ileal pouch–anal anastomosis
NSAIDs	nonsteroidal antiinflammatory drugs
PSC	primary sclerosing cholangitis
UC	ulcerative colitis
VCE	video capsule endoscopy

Introduction

Disease spectrum of Crohn's disease (CD) and ulcerative colitis (UC) is beyond the two classic phenotypes. Differential diagnosis between CD and UC is important for medical and surgical management and prognosis. For example, oral antibiotics have been used as a first-line therapy for perianal CD, Crohn's colitis, and prophylaxis and treatment of postoperative CD, the agents are not recommended in the treatment of UC, with except in superimposed *Clostridium difficile* infection. Restorative proctocolectomy with ileal pouch–anal anastomosis (IPAA) is the surgical

Atlas of Endoscopy Imaging in Inflammatory Bowel Disease. DOI: https://doi.org/10.1016/B978-0-12-814811-2.00010-4
© 2020 Elsevier Inc. All rights reserved.

treatment of choice for patients with classic UC who require colectomy, whereas the procedure is contraindicated in the CD patients with ileitis, ileocolitis, transmural colitis,perianal disease, or upper gastrointestinal (GI) disease.

Differential diagnosis of CD and UC can be difficult. Endoscopy together with other diagnostic modalities may only differentiate CD from UC in ≥ 85% of patients [1]. In a prospective study of more than 350 patients with inflammatory bowel disease (IBD) followed up for more than 22 months, index colonoscopy and biopsy were accurate in distinguishing CD from UC in 89% of cases. IBD diagnosis was revised in 4% of cases, and the diagnosis of indeterminate colitis (IC) remained in 7% of cases [2]. A range of clinical, endoscopic, and histologic phenotypes cannot be easily classified into CD or UC. Two terms have been widely used in clinical practice, that is, IC and IBD-type unclassified (IBD-U). They may represent phenotypes of the wide spectrum of IBD (Chapter 1: Introduction and classification of inflammatory bowel diseases). However, the characterization of endoscopic features of IC and IBD-U will help guide clinical diagnosis, differential diagnosis, and management of the phenotypes of IBD.

Classic Crohn's disease and ulcerative colitis

Crohn's disease can affect any part of the GI tract, from the mouth to anus or perianal area, although it is predominantly located at the distal ileum (L1 in the Montreal classification) and/or colon (L2 and L3) [3]. The disease distribution pattern is typically discontinuous or segmental with skip lesions and rectal sparing. Anal or perianal lesions, such as fissures, fistulas, skin tags, abscesses, and perianal dermatitis, are common in CD. It appears that the disease location and length of bowel involved of a given patient remain relatively stable over the natural course. However, the transmural inflammation in CD dictates its progression to both luminal (i.e., formation of strictures) and extraluminal (i.e., formation of fistulae and abscesses). Histologic features of CD are characterized by granulomatous infiltration and transmural disease, in addition to chronic structural changes [e.g., crypt distortion, crypt branching, infiltration of mononuclear cells of any layers of the bowel wall, fibrosis, muscular hyperplasia, and Paneth cells metaplasia (of the large bowel), and pyloric gland metaplasia (of small and large bowel)]. The granulomas in CD are usually small in number and size without central necrosis, that is, epithelioid noncaseating granulomas.

The disease process of classic UC is confined to the large bowel, that is, the rectum with or without the involvement of the left part or majority of the colon, corresponding to E1, E2, and E3 phenotypes in the Montreal classification [1]. In the time of the initial diagnosis, before the exposure to medical therapy, all patients with UC have rectal involvement. Patients with UC can have proctitis, left-sided colitis, or extensive colitis at presentation or initially have E1 or E2 disease with subsequent proximal extension. The disease distribution pattern in UC is typically continuous, with a sharp demarcation between the diseased and nondiseased segments (Chapter 9: Ulcerative colitis). On histology, the depth of inflammation in UC is the mucosa, muscularis mucosae, and superficial submucosa. Like CD, UC is characterized by the presence of chronic structural changes with crypt distortion, infiltration of chronic inflammatory cells in the lamina propria and superficial submucosa, basal lymphoplasmacytosis, and Paneth cells or pyloric gland metaplasia.

On histology, acute disease flare-up of CD or UC is featured with the infiltration of acute inflammatory cells such as neutrophils and eosinophils, and the development of erosions, ulcers, and microabscesses. These histologic features of acute disease activity are superimposed on chronic inflammatory structural changes. In clinical practice, mucosal healing or transmural healing after effective medical therapy is defined the absence of these acute histologic disease activities.

Distinguishing endoscopic features of CD and UC are listed in Table 10.1.

Definitions of indeterminate colitis and inflammatory bowel disease unclassified

The terms of IC and IBD-U are used loosely and almost interchangeably in clinical practice. The working group for the Montreal classification recommended that the term of IC is limited to the setting of surgical histopathology of colectomy specimens [1,4–6]. Patients with severe UC or Crohn's colitis and deep ulcers may have transmural inflammation with lymphoid aggregates or microabscesses, which precludes a clear diagnosis of either disease. The term of colonic IBD-U is applied in the situation in which ileocolonoscopy is inconclusive, and endoscopic inflammation and histologic features of chronic inflammatory changes limited to the colon with the absence of diagnostic features of either CD or UC [1]. Infectious etiologies have to be excluded. IBD-U may be applied to other situations, such as duodenitis, gastritis, aphthous small bowel ulcers, and anal fistulae in UC and Crohn's colitis with diffuse, continuous proctocolitis. Those conditions may be categorized into CD variants or UC variants.

TABLE 10.1 Endoscopic differential diagnosis between *classic* Crohn's disease and ulcerative colitis.

	Crohn's disease	Ulcerative colitis
Disease location	Any parts of the gastrointestinal tract	Colon and rectum
Disease distribution pattern	Segmental with skip lesions	Continuous, extension proximally
Ileocecal valve	Often strictured or deformed in ileitis or ileocolitis	Patulous in patients with primary sclerosing cholangitis or backwash ileitis
Ulcers in the colon	Predominantly along the mesentery edge	Small or large ulcers evenly distributed circumferentially even with multiple linear ulcers
Stricture and fistula	Common	Fibrosis common; stricture rare; anal fistula rare
Perianal disease	Common	Rare

Crohn's disease variants

CD may present with superficial lesions in a continuous distribution pattern.

Superficial or continuous inflammation in Crohn's disease

Transmural lymphoid aggregates, one of the histologic hallmarks of CD more common in the ileum than colon, which may explain the higher risk of strictures and fistulae in small bowel CD than CD colitis [7,8]. The disease process can affect superficial layers of the bowel or even in a continuous pattern, like that in UC [9]. Diagnosis of CD is established by the detection of noncaseating, nonmucinous granulomas, in the absence of evidence of intestinal tuberculosis.

Nonspecific small bowel erosions and ulcers

The extensive application of video capsule endoscopy (VCE) and device-assisted enteroscopy in small bowel disease has opened "the can of worms." In the absence of the use of nonsteroidal antiinflammatory drugs (NSAIDs), ischemia, or infectious or neoplastic etiology, the finding of erosions or ulcers in the small bowel raises the possibility of CD or immune-mediated enteritis. Persistent nonspecific isolated ileitis or ileal ulcers also exist (Fig. 10.1). The presence of erosions or ulcers, especially without chronic inflammatory structural changes, is not necessarily pointing the diagnosis of small bowel CD [10]. Although VCE is more sensitive in the diagnosis of small bowel CD than computed tomography enterography or small bowel follow-through, it suffers from a lower specificity [8,11]. The role of VCE and enteroscopy in the diagnosis and differential IBD are discussed in Chapter 17, Capsule endoscopy in inflammatory bowel disease, and Chapter 18, Enteroscopy in inflammatory bowel disease and inflammatory bowel disease—like conditions.

Circumferential erosions or ulcers

Erosions and ulcers in a longitudinal fashion, especially those along the mesenteric edge is one of the hallmarks of CD. The presence of longitudinal ulcers has been used to distinguish CD from intestinal tuberculosis. However, circumferential ulcers can occur in patients with small or large bowel CD, with more often being seen in in the upper small bowel [12] or the neo-terminal ileum after ileocolonic resection and ileocolonic anastomosis (Fig. 10.2).

Ulcerative colitis variants

Ulcerative colitis may also present in nontraditional patterns, such as rectal sparing and patchy disease distribution. Occasionally, there are backwash ileitis (BWI) and nonspecific gastritis, duodenitis, or enteritis in patients with UC.

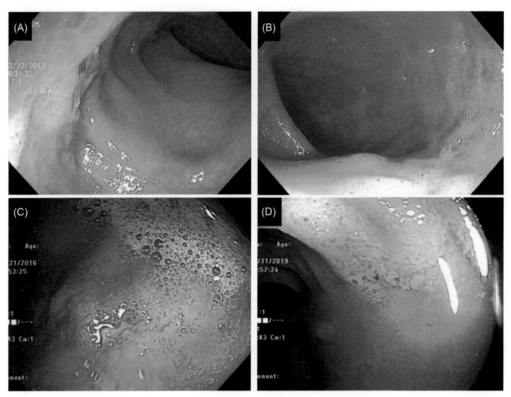

FIGURE 10.1   CD-like, persistent small bowel ulcers. (A−D) Persistent small ulcers in the terminal ileum and normal colon in a 36-year-old male. The latest medicine was only colchicine. Colonoscopy was performed in 2012 (A), 2013 (B), 2014 (not shown), 2016 (C), and 2019 (D). Histology revealed chronic active enteritis. The disease did not progress or became structuring disease in the absence of traditional therapy for CD. *CD*, Crohn's disease.

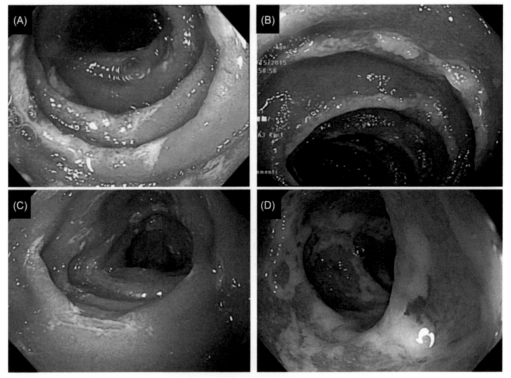

FIGURE 10.2   CD with circumferential ulcers in the small and large bowels, in contrast to classic linear, longitudinal ulcers: (A and B) circumferential ulcers in the neo-terminal ileum after ileocolonic resection and (C and D) circumferential ulcers in the colon. *CD*, Crohn's disease.

Backwash ileitis

Patients with UC pancolitis may present with concurrent endoscopic and/or histologic inflammation in the terminal ileum. The term of backwash ileitis (BWI) has been used. Various definitions of BWI have been used, based on the length [5 cm [13] or 10 cm, or any length from 0.5 cm above ileocecal valve (ICV) [14]] of ileal involvement, way of diagnosis (macroscopic, endoscopic, or histologic, or combinations [4,5,15]). Approximately 10%–30% of patients with UC pancolitis have concurrent BWI [4–6,16]. Patients with UC pancolitis and BWI carry a higher risk for colitis-associated neoplasia than those with UC pancolitis without BWI, or left-sided UC [4]. UC patients with primary sclerosing cholangitis (PSC) have a significantly higher risk for endoscopic [17] or histologic BWI [8,18].

On ileocolonoscopy, BWI is characterized by diffuse erythema, edema, granularity, loss of vascularity of mucosa with or without superficial ulceration. The endoscopic pattern of mucosal inflammation of the ileum is similar to that in the right colon. In contrast to CD ileitis, ICV in UC–BWI is often widely patent or patulous, without ulceration, stenosis, or deformity (Figs. 10.3 and 10.4). On ileocolonoscopy and contrasted abdominal imaging, the terminal ileum and ICV often show smooth and tubular appearance. On the contrary, patients with CD ileitis (L1) or CD ileocolitis (L3) usually present with strictured or deformed ICV and discrete lesions or cobblestoning of the distal ileum mucosa (Fig. 10.5).

The distribution of histologic inflammation can be focal, patchy, or diffuse. Histologic inflammation has even been graded into: Grade 1, cryptitis; Grade 2, scattered crypt abscesses; Grade 3, numerous crypt abscesses; and Grade 4, ulcer [5]. Normally, there is the absence of granulomas, transmural lymphoid aggregates, fissuring ulcers. Pyloric gland metaplasia may be present in UC with BWI.

Rectal sparing and patchy colitis

Topical or systemic medical therapy may result in endoscopic and/histologic rectal sparing and patchy colitis in adult patients [19,20]. It appears that rectal sparing and/or patchiness can occur in untreated pediatric UC patients on the index colonoscopy [21,22]. Rectal sparing is common in patients with PSC–UC than those with UC only

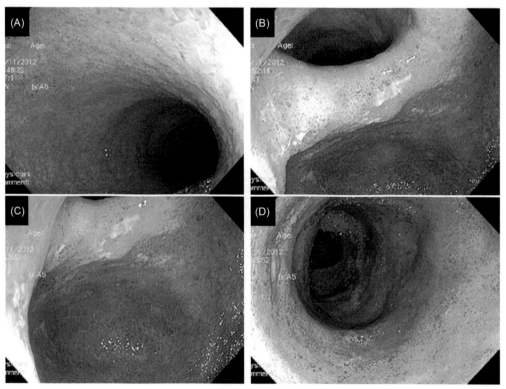

FIGURE 10.3   UC with BWI: (A) diffuse mild enteritis with loss of normal vascularity and "ground glass" appearance; (B) patulous, "fish mouth" ileocecal valve; and (C and D) diffuse mild colitis of the cecum and ascending colon, with a similar mucosal appearance to that in the distal ileum. *BWI*, Backwash ileitis; *UC*, ulcerative colitis.

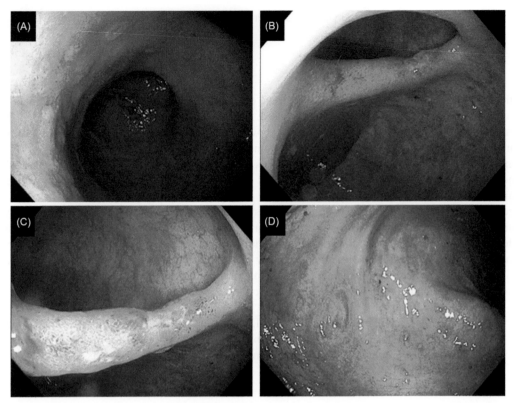

FIGURE 10.4 UC with BWI in a patient with concurrent PSC: (A) again, diffuse mild enteritis with loss of normal vascularity and "ground glass" appearance; (B and C) patulous, "fish mouth" ileocecal valve; and (D) diffuse mild colitis of the cecum, with a similar mucosal appearance to that in distal ileum. *BWI*, Backwash ileitis; *PSC*, primary sclerosing cholangitis; *UC*, ulcerative colitis.

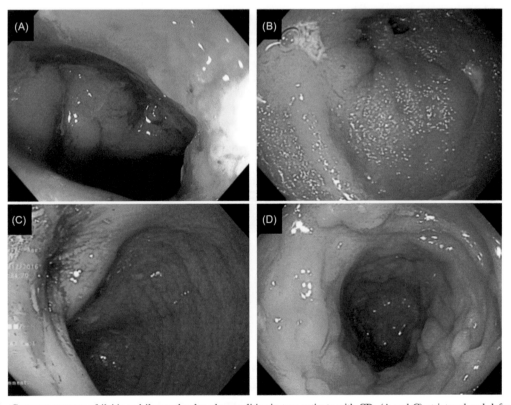

FIGURE 10.5 Common patterns of ileitis and ileocecal valve abnormalities in two patients with CD: (A and C) strictured and deformed ileocecal valves and (B and D) discrete ulcers or nodular mucosa (i.e., cobblestoning) in the terminal ileum. The patterns are different from UC with BWI. *BWI*, Backwash ileitis; *CD*, Crohn's disease; *UC*, ulcerative colitis.

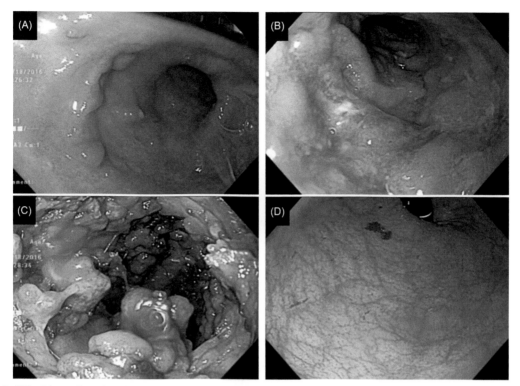

FIGURE 10.6 UC with rectal sparing after medical therapy in a patient: (A–C) diffuse colitis with multiple inflammatory pseudopolyps in the ascending, transverse, descending, and sigmoid colon and (D) normal rectal mucosa. *UC*, Ulcerative colitis.

(52% vs 6%) [15]. The rectum in these patients typically show normal mucosa and distensibility, which can be difficult to distinguish from rectal sparing and segmental colitis in CD (Figs. 10.6 and 10.7). The diagnosis of UC with rectal sparing has clinical implications. For example, colectomy with ileorectal anastomosis, instead of standard total procto-colectomy with IPAA, may be performed in selected patients [23].

Periappendiceal or cecal patch

Periappendiceal or cecal patch occasionally can be found in patients with UC, which is found more common in male patients with left-sided colitis than female patients or these with pancolitis [24]. Histologic inflammation of the periap-pendiceal area in colectomy specimens of UC patients appears to be common. On colonoscopic, periappendiceal patch presents with erythema around the orifice and cecal bases. In contrast, CD ileocolitis can also involve the cecal base, typically presenting with ulcers, nodularity, or deformity, along with diseases in ICV and terminal ileum (Fig. 10.8). The presence of cecal patch does not seem to impact the disease course or prognosis of left-sided UC [25].

Aphthous and longitudinal ulcers and other uncommon mucosal diseases

Aphthous erosions or ulcers in the distal ileum and proximal colon are common in CD. Histologic aphthous ulcers occa-sionally can be seen in UC [26], while the lesions on endoscopy are rare, mainly representing mild colitis (Fig. 10.9A and B). The use of NSAIDs or sodium phosphate-based bowel preparation should be excluded.

Longitudinal ulcers, especially those along the mesenteric edge, are an important endoscopic feature of CD. However, these can also be encountered in UC (Fig. 10.9C; see Fig. 10.10). In this author's experience, multiple longi-tudinal ulcers in UC are often evenly distributed in the 360 degrees circumferential. In contrast, the distribution of multiple longitudinal ulcers in CD is asymmetric, with deeper or more prominent ulcer(s) along the mesenteric edge (Fig. 10.10).

Partial colectomy with colo-colonic anastomosis is not recommended for the treatment of left-sided UC. As the disease recurrence after the surgery is a norm (Fig. 10.9D).

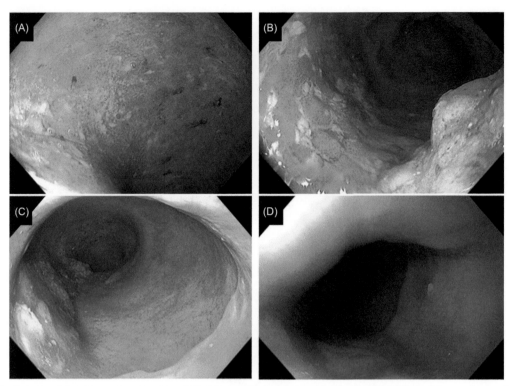

FIGURE 10.7 Crohn's colitis with rectal sparing: (A and B) diffuse colitis in the sigmoid and descending colons; (C) normal appearing rectum; and (D) normal anal canal.

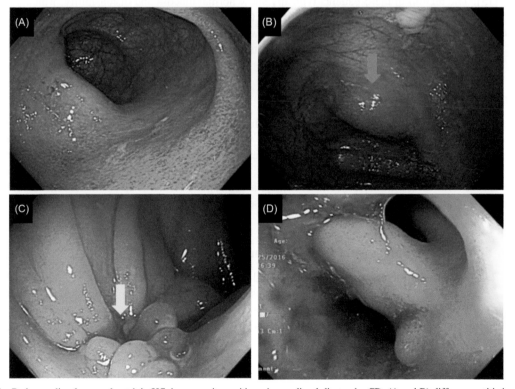

FIGURE 10.8 Periappendiceal or cecal patch in UC, in comparison with periappendiceal disease by CD: (A and B) diffuse proctitis in a patient with UC, who also had periappendiceal inflammation with erythema (*green arrow*) and (C and D) an exit opening of CD ileosigmoid fistula at the distal sigmoid colon (*yellow arrow*). *CD*, Crohn's disease; *UC*, ulcerative colitis.

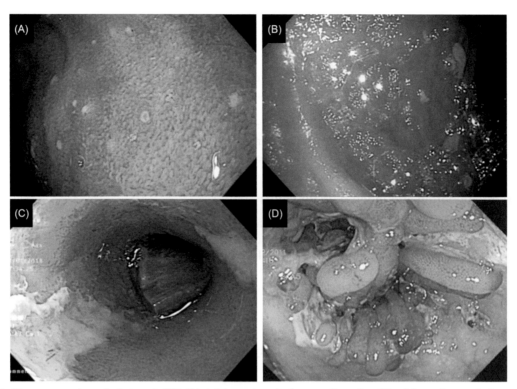

FIGURE 10.9 Uncommon endoscopic features in UC: (A) Aphthous ulcers in the ascending colon; (B) aphthous ulcers and edematous mucosa in the descending colon; (C) superficial longitudinal ulcers in the transverse colon; and (D) inappropriate partial colectomy in a patient with UC. Multiple pedunculated inflammatory polyps at the colo-colonic anastomosis. *UC*, Ulcerative colitis.

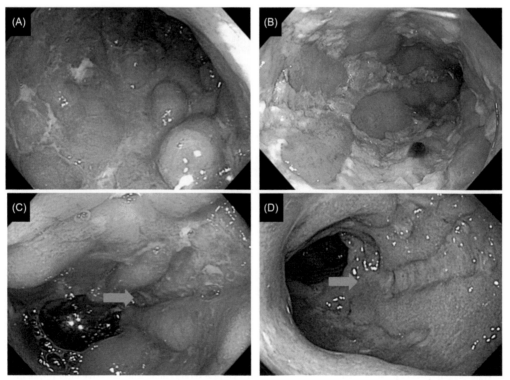

FIGURE 10.10 Longitudinal ulcers in UC as compared with CD: (A and B) Multiple longitudinal ulcers in the descending colon in two patients with UC. The ulcers evenly involved all aspects of circumferences and (C and D) longitudinal ulcers in CD predominantly at the mesentery edge of the descending colon (*green arrows*). *CD*, Crohn's disease; *UC*, ulcerative colitis.

Pseudogranulomas

Granulomatous inflammation in response to ruptured crypts can occur in patients with UC. This type of granulomas has been termed pseudogranulomas, mucinous granuloma, or foreign body–associated granulomas, which are normally located next to the ruptures crypts or ulcer beds. Since nonspecific gastritis or duodenitis can occur in patients with UC, especially in the pediatric population, the pseudogranulomas or clusters of giant histiocytes have been detected near a destructed gland in the gastric or duodenal mucosa [27].

Histologic features of UC-associated pseudogranulomas and CD-associated noncaseating granulomas are further discussed in Chapter 15, Histologic evaluation of disease activity in inflammatory bowel disease, and Chapter 34, Histology correlation with common endoscopic abnormalities.

Transmural inflammation and indeterminate colitis

Transmural lymphoid aggregates or microabscesses can be found underneath fissuring ulcer bed in severe fulminant UC [28,29]. This pattern of injury precludes a definitive diagnosis of UC or CD, leading to the pathological diagnosis of IC. Nonetheless, transmural lymphoid aggregates or granulomas in CD may be present underneath the intact mucosa (Fig. 10.11). The diagnosis of IC, however, may have clinical implications, especially the impact on the outcome of restorative proctocolectomy with IPAA. Patients with IC undergoing the surgery carry a numerically higher risk for CD of the pouch, complex perianal fistulas, pelvic abscess [26], anastomotic leaks, chronic pouchitis, and pouch strictures than these with classic UC undergoing the same procedure [30]. On the other hand, extensive use of corticosteroids and various biological agents to rescue the patients from colectomy may result in an increased proportion of individuals to have IC who fail the medical therapy and undergo colectomy.

Strictures in the large bowel

Colon strictures in patients with UC should be carefully evaluated for malignancy. However, benign strictures can occur in these patients. The underlying disease process includes fibrosis at the base of the lamina propria between the base of

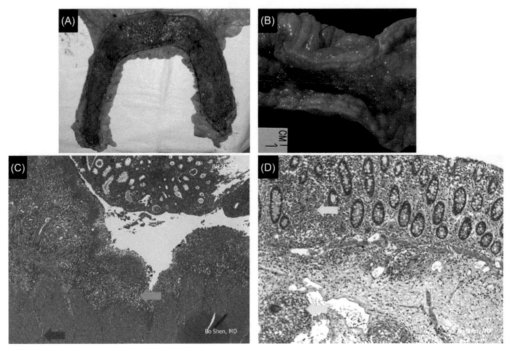

FIGURE 10.11 Transmural disease in IC and Crohn's colitis. (A and C) IC: Diffuse severe colitis with extensive ulcer (*green arrow*) and denuded mucosa on macroscopic and microscopic examination and transmural lymphoid aggregates (*blue arrow*) and (B and D) in contrast, Crohn's colitis with luminal stricture (B) is characterized by transmural lymphoid aggregates and granulomas (*yellow arrows*) with intact mucosa. *IC*, Indeterminate colitis. *Photo (B) Courtesy Lisa Yerian, MD, of Cleveland Clinic.*

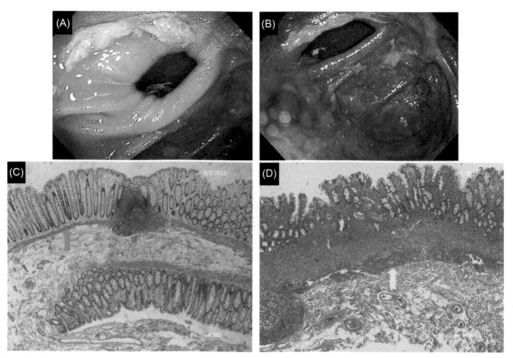

FIGURE 10.12 **Fibrosis and stricture of the colon in UC:** (A and B) mild nonulcerated stricture in the ascending colon, on white-light colonoscopy and narrow banding imaging; (C and D) surgical colectomy specimen of diverticulitis, showing normal epithelia, lamina propria, muscularity mucosae (*green arrow*); (D) colectomy specimen of UC, showing infiltration of acute and chronic inflammatory cells in the lamina propria, and basal space. There were yperplasia and thickening of the muscularis mucosae causing stricture (*yellow arrow*). *UC*, Ulcerative colitis.

the crypts and the muscularis mucosae, hypertrophic muscularis mucosae, thickening and splaying of the muscularis mucosae, and fibrosis involving the lamina propria, the muscularis mucosae, and the luminal aspect of the submucosa (Fig. 10.12) [31].

Chronically active UC with deep ulcerations can result in transmural inflammation and associated tissue damage, and fibrosis [5,6,23,32]. In a large series of 1156 UC cases, 42 (5.1%) were found to have endoscopic or radiographic bowel narrowing [33]. Reported frequencies of benign strictures in UC ranged from 1% to 11% [29,34−37]. UC-associated benign strictures are usually nonulcerated and circumferential (Figs. 10.13 and 10.14). They may affect any parts of the colon or rectum. However, UC-associated anal or anal canal strictures are rare (Fig. 10.13). Tissue biopsy should still be taken to rule out malignancy at the index colonoscopy and periodic follow up colonoscopy.

Upper gastrointestinal tract involvement

Upper endoscopy should be performed in patients as a part of workup for IBD [3]. In pediatric populations [38−40] and Asian adult populations [41], the reported frequencies ranged from 33% to 66% in UC and 51% to 78% in CD; and reported frequencies of endoscopic duodenitis ranged from 13% to 23% in UC and from 27% to 42% in CD. In UC patients, gastritis or duodenitis can present with erythema, granularity, erosions, and ulcers (Fig. 10.15). Granulomas of the stomach or duodenum on histology have been reported in 12%−28% of patients with IBD [3]. Other histologic changes include acute or chronic, and focal or diffuse gastritis or duodenitis with crypt abscesses, lymphoid aggregates, and ulcers [42]. Our group performed a retrospective study of 98 patients with UC who had both upper endoscopy and colonoscopy, with 38 (38.8%) had biopsy of stomach and 36 (36.7%) had a biopsy of the duodenum. Thirty patients (30.6%) had some forms of upper GI tract involvement. On histology, eight had active duodenitis and four had chronic duodenitis. Prior exposure to immunomodulator therapy in UC patients was associated with a lower risk for duodenal inflammation than those without. Of the 30 patients, 6 patients had stained positively for IgG4 immunohistochemistry (Fig. 10.16) [43]. Besides, patients with UC−PSC may have endoscopic or histologic duodenitis, especially in the second part of the duodenum (Fig. 10.17).

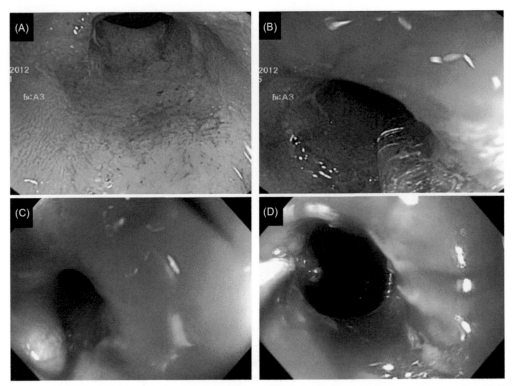

FIGURE 10.13 Anorectal ring strictures in UC versus CD: (A and B) nonulcerated stricture at the distal rectal in a patient with UC on long-term antitumor necrosis factor therapy. The stricture was dilated with endoscopic balloon and (C and D) tight ulcerated stricture at the anal canal, which was treated with endoscopic needle knife stricturotomy. *CD*, Crohn's disease; *UC*, ulcerative colitis.

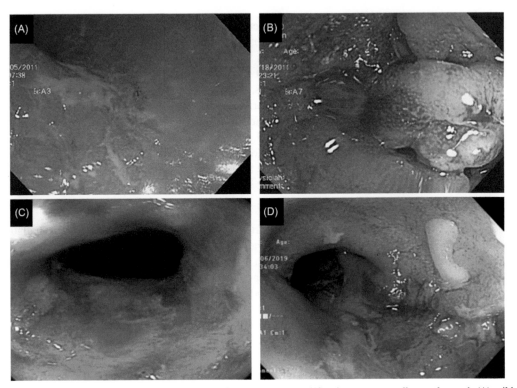

FIGURE 10.14 Benign colon strictures in UC as compared with that in CD. Benign UC strictures are usually nonulcerated: (A) mild smooth stricture in the rectosigmoid junction in UC; (B) inflammatory strictures from edema and pseudopolyps in UC; and (C and D) ulcerated strictures of the left colon in CD. *CD*, Crohn's disease; *UC*, ulcerative colitis.

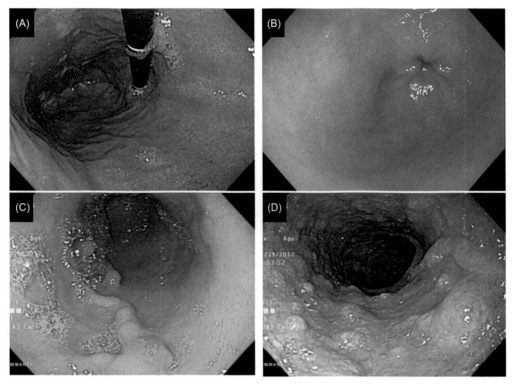

FIGURE 10.15   Gastritis, gastropathy, and duodenopathy in UC: (A and B) antral and body gastritis with linear erythema in two UC patients; (C) linear nodularity in the gastric body and antrum in a UC patient with ileal pouch−anal anastomosis and sleeve gastrectomy; and (D) the same patient had diffuse erythema and nodularity of entire duodenum. *UC*, Ulcerative colitis.

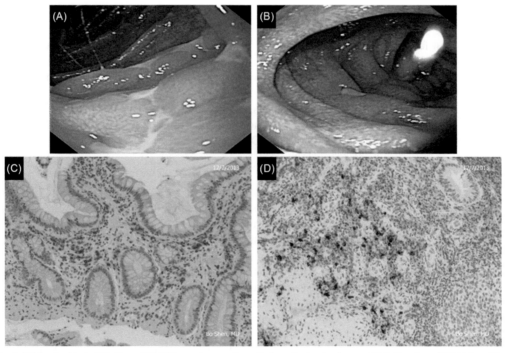

FIGURE 10.16   IgG4 duodenitis in UC: (A and B) diffuse edema and small ulcers in the duodenum; (C) infiltration of CD138-positive (pink) plasma cells of the lamina propria of duodenal biopsy; and (D) IgG4-positive (brown) plasma cells in the lamina propria. *UC*, Ulcerative colitis.

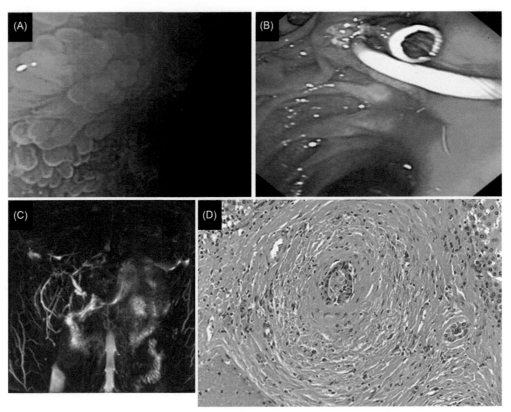

FIGURE 10.17   Duodenitis in concurrent UC and PSC: (A and B) granular duodenum mucosa, a transpapilla stent was in place for the treatment of PSC; (C) dilated and strictured intrahepatic bile ducts on magnetic resonance cholangiopancreatography; and (D) classic histologic features of PSC with "onion skin-like" concentric periductal fibrosis involving interlobular bile ducts. *PSC*, Primary sclerosing cholangitis; *UC*, ulcerative colitis.

Postcolectomy enteritis syndrome

This phenotype occurs in few UC patients after colectomy and ileostomy with or without the construction of the ileal pouch (Fig. 10.18) [44–47]. The etiology of this syndrome is unknown. The patients present with increased ileostomy output, severe dehydration, malnutrition, and fever; and poorly respond to corticosteroids or immunosuppressive medications. Early reestablishment of bowel continuity may be helpful. Upper endoscopy or ileoscopy via stoma typically show diffuse enteritis with granularity, nodularity, exudates, edema, and ulcers. Histology often shows diffuse superficial acute and chronic inflammation, crypt distortion, with no skipping lesions or noncaseating granulomas. These endoscopic and histologic patterns are different from these of CD enteritis.

Anal and perianal diseases

Anorectal dysfunction has been listed as one of signs for disease progression in UC, along with the proximal extension, stricturing, pseudopolyposis, dysmotility, and impaired permeability [48]. Distal rectal prolapse can occur in both UC and CD patients. The prolapsed rectum may present as distal anterior or circumferential proctitis or rectal cuffitis (in patients with IPAA) or even fistulae (Fig. 10.19). The main complaint is dyschezia.

Chronic anal fissures can occur in patients with UC as well as CD (3% vs. 6%) [49]. African American patients were shown to be at risk [45]. UC may also been associated with Type 1 (i.e., large, hard, edematous, cyanotic, pain near or from healed fissures or ulcers) or Type 2 (i.e., "elephant ear" type, soft, flat, broad or narrow, various size, smooth, painless) skin tags [50]. In UC patients, Type 1 skin tags are more common than Type 2. Unlike CD, concurrent perianal fistulae or abscesses are rare (Fig. 10.20).

Perianal fistulae or abscesses can also occur in patients with UC [51]. The cumulative incidence rates of perianal sepsis (abscess or fistula) were 2.2% and 4.5% at 5 and 10 years, respectively [52]. The fistula and abscess are normally simple, small, and single track, largely resulting from cryptoglandular source (Fig. 10.21). The risk factors for perianal fistulas and abscesses included male gender and extensive disease [48].

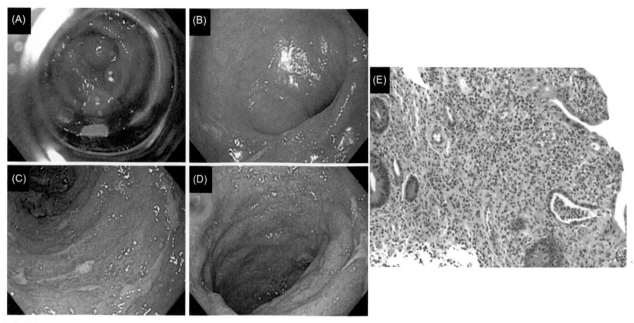

FIGURE 10.18   Postcolectomy (for ulcerative colitis) enteritis syndrome. Diffuse enteritis on endoscopy (via stoma) (A−D) and histology (E). *Courtesy Chang S, Shen B. In: Shen B, editor. Interventional inflammatory bowel disease. Elsevier; 2018; p. 27.*

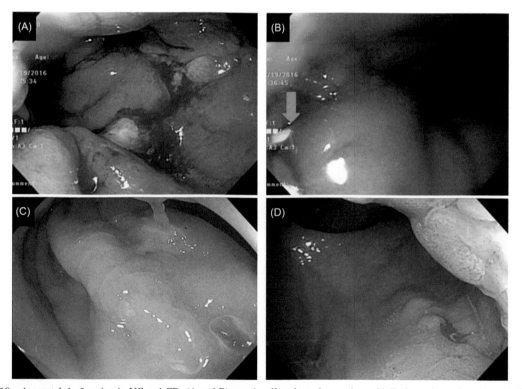

FIGURE 10.19   Anorectal dysfunction in UC and CD: (A and B) rectal cuff prolapse in a patient with ileal pouch−anal anastomosis for refractory UC. Prolapsed mucosa was edematous and ulcerated, with a fistula formed between folds on a guidewire (*green arrow*) and (C and D) distal anterior rectal prolapse with nodular mucosa in a patient with small bowel CD. *CD*, Crohn's disease; *UC*, ulcerative colitis.

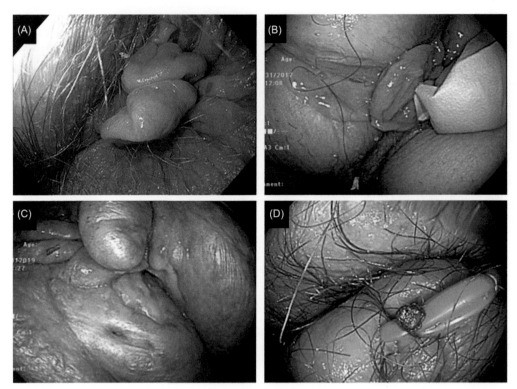

FIGURE 10.20 Comparison of skin tags in UC and CD: (A and B) skin tags in UC are typical of Type 1 or nontender "elephant ear" type and (C and D) skin tags of CD can be Type 1 or Type 2, which are often associated with other perianal diseases, such as perianal fistula or abscess, as shown here. *CD*, Crohn's disease; *UC*, ulcerative colitis.

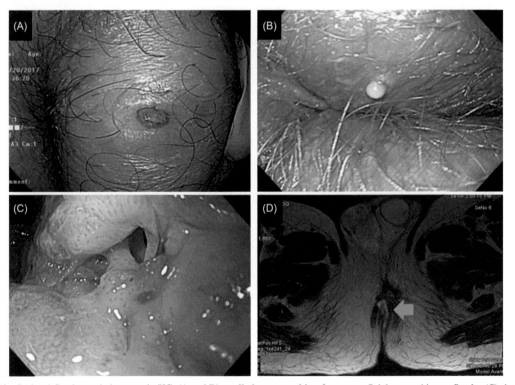

FIGURE 10.21 Perianal fistulae and abscesses in UC: (A and B) small abscess resulting from superficial extrasphincter fistula; (C) the fistula opening at the dentate line, suggesting cryptoglandular source; and (D) the simple extrasphincter on magnetic resonance imaging. *UC*, Ulcerative colitis.

Summary and recommendations

Patients with IBD lack of classic features of CD or UC have been classified into IC or IBD-U. These patients often have overlapping clinical, endoscopic, histologic, and radiographic features of CD and UC. The terms of CD variants and UC variants are used. CD variants and UC variants may represent phenotypes within a wide spectrum of IBD. The impact of the diagnoses on the management and prognosis warrants further studies. We anticipate that further classification of IC, IBD-U, or CD or UC variants is possible with the incorporation of clinical, endoscopic, radiographic, histologic, serological and genetic markers.

References

[1]   Chutkan RK, Scherl E, Waye JD. Colonoscopy in inflammatory bowel disease. Gastrointest Endosc Clin N Am 2002;12:463–83.

[2]   Pera A, Bellando P, Caldera D, Ponti V, Astegiano M, Barletti C, et al. Colonoscopy in inflammatory bowel disease. Diagnostic accuracy proposal an endoscopic score. Gastroenterology 1987;92:181–5.

[3]   Silverberg MS, Satsangi J, Ahmad T, Arnott ID, Bernstein CN, Brant SR, et al. Toward an integrated clinical, molecular and serological classification of inflammatory bowel disease: report of a Working Party of the 2005 Montreal World Congress of Gastroenterology. Can J Gastroenterol 2005;19(Suppl. A):5A–36A.

[4]   Satsangi J, Silverberg MS, Vermeire S, Colombel JF. The Montreal classification of inflammatory bowel disease: controversies, consensus, and implications. Gut 2006;55:749–53.

[5]   North American Society for Pediatric Gastroenterology, Hepatology, and Nutrition, Colitis Foundation of America, Bousvaros A, Antonioli DA, Colletti RB, Dubinsky MC, Glickman JN, Gold BD, et al. Differentiating ulcerative colitis from Crohn disease in children and young adults: report of a working group of the North American Society for Pediatric Gastroenterology, Hepatology, and Nutrition and the Crohn's and Colitis Foundation of America. J Pediatr Gastroenterol Nutr 2007;44:653–74.

[6]   Geboes K, Colombel JF, Greenstein A, Jewell DP, Sandborn WJ, Vatn MH, et al. Pathology Task Force of the International Organization of Inflammatory Bowel Diseases. Indeterminate colitis: a review of the concept—what's in a name? Inflamm Bowel Dis 2008;14:850–7 Review.

[7]   Yantiss RK, Odze RD. Diagnostic difficulties in inflammatory bowel disease pathology. Histopathology 2006;48:116–32 Review.

[8]   Yantiss RK, Farraye FA, O'Brien MJ, Fruin AB, Stucchi AF, Becker JM, et al. Prognostic significance of superficial fissuring ulceration in patients with severe "indeterminate" colitis. Am J Surg Pathol 2006;30:165–70.

[9]   Odze R. Diagnostic problems and advances in inflammatory bowel disease. Mod Pathol 2003;16:347–58.

[10] Goldstein JL, Eisen GM, Lewis B, Gralnek IM, Zlotnick S, Fort JG. Investigators. Video capsule endoscopy to prospectively assess small bowel injury with celecoxib, naproxen plus omeprazole, and placebo. Clin Gastroenterol Hepatol 2005;3:133–41.

[11] Solem CA, Loftus Jr EV, Fletcher JG, Baron TH, Gostout CJ, Petersen BT, et al. Small-bowel imaging in Crohn's disease: a prospective, blinded, 4-way comparison trial. Gastrointest Endosc. 2008;68:255–66.

[12] Esaki M, Matsumoto T, Ohmiya N, Washio E, Morishita T, Sakamoto K, et al. Capsule endoscopy findings for the diagnosis of Crohn's disease: a nationwide case-control study. J Gastroenterol 2019;54:249–60.

[13] Heuschen UA, Hinz U, Allemeyer EH, Stern J, Lucas M, Autschbach F, et al. Backwash ileitis is strongly associated with colorectal carcinoma in ulcerative colitis. Gastroenterology 2001;120:841–7.

[14] Arrossi AV, Kariv Y, Bronner MP, Hammel J, Remzi FH, Fazio VW, et al. Backwash ileitis does not affect pouch outcome in patients with ulcerative colitis with restorative proctocolectomy. Clin Gastroenterol Hepatol 2011;9:981–8.

[15] Haskell H, Andrews Jr CW, Reddy SI, Dendrinos K, Farraye FA, Stucchi AF, et al. Pathologic features and clinical significance of "backwash" ileitis in ulcerative colitis. Am J Surg Pathol 2005;29:1472–81.

[16] Alexander F, Sarigol S, DiFiore J, Stallion A, Cotman K, Clark H, et al. Fate of the pouch in 151 pediatric patients after ileal pouch anal anastomosis. J Pediatr Surg 2003;38:78–82.

[17] Loftus Jr EV, Harewood GC, Loftus CG, Tremaine WJ, Harmsen WS, Zinsmeister AR, et al. PSC-IBD: a unique form of inflammatory bowel disease associated with primary sclerosing cholangitis. Gut 2005;54:91–6.

[18] Joo M, Abreu-e-Lima P, Farraye F, Smith T, Swaroop P, Gardner L, et al. Pathologic features of ulcerative colitis in patients with primary sclerosing cholangitis: a case-control study. Am J Surg Pathol 2009;33:854–62.

[19] Bernstein CN, Shanahan F, Anton PA, Weinstein WM. Patchiness of mucosal inflammation in treated ulcerative colitis: a prospective study. Gastrointest Endosc 1995;42:232–7.

[20] Kim B, Barnett JL, Kleer CG, Appelman HD. Endoscopic and histological patchiness in treated ulcerative colitis. Am J Gastroenterol 1999;94:3258–62.

[21] Glickman JN, Bousvaros A, Farraye FA, Zholudev A, Friedman S, Wang HH, et al. Pediatric patients with untreated ulcerative colitis may present initially with unusual morphologic findings. Am J Surg Pathol 2004;28:190–7.

[22] Markowitz J, Kahn E, Grancher K, Hyams J, Treem W, Daum F. Atypical rectosigmoid histology in children with newly diagnosed ulcerative colitis. Am J Gastroenterol 1993;88:2034–7.

[23] da Luz Moreira A, Kiran RP, Lavery I. Clinical outcomes of ileorectal anastomosis for ulcerative colitis. Br J Surg 2010;97:65–9.

[24] Rubin DT, Rothe JA. The peri-appendiceal red patch in ulcerative colitis: review of the University of Chicago experience. Dig Dis Sci 2010;55:3495–501.

[25] Byeon JS, Yang SK, Myung S, Pyo SI, Park HJ, Kim YM, et al. Clinical course of distal ulcerative colitis in relation to appendiceal orifice inflammation status. Inflamm Bowel Dis 2005;11:366−71.

[26] Yantiss RK, Sapp HL, Farraye FA, El-Zammar O, O'Brien MJ, Fruin AB, et al. Histologic predictors of pouchitis in patients with chronic ulcerative colitis. Am J Surg Pathol 2004;28:999−1006.

[27] Queliza K, Ihekweazu FD, Schady D, Jensen C, Kellermayer R. Granulomatous upper gastrointestinal inflammation in pediatric ulcerative colitis. J Pediatr Gastroenterol Nutr 2018;66:620−3.

[28] Odze RD. Pathology of indeterminate colitis. J Clin Gastroenterol 2004;38(Suppl. 1):S36−40.

[29] Gramlich T, Delaney CP, Lynch AC, Remzi FH, Fazio VW. Pathological subgroups may predict complications but not a late failure after ileal pouch-anal anastomosis for indeterminate colitis. Colorectal Dis 2003;5:315−19.

[30] Fazio VW, Kiran RP, Remzi FH, Coffey JC, Heneghan HM, Kirat HT, et al. Ileal pouch anal anastomosis: analysis of outcome and quality of life in 3707 patients. Ann Surg 2013;257:679−85.

[31] Gordon IO, Agrawal N, Goldblum JR, Fiocchi C, Rieder F. Fibrosis in ulcerative colitis: mechanisms, features, and consequences of a neglected problem. Inflamm Bowel Dis 2014;20:2198−206.

[32] Swan NC, Geoghegan JG, O'Donoghue DP, Hyland JM, Sheahan K. Fulminant colitis in inflammatory bowel disease: detailed pathologic and clinical analysis. Dis Colon Rectum 1998;41:1511−15.

[33] Gumaste V, Sachar DB, Greenstein AJ. Benign and malignant colorectal strictures in ulcerative colitis. Gut 1992;33:938−41.

[34] De Dombal FT, Watts JM, Watkinson G, Goligher JC. Local complications of ulcerative colitis: stricture, pseudopolyposis, and carcinoma of colon and rectum. Br Med J 1966;1:1442−7.

[35] Edwards FC, Truelove SC. The course and prognosis of ulcerative colitis. III. Complications. Gut 1964;5:1−22.

[36] Lashner BA, Turner BC, Bostwick DG, Frank PH, Hanauer SB. Dysplasia and cancer complicating strictures in ulcerative colitis. Dig Dis Sci 1990;35:349−52.

[37] Yamagata M, Mikami T, Tsuruta T, Yokoyama K, Sada M, Kobayashi K, et al. Submucosal fibrosis and basic fibroblast growth factor-positive neutrophils correlate with colonic stenosis in cases of ulcerative colitis. Digestion 2011;84:12−21.

[38] Ruuska T, Vaajalahti P, Arajärvi P, Mäki M. Prospective evaluation of upper gastrointestinal mucosal lesions in children with ulcerative colitis and Crohn's disease. J Pediatr Gastroenterol Nutr 1994;19:181−6.

[39] Abdullah BA, Gupta SK, Croffie JM, Pfefferkorn MD, Molleston JP, Corkins MR, et al. The role of esophagogastroduodenoscopy in the initial evaluation of childhood inflammatory bowel disease: a 7-year study. J Pediatr Gastroenterol Nutr 2002;35:636−40.

[40] Kovacs M, Muller KE, Arato A, Lakatos PL, Kovacs JB, Varkonyi A, , et al.Hungarian IBD Registry Group (HUPIR) Diagnostic yield of upper endoscopy in paediatric patients with Crohn's disease and ulcerative colitis. Subanalysis of the HUPIR registry. J Crohns Colitis 2012;6:86−94.

[41] Hori K, Ikeuchi H, Nakano H, Uchino M, Tomita T, Ohda Y, et al. Gastroduodenitis associated with ulcerative colitis. J Gastroenterol 2008;43:193−201.

[42] Lee H, Westerhoff M, Shen B, Liu X. Clinical aspects of idiopathic inflammatory bowel disease: a review for pathologists. Arch Pathol Lab Med 2016;140:413−28 Review.

[43] Kochhar G, Singh T, Navaneethan N, Singh G, Shrestha K, Shen B. Upper gastrointestinal tract involvement mediated by igg4 in patients with ulcerative colitis, a retrospective study. Inflamm Bowel Dis 2013;19(Suppl. 1):S75.

[44] Rubenstein J, Sherif A, Appelman H, Chey WD. Ulcerative colitis associated enteritis: is ulcerative colitis always confined to the colon? J Clin Gastroenterol 2004;38:46−51.

[45] Corporaal S, Karrenbeld A, van der Linde K, Voskuil JH, Kleibeuker JH, Dijkstra G. Diffuse enteritis after colectomy for ulcerative colitis: two case reports and review of the literature. Eur J Gastroenterol Hepatol 2009;21:710−15.

[46] Yang Y, Liu Y, Zheng W, Zhou W, Wu B, Sun X, et al. A literature review and case report of severe and refractory post-colectomy enteritis. BMC Gastroenterol 2019;19:61 e-journal.

[47] Feuerstein JD, Shah S, Najarian R, Nagle D, Moss AC. A fatal case of diffuse enteritis after colectomy for ulcerative colitis: a case report and review of the literature. Am J Gastroenterol 2014;109:1086−9.

[48] Torres J, Billioud V, Sachar DB, Peyrin-Biroulet L, Colombel JF. Ulcerative colitis as a progressive disease: the forgotten evidence. Inflamm Bowel Dis 2012;18:1356−63.

[49] Malaty HM, Sansgiry S, Artinyan A, Hou JK. Time trends, clinical characteristics, and risk factors of chronic anal fissure among a national cohort of patients with inflammatory bowel disease. Dig Dis Sci 2016;61:861−4.

[50] Bonheur JL, Braunstein J, Korelitz BI, Panagopoulos G. Anal skin tags in inflammatory bowel disease: new observations and a clinical review. Inflamm Bowel Dis 2008;14:1236−9.

[51] Zwintscher NP, Shah PM, Argawal A, Chesley PM, Johnson EK, Newton CR, et al. The impact of perianal disease in young patients with inflammatory bowel disease. Int J Colorectal Dis 2015;30:1275−9.

[52] Choi YS, Kim DS, Lee DH, Lee JB, Lee EJ, Lee SD, et al. Clinical characteristics and incidence of perianal diseases in patients with ulcerative colitis. Ann Coloproctol 2018;34:138−43.

Chapter 11

Ulcerative colitis: postsurgical

Bo Shen

Center for Inflammatory Bowel Diseases, Columbia University Irving Medical Center-New York Presbyterian Hospital, New York, NY, United States

Chapter Outline

Abbreviations

ATZ	anal transition zone
CD	Crohn's disease
IPAA	ileal pouch—anal anastomosis
IRA	ileorectal anastomosis
PDAI	the Pouchitis Disease Activity Index
PVF	pouch-vaginal fistula
RPC	restorative proctocolectomy
UC	ulcerative colitis

Introduction

For the past two decades, we have witnessed rapid progress in medical management of ulcerative colitis (UC). However, it is estimated that approximately 20%−25% of UC patients would eventually require surgical intervention [1,2]. There appears to be a trend in a reduced rate of colectomy [2], which may be attributed to a better disease control with a wide use of immunomodulators and biological agents. Medically refractory disease, poor tolerance of medicines, and colitis-associated neoplasia are the main indications for colectomy. A variety of surgical treatment modalities have been performed, resulting in an altered bowel anatomy, which can pose a challenge for postoperative assessment, monitoring, and management of the disease with endoscopy. In this chapter the author highlights common UC-associated surgical procedures, bowel anatomy after the surgery, and landmarks and diseased conditions on endoscopic evaluation.

Following colectomy, ileorectal anastomosis (IRA) (Fig. 11.1) may occasionally be performed in a highly selected patient population with refractory UC, such as female patients with limited inflammation of the rectum and concern on adverse impact of the ileal pouch surgery on fecundity and those with significant comorbidities [3]. However, restorative proctocolectomy (RPC) with construction of ileal pouch—anal anastomosis (IPAA) is the surgical treatment of choice for patients with UC or familial adenomatous polyposis who require colectomy. If a single-stage RPC and IPAA is to be performed, it would involve total proctocolectomy, construction of an ileal pouch reservoir, and anastomosis of the pouch and anal transition zone (ATZ). As the surgery is reconstructive in nature and the majority of patients with UC have inflamed large bowel, RPC is normally performed in either two or three stages. The two-stage surgery is normally consisted total proctocolectomy and construction of an ileal pouch with anastomosis to ATZ and diverting loop ileostomy, followed by ileostomy closure. In contrast, the three-stage surgery consists of colectomy, the Hartmann procedure (i.e., diverted rectal stump), diverting ileostomy as the first stage (Fig. 11.2); completion proctectomy, construction of an ileal pouch reservoir with anastomosis to ATZ, and diverting loop ileostomy as the second stage; and closure of loop ileostomy as the final stage. The three-stage RPC and pouch surgery is reserved for patients with risk for

Atlas of Endoscopy Imaging in Inflammatory Bowel Disease. DOI: https://doi.org/10.1016/B978-0-12-814811-2.00011-6
© 2020 Elsevier Inc. All rights reserved.

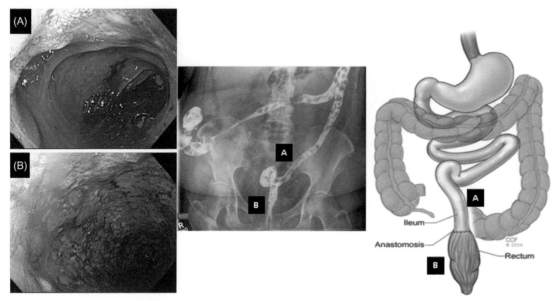

FIGURE 11.1 IRA for ulcerative colitis. The IRA procedure may be reserved for a highly selected patient population, such as gestational-age female patients who are concerned about the impact of pelvic pouch surgery on fecundity and have minimum inflammation of the rectum. (A) Normal neo-terminal ileum proximal to IRA; (B) diffuse proctitis, which often requires long-term medical therapy. The procedure will not burn the bridge for future pouch surgery, if indicated. The anatomy is illustrated by gastrografin enema (middle figure) and artist's depiction (right figure). *IRA*, Ileorectal anastomosis. *Courtesy Mr. Joel Pangrace of Cleveland Clinic.*

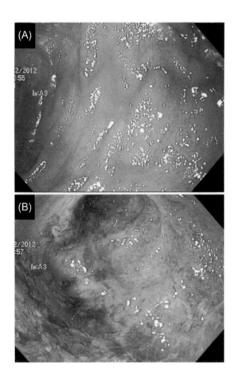

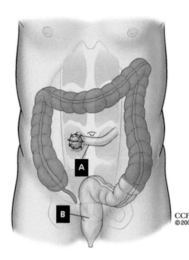

FIGURE 11.2 Colectomy with end ileostomy in ulcerative colitis as the initial stage of restorative proctocolectomy. (A) Brooke end ileostomy with normal mucosa of the neo-terminal ileum; (B) inflammation of Hartmann pouch (i.e., stapled rectal stump with proximal fecal diversion). The anatomy is illustrated by artist's depiction (right figure). *Courtesy Mr. Joel Pangrace of Cleveland Clinic.*

developing postoperative complications, such as those with severe colitis or under marked immunosuppression. In patients with severe colitis, mucous fistula is intentionally constructed with the rectal or sigmoid colon stump being sutured to the fascia of abdominal wall, which can lead to a colocutaneous fistula (Fig. 11.3) (Table 11.1).

A variety of configurations of the ileal pouch have been designed, with J pouch (Fig. 11.4), S pouch (Fig. 11.5), W pouch (Fig. 11.6), H pouch (Fig. 11.7), and the continent ileostomies, that is, Kock (K) pouch (Fig. 11.8) and Barnett continent ileal reservoir (Fig. 11.9). The choice among commonly performed pelvic pouches (J or S) versus

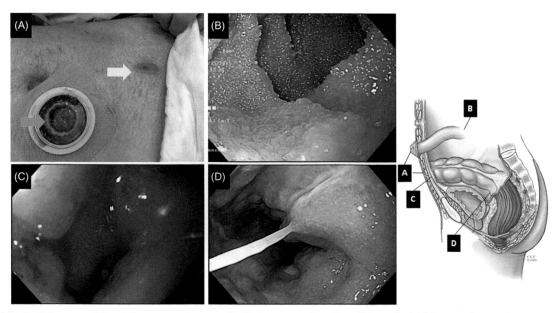

FIGURE 11.3   Colectomy with end ileostomy and mucus fistula for severe ulcerative colitis as the initial stage of restorative proctocolectomy. (A) Brooke end ileostomy (*green arrow*) and opening of mucus fistula (*yellow arrow*); (B) normal mucosa of the neo-terminal ileum; (C) internal opening of the mucus fistula; (D) Hartmann pouch, that is, colorectal stump with proximal fecal diversion. The anatomy is illustrated by the artist's depiction (right figure). *Courtesy Mr. Joel Pangrace of Cleveland Clinic.*

TABLE 11.1 Surgical modalities for ulcerative colitis.

	Subcategories	Examples of diseased conditions
Colectomy	Subtotal colectomy or colectomy	Diversion colitis
	Total proctocolectomy	Persistent perineal sinus
	Completion proctectomy	
	Mucus fistula	Mucus discharge
	Ileorectal anastomosis	Proctitis
Ileostomy	End ileostomy	Postcolectomy enteritis; stoma prolapse, stenosis, retraction; peristomal ulcer, or pyoderma gangrene nodosum
	Loop ileostomy	
Pelvic pouch configurations	J pouch	Various pouch disorders, such as pouchitis, Crohn's disease of the pouch
	S pouch	
	W pouch	
	H pouch	
Abdominal pouch or continent ileostomy configuration	Kock pouch	
	Barnett continent ileal reservoir	
Ileal pouch–anastomosis pattern	Handsewn	Stricture; anastomotic leak or sinus
	Stapled	
Management of distal rectum or anal transition zone mucosa	Mucosectomy	Cuffitis
	Without mucosectomy	

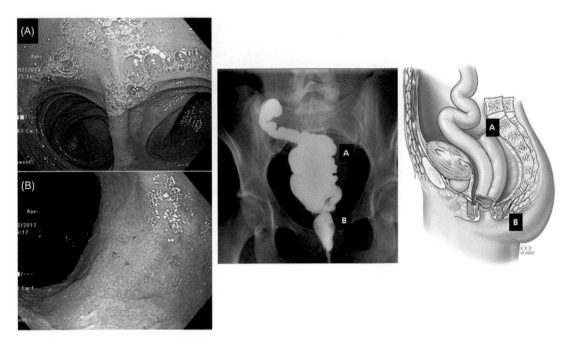

FIGURE 11.4   Anatomy of a J pouch. (A) Proximal part of the pouch body with an owl's eye configuration, one eye leading to the tip of the "J" and the other heading to the afferent limb; (B) distal part of the J pouch linked to the cuff with an anastomosis. The landmarks are highlighted on gastro-grafin enema (middle figure) and the artist's illustration (right figure). *Courtesy Mr. Joel Pangrace of Cleveland Clinic.*

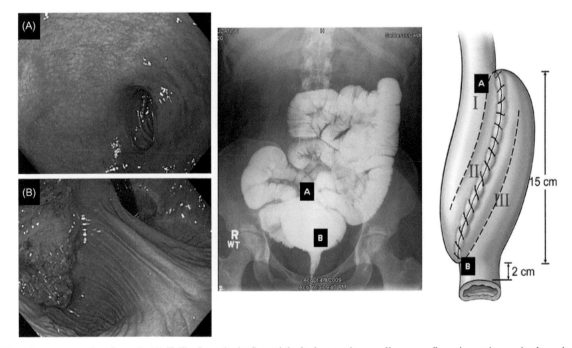

FIGURE 11.5   Anatomy of an S pouch. (A) Unlike J pouch, the S pouch body does not have owl's eye configuration at the proximal pouch and has a larger volume; (B) retroflex view of the pouch body, which is typically larger than that of the J pouch. The landmarks are highlighted on gastrografin enema (middle figure) and the artist's illustration (right figure). *Courtesy Mr. Joel Pangrace of Cleveland Clinic.*

abdominal pouches (K pouch) is determined by patient's anatomy and underlying disease (e.g., body mass index, length of mesentery, and function of anal sphincter) and surgical expertise. In addition, various techniques of construction of IPAA are used, including handsewn versus stapled and mucosectomy versus without mucosectomy (Fig. 11.10). J pouch with a stapled anastomosis and without mucosectomy is preferred surgical modality. The latter approach is normally

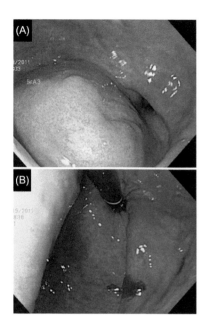

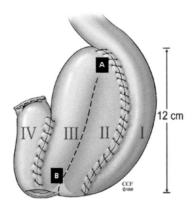

FIGURE 11.6 Anatomy of a W pouch. W pouch is created with an intention of augmentation of pouch volume and subsequent reduction in bowel frequency. The procedure has been largely abandoned in surgical practice due to high risk for complications. (A and B) Forward and retroflex view of the pouch body, which is typically larger than that of the J or S pouches. The landmarks are highlighted on the artist's illustration (right figure). *Courtesy Mr. Joel Pangrace of Cleveland Clinic.*

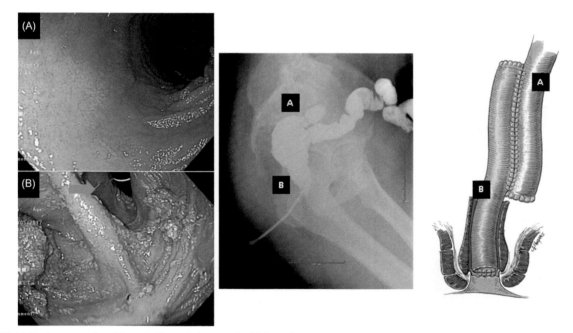

FIGURE 11.7 Anatomy of an H pouch. H pouch is created with isoperistaltic fashion, for technical reasons which are prohibitive for the J or S pouch. The efferent limb is made from a segment of distal ileum. (A and B) Forward and retroflex view of the pouch body. The long efferent limb is highlighted with a green arrow. An angulation of the efferent limb can cause obstruction. The landmarks are highlighted on gastrografin enema (middle figure) and artist's illustration (right figure). *Courtesy Mr. Joel Pangrace of Cleveland Clinic.*

performed in UC patients with colitis-associated neoplasia. J or S pouches with handsewn anastomosis with mucosectomy are usually performed in UC patients with colitis-associated neoplasia being indications for colectomy (Table 11.1).

The creation of a diverting end or loop ileostomy is often required in the staged RPC and IPAA for the purpose of maturity of the newly created pouch reservoir. In addition, temporary or permanent fecal diversion with an ileostomy can also be used for the treatment of medically or endoscopically refractory pouch disorders (Table 11.1). A rare, de novo disease entity, named postcolectomy enteritis, can occur in UC patients who undergo colectomy and ileostomy,

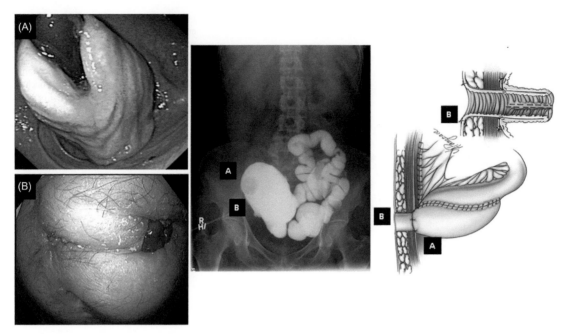

FIGURE 11.8   Kock pouch. Unique structure is the nipple valve through the abdominal wall, to keep continence and the patient does not need the use of external appliance. Fecal content of the pouch is emptied through self-catheterization. (A) Retroflex view of the nipple valve; (B) opening of the nipple valve, typically located at the right lower quadrant. The landmarks are highlighted on gastrografin enema (middle figure) and the artist's illustration (right figure). *Courtesy Mr. Joel Pangrace of Cleveland Clinic.*

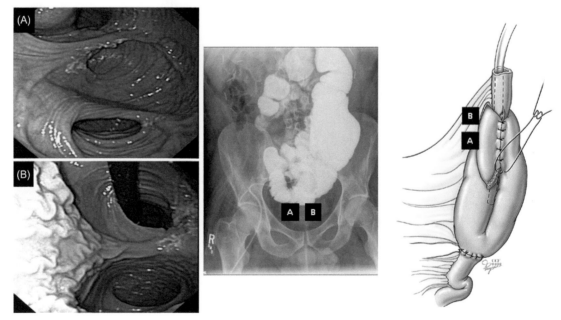

FIGURE 11.9   Variant of Barnett continent ileal reservoir. This pouch configuration is a form of continent ileostomy. The main structure which is different from a Kock pouch is a loop of small bowel wrapping around the valve. (A and B) Retroflex view of the pouch body showing a loop of bowel wrapping around the valve. The landmarks are highlighted on gastrografin enema via valve (middle figure) and the artist's illustration (right figure). *Courtesy Mr. Joel Pangrace of Cleveland Clinic.*

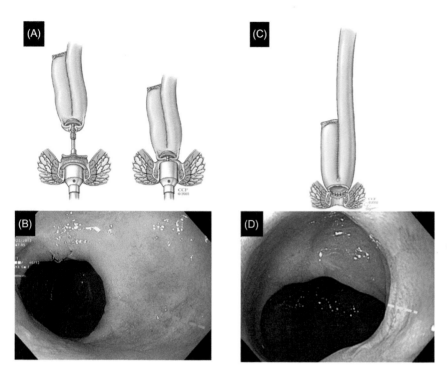

FIGURE 11.10 Stapled versus handsewn anastomosis at pelvic pouch construction. (A and B) Stapled anastomosis without mucosectomy on illustration and endoscopy. The cuff is highlighted with a green dot line; (C and D) handsewn anastomosis with mucosectomy on illustration and endoscopy. The anal transition zone is highlighted with a yellow dot line. *Courtesy Mr. Joel Pangrace of Cleveland Clinic.*

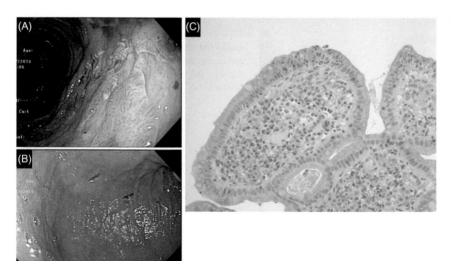

FIGURE 11.11 Postcolectomy enteritis in ulcerative colitis. With unknown etiology, this disease entity occurs in few ulcerative colitis patients immediately after colectomy and diverting ileostomy. (A) Diffuse enteritis with ulcers and granularity of ileal mucosa on ileoscopy via stoma; (B) normal ileal pouch on endoscopy with proximal fecal diversion; (C) histology of ileoscopy biopsy showed intact villi with neutrophil infiltration in epithelia, that is, cryptitis, and mononuclear cell infiltration in the lamina propria .

with or without creation of a pouch. Postcolectomy enteritis is characterized by diffuse mucosal inflammation of small bowel, starting from the stoma site, extending proximally (Fig. 11.11).

Temporary stoma would require eventual closure. Loop ileostomy can be closed with side-to-side (Fig. 11.12) or end-to-end anastomosis (Fig. 11.13). The bowel segment at the side-to-side anastomosis site is featured with a dilated lumen, making a false pouch on top of real pouch (Fig. 11.12) and causing symptoms of bloating and pain. In contrast, strictures are common patients with end-to-end anastomosis site (Fig. 11.13).

RPC with IPAA significantly improves patient's health-related quality of life, by preservation of the natural route of defecation. However, surgery-associated adverse sequelae can occur. Those adverse sequelae have been classified into

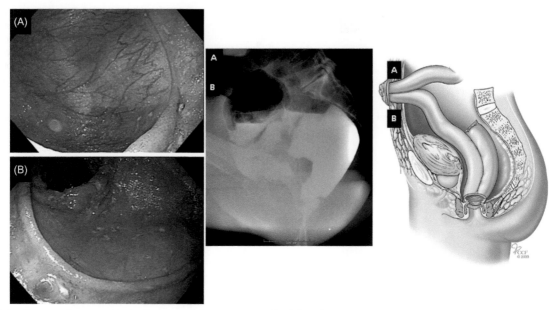

FIGURE 11.12 Loop ileostomy site in ulcerative colitis for temporary fecal diversion, as a part of staged restorative proctocolectomy. Side-to-side anastomosis is performed for ileostomy closure after ileal pouch is "matured." (A and B) Side-to-side anastomosis site with luminal dilation, making a false pouch body over real pouch body. The anatomy is also illustrated with gastrografin enema (middle figure) and the artist's depiction (right figure). *Courtesy Mr. Joel Pangrace of Cleveland Clinic.*

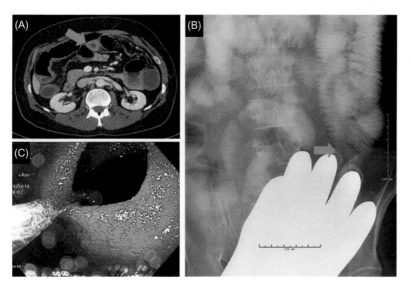

FIGURE 11.13 Loop ileostomy site closed with end-to-end anastomosis. (A and B) Strictured end-to-end anastomosis on CT and small bowel series (*green arrow*); (C) endoscopic balloon dilation of the stricture.

five categories: (1) structural complications, (2) inflammatory conditions, (3) functional disorders, (4) neoplastic conditions, and (5) metabolic abnormalities [4], which have been proposed in the Cleveland Clinic Classification of Ileal Pouch Disorders [5] (Table 11.2). The majority of those post−ileal pouch surgery-associated disorders can be reliably evaluated with endoscopy. Surgery-associated structural or mechanical complications can mainly be divided into two broad categories, obstructive lesions and anastomotic or suture-line leaks. Common obstructive entities include strictures, afferent limb syndrome (Fig. 11.14), efferent limb syndrome (Fig. 11.15), prolapse or pouchocele, pouch septum or bridge, and twisted pouch (Fig. 11.16). On the other hand, chronic suture-line or anastomosis leaks can lead to the formation of sinuses or fistulae. The most common location of the sinus is the presacral space, commonly resulting from anastomotic leak at posterior side. The sinus can be treated with endoscopic sinusotomy by cutting the pouch wall between the sinus and pouch body (Fig. 11.17) [6]. The leaks can also cause fistulae, including pouch-to-pouch, enterocutaneous, or perianal. To enhance the opportunity for the detection of fistula, various endoscopic techniques can be

TABLE 11.2 The Cleveland classification of ileal pouch disorders.

Surgical and mechanical disorders	Leak	Anastomotic or suture leak or separation
		Pelvic abscess or sepsis
		Pouch sinus
		Pouch fistula
	Obstruction	Stricture
		Afferent limb syndrome and efferent limb syndrome
		Twisted pouch, folded pouch
		Prolapse, pouchocele
		Pouch septum
	Others	Infertility and sexual dysfunction
		Portal−mesenteric vein thrombi
		Sphincter injury or dysfunction
Inflammatory disorders	Pouchitis	Microbiota associated
		Autoimmune associated
		Ischemia associated
	Cuffitis	Steroid/Mesalamine responsive
		Steroid/Mesalamine dependent
		Steroid/Mesalamine refractory
	Crohn's disease of the pouch	Inflammatory
		Stricturing
		Fistulizing
	Proximal small bowel bacterial overgrowth	
	Inflammatory polyps	
Functional disorders	Irritable pouch syndrome	
	Dyssynergic defecation	
	Saw-tooth contraction	
	Poucholgia fugax	
	Pseudo-obstruction or megapouch	
Neoplastic disorders	Neoplasia originated from intestinal epithelia	Neoplasia at the cuff or anal transitional zone
		Neoplasia at the pouch body
		Neoplasia at the afferent limb
	Neoplasia from squamous cell	
	Neoplasia of hematogenous origin	Lymphoma
		Systemic mastocytosis
	Metastatic cancer	Melanoma
Systemic and metabolic disorders	Anemia	Anemia of chronic disease
		Iron-deficiency anemia
	Bone loss	
	Vitamin D deficiency	
	Vitamin B12 deficiency	
	Bile acids	
	Nephrolithiasis	
	Celiac disease	Preexisting celiac disease
		De novo celiac disease

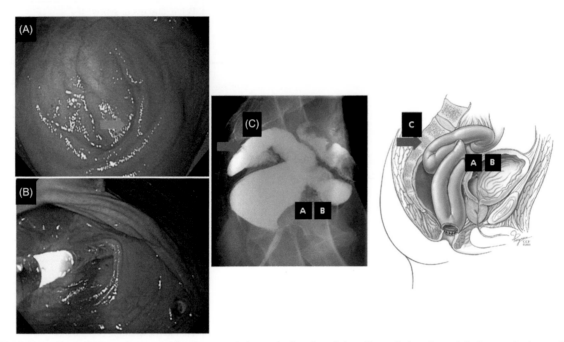

FIGURE 11.14 Afferent limb syndrome resulting from angulation at the junction of the afferent limb and pouch body or redundancy of afferent limb. (A) An angulation at the pouch inlet which prevents an easy intubation of the afferent limb (*green arrow*); (B) a through-the-scope balloon was used to guide the intubation; (C) a loop of the small bowel was trapped between the pouch body and tailbone, causing an angulated inlet (blue arrow). The landmarks are highlighted with barium defecography and and the trapped afferent limb (blue arrow) theartist's illustration (right figure). *Courtesy Mr. Joel Pangrace of Cleveland Clinic.*

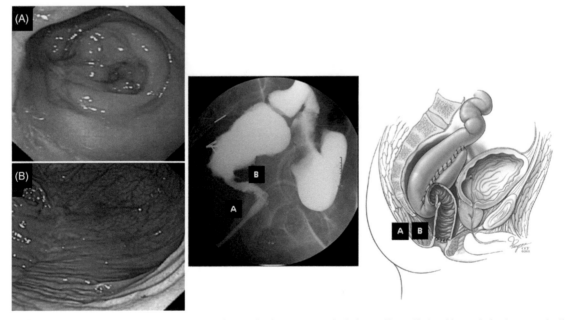

FIGURE 11.15 Efferent limb syndrome in an S pouch. It often results from an excessively long efferent limb, with angulation between the limb and pouch body. Common symptoms are bloating, nausea, and dyschezia. (A) Endoscopic view of the long efferent limb, with normal mucosa; (B) a retroflex view of pouch body. The landmarks are highlighted with barium defecography (middle figure) and the artist's illustration (right figure) . *Courtesy Mr. Joel Pangrace of Cleveland Clinic.*

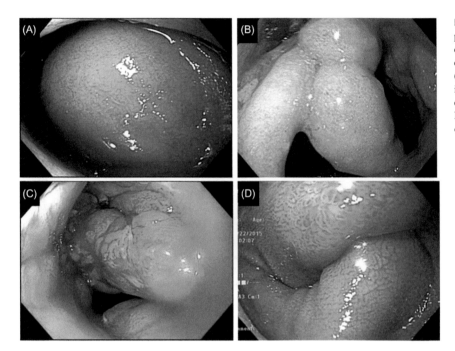

FIGURE 11.16 Obstructive disorders after pouch surgery. (A) Prolapse of anterior wall of the distal pouch body blocking the outlet, detected by endoscopic aspiration of air; (B) mucosal bridge in the pouch body, resulting in a compartmentalized pouch body; (C) distal pouch septum, blocking the pouch outlet; (D) twisted distal pouch body, causing obstruction.

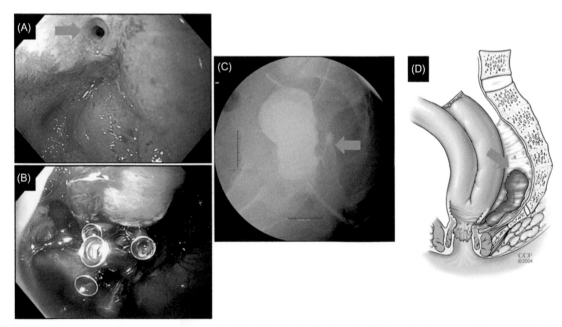

FIGURE 11.17 Presacral sinus from anastomotic or suture-line leaks in the ileal pouch. (A) Opening of presacral sinus resulting from chronic pouch–anal anastomosis leak (*green arrow*); (B) status of postendoscopic sinusotomy of the presacral sinus. Notice that endoclips were placed to keep the orifice of the incised sinus open; (C) the sinus is demonstrated on gastrografin enema; (D) illustration of presacral sinus (*green arrow*). *Courtesy Mr. Joel Pangrace of Cleveland Clinic.*

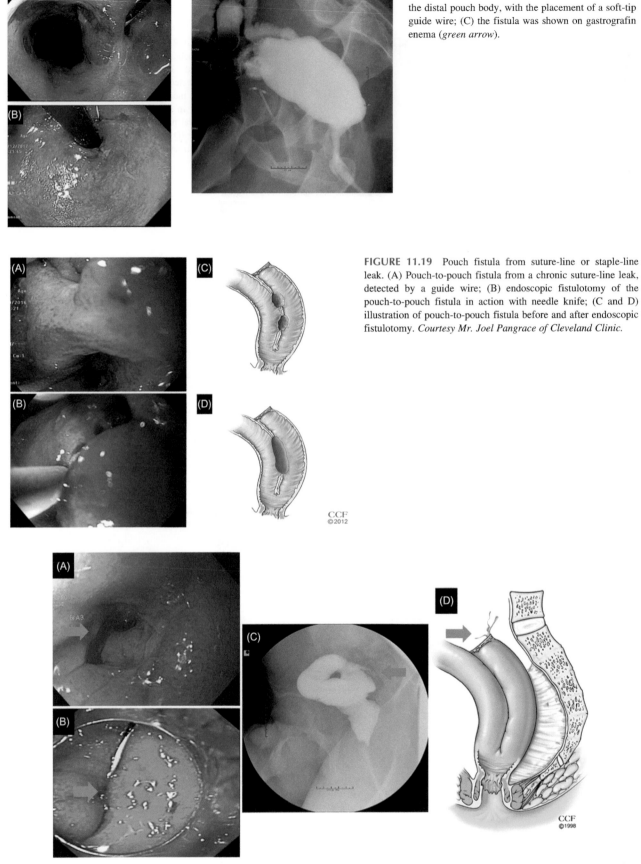

FIGURE 11.18 Pouch fistula. (A and B) Forward and retroflex view of a fistula at the anterior wall of the distal pouch body, with the placement of a soft-tip guide wire; (C) the fistula was shown on gastrografin enema (*green arrow*).

FIGURE 11.19 Pouch fistula from suture-line or staple-line leak. (A) Pouch-to-pouch fistula from a chronic suture-line leak, detected by a guide wire; (B) endoscopic fistulotomy of the pouch-to-pouch fistula in action with needle knife; (C and D) illustration of pouch-to-pouch fistula before and after endoscopic fistulotomy. *Courtesy Mr. Joel Pangrace of Cleveland Clinic.*

FIGURE 11.20 Leak at the tip of the "J" of J pouch. (A) A large defect or leak at the tip of the "J"; (B) a leak detected with a soft-tip guide wire, which was undergoing endoscopic closure with over-the-scope clipping; (C) a linear leak at the tip of the "J" detected with gastrografin enema; (D) illustration of the J pouch and the leak (*green arrows*). *Courtesy Mr. Joel Pangrace of Cleveland Clinic.*

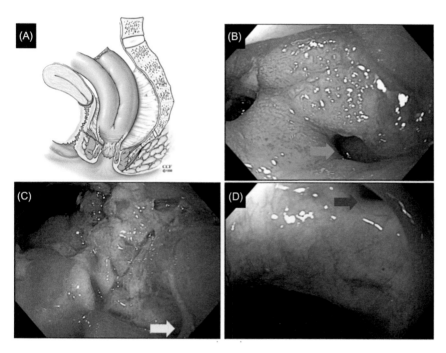

FIGURE 11.21 Etiology of PVF. (A) Illustration of PVF; (B) posterior wall of the vagina can be trapped between stapled heads during pouch construction, resulting in an iatrogenic PVF with normal mucosa around the fistula opening (*green arrow*); (C) Crohn's disease−associated fistula with inflamed anal transition zone (*yellow arrow*); (D) small cryptoglandular fistula opening at the dentate line with surrounding normal squamous mucosa (*blue arrow*). *PVF*, Pouch-vaginal fistula. *Courtesy Mr. Joel Pangrace of Cleveland Clinic.*

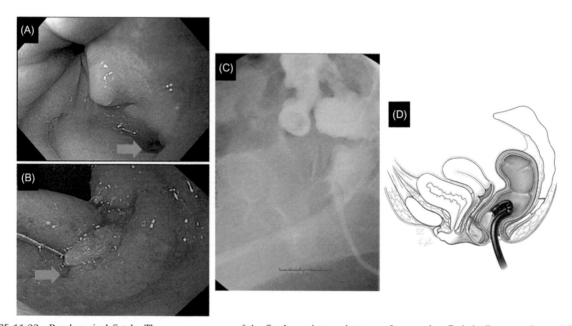

FIGURE 11.22 Pouch-vaginal fistula. The common causes of the fistula are iatrogenic trauma from stapler, Crohn's disease, and cryptoglandular abscess. (A) Forward view of the orifice of a pouch-vaginal fistula at the anal transition zone; (B) retroflex view of the fistula with a soft-tip guide wire; (C) the fistula is demonstrated on gastrografin enema; (D) illustration of pouch-vaginal fistula with a retroflexed endoscope. The vaginal fistula opening and tract are highlighted with green arrows. *Courtesy Mr. Joel Pangrace of Cleveland Clinic.*

TABLE 11.3 The pouchitis disease activity index [8].

Category	Subcategory	Score
Clinical	Stool frequency	
	Usual postoperative stool frequency	0
	1−2 stools/day > postoperative usual	1
	3 or more stools/day > postoperative usual	2
	Rectal bleeding	
	None or rare	0
	Present daily	1
	Fecal urgency	
	None	0
	Occasional	1
	Usual	2
	Fever (temperature >37.8°C)	
	Absent	0
	Present	1
Endoscopy	Endoscopic inflammation	
	Edema	1
	Granularity	1
	Friability	1
	Loss of vascular pattern	1
	Mucous exudate	1
	Ulceration	1
Acute histological inflammation	Polymorphonuclear leukocyte infiltration	
	Mild	1
	Moderate + crypt abscess	2
	Severe + crypt abscess	3
	Ulceration per low power field (mean)	
	<25%	1
	≥25%− ≤50%	2
	>50%	3

used, including retroflex of the scope and application of a guide wire (Figs. 11.18 and 11.19). Endoscopic fistulotomy can be attempted in the treatment of pouch fistulae (Fig. 11.19) [7]. Another common place prone to suture-line or staple-line leak is the tip of the "J" of a J pouch (Fig. 11.20).

Pouch-vaginal fistula (PVF) is a unique entity related to RPC and IPAA. PVF can result from iatrogenic injury from staple head, Crohn's disease (CD) of the pouch, or cryptoglandular abscess. The distinction among the etiology may be made on the basis of endoscopic features (Fig. 11.21). The internal opening of PVF is often better visualized with a retroflexed scope (Fig. 11.22).

Inflammatory disorders of the pouch are common, with the most common forms being pouchitis, CD of the pouch, and cuffitis. Endoscopic evaluation plays a key role in their diagnosis, differential diagnosis, disease monitoring, dysplasia surveillance, and therapy. Pouch endoscopy or pouchoscopy is routinely used for the quantification of pouch

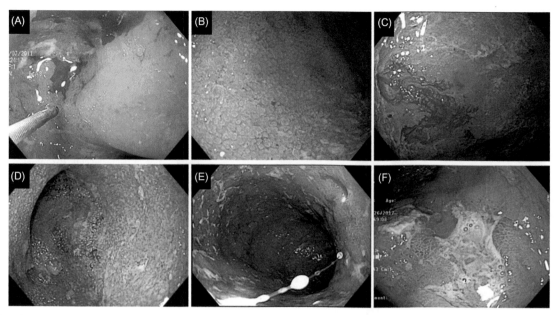

FIGURE 11.23 Endoscopic features reflecting the PDAI with a range 0−6. (A) Edema; (B) granularity; (C) friability; (D) loss of vascular pattern; (E) mucous exudates; (F) ulcers. Each feature contributes 1 point to the total of PDAI endoscopy subscores. *PDAI*, the Pouchitis Disease Activity Index.

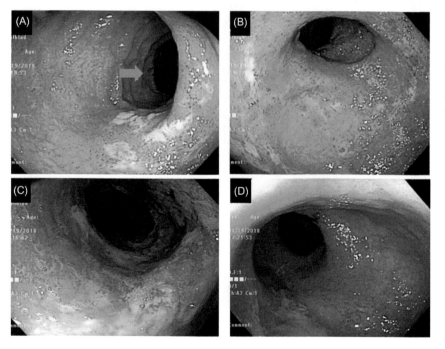

FIGURE 11.24 Classic primary microbiota-associated pouchitis. This phenotype of pouchitis is characterized by the presence of diffuse inflammation limited to the pouch body. (A) Normal afferent limb (*green arrow*) and patent pouch inlet; (B and C) diffuse moderate inflammation of whole pouch body with erythema, exudates, and loss of vascular pattern; (D) normal cuff.

inflammation. In fact, among various diagnostic instruments for pouchitis, endoscopy provides the most reliable approach for the measurement of pouch inflammation [8]. The 18-point Pouchitis Disease Activity Index (PDAI), consisting symptom, endoscopy, and histology subscores with each having 0−6 patients, is most commonly used (Table 11.3) [9]. The 6-point PDAI endoscopy subscores contain six endoscopic features for the description of inflammation, ranging from edema to ulceration (Fig. 11.23). Pouchitis is considered a disease spectrum with ranging etiology, disease course, and prognosis. Pouchitis may be broadly categorized into microbiota-associated (Fig. 11.24),

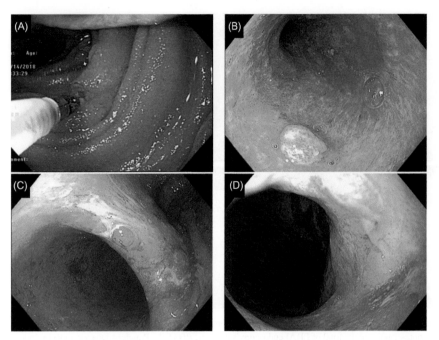

FIGURE 11.25 Patterns of immune-mediated pouchitis and enteritis. The classic example of this entity is PSC-associated pouchitis and enteritis. Sometimes, it can involve the duodenum. (A) Duodenum biopsy was taken in a patient with PSC and biliary stent to rule out duodenitis; (B) diffuse enteritis with a pseudopolyp in a long segment of the afferent limb above the pouch body; (C) diffuse inflammation at the pouch body with a widely patent pouch inlet; (D) mild diffuse inflammation of the rectal cuff. *PSC*, Primary sclerosing cholangitis.

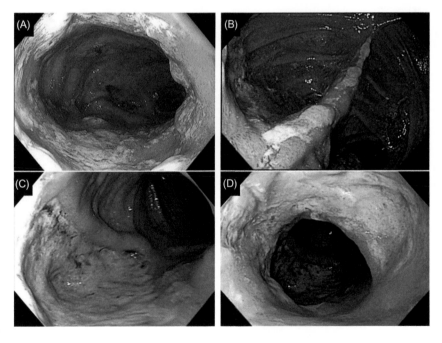

FIGURE 11.26 Patterns of ischemic pouchitis and cuffitis. The disease entities are diagnosed mainly on the basis of distribution of inflammation. (A) Inflammation is limited to the distal pouch body; (B) ulcers along the vertical staple line in the pouch body; (C) large, triangular-shaped ulcer at the distal pouch with a sharp demarcation; (D) the same patient as in (C) also had a semicircumferential ulcer in the cuff.

immune-mediated (Fig. 11.25), and ischemia-associated (Fig. 11.26) phenotypes, based on speculated etiopathogenetic pathways. A careful endoscopic evaluation of the distribution of pouch inflammation may provide clues for the diagnosis of the subtypes of pouchitis (Figs. 11.24—11.26).

CD of the pouch can occur in patients with RPC and IPAA who have a preoperative diagnosis of UC or indeterminate colitis, with the majority of the patients having de novo CD. CD of the pouch can be classified into inflammatory (Fig. 11.27), stricturing (Fig. 11.28), and fistulizing (Fig. 11.29) phenotypes. Endoscopic evaluation is an important modality for the diagnosis and differential diagnosis of phenotypes of CD along with other pouch disorders.

In some patients with IPAA, a phenotype of enteritiswas coined with the term "prepouch ileitis." The phenotype has been considered as a CD-like condition. Its representative endoscopic features are listed in Fig. 11.30.

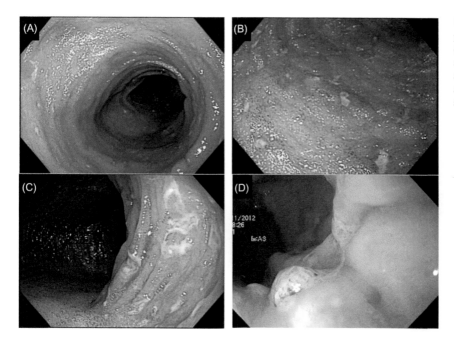

FIGURE 11.27 Patterns of inflammatory phenotype of Crohn's disease of the pouch. (A) Small erosions in a long segment of the afferent limb; (B) discrete small ulcers at the afferent limb; (C) patchy distribution of ulceration in the pouch body; (D) nodular anal transition zone.

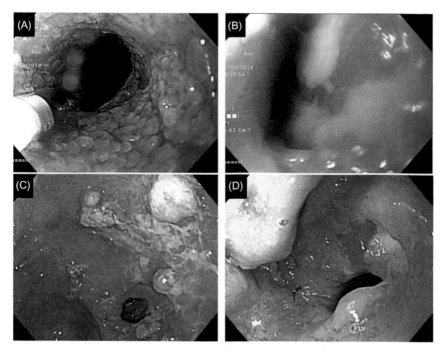

FIGURE 11.28 Patterns of stricturing phenotype of Crohn's disease of the pouch. (A) Stricture undergoing endoscopic dilation at the end-to-end anastomosis at the previous loop ileostomy site; (B) ulcerated stricture at the afferent limb; (C) stricture at the inlet with segmental inflammation at the pouch body; (D) stricture at the pouch–anal anastomosis with nodularity of the cuff.

Classic cuffitis has been considered a residual UC. Endoscopic evaluation is valuable for the diagnosis, disease monitoring, and differential diagnosis of cuffitis. The Cuffitis Activity Index, which is modified from PDAI, has been proposed and used for the diagnostic evaluation [10]. The endoscopy provides six features with each accounting for 1 point, a total of 6 points (Fig. 11.31). Pouchoscopy also provides an important modality for the identification of other cuff disorders, such as prolapse, fistulae, varices, and neoplasia (Fig. 11.32). Chapter 12 discusses pouch neoplasia in detail.

Fecal diversion with the creation of an ileostomy has been performed during and after RPC and IPAA for various indications. Long-term fecal diversion with pouch in situ may lead to diversion pouchitis and diversion-associated distal

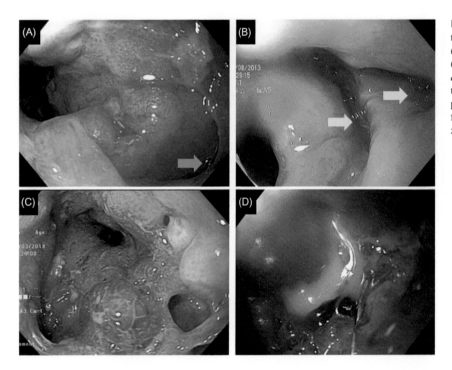

FIGURE 11.29 Patterns of fistulizing phenotype of Crohn's disease of the pouch. (A) Single fistula at ATZ with nodularity (*green arrow*); (B) two fistulae at ATZ (*yellow arrows*); (C) multiple fistulae along with cuffitis; (D) fistula with abscess (pus drainage from probing of a soft-tip guide wire) resulting from severe cuffitis. *ATZ*, Anal transition zone.

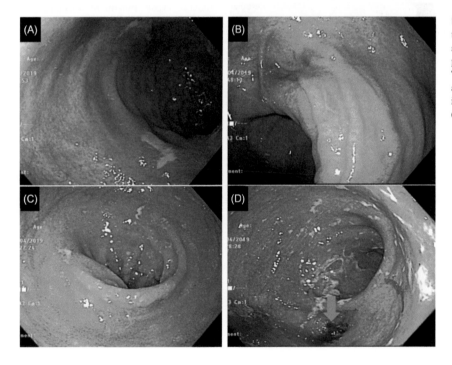

FIGURE 11.30 Phenotypes of prepouch ileitis. The first patient had discrete ulcers in a segment of 15 cm afferent limb above the pouch inlet (A) and a normal pouch body (B). The second patient had discrete ulcers in both afferent limb (C) and pouch body (D) with an inlet stricture, which was dilated with balloon (*green arrow*).

pouch or anastomotic stricture. In diversion pouchitis, mucosa of the pouch body is extremely friable, and mucosal bleeding often occurs even with a gentle air or carbon dioxide insufflation (Fig. 11.33).

Summary and recommendations

Surgical therapy of UC with RPC involves significant alterations in bowel anatomy. Pelvic ileal pouch with IPAA has become the surgical treatment of choice. Alternatively, patients may select to have permanent Brooke end ileostomy

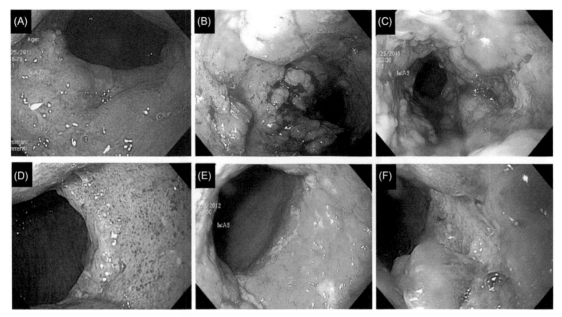

FIGURE 11.31 Endoscopic features reflecting the CAI with a range of 0–6. (A) Edema; (B and C) granularity and friability; (D) loss of vascular pattern and erythema; (E) mucous exudates over ulcers; (F) linear ulcers. Each feature contributes 1 point to the total of 6 CAI endoscopy subscores. *CAI*, the Cuffitis Activity Index.

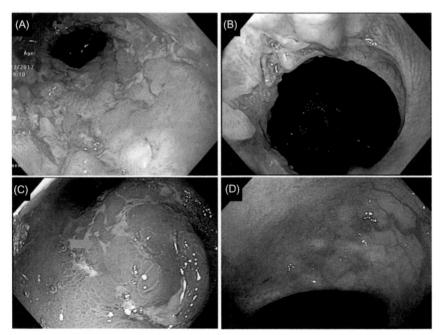

FIGURE 11.32 Cuffitis and cuff disorders. (A) Classic cuffitis with diffuse and circumferential inflammation, which has been considered residual ulcerative colitis; (B) nodular anal transition zone in a patient with mucosectomy during pouch surgery, mimicking cuffitis; (C) cuff inflammation is limited to the anterior wall (*green arrow*), from excessive straining and prolapse; (D) varices in the cuff in a patient with concurrent primary sclerosing cholangitis and portal hypertension.

and IRA. UC patients with poor anal-sphincter function may undergo continent ileostomies, such as Kock pouch, which may also be performed in some patients with failed J pouches. It is important for endoscopist to identify and photo document landmarks of various formats of the pouch.

There are a variety of pouch disorders, ranging from surgery-related structural abnormalities (including anastomotic strictures, afferent limb syndrome, and presacral sinus) to inflammatory disorders (such as pouchitis, CD of the pouch, and cuffitis). Endoscopic evaluation plays a key role in the diagnosis, differential diagnosis, disease monitoring,

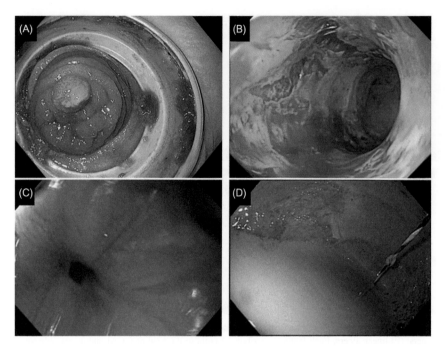

FIGURE 11.33 Disorders of the diverted ileal pouch. (A) Permanent diverting end ileostomy as a part of treatment for pouch failure; (B) friable mucosa in a diverted ileal pouch; (C and D) pinhole stricture at the anastomosis due to long-lasting fecal diversion. The stricture was treated with endoscopic stricturotomy over a guide wire. There is a large accumulation of secretions in the pouch body, which had resulted in patient's symptoms of pelvic discomfort and discharge of bloody mucus.

dysplasia surveillance, and treatment of those pouch disorders. It is important to document severity and distribution of inflammation, stricture, and fistula at topographic locations of IPAA or continent ileostomies.

References

[1] Samuel S, Ingle SB, Dhillon S, Yadav S, Harmsen WS, Zinsmeister AR, et al. Cumulative incidence and risk factors for hospitalization and surgery in a population-based cohort of ulcerative colitis. Inflamm Bowel Dis 2013;19(9):1858–66.

[2] Kaplan GG, Seow CH, Ghosh S, Molodecky N, Rezaie A, Moran GW, et al. Decreasing colectomy rates for ulcerative colitis: a population-based time trend study. Am J Gastroenterol 2012;107(12):1879–87.

[3] da Luz Moreira A, Kiran RP, Lavery I. Clinical outcomes of ileorectal anastomosis for ulcerative colitis. Br J Surg 2010;97(1):65–9.

[4] Shen B, Remzi FH, Lavery IC, Lashner BA, Fazio VW. A proposed classification of ileal pouch disorders and associated complications after restorative proctocolectomy. Clin Gastroenterol Hepatol 2008;6(2):145–58 quiz 124.

[5] Shen B. The Cleveland clinic classification of ileal pouch disorders. In: Shen B, editor. Pouchitis and ileal pouch disorders. Cambridge, MA: Elsevier; 2019. p. 80–112.

[6] Lan N, Hull TL, Shen B. Endoscopic sinusotomy versus redo surgery for the treatment of chronic pouch anastomotic sinus in ulcerative colitis patients. Gastrointest Endosc 2019;89(1):144–56.

[7] Kochhar G, Shen B. Endoscopic fistulotomy in inflammatory bowel disease (with video). Gastrointest Endosc 2018;88(1):87–94.

[8] Shen B, Achkar JP, Lashner BA, Ormsby AH, Remzi FH, Bevins CL, et al. Endoscopic and histologic evaluation together with symptom assessment are required to diagnose pouchitis. Gastroenterology 2001;121(2):261–7.

[9] Sandborn WJ, Tremaine WJ, Batts KP, Pemberton JH, Phillips SF. Pouchitis after ileal pouch-anal anastomosis: a pouchitis disease activity index. Mayo Clin Proc 1994;69(5):409–15.

[10] Shen B, Lashner BA, Bennett AE, Remzi FH, Brzezinski A, Achkar JP, et al. Treatment of rectal cuff inflammation (cuffitis) in patients with ulcerative colitis following restorative proctocolectomy and ileal pouch-anal anastomosis. Am J Gastroenterol 2004;99(8):1527–31.

Chapter 12

Inflammatory bowel disease–associated neoplasia

Bo Shen

Center for Inflammatory Bowel Diseases, Columbia University Irving Medical Center-New York Presbyterian Hospital, New York, NY, United States

Chapter Outline

Abbreviations

ATZ	anal transition zone
CAN	colitis-associated neoplasia
CD	Crohn's disease
CRC	colorectal cancer
EBV	Epstein–Barr virus
GI	gastrointestinal
HGD	high-grade dysplasia
HPV	human papillomavirus
IBD	inflammatory bowel disease
IPAA	ileal pouch–anal anastomosis
LGD	low-grade dysplasia
NHL	non-Hodgkin's lymphoma
PSC	primary sclerosing cholangitis
SCC	squamous cell cancer
UC	ulcerative colitis

Introduction

Chronic mucosal inflammation is a risk factor for the development of neoplasia in the gastrointestinal (GI) or extraintestinal organs. Inflammatory bowel disease (IBD) is a classic example. The natural history of colitis-associated neoplasia (CAN) in the majority of patients follows the sequence of chronic inflammation, low-grade dysplasia (LGD), high-grade dysplasia (HGD), and adenocarcinoma, which serves as a biological base for routine colonoscopy surveillance of long-standing ulcerative colitis (UC) or Crohn's disease (CD) of the colon. However, some patients may skip the stage(s) in the tumorigenic sequence or have endoscopically ill-defined dysplastic lesions, making the surveillance challeningg. Endoscopic differentiation between chronic inflammation, inflamma-tory polyps, LGD, and HGD is difficult, resulting in the development of various image-enhanced endoscopy techni-ques. Chromoendoscopy and other image-enhanced endoscopies in the diagnosis, surveillance, and disease monitoring are discussed in a separate chapter (Chapter 19: Chromoendoscopy and advanced endoscopy imaging in inflammatory bowel disease).

Endoscopic features of IBD-associated colorectal cancer (CRC) have not been well described in the literature. Common endoscopic features of IBD-associated adenocarcinoma are mass-like lesions, strictures, and ulcerative lesions. However, some of those endoscopic features overlap with is seen in CD or UC. Endoscopic features of colitis-associated dysplasia are described in Chapter 19, Chromoendoscopy and other advanced endoscopy imaging in inflammatory bowel disease.

The risk for neoplasia may persist in a minority of patients who undergo colectomy. Neoplasia can occur in the rectal cuff or pouch body in those with restorative proctocolectomy and ileal pouch−anal anastomosis (IPAA). The prognosis of IPAA-associated cancer is poor and surveillance endoscopy has been challenging due to the altered surgical anatomy and poorly defined disease course.

Patients with IBD can also develop squamous cell cancer (SCC) and non-Hodgkin's lymphoma (NHL). There are currently no published surveillance guidelines for SCC or NHL. However, common endoscopic features of SCC and NHL are discussed in this chapter.

Colitis- or enteritis-associated adenocarcinoma

CRC is a common cause of mortality for Crohn's colitis or UC. Colonoscopy plays a key role in diagnosis, surveillance, and in selected patients, endoscopic therapy (such as endoscopic mucosal resection).

Ulcerative colitis−associated cancer

Patients with UC carry a higher risk for CRC than the general population. A metaanalysis of 116 studies showed the probability of CRC in patients with UC was 1.6%, 8.3%, and 18.4% at 10, 20, and 30 years, respectively, after UC diagnosis. The incidence of UC-associated CRC is 4−10 times greater than that for sporadic CRC, with the average age of diagnosis of 20 years younger [1]. The presence of concurrent primary sclerosing cholangitis (PSC) with IBD further increases the risk for CRC [2]. A separate metaanalysis found that the cumulative prevalence of CRC was 21% of patients with UC−PSC and 4% of UC patients without, a 4.8-fold higher risk [3]. Therefore routine surveillance colonoscopy is recommended for patients with left-sided or extensive colitis for more than 8−9 years after UC diagnosis and yearly for all UC−PSC patients at the time of UC diagnosis [4−9].

Endoscopic features of UC-associated adenocarcinomas include ulcers, strictures, nodules, or masses or combinations. Endoscopically nonliftable lesions or lesions with central indentations should raise the suspicion for malignancy (Figs. 12.1 and 12.2). Patients with refractory to major available medicationsfor IBD) should be carefully evaluated for

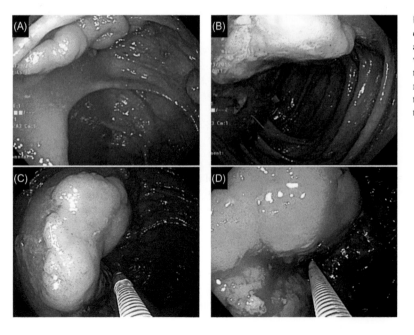

FIGURE 12.1 Nonliftable polypoid lesion in quiescent ulcerative colitis: (A) an elongated sessile polyp at the proximal ascending colon and (B−D) attempt was made to lift the polyp with 50% glucose. The feature of nonlifted central indentation raised concern for malignancy. En block endoscopic submucosal dissection was later performed. Fortunately, the patient has tubular adenoma with low-grade dysplasia.

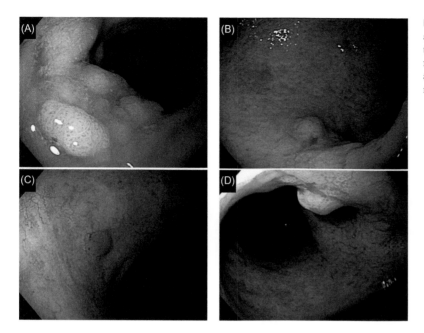

FIGURE 12.2 Invasive moderately differentiated adenocarcinoma in a patient with concurrent ulcerative colitis and primary sclerosing cholangitis: (A) malignant mass-type lesions with a central indentation and (B—D) quiescent extensive colitis in the with mucosal scars and inflammatory polyps.

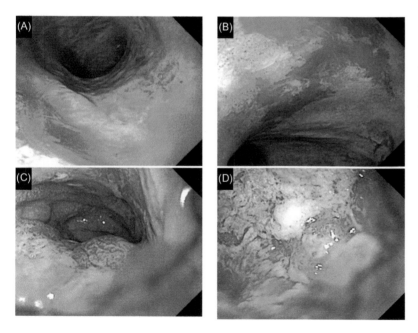

FIGURE 12.3 Adenocarcinoma of the rectum resulting from long-standing refractory ulcerative colitis in a 64-year-old male patient: (A and B) active colitis and proctitis with friable mucosa and spontaneous bleeding and (C and D) "inflammatory" stricture at the proximal rectum. Histology of biopsy of the stricture showed poorly differentiated adenocarcinoma.

malignancy. Endoscopic biopsy may cause excessive bleeding in the setting of active colitis and/or malignancy (Fig. 12.3). Differential diagnosis should be made between true refractory benign inflammation with ulcers or inflammatory polyps and malignant "inflammation" with ulcer or polypoid lesions

With the extensive use of biological agents in UC, benign strictures of the colon and rectum are increasingly identified [10]. The formation of benign strictures in this setting result from tissue healing and remodeling in the lamina propria, muscularis mucosae, and submucosa. Differential diagnosis should be made between benign and malignant strictures in IBD. UC-associated malignant strictures are often ulcerated, while long-standing UC-associated benign strictures are typically nonulcerated (Figs. 12.4—12.6). Tissue biopsy at the index and subsequent colonoscopies should be taken from all strictures in UC patients. Attempts should be made to traverse the stricture to assess the disease activity of IBD as well as synchronous neoplastic lesions in the proximal bowel, even if it may require endoscopic therapy, such as balloon dilation, stricturotomy, or temporary stenting (Figs. 12.4—12.6).

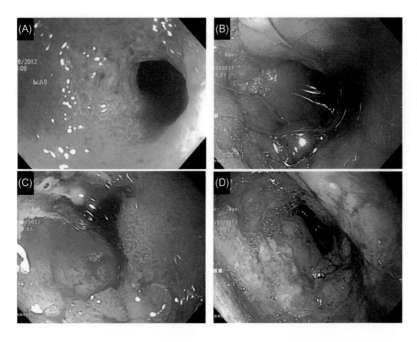

FIGURE 12.4 Benign and malignant strictures in four patients with ulcerative colitis: (A and B) inflammatory strictures at the sigmoid colon in two patients. Notice the strictures were not ulcered. (C and D) Ulcered malignant strictures in the transverse colon in other two patients. Tissue biopsy of the strictures for the differential diagnosis is imperative.

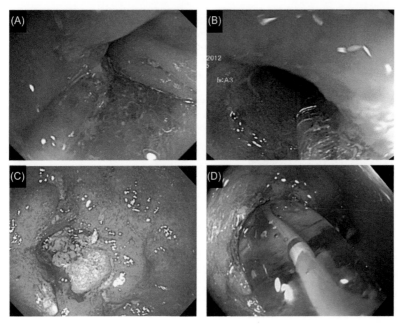

FIGURE 12.5 Endoscopic balloon dilation of non-traversable benign or malignant strictures in ulcerative colitis. The attempt was made to assess the disease activity and synchronous lesions in the proximal bowel. (A and B) Endoscopic dilation of a benign inflammatory stricture at the rectosigmoid junction and (C and D) endocopic dilation of an ulcerated malignant stricture in the transverse colon.

Flat and depressed lesions can also occur in CAN, which often harbor later-spreading malignancy (Fig. 12.7A and B).

Crohn's disease—associated cancer

Long-standing Crohn's colitis carries a similar risk for CRC to UC. In Crohn's colitis—associated dysplasia, approximately 50% were multifocal [11]. CD- as well as UC-associated cancer can be suspected or diagnosed based on endoscopic appearance or histologic evaluation of areas in active colitis. CD patients with CRC present with a more advanced cancer stage than that of UC [12]. Similar to UC, the presence of PSC in Crohn's colitis may increase the risk for CAN [13]. Interestingly, the presence of CD-like inflammatory changes in histology in colitis-associated CRC was shown to convey a better prognosis [14]. Endoscopic presentations of CD-associated CRC include deep ulcers,

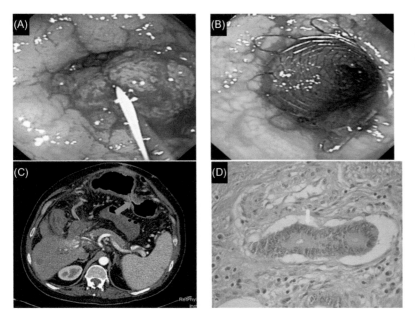

FIGURE 12.6 Malignant stricture in the ulcerative colitis: (A) a tight malignant stricture in the transverse colon detected with a soft-tip guidewire; (B) a non-covered self-expandable metallic stent was placed across the stricture for diagnosis and treatment; (C) the stricture on computed tomography (*green arrow*); and (D) vascular invasion of adenocarcinoma on surgical histopathology (*yellow arrow*).

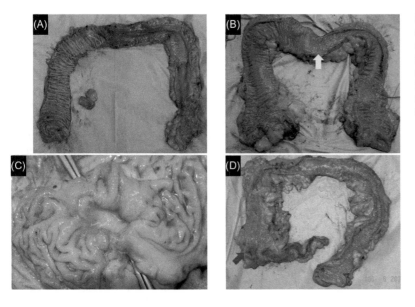

FIGURE 12.7 Colectomy specimens of colitis-associated neoplasia: (A) ulcerative colitis—associated malignant stricture at the sigmoid colon; (B) flat subtle lesions in the transverse colon with high-grade dysplasia; (C) malignant ulcer of the ascending colon in a patient with Crohn's disease and ileocecectomy; and (D) a malignant cecal mass in a patient with Crohn's colitis.

strictures, and masses (Fig. 12.7C and D). In contrast to the pattern in UC, patients with CD may develop adenoma-like lesions in previously inflamed or noninflamed colon [15].

Anorectal cancer in CD may be considered as a separate entity from classic CD-associated CRC (also see below). In a study of 116 patients with a duration of CD of ≥ 10 years, anorectal neoplastic lesions were detected through biopsy in 22 (19.0%), of whom 18 had carcinomas. Most patients were found to have poorly differentiated adenocarcinoma or mucinous adenocarcinoma [16]. CD-associated anorectal carcinoma was shown to have a poor prognosis due to the advanced stage and aggressive histology [17].

CD in the small intestine is rarely associated with nonfamilial small bowel adenocarcinomas which carries a poor prognosis. The investigation of histologic features (glandular vs diffuse poorly cohesive, mixed or solid), cell phenotype (intestinal vs gastric/pancreatobiliary duct type), and Wnt signaling activation showed that small bowel cancer from CD or from celiac disease are different [18]. Some small bowel cancers in the setting of CD were diagnosed incidentally during or after bowel resection surgery. Concurrent small bowel inflammation and adjacent dysplasia are

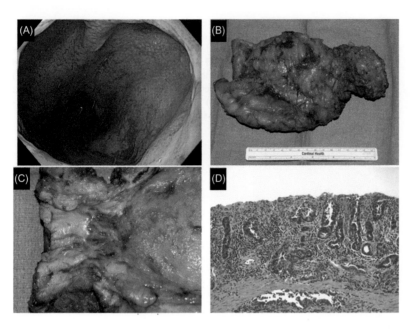

FIGURE 12.8 ATZ adenocarcinoma in a patient with restorative proctocolectomy, handsewn ileal pouch—anal anastomosis for colitis-associated neoplasia: (A) pouch endoscopy showed a normal-appearing ATZ with biopsy revealing high-grade dysplasia; (B) surgically excised pouch for the multifocal high-grade dysplasia; (C) macroscopic examination demonstrated no visible lesions in the ATZ; and (D) histology of the excised pouch reviewed invasive adenocarcinoma at the ATZ. *ATZ*, Anal transition zone.

common [19,20]. The prognosis of CD-associated small bowel adenocarcinoma is poor [19]. Endoscopic surveillance of small bowel cancer in CD has been challenging, mainly due to a low sensitivity, location, the presence of strictures [21], and limitation to the use of video capsule endoscopy.

Cuff and pouch cancer

UC patients undergoing proctocolectomy and IPAA still carry a small risk for the development of dysplasia or cancer in the cuff, anal transition zone (ATZ), or pouch body [22,23]. Patients with a precolectomy diagnosis of colorectal dysplasia or cancer are at a higher risk than those without the prior history of colorectal neoplasia. Additional risk factors include the presence of PSC, chronic pouchitis, cuffitis, CD of the pouch [22—24], or family history of CRC. The majority of IPAA-associated neoplasia is located at the cuff or ATZ; patients with mucosectomy are not immune to the development of neoplasia [22]. Of 11 patients with pouch cancer, 9 (82%) had annual pouch surveillance endoscopy in a study from the author's group. Three patients (27%) had no visible lesions on pouchoscopy at the time of cancer diagnosis (Fig. 12.8) [22]. Endoscopic features, if present, are masses or polypoid lesions (Figs. 12.9 and 12.10) or ulcer lesions (Fig. 12.11).

The prognosis of pouch cancer is poor. Of the 14 patients with pouch adenocarcinoma for a median follow-up of 2.1 years, 6 (42.9%) died from the pouch cancer [25]. The poor prognosis may be explained by the uncleared disease course, lateral spreading nature of cancer, or the lack of effective surveillance technology. Nonetheless, an endoscopic deep biopsy of the cuff or ATZ is recommended, especially in those with a precolectomy diagnosis of CAN.

Anorectal and perianal cancer

Patients with CD are at the risk for the development of anorectal cancer or perianal cancer. Malignant transformation of CD fistula has been reported [26]. Fistula-associated cancer can be adenocarcinoma from the rectal mucosa [27], or SCC [28,29]. In a population study of 19,486 patients with IBD with a median follow-up of 35 months (interquartile range, 29—40 months), 8 developed anal cancer and 14 had rectal cancer. In a subgroup analysis of 2911 patients with anal and/or perianal CD, 2 developed anal SCC, 3 developed perianal fistula—related adenocarcinoma, and 6 had rectal cancer. A history of anal and/or perianal lesions was found to be associated with the development of anal cancer (odds ratio = 11.2; 95% confidence interval, 1.18—551.51) [28]. In a retrospective study of 2382 patients with fistulizing perianal CD, cancer in a fistula tract was diagnosed in 19 (0.79%), with 9 having SCC, and 10 having adenocarcinoma. Common presenting symptoms in these patientswere drainage or abscess from perianal fistulas. The mean time from the fistula diagnosis to cancer diagnosis was 6.0 years [29].

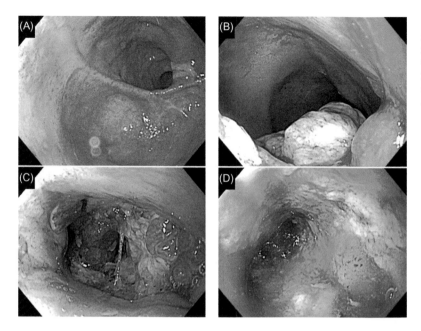

FIGURE 12.9   Ileal pouch cancer in a patient with a preoperative diagnosis of ulcerative colitis in a 43-year-old male: (A) normal proximal pouch body except superficial ulceration in the inlet and (B−D) malignant mass lesion in the distal pouch body and cuff with histology showing moderately differentiated adenocarcinoma.

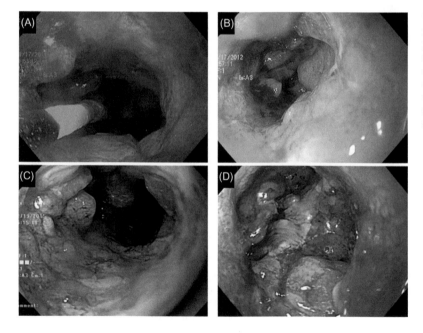

FIGURE 12.10   Progress of Crohn's disease of the ileal pouch to adenocarcinoma over 5 years in a 27-year-old patient with precolectomy diagnosis of ulcerative colitis: (A) inflammatory pouch—anal anastomosis stricture undergoing endoscopic balloon dilation; (B and C) persistent stricture and inflammation over the next 4 years, despite medical and endoscopic therapy; and (D) poorly differentiated adenocarcinoma developed at the anastomosis and cuff.

Endoscopic evaluation should include the inspection of colon, rectum, anal canal, anus, and perianal skin in patients with CD. If the rectum is not inflamed, a retroflex view should be performed. Endoscopic features of CD-associated anorectal cancer or fistula-associated cancer include diffusely friable and ulcerated mucosa or squamous epithelia, giving rise to a "dirty" appearance (Figs. 12.12 and 12.13). There may be polypoid lesions (Fig. 12.10) or strictures. The suspicion is even higher if these lesions do not respond to biological therapy or repeat endoscopic stricture dilations or endoscopic stricturotomy. These lesions require an extensive biopsy, even carrying a risk of biopsy-associated bleeding. A digital examination should be a part of the endoscopic evaluation. Anorectal cancer may be felt as hard and poor distensible tissue, strictures, or masses.

Patients with CD have a higher risk for melanoma or nonmelanoma skin cancer, especially in those using purine analogs or antitumor necrosis factor agents [30]. Perianal or anal skin cancer may be caused by chronic diarrhea or human papillomavirus (HPV) infection (Fig. 12.13).

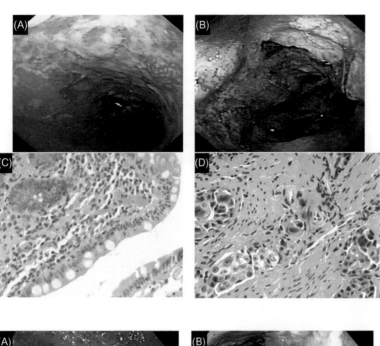

FIGURE 12.11 Aggressive form of pouch cancer in a 36-year-male with a preoperative diagnosis of ulcerative colitis: (A and C) endoscopic and histologic features of chronic active pouchitis with ulcers and mononuclear cell infiltration in the lamina propria and (B and D) deep ulcers developed 6 months later and histology of biopsy and pouch excision specimen showing poorly differentiated adenocarcinoma with metastasis to the lymph nodes and liver.

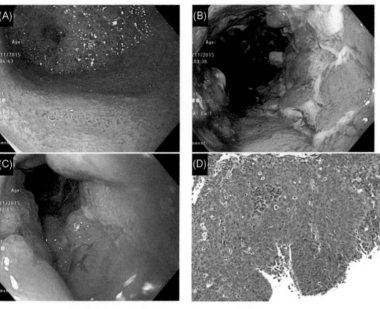

FIGURE 12.12 Squamous cell neoplasia in a patient with restorative proctocolectomy and ileal pouch—anal anastomosis ulcerative colitis: (A) normal afferent limb of the ileal pouch; (B) severe pouchitis with ulcers and nodularity; (C) severe cuffitis with ulcers and nodularity; and (D) histology of cuff biopsy showed superficial fragments of at least high-grade squamous intraepithelial lesion. *Histology photo: Courtesy of Dr. Ilyssa L. Gordon of Cleveland Clinic.*

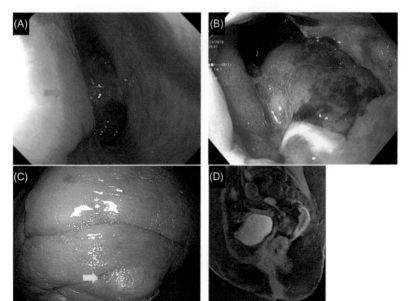

FIGURE 12.13 Human papillomavirus-associated squamous cell cancer in a 65-year-old patient with a history of hysterectomy for a benign uterine condition. The patient has had a long-standing ileal pouch—vaginal fistula with restorative proctocolectomy for underlying ulcerative colitis. (A) Vaginoscopy showed a mass with granular epithelia at the posterior wall (*green arrow*); (B) ulcerated and necrotic mass at the cuff; (C) small abscess at the labia (*yellow arrow*); and (D) the mass lesion between the cuff and vagina (*blue arrow*).

Gastrointestinal lymphoma

Endoscopic differential diagnosis of primary GI lymphoma and IBD is discussed in a separate chapter (Chapter 33: Inflammatory bowel disease—like conditions: gastrointestinal lymphoma and other neoplasms). Patients with IBD, especially CD [31], organ transplantation, chronic Epstein—Barr virus (EBV) infection, and the use of purine analogs or maybe antitumor necrosis factor agents have an increased risk for GI lymphoma [32,33]. The most prevalent form of NHL associated with IBD is diffuse large B-cell lymphoma. The common locations of lymphoma are the colorectumand small bowel [31]. GI lymphoma often presents with large ulcers, nodularity, masses, or "inflammatory" strictures on endoscopy (Figs. 12.14—12.16). The background IBD may be active or quiescent.

There are reports on lymphoma in ileal pouches [22,34,35], which may be associated with immunosuppressive therapy [36] or liver transplantation (Figs. 12.17 and 12.18).

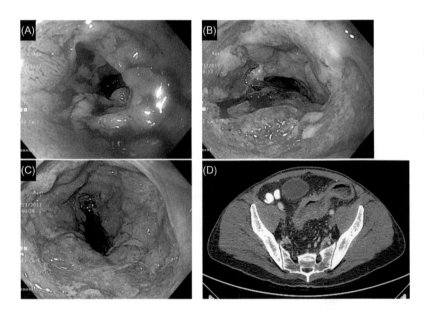

FIGURE 12.14 Diffuse large B-cell lymphoma of the rectosigmoid colon in a 30-year-old male with long-standing left-side ulcerative colitis and Epstein—Barr virus infection. The patient had been treated with long-term antitumor necrosis factor agents. (A—C) Circumferentially ulcerated stricture, edema, and nodularity in the sigmoid colon and (D) thickened rectosigmoid colon on computed tomography.

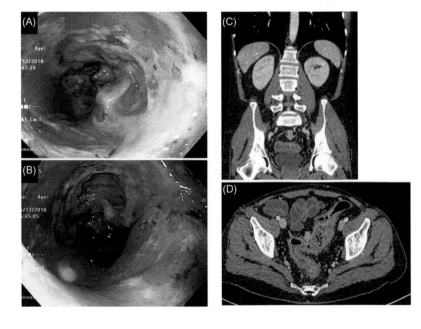

FIGURE 12.15 Diffuse large B-cell lymphoma of the colon with perforation in a 35-year-old male with diffuse ulcerative colitis. The patient had been treated with infliximab, vedolizumab, and 6-mercaptopurine and developed Epstein—Barr infection. (A and B) Diffuse colitis with "dirty" appearing large ulcers and exudates which was found to be malignant and (C and D) perforation at the sigmoid colon on computed tomography (*green arrows*).

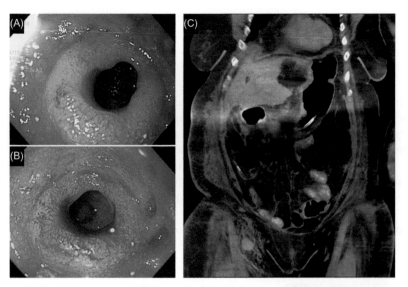

FIGURE 12.16 Diffuse large B-cell lymphoma of spleen and liver sparing gastrointestinal tract in a 58-year-old female patient with long-standing ileocolonic Crohn's disease resulting in proctocolectomy and end ileostomy. Her Crohn's disease at the distal 15 cm from the stoma was treated with azathioprine for 6 years. (A and B) Ileal stricture 15 cm from the stoma; (C) large liver mass found to be lymphoma (*green arrow*).

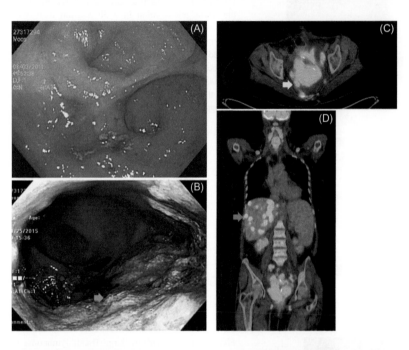

FIGURE 12.17 Monomorphic posttransplant, Epstein−Barr virus infection-negative lymphoproliferative disorder (diffuse large B-cell lymphoma, not otherwise specified) involving the ileal pouch, liver, and lymph nodes. The patient had had liver transplantation for primary sclerosing cholangitis: (A) initially normal pouch, (B) lymphoma with a large semicircumferential ulcerated mass in the pouch body (*green arrow*), and (C and D) Increased fludeoxyglucose uptake on positron emission tomography in the pouch (*blue arrow*), peripouch lymph nodes (*yellow arrow*), and liver (*red arrow*).

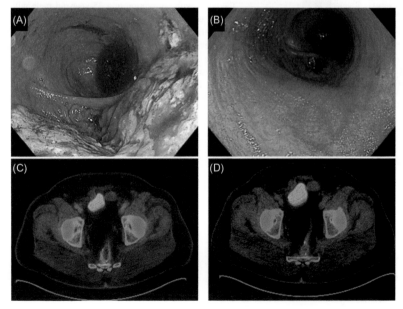

FIGURE 12.18 Diffuse large B-cell lymphoma of the ileal pouch in a 57-year-old male patient with restorative proctocolectomy for refractory ulcerative colitis. The patient presented with persistent anemia: (A and C) large nodular area almost covering half of the pouch body, which was highlighted with fludeoxyglucose uptake on positron emission tomography, and (B and D) resolution of lymphoma on endoscopy and positron emission tomography scan after effective chemotherapy.

Summary and recommendations

Chronic inflammation, immunosuppressive medications, and associated disease conditions (such as PSC and EBV virus or HPV infections) in IBD predispose the patients to the development of various forms of cancer, including adenocarcinoma, SCC, and lymphoma. The cancer risk may persist even after bowel resection or colectomy. Fortunately, colitis-associated with cancer is often preceded by LGD or HGD, which provides the rationale for routine surveillance colonoscopy for patients with long-standing UC or Crohn's colitis. However, the disease course of CAN does not necessarily follow the chronic inflammation—LGD—HGD—adenocarcinoma sequence; and some CAN lesions can be flat or slightly raised, posing challenge for the surveillance. In addition, there are no established protocols for endoscopic surveillance in small bowel adenocarcinoma, GI lymphoma, SCC in the anorectal area, and neoplasia in the ileal pouches.

References

[1] Eaden JA, Abrams KR, Mayberry JF. The risk of colorectal cancer in ulcerative colitis: a meta-analysis. Gut 2001;48:526—35.

[2] Molodecky NA, Kareemi H, Parab R, Barkema HW, Quan H, Myers RP, et al. Incidence of primary sclerosing cholangitis: a systematic review and meta-analysis. Hepatology 2011;53:1590—9.

[3] Soetikno RM, Lin OS, Heidenreich PA, Young HS, Blackstone MO. Increased risk of colorectal neoplasia in patients with primary sclerosing cholangitis and ulcerative colitis: a meta-analysis. Gastrointest Endosc 2002;56:48—54.

[4] Kornbluth A, Sachar DB. Ulcerative colitis practice guidelines in adults: American College Of Gastroenterology, Practice Parameters Committee. Am J Gastroenterol 2010;105:501—23 quiz 524.

[5] Farraye FA, Odze RD, Eaden J, Itzkowitz SH. AGA technical review on the diagnosis and management of colorectal neoplasia in inflammatory bowel disease. Gastroenterology 2010;138:746—74 774.e1-4; quiz e12-3.

[6] Leighton JA, Shen B, Baron TH, Adler DG, Davila R, Egan JV, et al. ASGE guideline: endoscopy in the diagnosis and treatment of inflammatory bowel disease. Gastrointest Endosc 2006;63:558—65.

[7] Cairns SR, Scholefield JH, Steele RJ, Dunlop MG, Thomas HJ, Evans GD, et al. Guidelines for colorectal cancer screening and surveillance in moderate and high risk groups (update from 2002). Gut 2010;59:666—89.

[8] Magro F, Gionchetti P, Eliakim R, Ardizzone S, Armuzzi A, Barreiro-de Acosta M, , et al.European Crohn's and Colitis Organisation [ECCO] Third European evidence-based consensus on diagnosis and management of ulcerative colitis. Part 1: Definitions, diagnosis, extra-intestinal manifestations, pregnancy, cancer surveillance, surgery, and ileo-anal pouch disorders. J Crohns Colitis 2017;11:649—70.

[9] Lamb CA, Kennedy NA, Raine T, Hendy PA, Smith PJ, Limdi JK, et al.; IBD Guidelines eDelphi Consensus Group. British Society of Gastroenterology consensus guidelines on the management of inflammatory bowel disease in adults. Gut 2019;68(Suppl. 3):s1—s106.

[10] Gordon IO, Agrawal N, Goldblum JR, Fiocchi C, Rieder F. Fibrosis in ulcerative colitis: mechanisms, features, and consequences of a neglected problem. Inflamm Bowel Dis 2014;20:2198—206.

[11] Kiran RP, Nisar PJ, Goldblum JR, Fazio VW, Remzi FH, Shen B, et al. Dysplasia associated with Crohn's colitis: segmental colectomy or more extended resection? Ann Surg 2012;256:221—6.

[12] Kiran RP, Khoury W, Church JM, Lavery IC, Fazio VW, Remzi FH. Colorectal cancer complicating inflammatory bowel disease: similarities and differences between Crohn's and ulcerative colitis based on three decades of experience. Ann Surg 2010;252:330—5.

[13] Lindström L, Lapidus A, Ost A, Bergquist A. Increased risk of colorectal cancer and dysplasia in patients with Crohn's colitis and primary sclerosing cholangitis. Dis Colon Rectum 2011;54:1392—7.

[14] Lewis B, Lin J, Wu X, Xie H, Shen B, Lai K, et al. Crohn's disease-like reaction predicts favorable prognosis in colitis-associated colorectal cancer. Inflamm Bowel Dis 2013;19:2190—8.

[15] Quinn AM, Farraye FA, Naini BV, Cerda S, Coukos J, Li Y, et al. Polypectomy is adequate treatment for adenoma-like dysplastic lesions (DALMs) in Crohn's disease. Inflamm Bowel Dis 2013;19:1186—93.

[16] Hirano Y, Futami K, Higashi D, Mikami K, Maekawa T. Anorectal cancer surveillance in Crohn's disease. J Anus Rectum Colon 2018;2:145—54.

[17] Ueda T, Inoue T, Nakamoto T, Nishigori N, Kuge H, Sasaki Y, et al. Anorectal cancer in Crohn's disease has a poor prognosis due to its advanced stage and aggressive histological features: a systematic literature review of Japanese patients. J Gastrointest Cancer 2018; Nov 26, [Epub ahead of print].

[18] Vanoli A, Di Sabatino A, Martino M, Klersy C, Grillo F, Mescoli C, et al. Small bowel carcinomas in celiac or Crohn's disease: distinctive histophenotypic, molecular and histogenetic patterns. Mod Pathol 2017;30:1453—66.

[19] Grolleau C, Pote NM, Guedj NS, Zappa M, Theou-Anton N, Bouhnik Y, et al. Small bowel adenocarcinoma complicating Crohn's disease: a single-centre experience emphasizing the importance of screening for dysplasia. Virchows Arch 2017;471:611—17.

[20] Svrcek M, Piton G, Cosnes J, Beaugerie L, Vermeire S, Geboes K, et al. Small bowel adenocarcinomas complicating Crohn's disease are associated with dysplasia: a pathological and molecular study. Inflamm Bowel Dis 2014;20:1584—92.

[21] Simon M, Cosnes J, Gornet JM, Seksik P, Stefanescu C, Blain A, , et al.GETAID Group Endoscopic detection of small bowel dysplasia and adenocarcinoma in Crohn's disease: A prospective cohort-study in high-risk patients. J Crohns Colitis 2017;11:47—52.

[22] Kariv R, Remzi FH, Lian L, Bennett AE, Kiran RP, Kariv Y, et al. Preoperative colorectal neoplasia increases risk for pouch neoplasia in patients with restorative proctocolectomy. Gastroenterology 2010;139:806—12.

[23] Derikx LAAP, Nissen LHC, Smits LJT, et al. Risk of neoplasia after colectomy in patients with inflammatory bowel disease: a systematic review and meta-analysis. Clin Gastroenterol Hepatol 2016;14:798—806.

[24] Derikx LA, Kievit W, Drenth JP, de Jong DJ, Ponsioen CY, Oldenburg B, et al. Dutch Initiative on Crohn and Colitis. Prior colorectal neoplasia is associated with increased risk of ileoanal pouch neoplasia in patients with inflammatory bowel disease. Gastroenterology 2014;146:119—28.

[25] Wu XR, Remzi FH, Liu XL, Lian L, Stocchi L, Ashburn J, et al. Disease course and management strategy of pouch neoplasia in patients with underlying inflammatory bowel diseases. Inflamm Bowel Dis 2014;20:2073—82.

[26] Bahadursingh AM, Longo WE. Malignant transformation of chronic perianal Crohn's fistula. Am J Surg 2005;189:61—2.

[27] Nishigami T, Kataoka TR, Ikeuchi H, Torii I, Sato A, Tomita N, et al. Adenocarcinomas associated with perianal fistulae in Crohn's disease have a rectal, not an anal, immunophenotype. Pathology 2011;43:36—9.

[28] Beaugerie L, Carrat F, Nahon S, Zeitoun JD, Sabaté JM, Peyrin-Biroulet L, , et al.Cancers et Surrisque Associé aux Maladies Inflammatoires Intestinales En France Study Group High risk of anal and rectal cancer in patients with anal and/or perianal Crohn's disease. Clin Gastroenterol Hepatol 2018;16:892—9.

[29] Shwaartz C, Munger JA, Deliz JR, Bornstein JE, Gorfine SR, Chessin DB, et al. Fistula-associated anorectal cancer in the setting of Crohn's disease. Dis Colon Rectum 2016;59:1168—73.

[30] Long MD, Martin CF, Pipkin CA, Herfarth HH, Sandler RS, Kappelman MD. Risk of melanoma and nonmelanoma skin cancer among patients with inflammatory bowel disease. Gastroenterology 2012;143:390—9.

[31] Holubar SD, Dozois EJ, Loftus Jr EV, Teh SH, Benavente LA, Harmsen WS, et al. Primary intestinal lymphoma in patients with inflammatory bowel disease: a descriptive series from the prebiologic therapy era. Inflamm Bowel Dis 2011;17:1557—63.

[32] Madanchi M, Zeitz J, Barthel C, Samaras P, Scharl S, Sulz MC, et al. Malignancies in patients with inflammatory bowel disease: a single-centre experience. Digestion 2016;94:1—8.

[33] Wang LH, Yang YJ, Cheng WC, Wang WM, Lin SH, Shieh CC. Higher risk for hematological malignancies in inflammatory bowel disease: a nationwide population-based study in Taiwan. Am J Gastroenterol 2016;111:1313—19.

[34] Smart CJ, Gibb A, Radford J. Burkitt's lymphoma of an ileal pouch following restorative proctocolectomy. Inflamm Bowel Dis 2012;18: E1596—7.

[35] Cosentino JS, Davies RJ, McLeod RS, Cohen Z. Lymphoma following ileal pouch anal anastomosis. Can J Surg 2009;52:E123—6.

[36] Schwartz LK, Kim MK, Coleman M, Lichtiger S, Chadburn A, Scherl E. Case report: lymphoma arising in an ileal pouch anal anastomosis after immunomodulatory therapy for inflammatory bowel disease. Clin Gastroenterol Hepatol 2006;4:1030—4.

Chapter 13

Diversion-associated bowel diseases

Bo Shen

Center for Inflammatory Bowel Diseases, Columbia University Irving Medical Center-New York Presbyterian Hospital, New York, NY, United States

Chapter Outline

Abbreviations

CD	Crohn's disease
CAN	colitis-associated neoplasia
DABD	diversion-associated bowel disease
IBD	inflammatory bowel disease
IPAA	ileal pouch—anal anastomosis
IRA	ileorectal anastomosis
PSC	primary sclerosing cholangitis
SCFAs	short-chain fatty acids
UC	ulcerative colitis

Introduction

Diversion colitis, proctitis, and pouchitis result from defunctionalized, surgically bypassed bowel segment. Diversion-associated bowel disease (DABD) occurs in fecal-stream bypassed segments of the bowel, following stoma surgery for congenital, inflammatory, or neoplastic disorders. The degree of inflammation of the bypassed segment of lower gastrointestinal track is largely determined by the extent of dependency of epithelia of the bowel on short-chain fatty acids (SCFAs), the fermented produces of gut luminal bacteria and dietary fibers. For example, the risk for diversion-associated mucosal injury of the ileum is different from that of the colon. Long-term fecal diversion often results in intrinsic luminal strictures in the bypassed bowel segment, with anastomosis, distal rectum, or anal canal carrying the highest risk. In addition, patients with a long history of inflammatory bowel disease (IBD) carry a risk for the development of colitis-associated neoplasia (CAN), and the diverted bowel is not an exception. The best treatment option for DABD is the closure of stoma and reestablishment of the natural route of defecation. However, closure of the stoma in patients is not possible or feasible, for personal and medical reasons. In addition, the excision of the diverted segment of bowel is often associated with complications, particularly persistent perineal sinus [1,2]. Therefore some patients may have to have long-term or permanent fecal diversion, which pose the risk for various forms of diversion-associated disorders (Table 13.1). Endoscopy plays a key role in the diagnosis, disease monitoring, surveillance, and treatment of diversion bowel disease.

Diversion-associated inflammatory disorders

DABD consists of diversion ileitis, diversion colitis, diversion proctitis, and diversion pouchitis, resulting from the lack of exposure to fecal stream by surgery, that is, the creation of temporary or permanent colostomy, ileostomy, or

Atlas of Endoscopy Imaging in Inflammatory Bowel Disease. DOI: https://doi.org/10.1016/B978-0-12-814811-2.00013-X
© 2020 Elsevier Inc. All rights reserved.

TABLE 13.1 Diversion-associated bowel diseases.

Category	Subcategory
Inflammatory disorders	Diversion ileitis
	Diversion colitis
	Diversion proctitis
	Diversion pouchitis
Stricturing disorders	Disease-associated or primary stricture
	Anastomotic or secondary stricture
Neoplasia	Colorectal neoplasia (dysplasia or cancer)
	Pouch neoplasia (dysplasia or cancer)

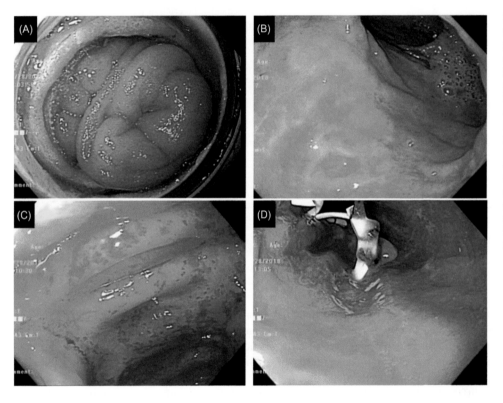

FIGURE 13.1 Different risks for diversion-associated injury between the ileum and colon. (A) Loop ileostomy constructed for the treatment of refractory perianal Crohn's disease; (B) intubation of the afferent limb of the loop ileostomy via stoma to the distal ileum showing mucus exudates; (C) subsequent passage of the endoscopy to the efferent limb showed the diverted colon with edematous and friable mucosa (diversion colitis); (D) setons seen in the rectum, which had been placed for the treatment of perianal disease.

jejunostomy (Table 13.1). Among them, the creation of a diverted rectum is also called Hartmann's procedure. It is estimated that 70%−100% of the patients developed either and/or endoscopic inflammation in the diverted bowel segment [3]. Diversion colitis or proctitis is more frequently seen in patients with IBD than those with colorectal cancer of diverticular diseases, suggesting the role of underlying disease in the development and process of DABD. In addition, it appears that the bypassed segment of the ileum carries a lower risk for diversion-associated mucosal injury than the bypassed segment of large bowel. For example, CD patients with diverting loop or end ileostomy and ileorectal anastomosis (IRA) following ileocolonic resection may have diversion proctitis in the absence of diversion ileitis (Figs. 13.1 and 13.2). Similar situation is also present in patients with underlying ulcerative colitis (UC) who had colectomy, loop ileostomy, and IRA (Fig. 13.3). The findings suggest that the epithelia of colon and rectum relies on SCFAs and other luminal nutrients (such as glutamine), than the epithelia of the ileum. Subsequently, the lack of those nutrients leads to alterations in both innate and adaptive mucosal immunity [4,5].

Common presenting symptoms of DABD are tenesmus, urgency, bloody or mucus discharge, and abdominal pain. Few patients with develop systemic symptoms such as fever, chills, and night sweats. This author has managed a patient with diversion

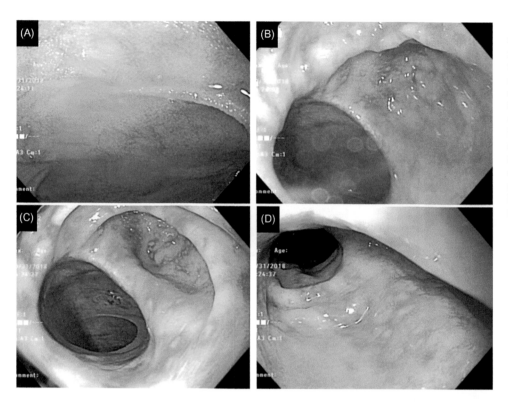

FIGURE 13.2 Diverted ileor-ectal anastomosis with loop ileostomy and ileocolonic resection for refractory Crohn's ileocolitis with relative rectal sparing. (A) Normal neoterminal ileum assessed with colonoscopy and (B—D) patent ileorectal anastomosis with patchy erythema and aphthous ulcers in the rectum, indicating diversion proctitis. Of note, the risk for diversion inflammation in the ileum is lower than that for the colon or rectum.

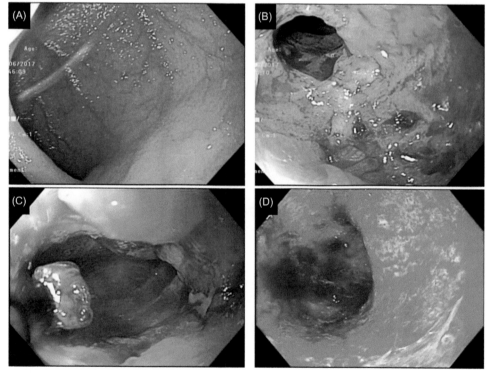

FIGURE 13.3 Diversion-associated colitis, but not ileitis, along with IRA stricture. Loop ileostomy with IRA following colectomy for refractory ulcerative colitis. (A) Normal diverted ileum proximal to IRA; (B and C) IRA stricture with endoscopic stricturotomy; and (D) diversion pouchitis with friable and edematous mucosa, in contrast to normal diverted ileum. *IRA*, Ileorectal anastomosis.

pouchitis and distal pouch stricture who develop liver abscess (Fig. 13.4). In addition, patients with severe diversion-associated bowel disorders may be prone to the development of bacteremia when endoscopic intervention is performed (Fig. 13.5). Patients with bowel obstruction may cause inflammation of downstream bowel, mimicking diversion colitis (Fig. 13.6).

The diagnosis of DABD is based on a combined assessment of surgical history, clinical presentations, and endoscopic and histologic features. The mucosa of diverted bowel is frequently friable, even with gentle air or carbon dioxide

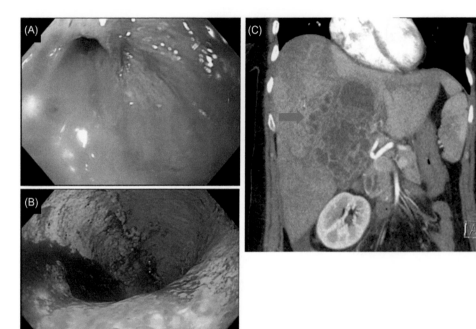

FIGURE 13.4 Diversion pouchitis and stricture resulting in liver abscess. Ileostomy was constructed for the treatment of Crohn's disease of the pouch. (A) Tight pouch—anal anastomosis, which was treated endoscopic stricturotomy; (B) severe diversion pouchitis with friable mucosa, spontaneous bleeding, and exudates; (C) multilobulated liver abscess resulting from severe diversion pouchitis and anastomotic stricture.

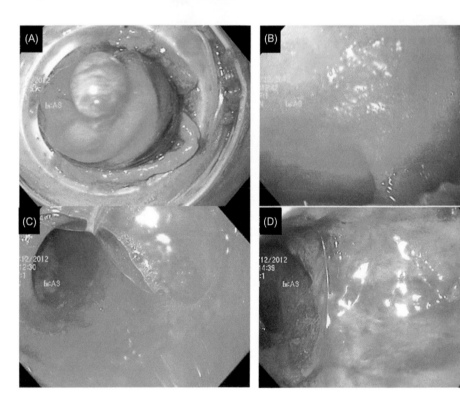

FIGURE 13.5 Diversion pouchitis, probiotic enema, and bacteremia. (A) End ileostomy created for the treatment of refractor pouchitis; (B and C) normal neo-distal ileum on ileoscopy via stoma; (D) diversion pouchitis with spontaneous bleeding which was biopsied. Probiotic enema was used to treat symptomatic diversion pouchitis, which later resulted in bacteremia. Prophylactic antibiotics may be used in patients with severe diversion pouchitis under-doing endoscopic intervention.

insufflation (Fig. 13.7). Biopsy-associated mucosal, even submucosal injury, is common (Fig. 13.7), which may be related to commonly present expanded lymphoid aggregates and inflammation in the lamina propria with lymphocytes and plasma cells on histology. Other endoscopic features include erythema, edema, granularity, "cat scratch" pattern with bright erythematous linear marks, the presence of inflammatory and filiform polyps, and pool of mucus or bloody mucus exudates.

Diversion pouchitis is a form of DABD. It is speculated that patients with long-term diverted ileal pouches have an intermediate risk for diversion-associated bowel inflammation, between diverted ileum and diverted colon. Restorative

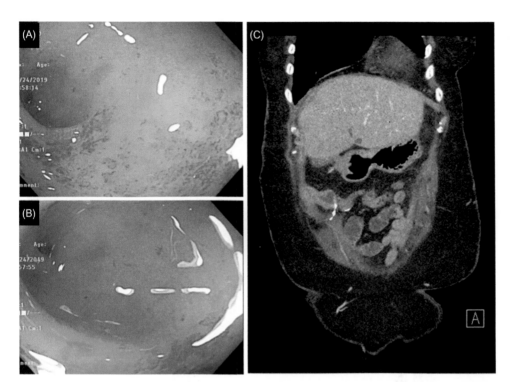

FIGURE 13.6 Severe anastomotic stricture leading to downstream colitis mimicking diversion colitis. (A and B) Severe ileocolonic anastomosis stricture with distal diversion colitis and (C) the anastomotic stricture on enterography.

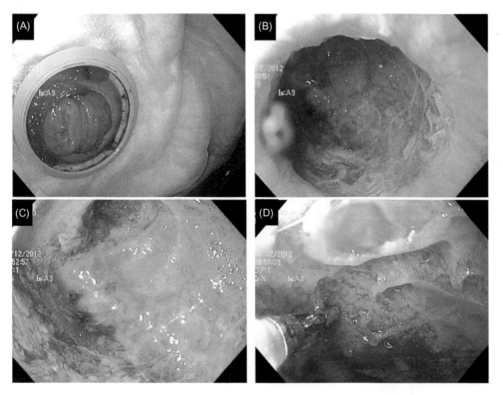

FIGURE 13.7 Diversion proctitis of the Hartmann pouch in staged restorative proctocolectomy. (A) End ileostomy; (B) inflammation in the distal rectum; (C) severe diversion proctitis with spontaneous bleeding; and (D) friable mucosa of the diverted rectum, with a high risk for bleeding and perforation on biopsy. Extreme caution should be taken when biopsy is taken.

proctocolectomy with ileal pouch−anal anastomosis (IPAA) has become the surgical treatment of choice for patients with UC or familial adenomatous polyposis who require colectomy. Alternatively, patients may undergo permanent Brooke end ileostomy with colectomy and Hartmann procedure. In patients with severe or fulminant UC, mucus fistula may be constructed during Hartmann procedure, to reduce the risk for intraabdominopelvic abscess (Fig. 13.8). Restorative proctocolectomy with IPAA procedures, however, are routinely performed in stages [6]. A temporary loop ileostomy is often constructed to allow for maturation of newly created ileal pouch. Patients with staged IPAA who are

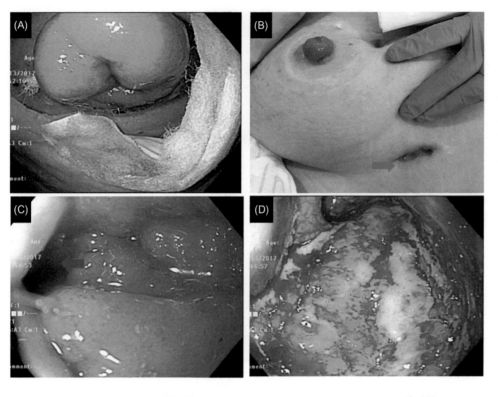

FIGURE 13.8 Ileostomy, mucus fistula, and Hartmann pouch or diverted rectum in staged proctocolectomy for ulcerative colitis. (A) End ileostomy; (B) a mucus fistula (*green arrow*) near the end ileostomy (*green arrow*); (C) the internal opening of the mucus fistula at the rectal stump (*blue arrow*); and (D) severe diversion proctitis with spontaneous bleeding and exudates.

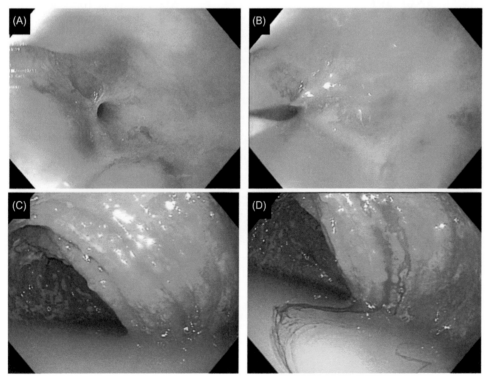

FIGURE 13.9 Diversion-associated inflammatory and stricturing pouch disorders. (A) A pinhole stricture at the ileal pouch—anastomosis, which was not traversable to a gastroscope; (B) a soft-tip guide wire was used to detect the lumen of the pouch body; and (C and D) passage of the endoscope, following endoscopic balloon dilation. Diffuse mucosal edema and copious exudates. Common presentations of patients with a fluid-loaded diverted bowel are urgency, odor, and pelvic pressure.

diverting loop ileostomy and diverted ileal pouch may develop mild diversion pouchitis. Diversion pouchitis from the temporary fecal diversion uniformly resolves after stoma closure and reestablishment of fecal stream. Temporary or permanent fecal diversion with the creation of an ileostomy has been used for the treatment of refractory pouch disorders, such as chronic pouchitis, CD of the pouch, and pouch vaginal fistula. Long-term fecal diversion, however, can result in various forms of pouchitis, with or without concurrent strictures (Fig. 13.9).

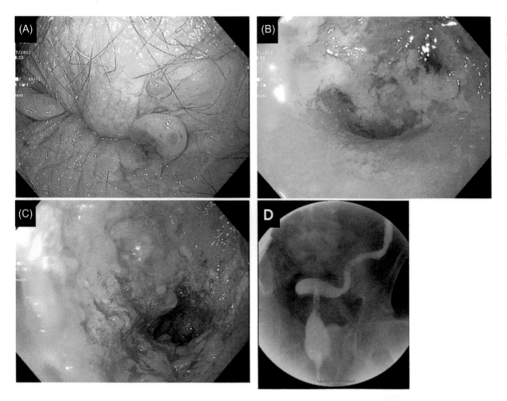

FIGURE 13.10 Long segment of diversion colitis with constricted bowel lumen in Crohn's disease. (A) Perianal disease, a major cause for the creation of an ileostomy; (B and C) diffuse severe diversion colitis and proctitis with granular and friable mucosa; and (D) diffuse luminal constriction of the diverted colon and rectum.

Endoscopic features of diversion pouchitis are similar to that in diversion colitis or diversion proctitis. In addition, diversion pouchitis is often coexistent with ileal pouch—anal anastomotic stricture.

Diversion-associated strictures

Due to long-term nonuse, diverted bowel carries a high risk for the development of strictures (Table 13.1). The strictures can be primary or disease associated versus secondary or anastomotic in nature. Although the stricture can occur in any segment of the diverted bowel, it is more often seen in distal rectum or anastomosis. The stricture can be short or long in length and/or can be fibrotic, inflammatory, or mixed type. Some patients may have completely sealed strictures (Figs. 13.9—13.11). Patients with strictures in the diverted bowel often present with symptoms of urgency, pelvic discomfort, continence, or seepage. The stricture frequently results in fecal stasis, exacerbating diversion-associated mucosal inflammation. The stricture can be treated with digital, bougie, or balloon dilation or endoscopic stricturotomy. Due to the presence of underlying etiological factors, the stricture often recurs, which requires periodic endoscopic therapy. Topical therapy with SCFA, mesalamine, and corticosteroids may also be beneficial.

Cancer surveillance in the diverted bowel

The incidence or prevalence of CAN in patients with a diverted bowel is not clear. However, the risk appears to be low in the absence of a history of colorectal cancer or IBD [7]. A study from The Netherlands comparing 12 IBD patients with cancer or high-grade dysplasia of the rectal stump and 18 IBD controls with rectal stump, but no neoplasia, showed that the presence of primary sclerosing cholangitis (PSC) or long duration of IBD before subtotal colectomy was risk factor for the cancer of the diverted large bowel [8]. Patients with restorative proctocolectomy and IPAA for CAN have been shown to have a higher risk for pouch neoplasia [9,10]. Therefore surveillance of the diverted colon, rectum, or ileal pouch is recommended for those with a history of colorectal cancer or underlying IBD or concurrent IBD and PSC (Table 13.1). The challenges of routine surveillance endoscopy in the diverted bowel include the presence of distal stricture and risk of bleeding or bacterial translocation of mucosal biopsy. In addition, the presence of diversion-associated inflammation may make an accurate diagnosis of dysplasia difficult, even with image-enhanced techniques, such as narrow banding imaging and chromoendoscopy (Fig. 13.12).

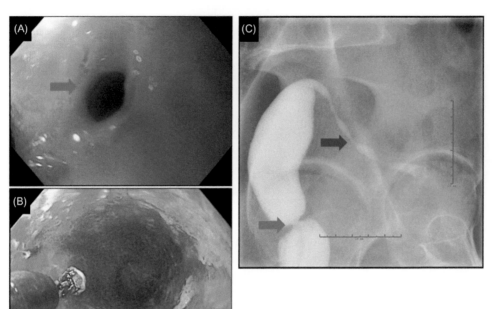

FIGURE 13.11 Diverted large bowel with primary and secondary strictures in Crohn's disease. (A) A tight anastomotic (or secondary) stricture (*green arrow*) at colorectal anastomosis; (B) endoscopic stricturotomy with insulated-tip knife for the treatment of the anastomotic stricture; and (C) long segment of diversion-associated primary stricture in a long segment of colon(*blue arrow*) in addition to the anastomotic stricture (*green arrow*), which is not accessible to endoscope.

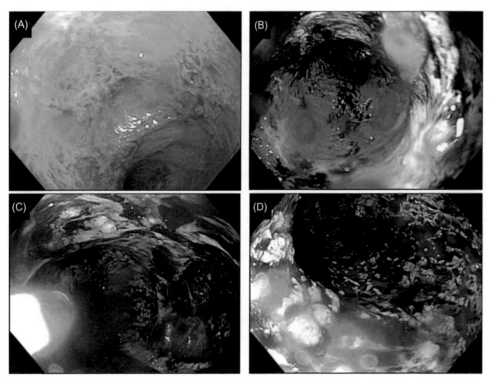

FIGURE 13.12 Challenges in surveillance endoscopy in the diverted rectum in ulcerative colitis. The patient had colectomy, permanent end ileostomy, and Hartmann pouch, for low-grade dysplasia of the sigmoid colon. (A) Diversion proctitis with friable mucosa and exudates, which prevent accurate identification of dysplasia; narrow band imaging (B). Chromoendoscopy (C and D) provided no additional benefits for the assessment of dysplastic lesions.

Summary and recommendations

Endoscopy plays a key role in the diagnosis, differential diagnosis, disease monitoring, and dysplasia surveillance in patients with diverted bowel. In addition, endoscopic balloon dilation and endoscopic stricture can safely performed in patients with diversion-associated strictures. However, diversion-associated bowel disorders are characterized with friable mucosa with a tendency for spontaneous bleeding. Patients with concurrent diversion-associated inflammatory and stricturing disorders may carry a risk for systemic septic complications, and endoscopic biopsy or intervention may further increase the risk. Therefore antibiotic prophylaxis is recommended for the patients undergoing endoscopic biopsy or endoscopic stricture therapy.

References

[1] Nisar PJ, Turina M, Lavery IC, Kiran RP. Perineal wound healing following ileoanal pouch excision. J Gastrointest Surg 2014;18:200−7.

[2] Shorthouse A. Perineal wound morbidity following proctectomy for inflammatory bowel disease (IBD). Colorectal Dis 2000;2:165−9.

[3] Shen B. Diversion colitis. Update®.

[4] Martinez CA, de Campos FG, de Carvalho VR, de Castro Ferreira C, Rodrigues MR, Sato DT, et al. Claudin-3 and occludin tissue content in the glands of colonic mucosa with and without a fecal stream. J Mol Histol 2015;46:183−94.

[5] Daferera N, Kumawat AK, Hultgren-Hörnquist E, Ignatova S, Ström M, Münch A. Fecal stream diversion and mucosal cytokine levels in collagenous colitis: a case report. World J Gastroenterol 2015;21:6065−71.

[6] Fazio VW, Kiran RP, Remzi FH, Coffey JC, Heneghan HM, Kirat HT, et al. Ileal pouch anal anastomosis: analysis of outcome and quality of life in 3707 patients. Ann Surg 2013;257:679−85.

[7] Winther KV, Bruun E, Federspiel B, Guldberg P, Binder V, Brynskov J. Screening for dysplasia and TP53 mutations in closed rectal stumps of patients with ulcerative colitis or Crohn disease. Scand J Gastroenterol 2004;39:232−7.

[8] Lutgens MW, van Oijen MG, Vleggaar FP, Siersema PD, Broekman MM, Oldenburg B, et al. Risk factors for rectal stump cancer in inflammatory bowel disease. Dis Colon Rectum 2012;55:191−6.

[9] Kariv R, Remzi FH, Lian L, Bennett AE, Kiran RP, Kariv Y, et al. Preoperative colorectal neoplasia increases risk for pouch neoplasia in patients with restorative proctocolectomy. Gastroenterology 2010;139:806−12.

[10] Derikx LA, Kievit W, Drenth JP, de Jong DJ, Ponsioen CY, Oldenburg B, et al. Prior colorectal neoplasia is associated with increased risk of ileoanal pouch neoplasia in patients with inflammatory bowel disease. Gastroenterology 2014;146:119−28.

Chapter 14

Endoscopic scores in inflammatory bowel disease

Jose Melendez-Rosado[1] and Bo Shen[2]

[1]*Department of Gastroenterology, Cleveland Clinic Florida, Weston, FL, United States,* [2]*Center for Inflammatory Bowel Diseases, Columbia University Irving Medical Center-New York Presbyterian Hospital, New York, NY, United States*

Chapter Outline

Abbreviations

CD Crohn's disease
CDAI the Crohn's Disease Activity Index
CDEIS the Crohn's Disease Endoscopic Index of Severity
CRP C-reactive protein
DAI the Diseases Activity Index
GI gastrointestinal
HBI the Harvey–Bradshaw Index
IBD inflammatory bowel disease
ICA ileocolonic anastomosis
ICR ileocolonic resection
IPAA ileal pouch-anal anastomosis
MES the Mayo Endoscopic Subscore
MH mucosal healing
MMES the modified Mayo Endoscopic score
MRE magnetic resonance enterography
PDAI the Pouchitis Disease Activity Index
RS the Rutgeerts Score
SES-CD the simple endoscopic score for Crohn's disease
STRIDE selecting therapeutic targets in inflammatory bowel disease
UC ulcerative colitis
UCEIS the Ulcerative Colitis Endoscopic Index of Severity
UCCIS the Ulcerative Colitis Colonoscopic Index of Severity

Atlas of Endoscopy Imaging in Inflammatory Bowel Disease. DOI: https://doi.org/10.1016/B978-0-12-814811-2.00014-1
© 2020 Elsevier Inc. All rights reserved.

Introduction

Endoscopy is a standard modality for diagnosis, differential diagnosis, assessment of disease activity, and response to treatment in Crohn's disease (CD) and ulcerative colitis (UC). Endoscopy provides the most reliable measurement of mucosal inflammation. The goal of management of inflammatory bowel disease (IBD) is constantly evolving. Earlier clinical trials have used clinical parameters to measure disease activity before and after therapeutic intervention. Recent studies have shown that endoscopic healing documented by high-definition colonoscopy is associated with less complication and better long-term outcomes [1]. The selecting therapeutic targets in IBD (STRIDE) initiated by the he International Organization for the Study of IBD proposed target goals for both CD and UC. The treat-to-target concept is being accepted in clinical practice. The targets are clinical, biological, and histologic. Endoscopy targets, that is, endoscopic mucosal healing (MH), are established with objective documentation of response to medical or surgical treatment. Endoscopic MH were defined as the absence of ulceration for CD or Mayo Endoscopic Subscore (MES) of 0 or 1 for UC [2]. Multiple endoscopic scores have been developed for CD, UC, and pouchitis, mainly based on features of white-light endoscopy. In this chapter, we present some of the most commonly used endoscopic scores with a wide range of illustrative images.

Components of endoscopic assessment of mucosal inflammation

The main components of endoscopic features shared by CD and UC are erythema, edema, loss of vascular pattern, granularity, nodularity, friability, spontaneous bleeding, erosions, ulcers, and pseudopolyps (Fig. 14.1). These inflammation conditions are often accompanied by the presence of mucopurulent exudates on the top of the erosion or ulcers or nonulcerated inflamed surface (Fig. 14.2).

There are endoscopic modifiers for CD, including cobblestoning and strictures. These modifiers can make an accurate measurement of disease activity complex. Cobblestoning is defined as serpiginous and linear ulcers intervened with nodular mucosa, which is commonly seen in the distal lymphoid tissue-predominant terminal ileum. This pattern of disease process gives rise to deep linear ulcer "cracks" in between the areas of inflamed or normal tissue with an appearance of "stones." The width of the ulcer cracks and the size of nodular mucosa vary (Fig. 14.3).

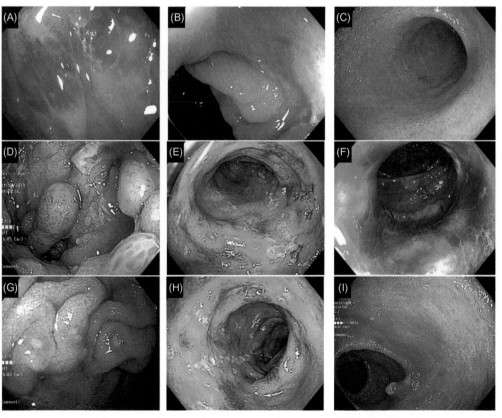

FIGURE 14.1 Common patterns of inflammation of Crohn's disease and ulcerative colitis: (A) erythema, (B) edema, (C) loss of vascular pattern, (D) nodularity, (E) friability, (F) spontaneous bleeding, (G) erosions, (H) ulcers, and (I) pseudopolyps.

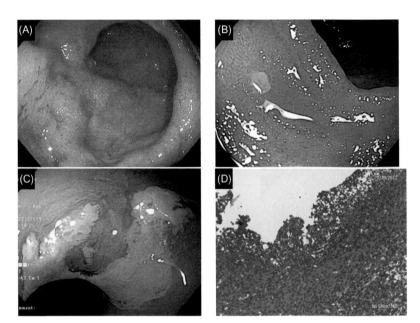

FIGURE 14.2 Additional patterns of mucosal inflammation in Crohn's disease: (A) exudates in the absence of mucosal inflammation and (B–D) exudates cover ulcers and exudates with histology.

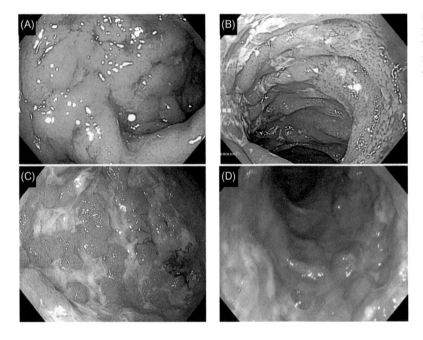

FIGURE 14.3 Spectrum of cobblestoning and nodular mucosa in Crohn's disease: (A and B) classic cobblestoning with networks of thin ulcers and nodular mucosa and (C and D) nodular mucosa with networks of deep and wider ulcers in between.

Another endoscopic modifier is the presence of stricture. Strictures can be observed in both CD and UC. Endoscopic appearance of strictures can be short or long, mild or severe, inflammatory or fibrotic, and ulcerated or nonulcerated. Nonulcerated strictures can be a part of mucosal or transmural healing or serosa- or mesentery-predominant transmural disease processwith normal overlying mucosa (Figs. 14.4 and 14.5C and D). On the other hand, the presence of ulcers on the anastomosis does not necessarily indicate active CD or UC. Rather they may result from surgery-associated ischemia or the use of nonsteroidal antiinflammatory drugs (Fig. 14.6).

Crohn's disease

The disease activity of CD has been measured with clinical, endoscopic, laboratory, and histologic features. Clinical and endoscopic disease activity instruments are routinely used in clinical trials and are increasingly used in clinical practices.

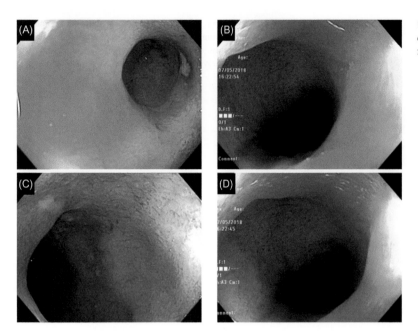

FIGURE 14.4 Nonulcerated strictures in Crohn' disease resulting from disease process per se or effective medical therapy with tissue healing (A—D).

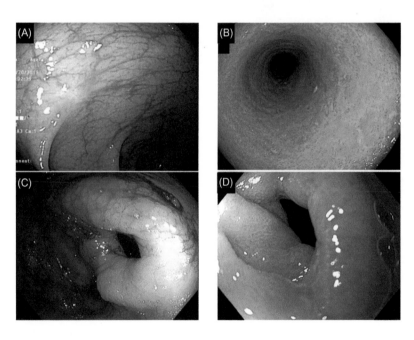

FIGURE 14.5 Forms of endoscopic mucosal healing: (A) mucosal scars, (B) "lead piping" of the colon, (C) nonulcerated stricture at the ileocecal valve, and (D) nonulcerated stricture at the duodenum.

Instruments for clinical disease activity

The 1100-point Crohn's Disease Activity Index (CDAI) was proposed and has been routinely used in clinical trials [3]. The items listed in the CDAI consist of the average number of liquid or soft stools per day over 7 days—using diphenoxylate or loperamide for diarrhea, average abdominal pain rating over 7 days; general well-being each day over 7 days; arthritis or arthralgia, iritis or uveitis, erythema nodosum, pyoderma gangrenosum, or aphthous stomatitis, anal fissure, fistula, or abscess, other fistula, temperature more than 100°F (37.8°C) in the last week; and finding of abdominal mass. CDAI has been used for quantification of clinical disease activity [4]: asymptomatic remission (CDAI 0—149), mild disease (CDAI 150—220), moderate-to-severe disease (CDAI 221—450), and severely active to fulminant disease (CDAI 451—1100). Patients requiring corticosteroids to remain asymptomatic should not be considered as being in remission. Rather, they are referred as being "steroid-dependent."

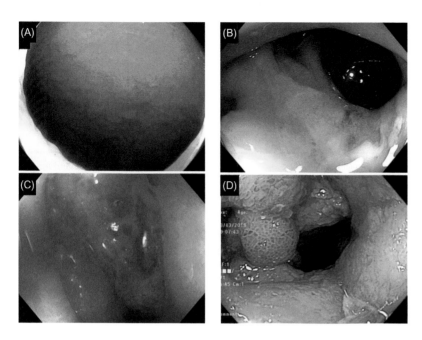

FIGURE 14.6 Inflammatory versus ischemic strictures in Crohn's disease patients with ileocolonic resection and anastomosis: (A) ulcerated stricture can be due to ischemia with a normal proximal segment of bowel, (B and C) inflammatory ulcerated strictures, and (D) inflammatory nodular stricture.

The Harvey—Bradshaw Index (HBI) of CD activity is a simplified clinical instrument [5]. Components of HBI consist of patient sense of general well-being (0—4) and abdominal pain (0—3) in the last 24 hours; number of liquid stools in last 24 hours, abdominal mass (0—4); and complications with arthralgias, uveitis, erythema nodosum, aphthous ulcers, pyoderma gangrenosum, anal fissures, newly discovered fistulae, or abscesses (0—8). An HBI score of <5 correlates with clinical remission.

The Crohn's Disease Endoscopic Index of Severity

Recently, the use of endoscopic instruments as outcome measures for clinical research has been more emphasized, as the endoscopic assessment is more objective than symptom-based indices. The most commonly used and validated endoscopic indices are the 44-point Crohn's Disease Endoscopic Index of Severity (CDEIS) [6], the Simple Endoscopic Score for Crohn's Disease (SES-CD) [7], and the Rutgeerts Score (RS) for postoperative recurrence in the neo-terminal ileum after ileocolonic resection and anastomosis [8]. These indices have been mainly applied for clinical trials. However, components of the indices are used for the diagnosis, differential diagnosis, disease monitoring, and assessment of response to therapy in clinical practice.

The first described CD endoscopic score is the 44-point CDEIS developed by Mary and Modigliani in 1989. Despite its correlation with clinical indices [9], this score involves a series of labor-intensive calculations. To calculate CDEIS, five segments of the intestine are assessed: the rectum, sigmoid and left colon, traverse colon, right colon, and ileum (Table 14.1). The definition of severity is based on the size and depth of ulceration, that is, superficial ulceration (Fig. 14.7) or deep ulceration (Figs. 14.8 and 14.9; Table 14.2). The aphthoid lesion, though mentioned in the instrument and evaluated during its initial development, is not a component for the final calculation of CDEIS (Fig. 14.10). In each colon segment CDEIS is calculated based on the presence of deep or superficial ulceration, the percentage of ulcerated surface; and the percentage of surface involved by CD are measured on a 10-cm visual analog scale (Table 14.1). In addition, the presence of ulcerated and nonulcerated stenoses is assessed.

The calculation of CDEIS is illustrated in Fig. 14.11A and B, corresponding to Tables 14.1 and 14.2.

The Simple Endoscopic Score for Crohn's Disease

The 56-point SES-CD (Table 14.3), was developed as a more user-friendly score that correlated with clinical parameters, and C-reactive protein (CRP) as well as CDEIS [7]. The score takes four segments of the colon along with the terminal ileum into consideration. Endoscopic variables in SES-CD measure the size of ulcer, proportion of ulcerated surface area, proportion of affected surface area, and the presence of stenosis with the range of score

TABLE 14.1 Crohn's Disease Endoscopic Index of Severity (CDEIS) [3].

	Rectum	Sigmoid and left colon	Transverse colon	Right colon	Ileum	Total
Deep ulceration quote 12 if present in the segment, 0 if absent						Total 1
Superficial ulceration quote 6 if present in the segment, 0 if absent						Total 2
Surface involved by the disease measured in cm[a]						Total 3
Ulcerated surface measured in cm[a]						Total 4

Total 1 + Total 2 + Total 3 + Total 4 = Total A. Number (n) of segments totally or partially explored (1−5) = n. Total A divided by n = Total B. Quote 3 if ulcerated stenosis anywhere, 0 if not = C. Quote 3 if nonulcerated stenosis anywhere, o if not = D. Total B + C + D = CDEIS.
[a]For partially explored segments and for the ileum, the 10 cm linear scale represents the surface effectively explored.

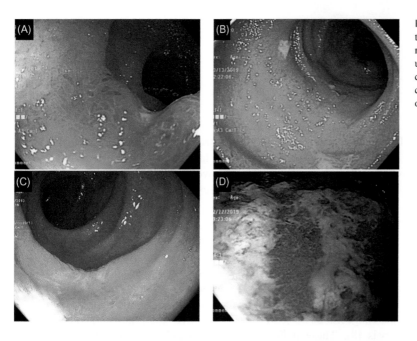

FIGURE 14.7 Superficial ulcers as a component in the CDEIS-"superficial ulceration". The ulcers are neither aphthoid nor deep: (A−C) small superficial ulcers in the terminal ileum and (D) diffuse superficial ulcers covered with mucopurulent exudates in the colon. *CDEIS*, the Crohn's Disease Endoscopic Index of Severity.

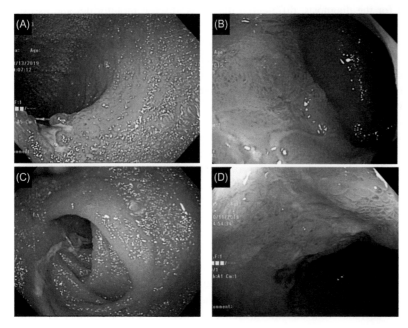

FIGURE 14.8 Deep ulceration as a component of the CDEIS: (A−D) deep ulcers in the terminal ileum. *CDEIS*, the Crohn's Disease Endoscopic Index of Severity.

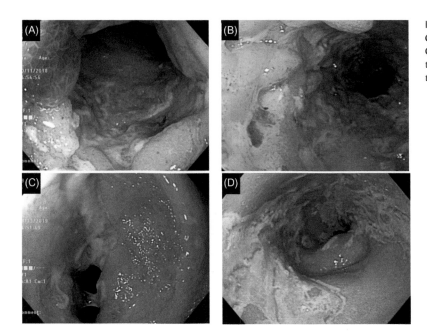

FIGURE 14.9 Deep ulceration as a component of CDEIS: (A) large ulcers in the terminal ileum, (B and C) large ulcers in the colon, and (D) ulcers in the rectum in a patient with colorectal anastomosis. *CDEIS*, the Crohn's Disease Dndoscopic Index of Severity.

TABLE 14.2 Further definitions of Crohn's Disease Endoscopic Index of Severity (CDEIS)[3].

Lesions	Definitions
Aphthoid ulceration	Tiny (2−3 mm), raised or flat red lesion with a white center
Superficial ulceration	Ulceration which was neither aphthoid nor deep
Deep ulceration	Only frankly deep ulceration

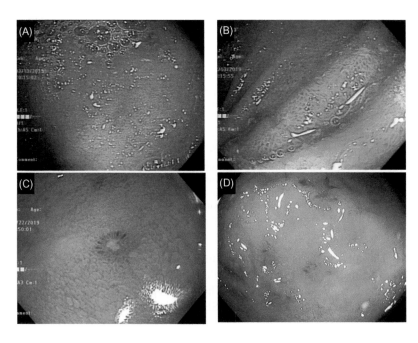

FIGURE 14.10 Various forms of aphthous ulceration in Crohn's disease. Of note, the presence of aphthous ulcers is not the component of CDEIS but is a component for SES-CD. Aphthous ulceration was defined as tiny (2−3 mm in CDEIS or 0.1−0.5 mm in SES-CD), raised or flat red lesion with a white center: (A−C) aphthous ulcers or erosions in the terminal ileum and (D) aphthous ulcers or erosions in the colon with edematous adjacent mucosa. *CDEIS*, the Crohn's Disease Endoscopic Index of Severity; *SES-CD*, the SEndoscopy Score of Crohn's Disease.

(A)

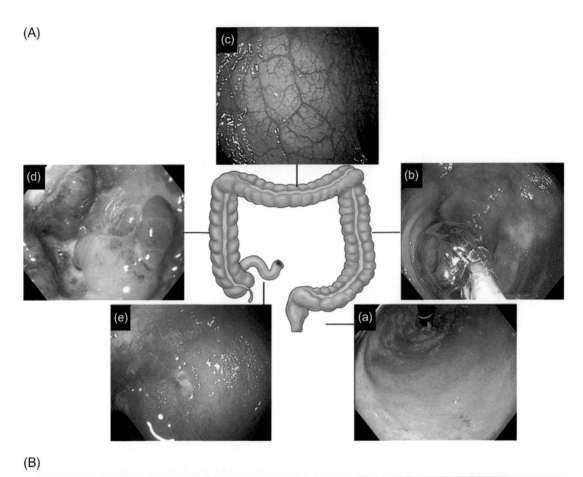

(B)

	Rectum	Sigmoid and left colon	Transverse colon	Right colon	Ileum	Total
Deep ulceration quote 12 if present in the segment, 0 if absent	12	12	0	12	0	Total 1=36
Superficial ulceration quote 6 if present in the segment, 0 if absent	0	0	0	0	6	Total 2=6
Surface involved by the disease measured in cm*	5	10	0	10	5	Total 3=30
Ulcerated surface measured in cm*	5	6	0	8	5	Total 4=24

Total A=Total 1 + Total 2 + Total 3 + Total 4 = 96
Number (*n*) of segments explored = 5
Total A divided by 5 (*n*) = Total B = 19.2
Quote 3 for ulcerated stenosis in left colon and right colon = C = 3
Quote 0 for absence of nonulcerated stenosis anywhere = D = 0
Total B + C + D = CDEIS = **22.2**

FIGURE 14.11 (A) Illustration of the calculation of CDEIS in a patient with inflammatory and fibrostenotic ileocolonic CD: (a) distal 5-cm rectum with deep ulcers, (b) a deeply ulcerated 6-cm-long stricture at the sigmoid colon with a 10-cm-long disease segment, (c) normal transverse, (d) an 8-cm-long deeply ulcerated area with stricture at the ascending colon with additional 2-cm-long inflamed area in the surrounding mucosa, and (e) a 5-cm-long area with superficial ulcers in the terminal ileum. (B) Calculation of CDEIS in the patient with inflammatory and fibrostenotic ileocolonic CD. *CD*, Crohn's disease; *CDEIS*, the Crohn's Disease Endoscopic Index of Severity.

from 0 to 3 (Table 14.4; Figs. 14.12−14.16A). The total score is calculated with the sum of each endoscopic characteristic in each segment of the colon plus the ileum. In contrast to CDEIS the presence of aphthous ulcers is a component of SES-CD, as being 1 in the 0−3 scale for ulcers (Fig. 14.11B). The term "aphthous ulcers" here may be better defined as aphthous erosions or aphthous lesions. The calculation of SES-CD is illustrated in Fig. 14.16A and B.

TABLE 14.3 The Simple Endoscopic Score for Crohn's Disease (SES-CD)* [7].

	Ileum	Right colon	Transverse colon	Left colon	Rectum	Total
Presence and size of ulcers (0–3)						
Extent of ulcerated surface (0–3)						
Extent of affected surface (0–3)						
Presence and type of narrowing (0–3)						
SES-CD						

*Ileum—full extent to which it is examined excluding ileocecal valve; right colon— ileocecal valve, ascending, and hepatic flexure; transverse colon— segment between the hepatic flexure to splenic flexure; left colon—descending and sigmoid colon up to rectosigmoid junction; and rectum—the portion distal to the rectosigmoid junction.

TABLE 14.4 Further definitions of the Simple Endoscopic Score for Crohn's Disease (SES-CD) [7].

Variable	0	1	2	3
Size of ulcer	None	Aphthous ulcers (ø 0.1–0.5 cm)	Large ulcers (ø 0.5–2 cm)	Very large ulcers (ø >2 cm)
Ulcerated surface	None	<10%	10%–30%	>30%
Affected surface	Unaffected segment	<50%	50%–75%	>75%
Presence of narrowing	None	Single, can be passed	Multiple, can be passed	Cannot be passed

ø = diameter.

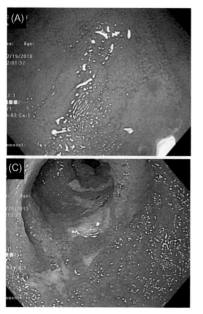
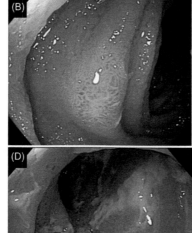

FIGURE 14.12 The size of ulcer as a component of SES-CD: (A) none (score = 0), (B) aphthous ulcers, 0.1–0.5 cm (score = 1), (C) large ulcers, 0.5–2 cm (score = 2), (D) very large ulcers <2 cm (score = 4). *SES-CD*, the Simple Endoscopic Score for Crohn's Disease.

The Rutgeerts Score in ileocolonic resection and ileocolonic anastomosis

The 4-point RS has been routinely used for the assessment of disease recurrence of the neo-terminal ileum in CD patients with ileocolonic resection (ICR) and ileocolonic anastomosis (ICA) in clinical practice as well as clinical trials.

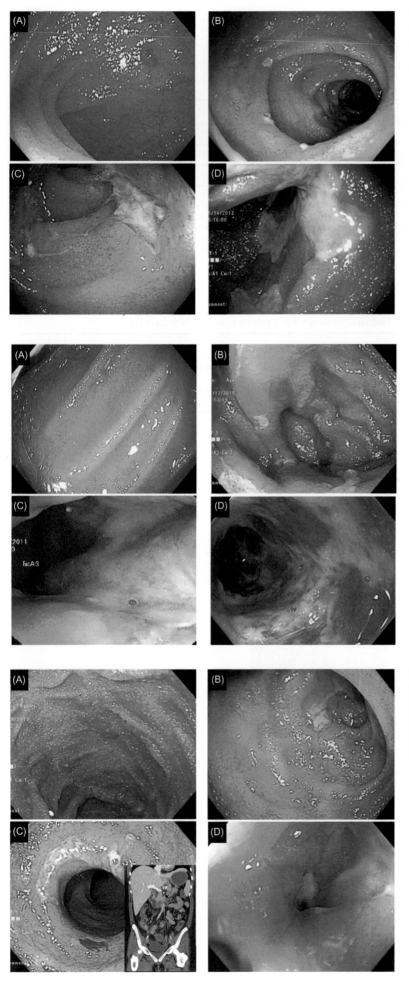

FIGURE 14.13 The ulcerated surface as a component of SES-CD: (A) none (score = 0), (B) <10% (score = 1), (C) 10%−30% (score = 2), and (D) >30% (score = 4). *SES-CD*, the Simple Endoscopic Score for Crohn's Disease.

FIGURE 14.14 The affected surface as a component of SES-CD: (A) unaffected segment (score = 0), (B) affected <50% (score = 1), (C) affected 50%−75% (score = 2), and (D) affected >75% (score = 4). *SES-CD*, the Simple Endoscopic Score for Crohn's Disease.

FIGURE 14.15 The presence of narrowing as a component of SES-CD: (A) none (score = 0); (B) single, can be passed (score = 1); (C) multiple, can be passed by an endoscope (scores = 2). Computed tomography highlights the strictures; and (D) cannot be passed (score = 3). *SES-CD*, the Simple Endoscopic Score for Crohn's Disease.

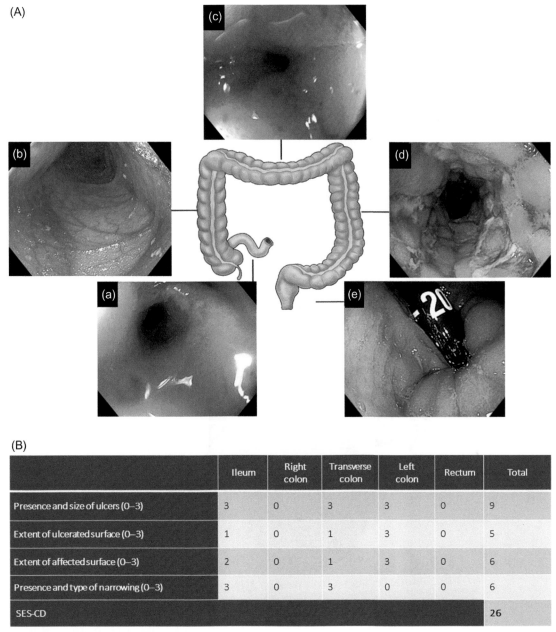

	Ileum	Right colon	Transverse colon	Left colon	Rectum	Total
Presence and size of ulcers (0–3)	3	0	3	3	0	9
Extent of ulcerated surface (0–3)	1	0	1	3	0	5
Extent of affected surface (0–3)	2	0	1	3	0	6
Presence and type of narrowing (0–3)	3	0	3	0	0	6
SES-CD						26

FIGURE 14.16   (A) Illustration of calculation of SES-CD in a patient with inflammatory and fibrostenotic ileocolonic CD: (a) a 2-cm-long ulcerated nontraversable stricture at the terminal ileum with a 5-cm segment of nonulcerated inflammation; (b) normal ascending colon; (c) a 3-cm long ulcerated nontraversable stricture at the transverse colon with 10-cm-long segment of disease; (d) 30-cm-long segment ulcerations at the left colon; (e) normal rectum. Of note, both strictures at the ascending colon and terminal ileum were treated with endoscopic balloon dilation, allowing for passage of the endoscope and assessment of the proximal bowel. (B) calculation of SES-CD in the patient with inflammatory and fibrostenotic ileocolonic CD. *CD*, Crohn's disease; *SES-CD*, the Simple Endoscopic Score for Crohn's Disease.

It is estimated that half of patients with CD will experience one surgery during their lifetime [10]. Surgery may be curative for some individuals; however, postoperative endoscopic recurrence occurs in 70%−90% of patients [8]. The RS was developed as a marker of clinical relapse after surgical resection. The score ranges from i,0 to i,4 and is determined by the number of ulcers in the neo-terminal ileum. Patients with ≥i,2 are at an increased risk of clinical recurrence at 4 years compared to those with <i,2. The description of the RS is detailed in Chapter 8, Crohn's disease: postsurgical, with Table 8.2 and Fig. 8.8 [8]. Additional figures are added in this chapter to further illustrate the endoscopic features of postoperative recurrence of CD in the neo-terminal ileum (Figs. 14.17−14.20).

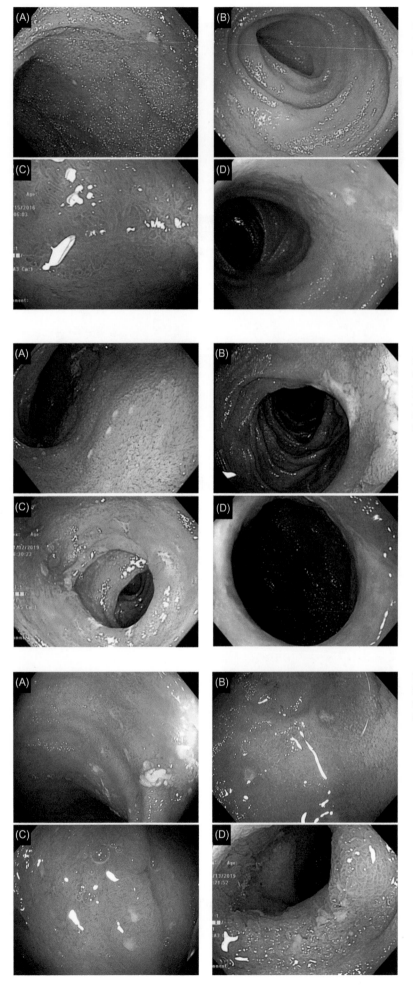

FIGURE 14.17 Rutgeerts Score i,1 lesions after ileocolonic resection and ileocolonic anastomosis for Crohn's disease: (A–D) discrete aphthous ulcers ≤ 5 in the neo-terminal ileum.

FIGURE 14.18 Rutgeerts Score i,2 lesions after ileocolonic resection and ileocolonic anastomosis for Crohn's disease: (A) >5 aphthous ulcers with normal intervening mucosa, (B and C) skip areas of larger lesions, (D) lesions (ulcers) confined to ileocolonic anastomosis. This can be labeled as an i,2a lesion in the modified Rutgeerts Score.

FIGURE 14.19 Rutgeerts Score i,3 lesions after ileocolonic resection and ileocolonic anastomosis for Crohn's disease: (A–D) diffuse aphthous ileitis with diffusely inflamed mucosa.

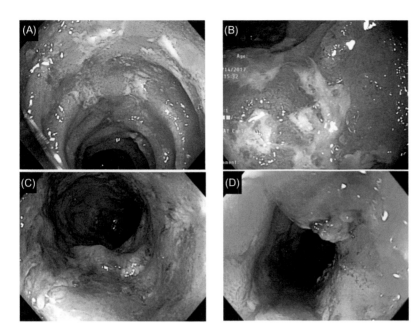

FIGURE 14.20 Rutgeerts Score i,4 after ileocolonic resection and ileocolonic anastomosis for Crohn's disease: (A and B) diffuse inflammation with larger ulcers, (C) diffuse inflammation with ulcers and nodularity, and (D) diffuse inflammation with strictures.

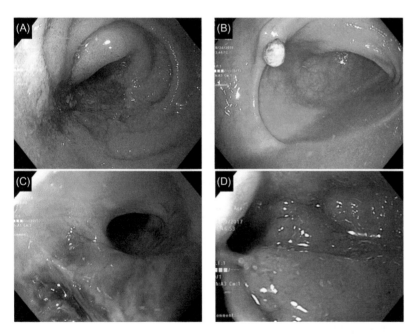

FIGURE 14.21 Other forms of bowel resection and anastomosis in Crohn's disease: (A) an asymmetric ulcerated stricture at the neo-terminal ileum in a patient with ileal resection and ileoileal anastomosis, (B) colo-colonic anastomosis with no signs of disease recurrence in a separate patient, (C) strictured ileorectal anastomosis with normal neo-terminal ileum viewed from distance , and (D) ileorectal anastomosis stricture in a patient with a diverting ileostomy.

Whether ulcer on the surface of ICA is considered as a part of the recurrent CD is controversial, as the surgery-associated ischemia can also cause ulcers. Nonetheless, ulcers on the surface of ICA should be separately labeled, which led to the development of the modified RS for i,2 lesions. The i,2a lesions are defined as the ulcer confined to the anastomosis with or without <5 isolated aphthous ulcers in the ileum, and i,2b lesions are defined as >5 aphthous ulcers in the ileum with normal mucosa in between, with or without anastomotic lesions. A recent retrospective study of 207 patients undergoing ICR and ICA showed that i,2a lesions were not associated with a higher rate of 1-year progression to more severe disease, while those with i,2b were [11]. This finding is consistent with the authors' clinical practice. Detailed information on the modified RS is discussed in Chapter 8, Crohn's disease: postsurgical, with Table 8.3 and Fig. 8.9 [12,13]. An additional figure is added to illustrate the concept of ulcers confined to ICA (Fig. 14.18D).

Variants of ICR and ICA for the treatment of CD include ileoileal anastomosis, ileorectal anastomosis, colo-colonic anastomosis, and colorectal anastomosis (Fig. 14.21). The role of RS in the diagnosis, differential diagnosis, and disease monitoring in these settings remains to be investigated.

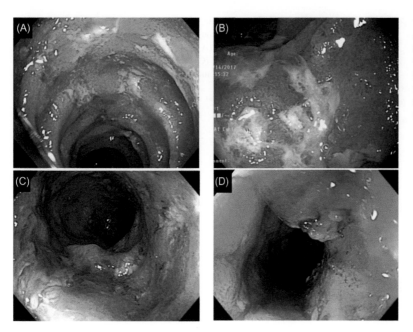

FIGURE 14.22 Representative endoscopic features of the Rutgeerts Score in the assessment of postoperative Crohn's disease in the neo-distal ileum in patients with ileostomies: (A) i,0-normal neo-distal ileum, (B) i,2 lesions (>5 aphthous lesions with normal mucosa), (C) diffuse aphthous ileitis with inflamed mucosa, and (D) diffuse inflammation with large ulcers and narrowed lumen.

TABLE 14.5 Severity scale for small bowel lesions according to the lesions or history of surgery or any other interventional procedure (the Lémann's Score) [15].

Grade	Stricturing lesions	Penetrating lesions	History of surgery or any other interventional procedure
0	Normal	Normal	None
1	Wall thickening <3 mm and/or segmental enhancement without prestenotic dilatation	NA	NA
2	Wall thickening ≥3 mm and/or mural stratification without prestenotic dilatation	Deep transmural ulceration	Bypass diversion or stricturoplasty
3	Stricture with prestenotic dilatation	Abscess or any type of fistula	Resection

The Rutgeerts Score in patients with ileostomy

There are limited data on the use of RS in CD with an ileostomy. This author's group demonstrated that RS is valuable the assessmentof recurrentCD in the neo-distal ileum in patients with ileostomies, as outlined in Chapter 8, Crohn's disease: postsurgical (Fig. 8.17) [14]. Additional figures are added to further illustrate the role of RS in the evaluation of the neo-distal ileum in patients with ileostomies (Fig. 14.22).

Endoscopy in the assessment of the Digestive Damage Score

The CDEIS and SES-CD only assess the disease activity of mucosal surface with or without luminal narrowing. The transmural or penetrating nature of the disease process requires a combined assessment of endoscopy, cross-sectional imaging, and histopathology. The "Lémann Score (LS)" also named as CD Digestive Damage Score was then developed. The LS evaluates CD at different gastrointestinal (GI) locations, strictures and penetrating disease, and the presence of surgical resection or bypass of the bowel, with upper GI endoscopy, colonoscopy, computed tomography enterography, or magnetic resonance enterography (MRE) [15]. The 10-point LS consists of three components, stricturing lesions (scores: 0−3), penetrating lesions (scores: 0−3), and history of surgery or any other interventional procedure (scores: 0−3) (Tables 14.5 and 14.6). The role of endoscopy in the calculation of the entire LS relies on the assessment of luminal narrowing with or without prestenotic dilation and the presence of surgical anastomosis or strictureplasty.

TABLE 14.6 Examinations required for inclusion in the study aimed to develop the Lémann score, according to Crohn's disease (CD) location [15].

CD location	Upper endoscopy	Colonoscopy	Abdominal MRI enterography	Pelvic MRI	Abdominopelvic CT enterography
Upper digestive tract	X		X		X
Small bowel			X		X
Colon and/or rectum		X	X		X
Perianal and anal			X	X	X

CT, Computed tomography; *MRI,* magnetic resonance imaging.

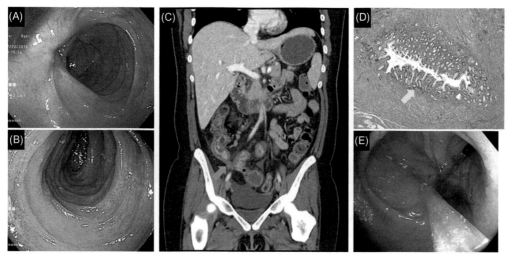

FIGURE 14.23   Calculation of the 10-point Lémann Score in a combined assessment of endoscopy, imaging, histopathology, and surgical history for Crohn's disease: (A and B) strictured ileocecal valve with prestenotic dilation of the lumen of the terminal lumen (score = 3), (C and D) penetrating disease with an ileo-sigmoid fistula on computed tomography (*green arrow*) and postoperative histopathology with epithelialized fistula tract (*yellow arrow*) (score = 3), (E) presence of surgical history with bowel resection (D) and strictureplasty (E) (score = 3). The strictured outlet of the strictureplasty was dilated with an endoscopic balloon (score = 3). *Histology photo: Courtesy of Dr. Shuyuan Xiao of University of Chicago Medical Center.*

Cross-sectional imaging can assess stricturing or penetrating disease and postsurgical anatomy. Post-surgical histopathology is used to evaluate transmural inflammation (Fig. 14.23).

Ulcerative colitis

Various clinical, endoscopic, or composite instruments have been developed for the assessment of disease activity, diagnosis, and monitoring of UC. The instruments integrated with endoscopic evaluation include the 3-point Truelove and Witts Score (for hyperemia and granularity) [16], the 3-point Baron Score (for the severity of mucosal bleeding and friability) [17], the 2-point Powell-Tuck Sigmoidoscopic Score (for mucosal bleeding) [18], the 2-point Sutherland Score (for severity of mucosal bleeding and friability) [19], and the 12-point Rachmilewitz Score (for vascular pattern, mucosal granularity, bleeding, mucus, fibrin, erosions, and ulcers) [20]. However, those scores are largely replaced by the intruments listed next.

The Mayo Endoscopy Subscore

The 3-point MES is one of the most commonly used endoscopic scores in clinical trials and practice given its simplicity and correlation with clinical outcomes (Table 14.7) [21,22]. MES is a part of a composite instrument, the Mayo Score,

TABLE 14.7 The Mayo Endoscopy Subscore (MES) [21].

Score	Descriptions
0	No friability or granularity, intact vascular pattern
1	Mild—erythema, diminished or absent vascular markings, mild granularity
2	Moderate—marked erythema, absent vascular marking, granularity, friability, no ulceration
3	Severe—marked erythema, absent vascular markings, granularity, friability, spontaneous bleeding in the lumen, ulcerations

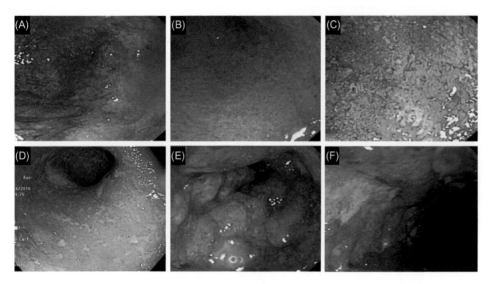

FIGURE 14.24 Representative features of components in MES in ulcerative colitis: (A) erythema, (B) loss of vascular pattern, (C) friability, (D) erosions (covered with exudates), (E) bleeding, and (F) ulcers. *MES*, the Mayo Endoscopic Score.

also known as the 12-point Disease Activity Index (DAI), which was developed for a clinical trial [21]. The MES or DAI consists of four 3-point subscores: stool frequency (0 = normal; 1 = 1−2 more than normal; 2 = 3−4 more than normal; 3 = ≥5 more than normal); rectal bleeding (0 = none; 1 = streaks <50% of time; 2 = obvious blood most of the time; 3 = blood alone passed); physician's global assessment (0 = normal; 1 = mild; 2 = moderate; 3 = severe); and endoscopy (0 = mild or inactive; 1 = mild disease; 2 = moderate disease; 3 = severe disease). The MES is calculated based on the features of erythema, vascular pattern, friability, erosions, spontaneous bleeding, and ulcerations (Table 14.7 and Fig. 14.24). Each category of the UC disease severity is illustrated (Figs. 14.25−14.28). An MES score of 0 or 1 has been considered endoscopic MH.

The MES is calculated with a mixture of various components of endoscopic inflammation. For example, an MES score of 3 is defined as the presence marked erythema, absent vascular markings, granularity, friability, spontaneous bleeding in the lumen, and ulcerations, with a total of six items. It is not clear whether the MES of 3 can be made based on meeting all 6-point criteria or meeting any of the 6-item criteria. In addition, the presence of erosions or pseudopolyps and extent of the disease are not listed in MES.

The modified Mayo Endoscopic Score

The modified 20-point Mayo Endoscopic Score uses the same definitions in terms of mucosal descriptions as MES, but applied to each segment of the colon individually, excluding the ileum (Table 14.8) [23]. This score did not only correlate with clinical activity (the partial Mayo Score), biological activity (CRP and fecal calprotectin) but also with histologic activity(Fig. 14.29A and B).

The Ulcerative Colitis Endoscopic Index of Severity

The Ulcerative Colitis Endoscopic Index of Severity (UCEIS) was the first validated endoscopic score to be applied to UC patients [24,25]. This score has three endoscopic descriptors (vascular pattern, bleeding, and erosion and ulcers) making it simple and capturing 90% of the variability in the overall disease severity (Table 14.9; Figs. 14.30−14.33).

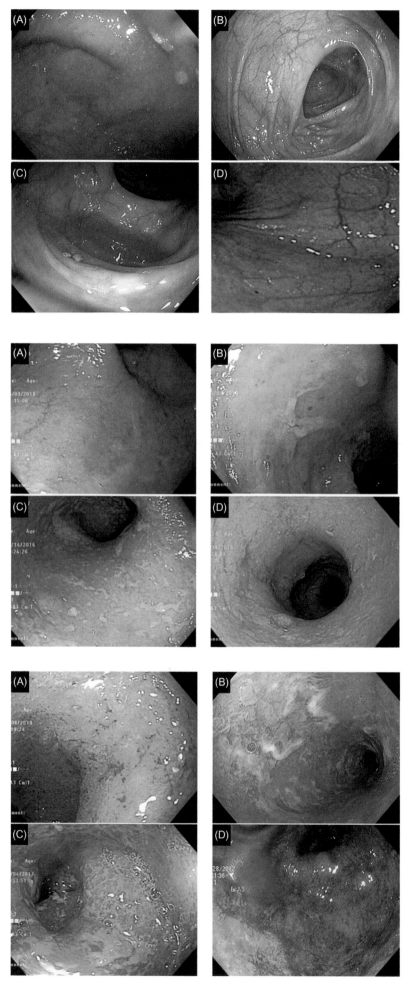

FIGURE 14.25 Mayo Endoscopy Subscore of 0: (A–D) no friability or granularity, intact vascular pattern at the cecum (A), ascending colon (B), splenic flexure (C), and rectum (D).

FIGURE 14.26 Mayo Endoscopy Subscore of 1: (A–D) mild-erythema, diminished or absent vascular markings, and mild granularity.

FIGURE 14.27 Mayo Endoscopy Subscore of 2: (A–D) marked erythema, absent vascular marking, granularity, friability, but no ulceration.

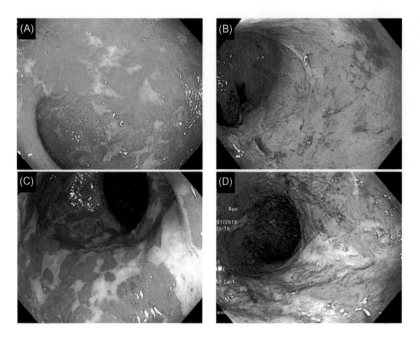

FIGURE 14.28 Mayo Endoscopy Subscore of 3: (A–D) marked erythema, absent vascular markings, granularity, friability, spontaneous bleeding in the lumen, and ulcerations.

TABLE 14.8 The modified Mayo Endoscopic Score (MMES) [23].

Colonic segments	Evaluated[a] (0 or 1)	Inflamed (0 or 1)[b]	MES (0–3)[c]
Rectum			
Sigmoid			
Descending colon			
Transverse colon			
Ascending colon			
Total (=Mayo score)			

[a]Evaluated: 1 if this segment was (completely or partially) evaluated.
[b]Inflamed: 1 if Mayo Endoscopic Subscore for this segment was not 0.
[c]MES: evaluated for the macroscopically most severely inflamed part; score 0 for a segment with normal or inactive disease; score 1 for a segment with erythema, decreased vascular pattern, mild friability; score 2 for a segment with marked erythema, absent vascular pattern, friability, erosions; score 3 for a segment with ulcerations or spontaneous bleeding.

Studies have shown that endoscopic features on UCEIS is reflective of clinical outcomes such as relapse and colectomy and predicts medium-to-long-term prognosis in patients undergoing induction medical therapy [26]. This score eliminates mucosal friability, which was used in the Baron Index to describe endoscopic severity [17]. One of the main disadvantages of UCEIS is its initial development with the use of flexible sigmoidoscopy and failure of taking rest of the colon into account. Examining the left side of the colon may not be sufficient to determine the severity and extent of UC [27,28].

Ulcerative Colitis Colonoscopic Index of Severity

The Ulcerative Colitis Colonoscopic Index of Severity (UCCIS) is a score, which overcomes this problem by taking into consideration the extent and severity of patient with UC in a full colonoscopy (Table 14.10; Figs. 14.34–14.38) [29]. In addition, UCCIS introduces friability as part of the endoscopic descriptors in comparison with UCEIS. The endoscopic features of UCCIS is correlated with different clinical indices of disease activity such as the Simple Clinical Colitis Activity Index, Lichtiger's Clinical Activity Index, as well as to laboratory parameters of active inflammation such as hemoglobin, CRP, and albumin [29].

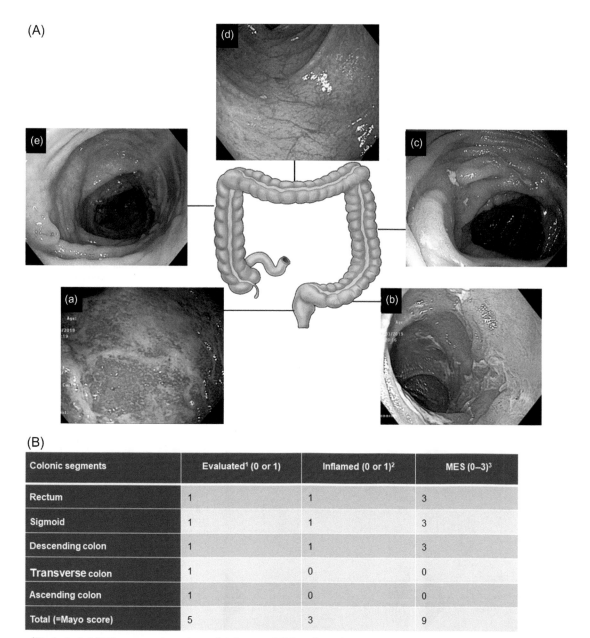

Colonic segments	Evaluated[1] (0 or 1)	Inflamed (0 or 1)[2]	MES (0–3)[3]
Rectum	1	1	3
Sigmoid	1	1	3
Descending colon	1	1	3
Transverse colon	1	0	0
Ascending colon	1	0	0
Total (=Mayo score)	5	3	9

[1]Evaluated: 1 if this segment was (completely or partially) evaluated.
[2]Inflamed: 1 if Mayo endoscopic subscore for this segment was not 0.
[3] MES: evaluated for the macroscopically most severely inflamed part; score 0 for a segment with normal or inactive disease; score 1 for a segment with erythema, decreased vascular pattern, mild friability; score 2 for a segment with marked erythema, absent vascular pattern, friability, erosions; score 3 for a segment with ulcerations or spontaneous bleeding.

FIGURE 14.29   (A) illustration of calculation of the 20-point MMES: (a) severe proctitis with ulcers and exudates (score = 3), (b) multiple ulcers in the sigmoid colon (score = 3), (c) discrete ulcers in the descending colon (score = 3), (d) normal transverse colon except subtle mucosal scars (score = 0), and (e) normal ascending colon except mucosal scars (score = 0). (B) Calculation of MMES in the patient with extensive colitis with active disease in the left colon and quiescent disease in the transverse and descending colon. *MMES*, the modified Mayo Endoscopy Subscore.

Inflammatory disorders of the ileal pouch

Restorative proctocolectomy with ileal pouch-anal anastomosis (IPAA) is the surgical treatment of choice for patients with refractory UC, UC-associated neoplasia, or familial adenomatous polyposis. Inflammatory conditions of the ileal pouch are common, which include pouchitis, cuffitis, and CD of the pouch. Endoscopy plays a key role in the

TABLE 14.9 The Ulcerative Colitis Endoscopic Index of Severity (UCEIS) [24].

Descriptor (score most severe lesion)	Likert scale anchor points	Definitions
Vascular pattern	Normal (1)	Normal vascular pattern with arborization of capillaries clearly defined, or with blurring or patchy loss of capillary margins
	Patchy obliteration (2)	Patchy obliteration of vascular pattern
	Obliterated (3)	Complete obliteration of vascular pattern
Bleeding	None (1)	No visible blood
	Mucosal (2)	Some spots or streaks of coagulated blood on the surface of the mucosa ahead of the scope, which can be washed away
	Luminal mild (3)	Some free liquid blood in the lumen
	Luminal moderate or severe (4)	Frank blood in the lumen ahead of endoscope or visible oozing from mucosa after washing intraluminal blood, or visible oozing from a hemorrhagic mucosa
Erosions and ulcers	None (1)	Normal mucosa, no visible erosions or ulcers
	Erosions (2)	Tiny (≤5 mm) defects in the mucosa, of a white or yellow color with a flat edge
	Superficial ulcers (3)	Large (>5 mm) defects in the mucosa, which are discrete fibrin-covered ulcers in comparison with erosions, but remain superficial
	Deep ulcers (4)	Deeper excavated defects in the mucosa, with a slightly raised edge

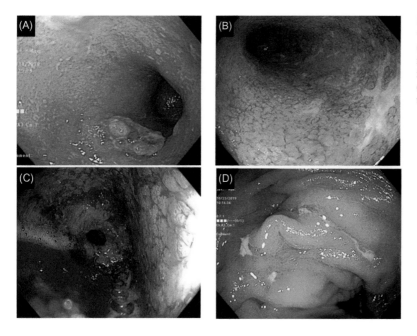

FIGURE 14.30 The 3-component, 10-point UCEIS: vascular pattern, bleeding, and erosions and ulcers: (A) obliterated vascular pattern (score = 3) with an inflammatory pseudopolyp, (B and C) mucosal bleeding (score = 1) (B) and severe luminal bleeding (score = 3) (C), and (D) erosions (score = 2). *UCEIS*, the Ulcerative Colitis Endoscopic Index of Severity.

diagnosis, differential diagnosis, disease monitoring, assessment of treatment response, and dysplasia surveillance in patients with IPAA. Pouchitis and other inflammatory disease conditions of the pouch are discussed in Chapter 11, "Ulcerative Colitis: Post-surgical". The 18-point Pouchitis Disease Activity Index (PDAI) contains clinical, endoscopy, and histology subscores, each having up to 6 points (Table 11.3) [30]. The PDAI endoscopy subscores consist of edema, granularity, friability, loss of vascular pattern, mucous exudates, and ulcers are highlighted in Fig. 11.23. One of the shortcomings of the PDAI endoscopy subscore is its failure to address the distribution of inflammation,

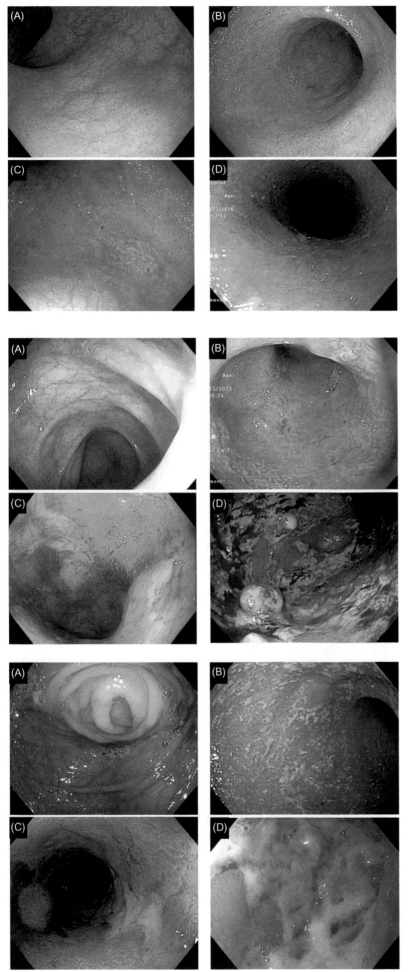

FIGURE 14.31 Vascular pattern component in UCEIS: (A) normal vascular pattern with capillaries clearly defined, or with blurring or patchy loss of capillary margins (score = 1), (B and C) patchy obliteration of vascular pattern (score = 2), and (D) complete obliteration of vascular pattern (score = 3). *UCEIS*, the Ulcerative Colitis Endoscopic Index of Severity.

FIGURE 14.32 Bleeding component in UCEIS: (A) normal no visible blood (score = 1), (B) mucosal bleeding-some spots or streaks of coagulated blood on the surface of the mucosa ahead of the scope, which can be washed away (score = 2), (C) mild luminal bleeding—some free liquid blood in the lumen (score = 3), and (D) moderate or severe luminal bleeding-frank blood in the lumen ahead of endoscope or visible oozing from mucosa after washing intraluminal blood, or visible oozing from a hemorrhagic mucosa (score = 4). *UCEIS*, the Ulcerative Colitis Endoscopic Index of Severity.

FIGURE 14.33 Erosion and ulcer component in the UCEIS: (A) none-normal mucosa of the cecum with no visible erosions or ulcers (score = 1), (B) erosions-tiny (≤5 mm) defects with of a white or yellow color and a flat edge (score = 2), (C) superficial ulcers-large (>5 mm) defects with discrete fibrin-covered ulcers (score = 3), and (D) deep ulcers-deeper excavated defects with a slightly raised edge (score = 4). *UCEIS*, the Ulcerative Colitis Endoscopic Index of Severity.

TABLE 14.10 The Ulcerative Colitis Colonoscopic Index of Severity (UCCIS) [29].

Lesion	Score	Definition
Vascular pattern	0	Normal, clear vascular pattern
	1	Partially visible vascular pattern
	2	Complete loss of vascular pattern
Granularity	0	Normal, smooth, and glistening
	1	Fine
	2	Coarse
Ulceration	0	Normal, no erosion or ulcer
	1	Erosions and pinpoint ulcerations
	2	Numerous shallow ulcer with mucus
	3	Deep, excavated ulcerations
	4	Diffusely ulcerated with >30% involvement
Bleeding/friability	0	Normal, no bleeding, no friability
	1	Friable, bleeding to light touch
	2	Spontaneous bleeding

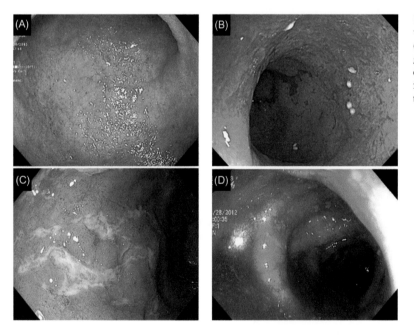

FIGURE 14.34 The 4-component, 10-point UCCIS: vascular pattern, granularity, and bleeding/friability: (A) vascular pattern (partially visible, score = 1), (B) granularity (fine, score = 1), (C) ulceration (deep, excavated ulcerations, score = 3), and (D) bleeding/ friability (spontaneous bleeding, score = 4). *UCCIS*, the Ulcerative Colitis Colonoscopic Index of Severity.

while disease distribution patterns may provide clues for the differential diagnosis of microbiota-associated pouchitis, immune-mediated pouchitis, and ischemic pouchitis [31]. However, the interobserver agreement is high only in the endoscopic items of ulceration and ulcerated surface in the pouch body, not in other items of endoscopic scores [32]. Future studies are needed to assess responsiveness to treatment and development of a weighed endoscopic pouchitis disease index.

The same endoscopic components have been adopted to measure the disease activity of cuffitis in a separate index, that is, the Cuffitis Activity Index (Fig. 11.31) [33].

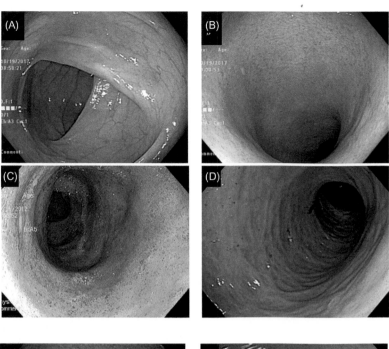

FIGURE 14.35 Vascular pattern component in the 10-point UCCIS: (A) normal vascular pattern (score = 1), (B) partially visible vascular pattern (score = 2), (C) complete loss of vascular pattern (score = 3), and (D) complete loss of vascular pattern with extensive mucosal scars (score = 3). The formation of mucosal scars is a sign of healing, which is often accompanied by loss of vascular pattern. The score of 3 here may be overrated. *UCCIS*, the Ulcerative Colitis Colonoscopic Index of Severity.

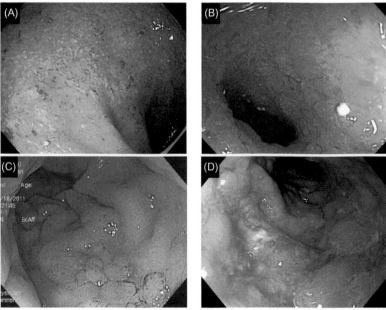

FIGURE 14.36 Granularity component in the 10-point UCCIS: (A and B) fine granularity of mucosa (score = 1), (C) coarse granularity of mucosa (score = 2), and (D) coarse granularity of mucosa with ulcers (score = 2). *UCCIS*, the Ulcerative Colitis Colonoscopic Index of Severity.

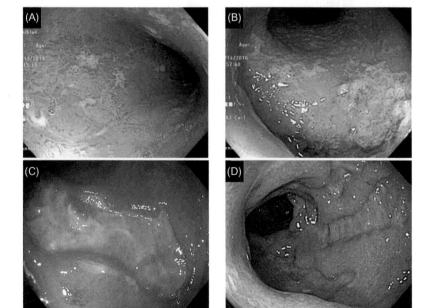

FIGURE 14.37 Ulceration component in the 10-point UCCIS: (A) erosions and pinpoint ulcerations (score = 1), (B) numerous shallow ulcer with mucus (score = 2), (C) deep, excavated ulcerations (score = 3), and (D) diffusely ulcerated with >30% involvement (score = 4). *UCCIS*, the Ulcerative Colitis Colonoscopic Index of Severity.

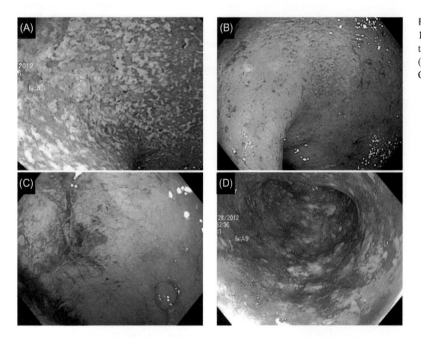

FIGURE 14.38 Bleeding/friability component in the 10-point UCCIS: (A and B) friable, bleeding to light touch (score = 1) and (C and D) spontaneous bleeding (score = 2). *UCCIS*, the Ulcerative Colitis Colonoscopic Index of Severity.

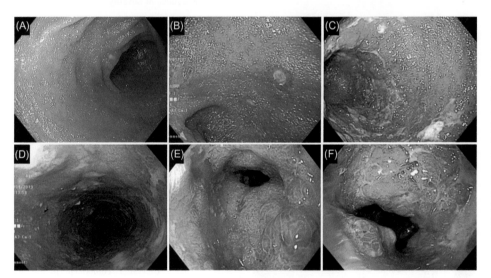

FIGURE 14.39 Assessment of Crohn's disease—associated *afferent limb* inflammation using the 6-point PDAI endoscopy subscores: (A and B) discrete ulceration, (C and D) edema, loss of vascular pattern, exudates, granularity, ulcers with mucopurulent cap, and (E and F) edema, granularity, loss of vascular pattern, ulceration, and luminal narrowing. Luminal narrowing or stricture is a part of fibrostenotic Crohn's disease of the pouch, which is not a component of the PDAI endoscopy subscore. *PDAI*, the Pouchitis Disease Activity Index.

The endoscopic scores have also been applied to calculate the degree of inflammation of the afferent limb (Fig. 14.39), pouch body (Fig. 14.40), and cuff (Fig. 14.41) in patients with CD of the pouch [34,35]. In addition to inflammatory phenotypes of CD of the pouch, pouchoscopy provides useful information on fibrostenotic and fistulizing CD, as outlined in Chapter 11, Ulcerative colitis: postsurgical (Figs. 11.27—11.29). Erosions, erythema, strictures, fistulae, and pseudopolyps are not listed in the PDAI endoscopy subscores.

Diverted bowel

Fecal diversion with an ileostomy, and occasionally colostomy, or jejunostomy are performed as a temporary or permanent measure for the treatment of refractory disease in the downstream segment of the bowel, anal, or perianal area. Diversion-associated mucosal inflammation often occurs in the diverted rectum (also called Hartmann's pouch), diverted colon, or a diverted ileal pouch. Patients with underlying IBD are at a higher risk for diversion bowel disease than those without. The common endoscopic features of diversion bowel disease are nodularity, friability, and spontaneous bleeding (Figs. 14.42—14.44), also as outlined in Chapter 13, Diversion-associated bowel diseases. Currently, there

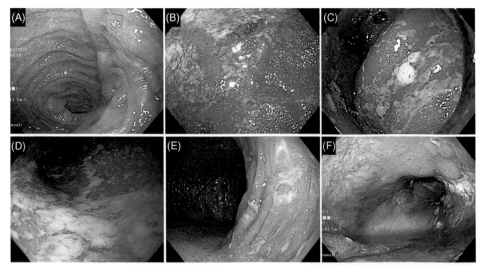

FIGURE 14.40 Assessment of Crohn's disease–associated *pouch body* inflammation using the 6-point PDAI endoscopy subscores: (A) edema, (B–D) edema, erythema, loss of vascular pattern, and exudates, (E) ulceration, and (F) loss of vascular pattern and ulceration. Erythema is not a component of the PDAI endoscopy subscores. *PDAI*, the Pouchitis Disease Activity Index.

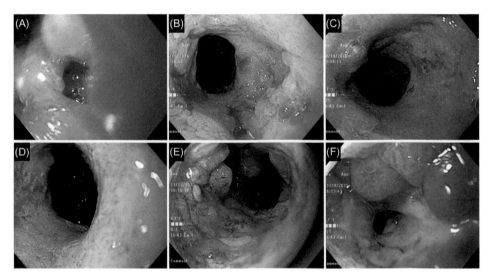

FIGURE 14.41 Assessment of Crohn's disease–associated *cuff* inflammation using the 6-point PDAI endoscopy subscores: (A) edema with a fistula opening, (B and C) granularity, loss of vascular pattern, and friability, (D) loss of vascular pattern and friability, (E and F) nodularity, mucous exudates on ulcers with a stricture, and (F) the presence of stricture, fistula, or stricture is not a component of the PDAI endoscopy subscores. *PDAI*, the Pouchitis Disease Activity Index.

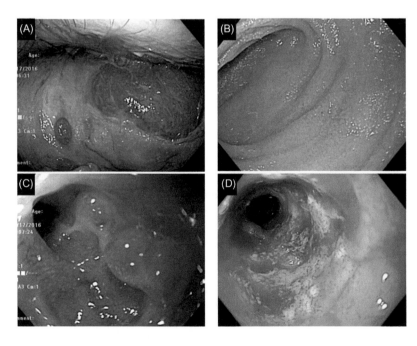

FIGURE 14.42 Endoscopic evaluation of the diverted colon or Hartmann pouch for refractory rectal and perianal Crohn's disease: (A) perianal disease with ulcers and fistula, (B) diverting ileostomy with normal mucosa of the neo-distal ileum, (C) ulcerated stricture at the distal rectum, and (D) diversion proctitis with nodular and friable mucosa with spontaneous bleeding.

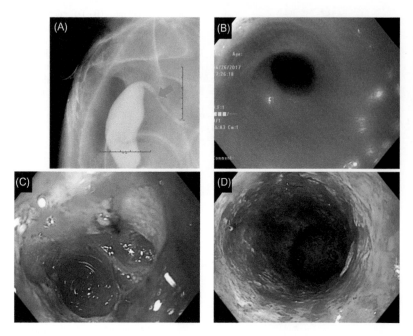

FIGURE 14.43 Endoscopic evaluation of diverted colon for refractory rectal and perianal Crohn's disease: (A) diverted colon with a long stricture (*green arrow*) on gastrograffin enema; (B) distal rectal stricture due to long-term fecal diversion; and (C and D) diverted colon with friable mucosa and spontaneous bleeding.

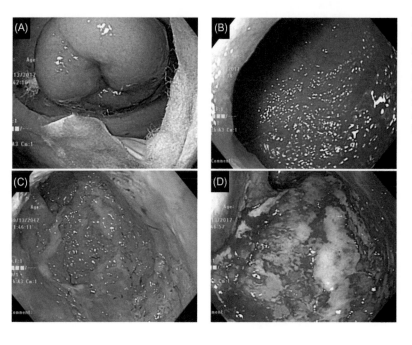

FIGURE 14.44 Endoscopic evaluation of diverted ileal pouch for refractory Crohn's disease of the pouch in a patient with a precolectomy diagnosis of ulcerative colitis: (A) end ileostomy, (B) normal neo-distal ileum assessed by ileoscopy via stoma, and (C and D) diverted pouch before and after air insufflation. Mucosa of the diverted pouch was extremely friable even with minimum air or carbon dioxide insufflation.

are no endoscopy disease activity instruments specifically designed for the measurement of inflammation of diversion bowel disease.

Strictureplasty

Strictureplasty is increasingly performed in patients with small bowel CD. To date, there are no published disease activity instruments for the measurement of inflammation in the strictureplasty (Fig. 14.45). The anatomy and CD activity in strictureplasty site are also discussed in Chapter 8, Crohn's disease: postsurgical, along with Figs. 8.10–8.14.

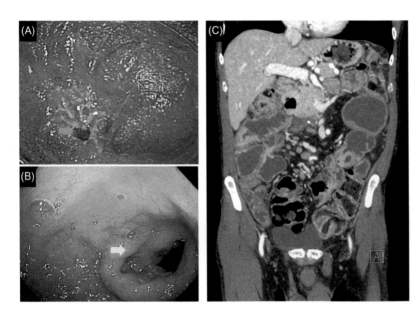

FIGURE 14.45 Endoscopic evaluation of stricture-plasty for Crohn's disease: (A) ulcerated outlet stricture in a strictureplasty site (*green arrow*) with normal mucosa of the body. (B) normal outlet (*yellow arrow*) of a strictureplasty site in a separate patient with loss of vascular pattern of mucosa of the body, and (C) multiple stricturoplasties in a patient with Crohn's disease on computed tomography enterography.

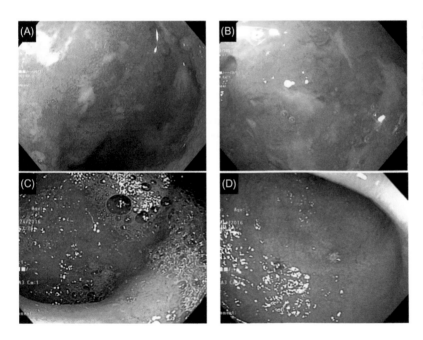

FIGURE 14.46 Assessment of treatment response with antiintegrin therapy in a Crohn's disease patient with ileorectal anastomosis: (A and B) severe inflammation with granularity, exudates, and ulcers, exudates in the rectum and (C and D) resolution of the inflammation of the rectum except for mild loss of vascular pattern.

Disease monitoring and endoscopic healing

Adjunct to clinical, biological, and radiographic evaluation, endoscopic assessment plays a key role in the disease monitoring and assessment of treatment response, and prediction of the prognosis. The endoscopic disease indices have been used to assess the response to medical and surgical treatment (Figs. 14.46–14.49).

Endoscopic mucosal healing has routinely been used as an endpoint of trials of immunomodulators, antitumor necrosis factor-α, antiintegrin [36], antiinterleukin [37], anti-Janus kinase [38] therapy for moderate-to-severe CD or UC. A precise term of MH has not clearly been defined, although the phrases of "absence of ulcers," "endoscopic scores improvement," or "complete absence of all inflammatory and ulcerative lesions in all segments of gut" have been used in the majority of clinical trials. MH is mainly measured by the white-light-endoscopy-based scores in various CD- or UC-related indices. For example, MH in CD has been defined as CDEIS 0 or <3, SES-CD 0 or <5, RS 0–1, and MES 0–1. It appears that the current phrase is the definition of exclusion, rather than a proactive term for diagnosis.

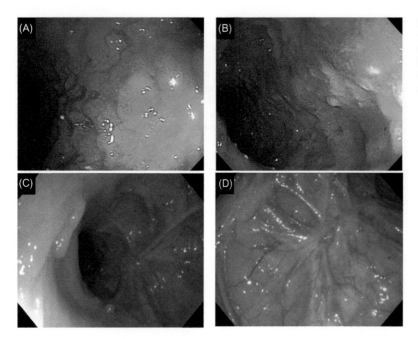

FIGURE 14.47 Assessment of treatment response with tumor-necrosis factor therapy in a patient with Crohn's colitis: (A and B) active colitis with nodularity and ulcers and (C and D) resolution of the inflammation of the colon with the formation of mucosal scars and pseudopolyps.

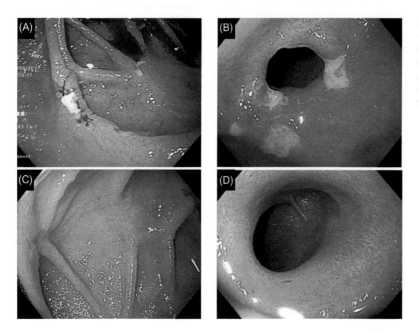

FIGURE 14.48 Assessment of treatment response with tumor-necrosis factor therapy in a patient with Crohn's disease of the duodenum: (A and B) ulcers in the second portion of the duodenum (A) and ulcers and ulcerated stricture at the duodenum bulb and cap (B),and (C and D) resolution of the ulcers, including ulcers on the surface of the stricture.

Image-enhanced endoscopy with the demonstration of a more detailed mucosal and vascular structure may further define MH, as outlined in Chapter 19, "Chromoendoscopy in Inflammatory Bowel Disease".

Tissue healing may not only be present with the absence of active inflammation, such as erosions, ulcers, friability, and spontaneous bleeding, but also with other forms, such as the formation of pseudopolyps (Fig. 14.50), mucosal scars (Fig. 14.51), mucosal ridges (Fig. 14.52), or mucosal bridges (Fig. 14.53). These endoscopic features can be seen in both CD and UC, and in both small bowel and large bowel CD. MH is often associated with the formation of pseudopolyps. Therefore isolated sparse, multiple, or numerous pseudopolyps with no active disease in surrounding mucosa can be a part of healing [39]. There is limited literature on mucosal scars, ridges, or bridges. These endoscopic features have not been listed in the current endoscopy disease activity scores. We suggest that these endoscopic findings be markers for quiescent or inactive disease.

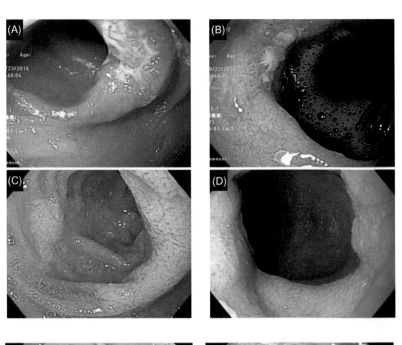

FIGURE 14.49 Assessment of treatment response with antiintegrin therapy in a patient with Crohn's disease of the pouch: (A and B) ulcers the afferent limb (the Pouchitis Disease Activity Index endoscopy subscore = 1) and (C and D) resolution of the ulcers (the Pouchitis Disease Activity Index endoscopy subscore = 0).

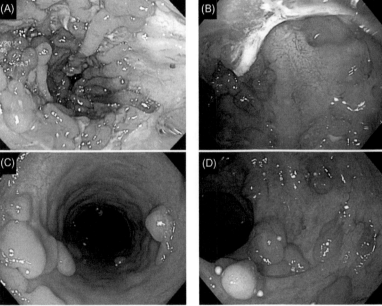

FIGURE 14.50 Pseudopolyps associated with quiescent disease in ulcerative colitis: (A–D) various forms of pseudopolyps associated with mucosal healing.

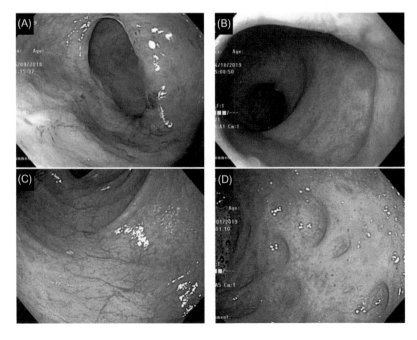

FIGURE 14.51 Mucosal scars in patients with Crohn's disease or ulcerative colitis: (A) mucosal scars with disease activity (bleeding) in a patient with Crohn's disease and (B–D) various forms of mucosal scars after effective medical therapy in patients with ulcerative colitis.

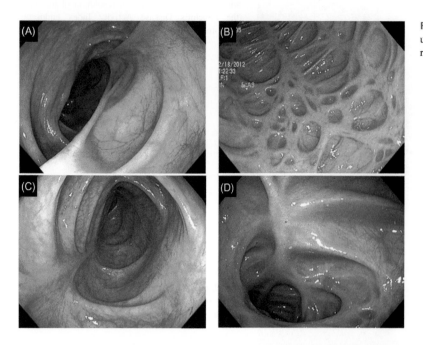

FIGURE 14.52 Mucosal ridges in patients with ulcerative colitis: (A−D) various forms of mucosal ridges without active disease activity.

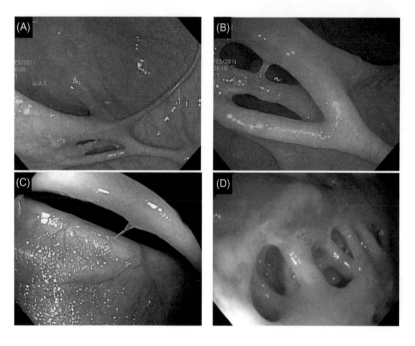

FIGURE 14.53 Mucosal bridges in patients with Crohn's disease: (A−C) various forms of mucosal bridges without active disease activity and (D) mucosal bridges with active disease.

Endoscopic documentation of MH has been used in conjunction with cross-sectional imaging evaluation for transmural healing in CD. Patients with endoscopic MH and transmural healing on MRE are considered as having deep healing. The features of transmural healing on MRE include the absence of abnormal bowel wall thickening (defined has a thickness >3 mm), increased contrast enhancement, and complications (stricture, abscess, or fistulae) [40,41]. Endoscopic MH and MRE's transmural healing do not necessarily correlate. In a study of 151 children with CD undergoing a clinical trial, 21 (14%) had deep healing (consisting of MH and transmural healing), 9 (6%) had MH without transmural healing, 38 (25%) had transmural healing without MH, and 83 (55%) had both mucosal and transmural inflammation [42].

Summary and recommendations

Endoscopy, upper endoscopy, colonoscopy, and ileoscopy have been routinely used to diagnose, differential diagnosis, disease monitoring, and assessment of treatment response in IBD. Various endoscopy disease activity instruments have been developed in CD, UC, and pouchitis. These disease activity indices are mainly used in clinical trials but are increasingly been used in routine clinical practice. Some of disease activity indices have been incorporated into the template of endoscopy reporting system. On the other hand, the calculation of these indices is still cumbersome. We are looking forward to seeing new, user-friendly, and statistically weighted instruments being developed.

References

[1] Rutgeerts P, Vermeire S, Van Assche G. Mucosal healing in inflammatory bowel disease: impossible ideal or therapeutic target? Gut 2007;56:453—5.

[2] Peyrin-Biroulet L, Sandborn W, Sands BE, Reinisch W, Bemelman W, Bryant RV, et al. Selecting therapeutic targets in inflammatory bowel disease (STRIDE): determining therapeutic goals for treat-to-target. Am J Gastroenterol 2015;110:1324—38.

[3] Best WR, Becktel JM, Singleton JW, Kern Jr. F. Development of a Crohn's disease activity index. National Cooperative Crohn's Disease Study. Gastroenterology 1976;70:439—44.

[4] Lichtenstein GR, Loftus EV, Isaacs KL, Regueiro MD, Gerson LB, Sands BE. ACG clinical guideline: management of Crohn's disease in adults. Am J Gastroenterol 2018;113:481—517.

[5] Harvey RF, Bradshaw JM. A simple index of Crohn's-disease activity. Lancet 1980;1(8167):514.

[6] Mary JY, Modigliani R. Development and validation of an endoscopic index of the severity for Crohn's disease: a prospective multicentre study. Groupe d'Etudes Therapeutiques des Affections Inflammatoires du Tube Digestif (GETAID). Gut 1989;30:983—9.

[7] Daperno M, D'Haens G, Van Assche G, Baert F, Bulois P, Maunoury V, et al. Development and validation of a new, simplified endoscopic activity score for Crohn's disease: the SES-CD. Gastrointest Endosc 2004;60:505—12.

[8] Rutgeerts P, Geboes K, Vantrappen G, Beyls J, Kerremans R, Hiele M. Predictability of the postoperative course of Crohn's disease. Gastroenterology 1990;99:956—63.

[9] D'Haens G, Van Deventer S, Van Hogezand R, Chalmers D, Kothe C, Baert F, et al. Endoscopic and histological healing with infliximab antitumor necrosis factor antibodies in Crohn's disease: a European multicenter trial. Gastroenterology 1999;116:1029—34.

[10] Jess T, Riis L, Vind I, Winther KV, Borg S, Binder V, et al. Changes in clinical characteristics, course, and prognosis of inflammatory bowel disease during the last 5 decades: a population-based study from Copenhagen, Denmark. Inflamm Bowel Dis 2007;13:481—9.

[11] Ollech JE, Aharoni-Golan M, Weisshof R, Normatov I, Sapp AR, Kalakonda A, et al. Differential risk of disease progression between isolated anastomotic ulcers and mild ileal recurrence after ileocolonic resection in patients with Crohn's disease. Gastrointest Endosc 2019;90:269—75.

[12] Domènech E, Mañosa M, Bernal I, Garcia-Planella E, Cabré E, Piñol M, et al. Impact of azathioprine on the prevention of postoperative Crohn's disease recurrence: results of a prospective, observational, long-term follow-up study. Inflamm Bowel Dis 2008;14:508—13.

[13] Ma C, Gecse KB, Duijvestein M, Sandborn WJ, Zou G, Shackleton LM, et al. Reliability of endoscopic evaluation of postoperative recurrent Crohn's disease. Clin Gastroenterol Hepatol 2019; Aug 29 [Epub ahead of print].

[14] Chongthammakun V, Fialho A, Fialho A, Lopez R, Shen B. Correlation of the Rutgeerts score and recurrence of Crohn's disease in patients with end ileostomy. Gastroenterol Rep (Oxf) 2017;5:271—6.

[15] Pariente B, Cosnes J, Danese S, Sandborn WJ, Lewin M, Fletcher JG, et al. Development of the Crohn's disease digestive damage score, the Lémann score. Inflamm Bowel Dis 2011;17:1415—22.

[16] Truelove SC, Witts IL. Cortisone in ulcerative colitis; final report on a therapeutic trial. Br Med J 1955;2:1041—8.

[17] Baron JH, Connell AM, Lennard-Jones JE. Variation between observers in describing mucosal appearances in proctocolitis. Br Med J 1964;189—92.

[18] Powell-Tuck J, Bown RL, Lennard-Jones JE. A comparison of oral prednisolone given as single or multiple daily doses for active proctocolitis. Scand J Gastroenterol 1978;13:971—6.

[19] Sutherland LR, Martin F, Greer S, Robinson M, Greenberger N, Saibil F, et al. 5-Aminosalicylic acid enema in the treatment of distal ulcerative colitis, proctosigmoiditis and proctitis. Gastroenterology 1987;92:1894—8.

[20] Rachmilewitz D. Coated mesalazine (5-aminosalicylic acid) versus sulphasalazine in the treatment of active ulcerative colitis: a randomised trial. Br Med J 1989;298:82—6.

[21] Schroeder KW, Tremaine WJ, Ilstrup DM. Coated oral 5-aminosalicylic acid therapy for mildly to moderately active ulcerative colitis. A randomized study. N Engl J Med 1987;317:1625—9.

[22] Colombel JF, Rutgeerts P, Reinisch W, Esser D, Wang Y, Lang Y, et al. Early mucosal healing with infliximab is associated with improved long-term clinical outcomes in ulcerative colitis. Gastroenterology 2011;141:1194—201.

[23] Lobaton T, Bessissow T, De Hertogh G, Lemmens B, Maedler C, Van Assche G, et al. The Modified Mayo Endoscopic Score (MMES): A new index for the assessment of extension and severity of endoscopic activity in ulcerative colitis patients. J Crohns Colitis 2015;9:846—52.

[24] Travis SP, Schnell D, Krzeski P, Abreu MT, Altman DG, Colombel JF, et al. Developing an instrument to assess the endoscopic severity of ulcerative colitis: the Ulcerative Colitis Endoscopic Index of Severity (UCEIS). Gut 2012;61:535—42.

[25] Travis SP, Schnell D, Krzeski P, Abreu MT, Altman DG, Colombel JF, et al. Reliability and initial validation of the ulcerative colitis endoscopic index of severity. Gastroenterology 2013;145:987—95.

[26] Ikeya K, Hanai H, Sugimoto K, Kawasaki S, Iida T, Maruyama Y, et al. The ulcerative colitis endoscopic index of severity more accurately reflects clinical outcomes and long-term prognosis than the Mayo endoscopic score. J Crohns Colitis 2016;10:286—95.

[27] Bernstein CN, Shanahan F, Anton PA, Weinstein WM. Patchiness of mucosal inflammation in treated ulcerative colitis: a prospective study. Gastrointest Endosc 1995;42:232—7.

[28] Kim B, Barnett JL, Kleer CG, Appelman HD. Endoscopic and histological patchiness in treated ulcerative colitis. Am J Gastroenterol 1999;94:3258−62.

[29] Samuel S, Bruining DH, Loftus Jr. EV, et al. Validation of the ulcerative colitis colonoscopic index of severity and its correlation with disease activity measures. Clin Gastroenterol Hepatol 2013;11:49−54.

[30] Sandborn WJ, Tremaine WJ, Batts KP, Pemberton JH, Phillips SF. Pouchitis after ileal pouch-anal anastomosis: a pouchitis disease activity index. Mayo Clin Proc 1994;69:409−15.

[31] Shen B. Problems after restorative proctocolectomy: assessment and therapy. Curr Opin Gastroenterol 2016;32:49−54.

[32] Samaan MA, Shen B, Mosli MH, Zou G, Sandborn WJ, Shackelton LM, et al. Reliability among central readers in the evaluation of endoscopic disease activity in pouchitis. Gastrointest Endosc 2018;88:360−9.

[33] Shen B, Lashner BA, Bennett AE, Remzi FH, Brzezinski A, Achkar JP, et al. Treatment of rectal cuff inflammation (cuffitis) in patients with ulcerative colitis following restorative proctocolectomy and ileal pouch-anal anastomosis. Am J Gastroenterol 2004;99:1527−31.

[34] Shen B, Remzi FH, Lavery IC, Lopez R, Queener E, Shen L, et al. Administration of adalimumab in the treatment of Crohn's disease of the ileal pouch. Aliment Pharmacol Ther 2009;29:519−26.

[35] Li Y, Lopez R, Queener E, Shen B. Adalimumab therapy in Crohn's disease of the ileal pouch. Inflamm Bowel Dis 2012;18:2232−9.

[36] Cholapranee A, Hazlewood GS, Kaplan GG, Peyrin-Biroulet L, Ananthakrishnan AN. Systematic review with meta-analysis: comparative efficacy of biologics for induction and maintenance of mucosal healing in Crohn's disease and ulcerative colitis controlled trials. Aliment Pharmacol Ther 2017;45:1291−302.

[37] Rutgeerts P, Gasink C, Chan D, Lang Y, Pollack P, Colombel JF, et al. Efficacy of ustekinumab for inducing endoscopic healing in patients with Crohn's disease. Gastroenterology 2018;155:1045−58.

[38] Ma C, Lee JK, Mitra AR, Teriaky A, Choudhary D, Nguyen TM, et al. Systematic review with meta-analysis: efficacy and safety of oral Janus kinase inhibitors for inflammatory bowel disease. Aliment Pharmacol Ther 2019;50:5−23.

[39] Rezapour M, Quintero MA, Khakoo NS, Sussman DA, Barkin JA, Clarke J, et al. Reclassifying pseudopolyps in inflammatory bowel disease: histologic and endoscopic description in the new era of mucosal healing. Crohns Colitis 360 2019;1(3):otz033 Epub 2019 Oct 15.

[40] Fernandes SR, Rodrigues RV, Bernardo S, Cortez-Pinto J, Rosa I, da Silva JP, et al. Transmural healing is associated with improved long-term outcomes of patients with Crohn's disease. Inflamm Bowel Dis 2017;23:1403−9.

[41] Castiglione F, Mainenti P, Testa A, Imperatore N, De Palma GD, Maurea S, et al. Cross-sectional evaluation of transmural healing in patients with Crohn's disease on maintenance treatment with anti-TNF alpha agents. Dig Liver Dis 2017;49:484−9.

[42] Weinstein-Nakar I, Focht G, Church P, Walters TD, Abitbol G, Anupindi S, et al. Associations among mucosal and transmural healing and fecal level of calprotectin in children with Crohn's disease. Clin Gastroenterol Hepatol 2018;16:1089−97.

Chapter 15

Histologic evaluation of disease activity in inflammatory bowel disease

Chanqing Ma[1] and Xiuli Liu[2]

[1]Department of Pathology, University of Pittsburgh Medical Center, Pittsburgh, PA, United States, [2]Department of Pathology, Immunology, and Laboratory Medicine, University Florida School of Medicine, Gainesville, FL, United States

Chapter Outline

Abbreviations

CD Crohn's disease
GHAS the Global Histologic Disease Activity Score
GI gastrointestinal
IBD inflammatory bowel disease
IPAA ileal pouch−anal anastomosis
IOIBD the International Organization for the Study of Inflammatory Bowel Disease
STRIDE the Selecting Therapeutic Targets in Inflammatory Bowel Disease
UC ulcerative colitis

Introduction

Histologic assessment of biopsy and resection specimens of the small and large intestine has always been an integral part of the diagnosis and management of idiopathic inflammatory bowel disease (IBD). Most IBD cases can be classified into two major subtypes: ulcerative colitis (UC) and Crohn's disease (CD). UC is a chronic inflammatory disease of the colon, while CD is an inflammatory disease with a propensity to involve the gastrointestinal (GI) tract [1−4]. From a histopathologic standpoint, despite many shared characteristics, each subtype has unique macroscopic (gross) and microscopic (histologic) features that can help distinguish one from another.

Macroscopic features of UC and CD and guidelines of standard endoscopy sampling with adequate specimens for histopathology evaluation will be discussed first in this chapter. This is followed by discussions of histologic features of each entity and histologic grading and scoring systems that have been used in clinical trials of UC and CD. The remainder of the chapter is focused on histologic mucosal healing in IBD.

Atlas of Endoscopy Imaging in Inflammatory Bowel Disease. DOI: https://doi.org/10.1016/B978-0-12-814811-2.00015-3
© 2020 Elsevier Inc. All rights reserved.

Macroscopic features of ulcerative colitis and Crohn's disease

UC classically starts in the rectum and can extend proximally to involve other segments or even the entire colon. The extent of involvement is determined by the severity of the disease. Small intestine or upper GI tract is usually spared in UC. Inflammation in UC is contiguous and homogenous in the affected segment of the colon without any uninflamed areas. Further, inflammation in UC is superficial and predominantly confined to the mucosa muscularis mucosae, or superficial submucosa. Ulcers in UC are shallow and typically do not involve the muscle wall. The deep layers of the bowel wall and mesentery are intact in UC [1] (Fig. 15.1A and Table 15.1).

In contrast, CD can affect any part of the GI tract, though the small intestine and proximal colon being the most often-affected segments. Inflammation in CD is segmental and discontinuous. Inflamed areas are intermixed with unin-flamed areas, resulting in an appearance of "skip lesions." In addition, inflammation in CD is transmural and, therefore,

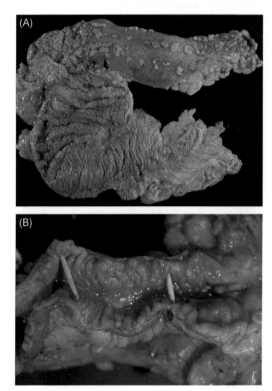

FIGURE 15.1   Macroscopic features of ulcerative colitis and Crohn's disease. (A) Macroscopic examination of a colectomy specimen reveals a dif-fuse and homogenous colitis with erythema and ulceration. The colon lies flat on the examination table upon opening. (B) Macroscopic examination of a segmental colectomy specimen reveals segmental colitis with strictures and longitudinal ulceration. The colon does not lie flat on the examination table upon opening.

TABLE 15.1 Macroscopic features of ulcerative colitis and Crohn's disease.

Macroscopic appearance	Ulcerative colitis	Crohn's disease
Rectum involvement	Yes	No
Small intestine involvement	No	Yes
Diffuse and contiguous	Yes	No
Segmental disease with skip lesions	No	Yes
Penetrating ulcer	No	Yes
Fistula	No	Yes
Fat wrapping in mesentery	No	Yes

can cause injuries to the muscle wall, serosa, or mesentery. Deep-penetrating ulcers, fissures, fistula tracts, strictures, and fat wrapping along the antimesenteric border are hallmark features of CD [2] (Fig. 15.1B and Table 15.1).

Endoscopic tissue sampling in inflammatory bowel disease

In clinical practice, histologic evaluation of biopsy specimens obtained through colonoscopy or ileocolonoscopy in UC or CD patients is performed to establish IBD diagnosis, evaluate for dysplasia, and confirm the presence of active disease during an episode of clinical flare [5−8]. In clinical trials, the histologic evaluation of biopsy specimens is predominantly used to assess changes in histopathology over a period of time and/or before and after the administration of treatment. Regardless of the purpose, the diagnostic yield of biopsies is determined by appropriate sampling of colorectal and ileal mucosa [5−8].

It is recommended that at least two, preferably multiple, biopsy fragments should be obtained from each segment of the colon and from the segment of ileum accessible through ileocolonoscopy [7]. The use of large-cup or large-capacity forceps is recommended to obtain biopsy samples at least 4 mm in size. It is recommended to take an additional biopsy for each biopsy that measures less than 4 mm [7]. Large biopsies are not necessarily deep ones. Since certain histologic features, such as basal plasmacytosis which will be defined and discussed later in the chapter, involve the space right above *muscularis mucosae*, deep biopsy should be obtained if possible [6,7].

Both endoscopically normal and abnormal locations or distinctive areas in the same anatomic location should be sampled [6,7]. When sampling endoscopically abnormal or inflamed mucosa, such as ulcers, biopsies should be taken from the most severely affected or inflamed region [6,7]. Random biopsies from the same segment presumably with similar endoscopic appearances can be collected in one single specimen container. However, biopsies from endoscopically abnormal areas with different appearances in the same segment should be submitted separately [5]. Containers should be labeled by anatomic segments [5]. Biopsy specimens should be immediately fixed in 10% neutral buffer formalin [5,7] or other fixatives.

In surgery-naïve patients a standardized biopsy procurement protocol may include sampling each segment of the colon, including the rectum, left colon, transverse colon, right colon, and terminal ileum [8]. When flexible sigmoidoscopy is used in patients with severe colitis or megacolon, biopsies of the descending colon, sigmoid colon, and rectum are recommended as the basic minimum series [7]. In patients with previous surgery or CD patients with severe strictures, all accessible segments should be biopsied and submitted separately [8]. It is additionally recommended to always take two biopsies from the rectum [7,8].

In addition to providing a sufficient number of good-quality biopsies, pertinent clinical information and endoscopic findings preferably together with endoscopic pictures should be made available to pathologists.

Microscopic features of ulcerative colitis

UC and CD have substantial overlapping histologic features. Upon microscopic examination, histologic findings of UC and CD as a whole can be subclassified as features of activity and features of chronicity (Fig. 15.2 and Table 15.2). Activity is defined as neutrophilic epithelial injury. Histologic features of activity include cryptitis, crypt abscesses, erosions, and ulcers. Chronicity refers to the presence of a constellation of the following histologic features, including crypt architecture distortion, infiltration of chronic inflammatory cells (such as lymphocytes and monocytes), and metaplasia. Chronic inflammation consists of both expansions of lamina propria by lymphoplasmacytic inflammation and basal plasmacytosis, which is defined as plasma cells filling the space between the bases of the crypts and *muscularis mucosae* [9]. Metaplasia typically includes Paneth cell form in the left colon and pyloric gland form throughout colon.

It is important to note that the aforementioned features of chronicity are morphological findings observed in the mucosa. Therefore they can be appreciated in both UC and CD and by the same token in both biopsy and resection specimens. Other histologic features of chronicity, including *muscularis mucosae* hyperplasia and submucosal fibrosis, may not be seen in biopsy specimens. Furthermore, histologic features associated with inflammation and injury deep in the bowel wall, which are characteristically seen in CD, will be discussed in the microscopic features of resected surgical specimens of CD.

Features of activity and chronicity are not mutually exclusive. Rather, they frequently coexist. In other words, both sets of features can be appreciated when evaluating a specimen. In such instance the disease in a colon biopsy, for example, can be referred to active chronic colitis. Along these lines, active colitis in biopsy is a diagnosis referring to the presence of features of activity only without those of chronicity, while inactive chronic colitis, features of chronicity only without evidence of activity.

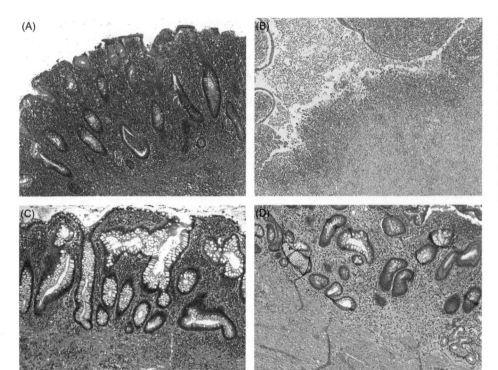

FIGURE 15.2 Microscopic features of activity and chronicity in ulcerative colitis and Crohn's disease. Activity is defined as neutrophil-mediated epithelial injury, including cryptitis, crypt abscesses (A); erosions and ulcerations (B). Chronicity refers to the presence of a constellation of the following features, including crypt architectural distortion (C); basal lymphoplasmacytosis, Paneth cell metaplasia in the left colon, and pyloric gland metaplasia in the colon (D).

TABLE 15.2 Histologic features of ulcerative colitis and Crohn's disease.

Histopathology features	Ulcerative colitis	Crohn's disease
Small intestine involvement	No	Yes
Rectum involvement	Yes	No (yes in some cases)
Skip lesions	No	Yes
Features of activity		
Cryptitis	Yes	Yes
Crypt abscess	Yes	Yes
Erosion	Yes	Yes
Ulceration	Yes, but shallow	Yes
Aphthous ulcer	No	Yes
Features of chronicity		
Crypt architecture distortion	Yes	Yes
Basal plasmacytosis	Yes	Yes
Increased lamina propria lymphocytes and plasma cells	Yes	Yes
Paneth cell metaplasia in left colon	Yes	Yes
Pyloric gland metaplasia	Yes	Yes
Epithelioid granuloma	No	Yes
Mural changes		
Muscularis mucosae hyperplasia	Yes (when severe)	Yes
Submucosal fibrosis	Yes (when severe)	Yes
Fissuring ulcers	No	Yes
Fistula/Sinus tract	No	Yes
Transmural lymphoid aggregates	No	Yes
Neural hypertrophy	No (yes when severe)	Yes
Muscularis propria hypertrophy	No	Yes

Histologic evaluation of biopsy specimen has been used to determine response to medical therapy in clinical trials for UC. To accomplish such a task a histopathology scoring index is often used to evaluate and compare the degree of active inflammation, chronic inflammation, and/or chronic injury, such as crypt architecture distortion in biopsy specimens obtained from colonoscopy before and after treatment. The first histologic scoring system used for UC was developed in 1956 [10]. To date, at least 30 histologic scoring systems have been developed for the disease [11]. Despite the large number and wide use in clinical trials, only a small subset of the existing scoring indexes have been validated to a certain extent for reproducibility and reliability in evaluating histologic findings in UC [7,8,11−14]. Furthermore, an even smaller number of scoring systems have been evaluated for responsiveness, which is defined as the ability to detect changes following a period of known histopathological alterations following known efficacious treatment [8,11].

The Riley score [15,16], modified Riley score [17], and Geboes score [18] are the most commonly used histopathology indexes among those validated scoring systems. All three scoring systems are shown to have substantial intraobserver agreements, while their interobserver agreements are less optimal [7,11,14,19]. However, none of the three scoring systems has been subjected to responsiveness testing. Two recently developed histopathology indexes, i.e. the Nancy histological index [19] and the Robarts histopathology index [20], are reported to have almost perfect intra- and interobserver agreements [11]. Both are shown to be responsive to histopathology changes [19,20]. Essential details of these histopathology indexes are listed in Table 15.3 and discussed as follows.

The Riley score was initially developed in 1988 [16] and subsequently described in 1991 [15] in a clinical study in investigation the role of microscopic inflammation in the prediction of relapse in UC patients. The patients were in symptomatic and endoscopic remission. The Riley score incorporates six histologic features, including characteristics of both activity (features 1−4) and chronicity (features 5 and 6). Each feature is graded on a four-point scale and is given an equal weight. Features of activity in the Riley score were shown to be predictive of the relapse in the following year [15].

The modified Riley score was described in 2005 in a multicenter, randomized, placebo-controlled trial for the treatment of active UC with vedolizumab [17]. In this study, rectal biopsies were graded with the subset of grading scores for activity in the Riley score (features 1−4 of the Riley score; Table 15.3). Active inflammation was graded on an eight-point scale from no inflammation to severe acute inflammation. This was designated as the modified Riley score. Details in histologic findings for each score in the modified Riley score were not discussed explicitly in the 2005 study. Key features of the modified Riley score listed in Table 15.3 are summarized on the basis of recent articles [8,20]. The rationale of discarding features of chronicity in the Riley score as discussed [11] is based on the observation that features of chronicity were resistant to short-term treatments.

The Geboes score is a comprehensive scoring system [18]. This scoring system evaluates multiple histologic features of activity and chronicity separately. Each feature is assigned to a major grade and is further graded on either a four-point or a five-point scale into subgrades. The assumption as discussed in the study is that the progression through each grade and subgrade correlates with increased disease severity or activity. The grade of the most affected biopsy fragment/the worse area should be used. In contrast to other scoring systems the Geboes score includes eosinophils in the lamina propria as one feature for disease activity in UC. In addition, the Geboes score has been shown to predict disease relapse in UC patients who were clinically in remission and had endoscopically inactive disease [21,22].

The Nancy histological index [19] and Robarts histopathology index [20] are new scoring systems developed around a similar time. Both used histologic features in existing histopathology indexes that were highly reliable in evaluating histologic disease activity. The Nancy histological index was built with statistical modeling using candidate histologic features, including those (1) that were reliable for the evaluation of histologic disease activity based on their preliminary work in UC and (2) selected based on expert opinion and literature review. The final Nancy index uses three histologic features, i.e. ulceration, acute and chronic inflammatory cell infiltration. The scoring algorithm is a stepwise, five-grade classification system detailed in Table 15.3.

The authors of the Robarts histopathology index first evaluated the modified Riley score and Geboes score and determined which features were reproducible and reliable for the evaluation of histologic disease activity [20]. Four features of the Geboes score, including chronic inflammatory infiltrates, lamina propria neutrophils, neutrophils in the epithelium, and erosion/ulceration, showed the best reliability that were used in the Robarts histopathology index. The total scores range from no disease activity to severe disease activity.

In pathology practice the severity of UC is usually assessed qualitatively or descriptively by pathologists using a grading system resembling the Geboes score [18]. Examples of quiescent colitis (Geboes score: grade 0), chronic inactive colitis (Geboes score: grade 1), mildly active colitis (Geboes score: grade 2), moderately active colitis (Geboes score: grades 3/4), and severely active colitis (Geboes score: grade 5) are illustrated in Fig. 15.3A−E. Additional

TABLE 15.3 Review of selected histopathology scoring systems in ulcerative colitis.

Score	Key features
Riley score, 1991 [15]	Six histological features graded on a four-point scale (0—none, 1—mild, 2—moderate, and 3—severe): 1. Neutrophil infiltration in lamina propria 2. Neutrophilic epithelial injury as crypt abscesses 3. Mucin depletion 4. Surface epithelial integrity 5. Chronic inflammatory cell infiltrate in lamina propria 6. Crypt architecture abnormalities
Modified Riley score, 2005 [17]	Scores only active inflammation: Score 0: normal mucosa or inactive colitis Score 1: scattered individual neutrophils in lamina propria Score 2: patchy collections of neutrophils in lamina propria Score 3: diffuse neutrophilic infiltrate in lamina propria Score 4: neutrophils in epithelium, <25% of crypts involved Score 5: neutrophils in epithelium, 25%–75% of crypts involved Score 6: neutrophils in epithelium >75% crypts involved Score 7: severe acute inflammation including erosion and ulceration

Geboes score, 2001 [18]	Evaluate multiple features of activity and chronicity; each major grade is subdivided into subgrades based upon severity or activity:
	Grade 0: structural changes only
	    Subgrades 0.0 (no abnormality)–0.3 (severe diffuse or multifocal abnormality)
	Grade 1: chronic inflammation
	    Subgrades 1.0 (no increase)–1.3 (marked increase)
	Grade 2: lamina propria eosinophils and neutrophils
	Grade 2A: eosinophils
	    Subgrades 2A.0 (no increase)–2A.3 (marked increase)
	Grade 2B: neutrophils
	    Subgrades 2B.0 (none)–2B.3 (marked increase)
	Grade 3: neutrophils in epithelium
	    Subgrades 3.0 (none)–3.3 (>50% crypts involved)
	Grade 4: crypt destruction
	    Subgrades 4.0 (none)–4.3 (unequivocal crypt destruction)
	Grade 5: erosions or ulcers
	    Subgrades 5.0 (no erosion, ulceration, or granulation tissue)–5.4 (ulcer or granulation tissue)

The Nancy index, 2017 [19]	Three histologic features and a five-grade, stepwise, classification scheme:
	Grade 4: severely active disease (ulceration)
	    If no ulceration, acute inflammatory cell infiltrate (presence of neutrophils) is assessed
	Grade 3: moderately active disease (moderate or severe acute inflammatory cells infiltrate)
	Grade 2: mildly active disease (mild acute inflammatory cells infiltrate)
	    If no acute inflammatory cells infiltrate, assessment of chronic inflammatory infiltrate (lymphocytes and plasma cells) is made
	Grade 1: chronic inflammatory cell infiltrate with no acute inflammatory infiltrate (moderate to marked increase in chronic inflammatory infiltrate)
	Grade 0: no histological significant disease (no or mild increase in chronic inflammatory infiltrate)

Robarts histopathology index, 2017 [20]	Used reproducible and reliable features in Geboes score; total score ranges from 0 (no disease activity) to 33 (severe disease activity):
	Robarts histopathology index
	    = 1 × chronic inflammatory infiltrate level (four levels)
	    + 2 × lamina propria neutrophils (four levels)
	    + 3 × neutrophils in epithelium (four levels)
	    + 5 × erosion or ulceration (four levels after combining Geboes 5.1 and 5.2)

descriptors are often provided for (1) the presence and absence of dysplasia, (2) severity of dysplasia (low-grade or high-grade) when present, (3) the presence and absence of granuloma, and (4) the quantity (one/single, rare, a few, or abundant) of granulomas when present. Histologically, the presence of granuloma may raise the suspicion for a diagnosis of CD. However, collections of histiocytes adjacent to ruptured crypts can be seen in UC.

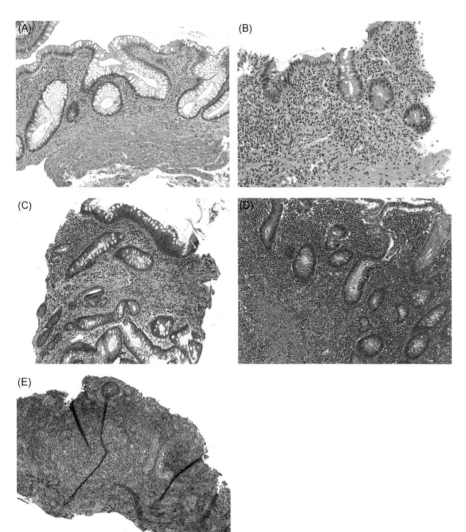

FIGURE 15.3 Examples of quiescent colitis (A), chronic inactive colitis (Geboes score: grade 1) (B), mildly active colitis (Geboes score: grade 2) (C), moderately active colitis (Geboes score: grade 3/4) (D), and severely active colitis (Geboes score: grade 5) (E).

Microscopic features of Crohn's disease

Inflammation in CD, as previously discussed, often involves small intestine and proximal colon; it is segmental and transmural. Histologically, CD involvement of mucosa in either small intestine or colon classically demonstrates "skip lesions," in which normal mucosa is alternating with areas of activity and chronic injury. Mucosal disease in CD shares many histologic features of activity and chronicity with UC, such as cryptitis, crypt abscess, crypt architecture distortion, and basal plasmacytosis (Table 15.2). Histologic features of mucosal disease in CD are aphthous ulcers and epithelioid granulomas. Histologically, aphthous ulcers are small erosions with neutrophilic infiltrates of the surface epithelium overlying lymphoid aggregates. Epithelioid granulomas are composed of aggregates of epithelioid histiocytes intermixed with lymphocytes and neutrophils without central necrosis (Fig. 15.4A and Table 15.2). In addition to mucosal disease, CD is characterized by mural changes, including deep fissuring ulcers, fistula and sinus tracts, transmural lymphoid aggregates, neural hypertrophy, *muscularis propria* hypertrophy, inflammation and fibrosis of mesentery, subserosal tissue, and serosa (Fig. 15.4B−D). Epithelioid granulomas can also be present in deep layers of the bowel wall (Table 15.2). Histologic features of mural changes are typically seen in CD resection specimens, not in UC.

Similar to UC, many histologic scoring systems (at least 13) have been developed for CD to evaluate disease activity in clinical trials (Table 15.4) [23,24]. Only a few indexes have been validated to a certain degree for their reproducibility, reliability, and responsiveness [8,23,24]. The most commonly used histopathology index is the Global Histologic Disease Activity Score (GHAS) [25,26]. This scoring index evaluates eight items in total. Seven items are histopathology features of active inflammation, chronic inflammation, architectural changes, and the presence of

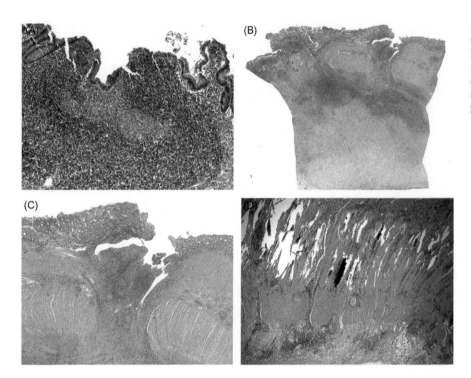

FIGURE 15.4 Features characteristics to Crohn's disease are epithelioid granuloma (A); sinus, and inflammation and fibrosis of mesentery, subserosal, and serosa (B). Other features are deep fissuring ulcers (C); and transmural lymphoid aggregates (D).

granulomas. The eighth item is the number of biopsy specimens affected [25]. GHAS has been modified many times to score colonic and ileal biopsies separately or to score only inflammation without architectural changes [27−30]. The details of these modified indexes are beyond the scope of this chapter. Readers can find detail information in the original publications [27−30] and review articles [8,23]. GHAS and indexes modified from GHAS have been shown to be responsive to histopathological changes due to known effective therapy [26].

The index by Naini and Cortina was built in an attempt to standardize pathology diagnosis of CD [31]. This index scores the ileum and colon separately, each by a set of histopathological features (Table 15.4). Both sets of features contain items for evaluating active and chronic inflammation, as well as chronic injury, such as architecture distortion and *muscularis mucosae* hyperplasia. Validation for reliability was performed during the development of this histopathology index. The interobserver agreement was excellent (correlation coefficient between pathologists: 0.94−0.96).

In pathology practice the severity of CD is usually assessed qualitatively or descriptively by pathologists using a grading system resembling the Geboes score [18], in a similar fashion to UC. Comparable descriptors for dysplasia and granulomas are also provided. The diagnostic categories include quiescent enteritis/colitis, chronic inactive enteritis/colitis, mildly active enteritis/colitis, moderately active enteritis/colitis, and severely active enteritis/colitis. Obviously, this assessment is not optimal, as it does not take changes in the deep layers of bowel, such as *muscular mucosae* hypertrophy and submucosal fibrosis, into consideration.

Histologic grading systems in pouchitis and cuffitis

Pouchitis is defined as a nonspecific inflammation of the ileal reservoir in IBD patients, who underwent ileal pouch−anal anastomosis (IPAA) after proctocolectomy [32]. The most commonly used disease activity index for pouchitis was first described in 1994 [33]. This index evaluates pouchitis through a combined assessment of clinical symptoms, endoscopic findings, and histopathology. The histopathology subscore evaluates acute inflammation with neutrophilic infiltration and ulceration separately. Each is graded on a three-point scale as follows. For acute inflammation, mild neutrophilic inflammation is assigned score 1; moderate neutrophilic inflammation with crypt abscess is assigned score 2; and severe neutrophilic inflammation with crypt abscess, score 3. For ulceration the average percentage of a biopsy involved by ulceration evaluated at low magnification is reported. Ulceration in less than 25% of the biopsy is graded as score 1; ulceration involving 25%−50% of the biopsy is graded as score 2; and ulceration in more than 50% of the biopsy, score 3. The maximum score of the histopathology component is 6.

TABLE 15.4 Review of selected histopathology scoring indexes in Crohn's disease.

Score	Key features

GHAS, 1998 [25]	Each feature is scored independently
	A moderate increase suggests up to twice the number of cells that can normally be expected
	A severe increase suggests more than twice the normal number of cells
	    Epithelial damage: 0—normal, 1—focal pathology, and 2—extensive pathology
	    Architectural changes: 0—normal, 1—moderately disturbed (<50%), 2—severely disturbed (>50%)
	    Infiltration of mononuclear cells in lamina propria: 0—normal, 1—moderate increase, 2—severe increase
	    Infiltration of polymorphonuclear cells in the lamina propria: 0—normal, 1—moderate increase, and 2—severe increase
	    Polymorphonuclear cells in epithelium: 1—in surface epithelium, 2—cryptitis, and 3—crypt abscess
	    Presence of erosion and/or ulcers: 0—no and 1—yes
	    Presence of granuloma: 0—no and 1—yes
	    Number of biopsy specimens affected: 0—none, 1—≤33%, 2—>33%–≤66%, and 3—>66%

Naini and Cortina, 2012 [31]	Ileum and colon are scored separately. Total ileitis score ranges from 0 to 10; total colitis score ranges from 0 to 17.
	Ileum:
	    Architecture distortion: 0—absent, 1—mild, and 2—conspicuous
	    Increased lymphoplasmacytic inflammation in lamina propria: 0—absent, 1—mild, and 2—conspicuous
	    Neutrophilic inflammation, including erosions/ulcerations: 0—absent, 1—mild, and 2—conspicuous
	    Granulomas: 0—absent and 1—present
	    Pyloric gland metaplasia: 0—absent and 1—present
	Colon:
	    Crypt architecture distortion: 0—absent, 1—mild, and 2—conspicuous
	    Basal lymphoplasmacytosis: 0—absent, 1—mild, and 2—conspicuous
	    Cryptitis and crypt abscesses: 0—absent, 1—mild, and 2—conspicuous
	    Ulcers: 0—absent and 1—present
	    Granulomas: 0—absent and 1—present
	    Increased lamina propria eosinophils: 0—absent, 1—mild, and 2—conspicuous
	    Paneth cell or pyloric gland metaplasia: 0—absent and 1—present
	    Lymphoid nodules at base: 0—absent and 1—present
	    Muscularis mucosae hyperplasia, splaying of fibers, adipose tissue in mucosa, and base of lamina propria fibrosis: 0—absent and 1—present
	    Hyperplasia of endocrine cells: 0—absent and 1—present

GHAS, Global histologic activity score.

The rectal cuff refers to the small segment of the distal rectal mucosa, typically 1.5—2 cm in length, between the stapled IPAA anastomosis line and the anal transitional zone [34]. Cuffitis is defined as inflammation in this segment of rectal mucosa on endoscopy and histopathology, with or without inflammation in the body of the ileal pouch [35,36]. Cuffitis has been considered a form of residual UC. Histologic features of cuffitis are similar to UC with active or acute inflammation (i.e., neutrophil infiltration) and chronic changes (e.g., crypt distortion) [35]. To the best of our knowledge the histologic scoring system has not been developed specifically for grading disease activity in the rectal cuff. However, it is likely that histologic disease activity indexes used for UC may be applicable in grading inflammation in the rectal cuff.

Histologic mucosal healing in ulcerative colitis and Crohn's disease

Advances in the medical treatment of IBD in the past two decades have made sustained mucosal healing possible [29,37—41]. As discussed in previous chapters, endoscopic mucosal healing is currently considered as the primary endpoint in many clinical trials as achieving endoscopic mucosal healing is associated with sustained clinical remission, reduced hospitalization rate, and decreased colectomy [37—39,42,43]. A recent consensus paper by the Selecting Therapeutic Targets in Inflammatory Bowel Disease (STRIDE) program recommends both clinical and endoscopic remissions as targets for routine clinical practice in patients with IBD [44]. This consensus paper acknowledges histopathology as a sensitive measure of inflammation in UC. However, due to insufficient evidence of its clinical utility in IBD patients, the achievement of histologic mucosal healing is not recommended as a therapeutic target for IBD at this time, but it is considered as an adjunctive goal for clinical management of UC. As also discussed in this consensus, a paradigm shift will be required when it is time to incorporate histopathology into routine patient management.

Currently, there is no widely accepted, standardized, definition of histologic healing for either UC or CD [6,8,44]. The term "healing" has been used interchangeably with the term "remission" in the literature, although they may not be synonymous [6,8]. In the various histopathology indexes discussed above, the definition of histologic remission ranges from persistent architecture changes without inflammation to complete normalization of the mucosa [7,13,15,17−20,31,45]. The general consensus of histologic remission is the lack of neutrophils in both epithelium and lamina propria [6,8]. The definition of histologic remission proposed by the International Organization for the Study of Inflammatory Bowel Disease (IOIBD) includes three components: (1) absence of neutrophils both in the crypts (epithelium) and lamina propria, (2) absence of basal plasma cells and ideally normal amount of lamina propria plasma cells, and (3) a normal amount of lamina propria eosinophils [6].

Histologic mucosal healing and endoscopic mucosal healing do not always correlate with each other. Many studies have shown that endoscopically normal mucosa, that is, the Mayo endoscopy score [46] of 0, can have persistent microscopic inflammation in UC [18,22,47−51]. Active inflammation (neutrophilic infiltration), chronic inflammation, and basal plasmacytosis have been seen in up to 40%, 29%, and 48% of endoscopically normal mucosa in UC, respectively. In addition, mild disease on endoscopy with a Mayo endoscopy score [46] of 1 (mild disease, erythema, and decreased vascularity) can have microscopic erosions and ulcers in up to 14% of UC cases [48]. In CD the assessment of disease activity using mucosal biopsies is limited due to the segmental and transmural nature of CD. However, observational findings in resection specimens show that microscopic mucosal inflammation in the absence of deep inflammation is common in CD, while deep inflammation without mucosal disease is rare [8]. These findings support the use of mucosal biopsies for the measurement of histologic disease activity in CD. Nevertheless, microscopic disease activity in CD mucosal biopsies, including active inflammation, chronic inflammation, architecture changes, or granulomas, has been reported in 25%−40% of endoscopically normal mucosa [45,52,53]. These results suggest that histopathology is more sensitive than endoscopy at detecting disease activity or inflammation in both UC and CD.

Persistent microscopic disease activity has been shown to be associated with an increased risk for clinical relapse in UC [15,22,47,49,50,54,55]. Multiple clinical studies have demonstrated that histologic mucosal healing/remission at baseline is associated with significantly decreased risk of clinical relapse/exacerbation. Histologic mucosal healing at baseline has also been shown to be superior to endoscopic mucosal healing and clinical remission in predicting clinical outcomes, such as clinical relapse, corticosteroid use, and hospitalization [49,50]. In a metaanalysis of 15 studies consisting of 1573 patients, histologic features associated with improved clinical outcomes include the absence of neutrophils in the epithelium or lamina propria and absence of increased eosinophils and chronic inflammatory cell infiltrates [55]. A recent clinical study demonstrates that complete histologic normalization of the mucosa can be achieved in UC and, when achieved, is a better predictor for relapse-free survival than either endoscopic mucosal healing or histologic quiescent disease (i.e., histologic findings of architecture changes only without neutrophilic inflammation) [56].

Histologic disease activity is also linked to the development of colitis-associated dysplasia or colorectal carcinoma in UC [57−59]. UC patients with persistent microscopic inflammation with neutrophilic inflammation in the epithelium over a long period of time (up to 10 years) carry an increased risk in developing neoplasia, including low-grade, high-grade, or colorectal carcinoma. The severity of inflammation over time has also been shown to be an independent risk factor for the progression to high-grade dysplasia and/or carcinoma [58].

The prognostic value of histologic mucosal healing for clinical relapse or neoplasia in CD is much less studied compared with that in UC [51,60]. The presence of active inflammation at baseline is shown to be predictive of subsequent disease flares at 6, 12, and 24 months, while endoscopic remission at baseline is not [60]. In CD resection specimens, inflammation of the submucosal plexus and lymphatic vessel density at resection margin are associated with clinical outcomes and/or postoperative recurrence [61−65]. There is conflicting literature regarding the significance of histologically active disease and the presence of granulomas at the resection margin [8,61]. CD patients have an increased risk in developing small intestinal and colorectal carcinoma [66,67] likely as a result of persistent mucosal inflammation. However, the correlation between histologic disease activity in mucosa and risk of neoplasia in CD has not been studied.

Lastly, the assessment of histologic mucosal healing in IBD is limited by the lack of a wildly accepted definition of histologic mucosal healing and the lack of adequately validated, robust histopathology scoring systems. The definition proposed by the IOIBD may be considered as a good starting point toward a unified definition. However, the precise definition of basal plasmacytosis remains unclear and currently, basal plasmacytosis is not graded by most histologic scoring indexes. Data available in the literature regarding the association between eosinophils and histologic remission are limited. Additional studies are needed to investigate the role of eosinophils in histologic remission. Further, clinical studies should also be performed to explore the benefit of achieving histologic mucosal healing/remission in CD. The application and assessment of histologic mucosal healing are further limited by the availability and quality of tissue samples. The number of biopsies needed for an accurate measurement of histologic disease activity remains to be determined. Thus studies and validations are also required to optimize biopsy protocol and sampling.

Summary and recommendations

Histopathological examination of intestinal specimens, either biopsy or resection, remains an essential component of patient care in IBD. Pathology practice in this field is advancing into achieving the ability to reliably and thoroughly measure disease activity and predict the clinical outcome of IBD patients. In addition, there is a need for achieving a uniformed definition of histologic mucosal healing, which indeed poses a great challenge to pathologists in an era when they are face dilemmas of increased clinical demands in conjunction with decreased reimbursements.

Grant support

None.

Disclosure

None.

References

[1] Patil DT, Greenson JK, Odze RD. Inflammatory disorders of the large intestine. In: Odze RD, Goldblum JR, editors. Surgical pathology of the GI tract, liver, biliary tract, and pancreas. 3rd ed. Elsevier Saunders; 2015. p. 436−511.

[2] Robert ME, Gibson JA. Inflammatory disorders of the small intestine. In: Odze RD, Goldblum JR, editors. Surgical pathology of the GI tract, liver, biliary tract, and pancreas. 3rd ed. Elsevier Saunders; 2015. p. 402−35.

[3] Ordas I, Eckmann L, Talamini M, Baumgart DC, Sandborn WJ. Ulcerative colitis. Lancet 2012;380:1606−19.

[4] Baumgart DC, Sandborn WJ. Crohn's disease. Lancet 2012;380:1590−605.

[5] Abreu MT, Harpaz N. Diagnosis of colitis: making the initial diagnosis. Clin Gastroenterol Hepatol 2007;5:295−301.

[6] Bryant RV, Winer S, Travis SP, Riddell RH. Systematic review: histological remission in inflammatory bowel disease. Is 'complete' remission the new treatment paradigm? An IOIBD initiative. J Crohns Colitis 2014;8:1582−97.

[7] Marchal Bressenot A, Riddell RH, Boulagnon-Rombi C, Boulagnon-Rombi C, Reinisch W, Danese S, et al. Review article: the histological assessment of disease activity in ulcerative colitis. Aliment Pharmacol Ther 2015;42957−67.

[8] Pai RK, Geboes K. Disease activity and mucosal healing in inflammatory bowel disease: a new role for histopathology? Virchows Arch 2018;472:99−110.

[9] Nostrant TT, Kumar NB, Appelman HD. Histopathology differentiates acute self-limited colitis from ulcerative colitis. Gastroenterology 1987;92:318−28.

[10] Truelove SC, Richards WC. Biopsy studies in ulcerative colitis. Br Med J 1956;1:1315−18.

[11] Mosli MH, Parker CE, Nelson SA, Baker KA, MacDonald JK, Zou GY, et al. Histologic scoring indices for evaluation of disease activity in ulcerative colitis. Cochrane Database Syst Rev 2017;5:Cd011256.

[12] Mosli MH, Feagan BG, Sandborn WJ, D'haens G, Behling C, Kaplan K, et al. Histologic evaluation of ulcerative colitis: a systematic review of disease activity indices. Inflamm Bowel Dis 2014;20:564−75.

[13] Mosli MH, Feagan BG, Zou G, Sandborn WJ, D'Haens G, Khanna R, et al. Reproducibility of histological assessments of disease activity in UC. Gut 2015;64:1765−73.

[14] Bressenot A, Salleron J, Bastien C, Danese S, Boulagnon-Rombi C, Peyrin-Biroulet L. Comparing histological activity indexes in UC. Gut 2015;64:1412−18.

[15] Riley SA, Mani V, Goodman MJ, Dutt S, Herd ME. Microscopic activity in ulcerative colitis: what does it mean? Gut 1991;32:174−8.

[16] Riley SA, Mani V, Goodman MJ, Herd ME, Dutt S, Turnberg LA. Comparison of delayed release 5 aminosalicylic acid (mesalazine) and sulphasalazine in the treatment of mild to moderate ulcerative colitis relapse. Gut 1988;29:669−74.

[17] Feagan BG, Greenberg GR, Wild G, Fedorak RN, Paré P, McDonald JW, et al. Treatment of ulcerative colitis with a humanized antibody to the alpha4beta7 integrin. New Engl J Med 2005;352:2499−507.

[18] Geboes K, Riddell R, Ost A, Jensfelt B, Persson T, Lofberg R. A reproducible grading scale for histological assessment of inflammation in ulcerative colitis. Gut 2000;47:404−9.

[19] Marchal-Bressenot A, Salleron J, Boulagnon-Rombi C, Boulagnon-Rombi C, Bastien C, Cahn V, et al. Development and validation of the Nancy histological index for UC. Gut 2017;66:43–9.

[20] Mosli MH, Feagan BG, Zou G, D'Haens G, Khanna R, Shackelton LM, et al. Development and validation of a histological index for UC. Gut 2017;66:50–8.

[21] Azad S, Sood N, Sood A. Biological and histological parameters as predictors of relapse in ulcerative colitis: a prospective study. Saudi J Gastroenterol 2011;17:194–8.

[22] Bessissow T, Lemmens B, Ferrante M, Bisschops R, Van Steen K, Geboes K, et al. Prognostic value of serologic and histologic markers on clinical relapse in ulcerative colitis patients with mucosal healing. Am J Gastroenterol 2012;107:1684–92.

[23] Mojtahed A, Khanna R, Sandborn WJ, D'Haens GR, Feagan BG, Shackelton LM, et al. Assessment of histologic disease activity in Crohn's disease: a systematic review. Inflamm Bowel Dis 2014;20:2092–103.

[24] Novak G, Parker CE, Pai RK, MacDonald JK, Feagan BG, Sandborn WJ, et al. Histologic scoring indices for evaluation of disease activity in Crohn's disease. Cochrane Database Syst Rev 2017;7:CD012351.

[25] D'Haens GR, Geboes K, Peeters M, Baert F, Penninckx F, Rutgeerts P. Early lesions of recurrent Crohn's disease caused by infusion of intestinal contents in excluded ileum. Gastroenterology 1998;114:262–7.

[26] D'Haens G, Geboes K, Rutgeerts P. Endoscopic and histologic healing of Crohn's (ileo-) colitis with azathioprine. Gastrointest Endosc 1999;50:667–71.

[27] Agnholt J, Dahlerup JF, Buntzen S, Tottrup A, Nielsen SL, Lundorf E. Response, relapse and mucosal immune regulation after infliximab treatment in fistulating Crohn's disease. Aliment Pharmacol Ther 2003;17:703–10.

[28] Geboes K, Rutgeerts P, Opdenakker G, Olson A, Patel K, Wagner CL, et al. Endoscopic and histologic evidence of persistent mucosal healing and correlation with clinical improvement following sustained infliximab treatment for Crohn's disease. Curr Med Res Opin 2005;21:1741–54.

[29] Mantzaris GJ, Christidou A, Sfakianakis M, Roussos A, Koilakou S, Petraki K, et al. Azathioprine is superior to budesonide in achieving and maintaining mucosal healing and histologic remission in steroid-dependent Crohn's disease. Inflamm Bowel Dis 2009;15:375–82.

[30] Sipponen T, Karkkainen P, Savilahti E, et al. Correlation of faecal calprotectin and lactoferrin with an endoscopic score for Crohn's disease and histological findings. Aliment Pharmacol Ther 2008;28:1221–9.

[31] Naini BV, Cortina G. A histopathologic scoring system as a tool for standardized reporting of chronic (ileo)colitis and independent risk assessment for inflammatory bowel disease. Hum Pathol 2012;43:2187–96.

[32] Gonzalo DH, Collinsworth AL, Liu X. Common inflammatory disorders and neoplasia of the ileal pouch: a review of histopathology. Gastroenterology Res 2016;9:29–38.

[33] Sandborn WJ, Tremaine WJ, Batts KP, Pemberton JH, Phillips SF. Pouchitis after ileal pouch-anal anastomosis: a Pouchitis Disease Activity Index. Mayo Clin Proc 1994;69:409–15.

[34] Thompson-Fawcett MW, Mortensen NJ. Anal transitional zone and columnar cuff in restorative proctocolectomy. Br J Surg 1996;83:1047–55.

[35] Thompson-Fawcett MW, Mortensen NJ, Warren BF. "Cuffitis" and inflammatory changes in the columnar cuff, anal transitional zone, and ileal reservoir after stapled pouch-anal anastomosis. Dis Colon Rectum 1999;42:348–55.

[36] Wu B, Lian L, Li Y, Remzi FH, Liu X, Kiran RP, et al. Clinical course of cuffitis in ulcerative colitis patients with restorative proctocolectomy and ileal pouch-anal anastomoses. Inflamm Bowel Dis 2013;19:404–10.

[37] Colombel JF, Rutgeerts P, Reinisch W, Schnitzler F, Noman M, Van Assche G, et al. Early mucosal healing with infliximab is associated with improved long-term clinical outcomes in ulcerative colitis. Gastroenterology 2011;141:1194–201.

[38] Ferrante M, Vermeire S, Fidder H, et al. Long-term outcome after infliximab for refractory ulcerative colitis. J Crohns Colitis 2008;2:219–25.

[39] Schnitzler F, Fidder H, Ferrante M, Noman M, Arijs I, Van Assche G, et al. Mucosal healing predicts long-term outcome of maintenance therapy with infliximab in Crohn's disease. Inflamm Bowel Dis 2009;15:1295–301.

[40] Hanauer SB, Feagan BG, Lichtenstein GR, Mayer LF, Schreiber S, Colombel JF, et al. Maintenance infliximab for Crohn's disease: the ACCENT I randomised trial. Lancet 2002;359:1541–9.

[41] Rutgeerts P, Sandborn WJ, Feagan BG, Reinisch W, Olson A, Johanns J, et al. Infliximab for induction and maintenance therapy for ulcerative colitis. New Engl J Med 2005;353:2462–76.

[42] Baert F, Moortgat L, Van Assche G, Caenepeel P, Vergauwe P, De Vos M, et al. Mucosal healing predicts sustained clinical remission in patients with early-stage Crohn's disease. Gastroenterology 2010;138:463–8.

[43] Froslie KF, Jahnsen J, Moum BA, Vatn MH. Mucosal healing in inflammatory bowel disease: results from a Norwegian population-based cohort. Gastroenterology 2007;133:412–22.

[44] Peyrin-Biroulet L, Sandborn W, Sands BE, Reinisch W, Bemelman W, Bryant RV, et al. Selecting therapeutic targets in inflammatory bowel disease (STRIDE): determining therapeutic goals for treat-to-target. Am J Gastroenterol 2015;110:1324–38.

[45] D'Haens G, Van Deventer S, Van Hogezand R, Chalmers D, Kothe C, Baert F, et al. Endoscopic and histological healing with infliximab anti-tumor necrosis factor antibodies in Crohn's disease: a European multicenter trial. Gastroenterology 1999;116:1029–34.

[46] Schroeder KW, Tremaine WJ, Ilstrup DM. Coated oral 5-aminosalicylic acid therapy for mildly to moderately active ulcerative colitis. New Engl J Med 1987;317:1625–9.

[47] Calafat M, Lobaton T, Hernandez-Gallego A, Mañosa M, Torres P, Cañete F, et al. Acute histological inflammatory activity is associated with clinical relapse in patients with ulcerative colitis in clinical and endoscopic remission. Dig Liver Dis 2017;49:1327–31.

[48] Lemmens B, Arijs I, Van Assche G, Sagaert X, Geboes K, Ferrante M, et al. Correlation between the endoscopic and histologic score in assessing the activity of ulcerative colitis. Inflamm Bowel Dis 2013;19:1194–201.

[49] Bryant RV, Burger DC, Delo J, Walsh AJ, Thomas S, von Herbay A, et al. Beyond endoscopic mucosal healing in UC: histological remission better predicts corticosteroid use and hospitalisation over 6 years of follow-up. Gut 2016;65:408—14.

[50] Zenlea T, Yee EU, Rosenberg L, Boyle M, Nanda KS, Wolf JL, et al. Histology grade is independently associated with relapse risk in patients with ulcerative colitis in clinical remission: a prospective study. Am J Gastroenterol 2016;111:685—90.

[51] Baars JE, Nuij VJ, Oldenburg B, Kuipers EJ, van der Woude CJ. Majority of patients with inflammatory bowel disease in clinical remission have mucosal inflammation. Inflamm Bowel Dis 2012;18:1634—40.

[52] Molander P, Sipponen T, Kemppainen H, Jussila A, Blomster T, Koskela R, et al. Achievement of deep remission during scheduled mainte-nance therapy with TNFalpha-blocking agents in IBD. J Crohns Colitis 2013;7(9):730—5.

[53] Regueiro M, Kip KE, Schraut W, Baidoo L, Sepulveda AR, Pesci M, et al. Crohn's disease activity index does not correlate with endoscopic recurrence one year after ileocolonic resection. Inflamm Bowel Dis 2011;17:118—26.

[54] Bitton A, Peppercorn MA, Antonioli DA, Niles JL, Shah S, Bousvaros A, et al. Clinical, biological, and histologic parameters as predictors of relapse in ulcerative colitis. Gastroenterology 2001;120:13—20.

[55] Park S, Abdi T, Gentry M, Laine L. Histological disease activity as a predictor of clinical relapse among patients with ulcerative colitis: system-atic review and meta-analysis. Am J Gastroenterol 2016;111:1692—701.

[56] Christensen B, Hanauer SB, Erlich J, Kassim O, Gibson PR, Turner JR, et al. Histologic normalization occurs in ulcerative colitis and is associ-ated with improved clinical outcomes. Clin Gastroenterol Hepatol 2017;15:1557—1564.e1551.

[57] Choi CR, Al Bakir I, Ding NJ, Lee GH, Askari A, Warusavitarne J, et al. Cumulative burden of inflammation predicts colorectal neoplasia risk in ulcerative colitis: a large single-centre study. Gut 2019;68:414—21.

[58] Gupta RB, Harpaz N, Itzkowitz S, Hossain S, Matula S, Kornbluth A, et al. Histologic inflammation is a risk factor for progression to colorectal neoplasia in ulcerative colitis: a cohort study. Gastroenterology 2007;133:1099—105 quiz 1340—1091.

[59] Rutter M, Saunders B, Wilkinson K, Rumbles S, Schofield G, Kamm M, et al. Severity of inflammation is a risk factor for colorectal neoplasia in ulcerative colitis. Gastroenterology 2004;126:451—9.

[60] Brennan GT, Melton SD, Spechler SJ, Feagins LA. Clinical implications of histologic abnormalities in ileocolonic biopsies of patients with Crohn's disease in remission. J Clin Gastroenterol 2017;51:43—8.

[61] Bressenot A, Peyrin-Biroulet L. Histologic features predicting postoperative Crohn's disease recurrence. Inflamm Bowel Dis 2015;21:468—75.

[62] Decousus S, Boucher AL, Joubert J, Pereira B, Dubois A, Goutorbe F, et al. Myenteric plexitis is a risk factor for endoscopic and clinical post-operative recurrence after ileocolonic resection in Crohn's disease. Dig Liver Dis 2016;48:753—8.

[63] Lemmens B, de Buck van Overstraeten A, Arijs I, Sagaert X, Van Assche G, Vermeire S, et al. Submucosal plexitis as a predictive factor for postoperative endoscopic recurrence in patients with Crohn's disease undergoing a resection with ileocolonic anastomosis: results from a pro-spective single-centre study. J Crohns Colitis 2017;11:212—20.

[64] Sokol H, Polin V, Lavergne-Slove A, Panis Y, Treton X, Dray X, et al. Plexitis as a predictive factor of early postoperative clinical recurrence in Crohn's disease. Gut 2009;58:1218—25.

[65] Rahier JF, Dubuquoy L, Colombel JF, Jouret-Mourin A, Delos M, Ferrante M, et al. Decreased lymphatic vessel density is associated with post-operative endoscopic recurrence in Crohn's disease. Inflamm Bowel Dis 2013;19:2084—90.

[66] Munkholm P. Review article: the incidence and prevalence of colorectal cancer in inflammatory bowel disease. Aliment Pharmacol Ther 2003;18(Suppl. 2):1—5.

[67] Shaukat A, Virnig DJ, Howard D, Sitaraman SV, Liff JM, Lederle FA. Crohn's disease and small bowel adenocarcinoma: a population-based case-control study. Cancer Epidemiol Biomarkers Prev 2011;20:1120—3.

Chapter 16

Postoperative complications in Crohn's disease

Bo Shen

Center for Inflammatory Bowel Diseases, Columbia University Irving Medical Center-New York Presbyterian Hospital, New York, NY, United States

Chapter Outline

Abbreviations

CD Crohn's disease
ECF enterocutaneous fistula
ICA ileocolonic anastomosis
ICR ileocolonic resection

Introduction

Bowel resection and anastomosis, strictureplasty, and fecal diversion with ileostomy are common surgical treatment modalities for Crohn's disease (CD). Postoperative complications after any surgery for CD are common, ranging from anastomotic stricture to anastomotic leak. The high risk for the surgery-associated complications may be related to underlying disease process of CD, concurrent immunosuppressive medications, poor nutrition status, and mesenteric fat [1,2]. In fact, the majority of complications are located at or around anastomosis. The focus of this chapter is long-term complications after surgery. Endoscopic evaluation, along with abdominal imaging, plays an important role in the diagnosis and management of those complications.

Immediate postoperative complications, such as wound infection, ileus, pelvic abscess, or sepsis, are not discussed.

Postsurgical complications after restorative proctocolectomy and ileal pouch for ulcerative colitis are discussed in Chapter 11, Ulcerative colitis: postsurgical.

Obstruction

Partial small bowel obstruction is common in CD patients who have undergone surgery. The obstruction is commonly caused by intrinsic anastomotic stricture or extrinsic luminal angulation from adhesions or compression. Endoscopy plays a key role in the identification, measurement, and treatment of anastomotic stricture. Endoscopy can be reliable in characterizing end-to-end, end-to-side, and side-to-side anastomoses. In addition to detect luminal narrowing at the anastomosis, endoscopy may also diagnose concurrent structural abnormalities in the anastomosis area (Fig. 16.1). The anastomotic stricture can be treated with endoscopic balloon dilation (Fig. 16.2) [3] or endoscopic stricturotomy or endoscopic strictureplasty with needle knife or insulated-tip knife (Fig. 16.3) [4]. Endoscopic treatment of anastomotic stricture has been proven to be safe and effective. The response to endoscopic therapy is determined by multiple factors,

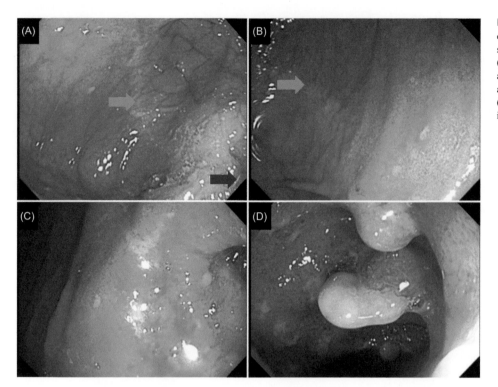

FIGURE 16.1 Side-to-side ileo-colonic anastomosis. (A and B) A sharp demarcation dividing ileum (*green arrow*) and colon, with anastomosis highlighted in blue arrow; (C) anastomotic stricture; (D) inflammatory polyps just proximal to the stricture.

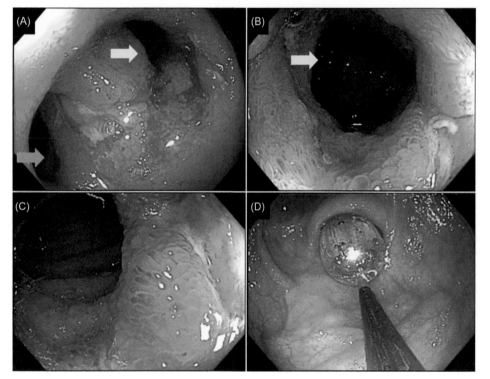

FIGURE 16.2 Side-to-side ileo-colonic anastomosis with stricture in Crohn's disease. (A) There are two orifices viewed from the colon side, one leading to the neo-terminal ileum (*green arrow*) with mild ulcerated stricture and another leading to a bind end of transverse staple line (*yellow arrows*). (B) The blind end side; (C) anasto-motic stricture with normal neo-terminal ileum; (D) endoscopic balloon dilation of the anastomotic stricture.

such as degree and length of strictures, size of endoscopic balloon, current medical therapy, and consumption of tobacco [5]. This author found that the poor response of anastomotic anastomosis to the endoscopic therapy in those with concurrent anastomotic stricture and mucosal prolapse in the segment of bowel just proximal to the anastomosis (Figs. 16.4 and 16.5). Sometimes, the anastomotic stricture may be covered by inflammatory polyps (Fig. 16.6).

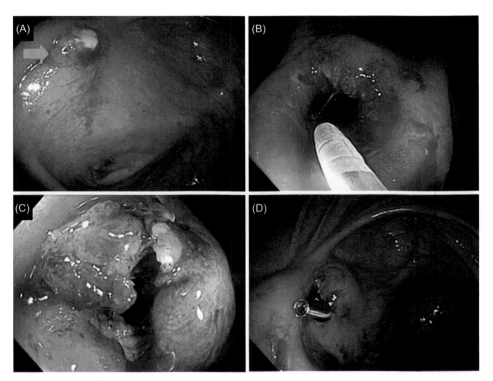

FIGURE 16.3 End-to-side ileo-colonic anastomotic stricture in Crohn's disease. (A) Stricture was covered by an inflammatory polyp (*green arrow*); (B) endoscopic stricturotomy with needle knife; (C) electroincised stricture; (D) placement of endoclips after stricturotomy to serve as spacers and to prevent bleeding and perforation. The combined endoscopic stricturotomy and placement of spacers is also termed endoscopic strictureplasty.

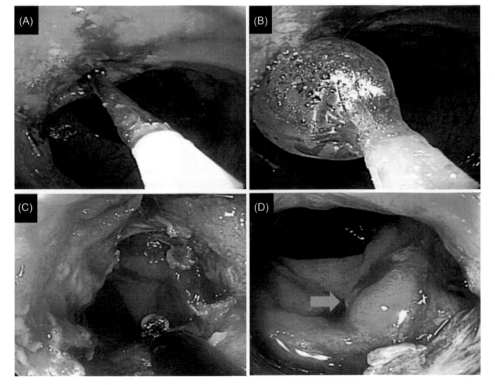

FIGURE 16.4 Refractory end-to-side anastomotic stricture in Crohn's disease. (A and B) The stricture was dilated with through-the-scope balloon; (C) the stricture was further treated with insulated-tip knife; (D) partial prolapse (*green arrow*) of the ileum just proximal to the strictured anastomosis, which resulted in the patient's poor response to the endoscopic therapy.

Therefore diagnostic and therapeutic endoscopy can be combined with other modalities, such as the use of soft-tip guide wire. Endoscopic therapy has become a standard treatment modality in the treatment of anastomotic strictures. A salvage plan for damage control of bleeding or perforation should be readily available and executed when endoscopic therapy is performed (Fig. 16.7).

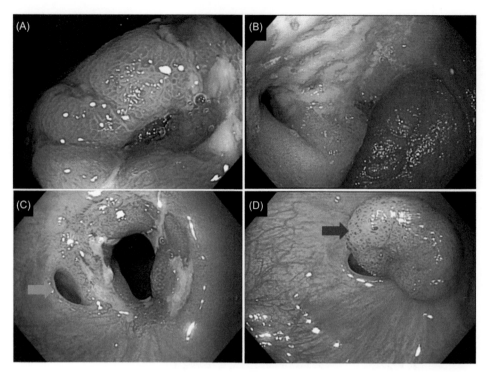

FIGURE 16.5 Concurrent stricture and prolapse at the anastomosis in two patients with Crohn's disease. (A and B) Circumferential mucosal prolapse blocking ileocolonic anastomosis. An ulcerated anastomotic stricture underneath prolapse (B); (C and D) A separate patient had a stricture at stapled ileorectal anastomosis with a small diverticulum nearby (*green arrow*). A prolapse ileum mucosa (*blue arrow*) intermittently blocked the anastomosis, contributing to his symptoms.

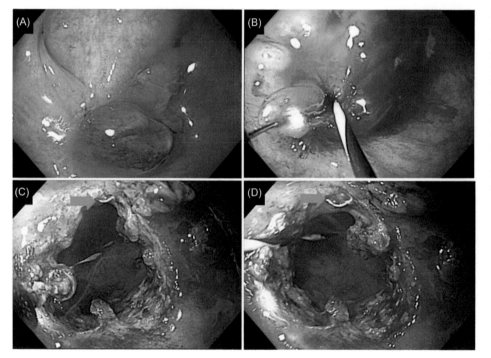

FIGURE 16.6 Sealed colorectal stapled anastomosis in Crohn's disease. (A) A short 1 cm tight stricture was blocked by granular tissue; (B and C) a soft-tip wire being used to guide endoscopic stricturotomy with an insulated-tip knife; (D) status after the endoscopic therapy. A dislodged staple was heighted by green arrow. In the absence of mucosal inflammation at the adjacent bowels, the stricture likely resulted from surgical ischemia.

   While the most common anastomotic strictures are located at the junction between small bowel and colon or rectum, strictures can occur in small-bowel-to-small-bowel anastomosis (Fig. 16.8), large-bowel-to-large-bowel anastomosis (Fig. 16.9), and the strictureplasty site (Fig. 16.10). Surgical ischemia-associated stricture can also occur in the ileostomy site, at the fascia (Fig. 16.11). A stricture can also occur in the anus, distal rectum, or anastomosis close to the anus, in patients with long-term fecal diversion (Fig. 16.12). Small-intestinal bacterial overgrowth is considered a "norm" in patients with strictureplasty or prestenotic luminal dilation (Table 16.1).

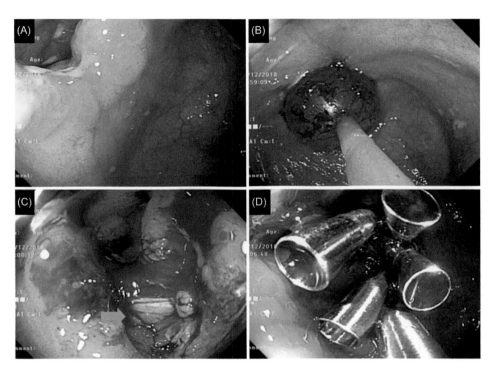

FIGURE 16.7 Stricture at the ileocolonic end-to-end anastomosis in Crohn's disease. (A) Ulcerated anastomotic stricture; (B) endoscopic balloon dilation of the stricture; (C) an iatrogenic perforation resulting from the balloon dilation (*green arrow*); (D) the perforation was sealed by the deployment of multiple endoclips.

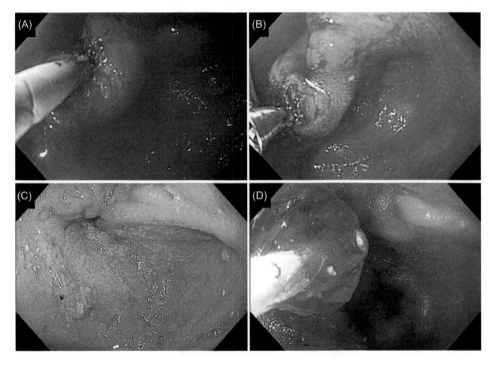

FIGURE 16.8 Strictures at ileoileal hand-sewn anastomoses in Crohn's disease. (A and B) Angulated stricture being treated with endoscopic needle knife strictureplasty with the placement of endoclips to prolong the patency of lumen and to prevent bleeding and perforation; (C and D) nonulcerated stricture being dilated with endoscopic balloon.

Acute and chronic sutures or staple line leaks

Acute or chronic leaks at the anastomosis, staple line, or suture line are common in CD patients undergoing surgery. Those leaks can cause wound infection, abscess, sepsis, sinus, and fistula. Both hand-sewn and stapled anastomosis or suturing can be associated with anastomotic complications. While acute leaks with abscess or sepsis are mainly diagnosed with abdominal and pelvic imaging, chronic leaks with sinus or fistula can be detected with a combined assessment of endoscopy and abdominal imaging (Table 16.1).

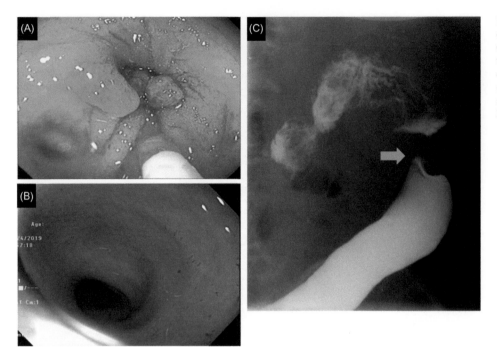

FIGURE 16.9 Severe stricture at colo-colonic anastomosis in Crohn's disease. (A) Pinhole stricture at the anastomosis being dilated with endoscopic balloon; (B) mild colitis proximal to the stricture; (C) the stricture is highlighted with green arrow.

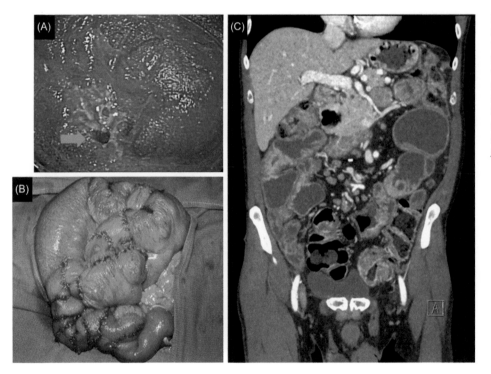

FIGURE 16.10 Strictures at jejunum strictureplasty in CD. (A) Stricture at the inlet or outlet of the strictureplasty site is common, which may be associated to CD or surgical ischemia (*green arrow*); strictureplasties with dilated bowel lumen shown in the operating room (B) and on CT enterography (C). *CD*, Crohn's disease. *Surgical photo: Courtesy Dr. Victor Fazio.*

Leaks at ileocolonic anastomosis and ileocolonic resection are most commonly studied. The leak can occur at the anastomosis or at the transverse staple line. On abdominal imaging the leak can present with enterocutaneous fistula (ECF) and intraabdominal abscess from transverse staple line leak. Anastomotic leak, however, rarely, leads to entero-oenteric fistula (Figure 8.5). While acute and chronic anastomotic leaks and their septic sequelae have traditionally been managed with interventional radiology or surgical intervention, endoscopy plays a growing role in the identification of the source of leak and possible delivery of therapy (Fig. 16.13). Endoscopy, mainly colonoscopy, may be able to

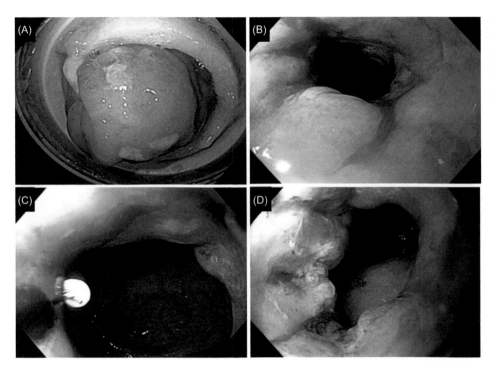

FIGURE 16.11 Stricture of the distal ileum in a Crohn's disease patient with an ileostomy and stoma site stricture. (A) Ileostomy; (B) a tight ischemic stricture at the fascia level, 10 cm from the stoma; (C and D) the stricture was treated with endoscopic stricturotomy with an insulated-tip knife.

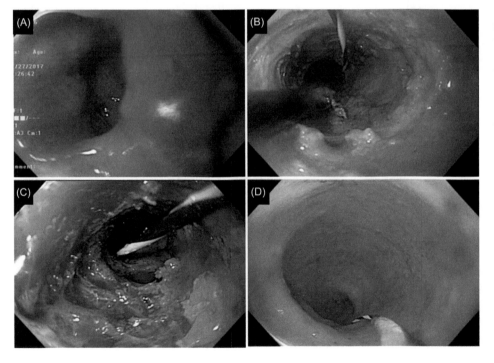

FIGURE 16.12 Anal stricture from long-term fecal diversion in Crohn's disease. (A) Severe anal stricture that was not traversable to gastroscope; (B) wire-guided endoscopic stricturotomy with insulated-tip knife; (C) the treated stricture; (D) endoscope was passed through the stricture, with a view of the diverted colon.

identify the leak at the staple or suture line. In patients with ECF the administration of hydrogen peroxide or betadine through the external opening at the skin or guide line side may help identify the internal opening at the bowel side (Figs. 16.14 and 16.15). Endoscopic closure with through-the-scope clip or over-the-scope clips may be attempted for suture or staple line leaks (Fig. 16.14) [5,6]. Other common locations of suture or staple line leak include one at the rectum (i.e.Hartmann pouch) or colon stump in patients with ostomies (Fig. 16.15). Patients with ostomies are at particular

TABLE 16.1 Postoperative complications in Crohn's disease.

Category	Subcategory	Examples
Obstruction	Anastomotic stricture	Ileocolonic anastomosis stricture
	Extrinsic obstruction	By adhesion and mass lesions, such as abscess
Acute and chronic suture or staple line leaks	Abscess	Psoas muscle abscess from transverse staple line leak after ileocolonic resection and anastomosis
	Fistula	Enterocutaneous fistula
	Sinus	Persistent perineal sinus
Anastomotic bleeding	Anastomosis ulcer bleeding	
	Anastomosis polyp bleeding	
	Friable anastomosis bleeding	
Stoma complications	Peristomal skin lesions	Peristomal pyoderma
	Stoma complications	Stoma prolapse or retraction
	Bowel complication proximal to the stoma	Parastomal hernia

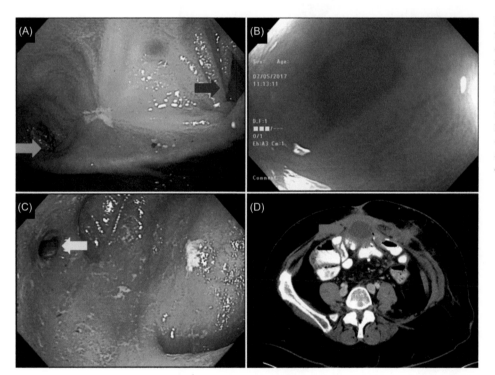

FIGURE 16.13 Ileocolonic side-to-side anastomosis in healthy and diseased. (A) Side-to-side anastomosis with a blind stump at transverse staple line (*green arrow*) and a lumen to the neo-terminal ileum (*blue arrow*); (B) staple seen at the transverse staple line; (C) a large leak at the transverse staple line (*yellow arrow*); (D) an abscess associated with the leak (*red arrow*).

risk for developing ECF, originating from a pinch of bowel by the fascia. Attention should be paid to the fascia level when conducting ileoscopy or colonoscopy via stoma (Fig. 16.16).

Patients with CD undergoing completion proctectomy are at risk for the development of persistent perineal sinus from anal stump leak, which has been difficult to treat [7]. Endoscopy or sinoscopy can easily detect the sinus cavity (Fig. 16.17). Vacuum suction, muscle flap, and sinoscopic therapy have been described.

Anastomosis bleeding

Brisk bleeding at the anastomosis after the surgery is a rare complication of bowel resection [8,9]. Anastomotic bleeding can also present with chronic iron-deficiency anemia. Colonoscopy or endoscopy may reveal active bleeding or stigmata of bleeding along the staple or suture lines. Bleeding can result from anastomosis ulcer, friable surface, or even anastomosis polyps (Table 16.1). This author has used endoclips to control anastomosis bleeding (Figs. 16.18 and 16.19).

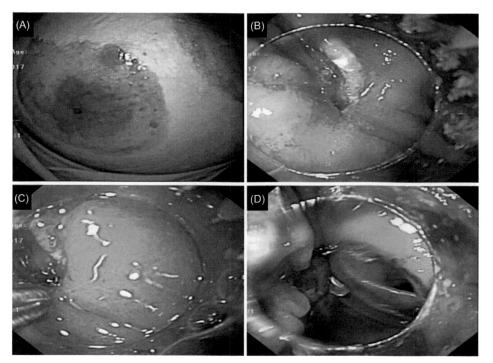

FIGURE 16.14 Enterocutaneous fistula from ileocolonic anastomosis leak in Crohn's disease. (A) Fistula opening on the skin; (B) the internal opening was identified by the administration of betadine through the external opening on the skin; (C) anchor was placed through endoscope with a cap for the subsequent endoscopic therapy; (D) deployment of over-the-scope clip at the internal opening of the fistula.

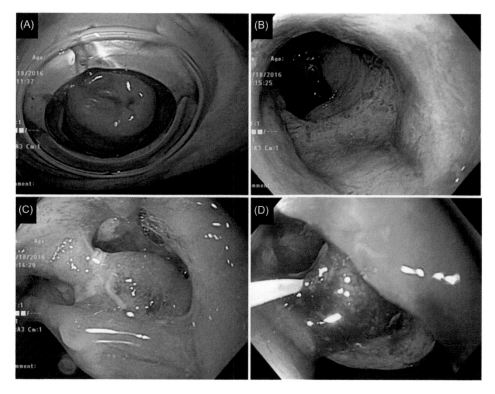

FIGURE 16.15 Stump leak at Hartmann pouch in Crohn's disease. (A) Diverting ileostomy; (B) diverted rectum with erythema and friable mucosa; (C and D) a large defect at the staple line at the stump, which was detected by a guide wire.

Stoma complications

Ileostomy, and less often colostomy and jejunostomy, are performed in patients with CD. The ostomy can be temporary or permanent; and with loop or end configuration. The ostomy procedure may be performed in a combination with

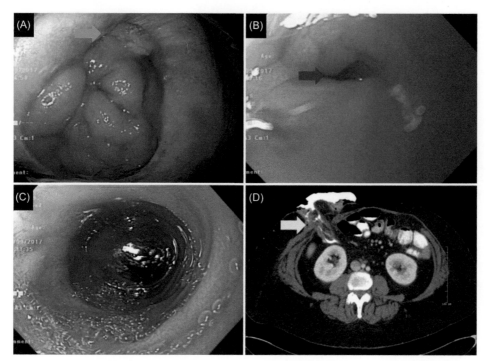

FIGURE 16.16 Peri-colostomy enterocutaneous fistula in Crohn's disease. (A) Fistula opening at the edge of colostomy (*green arrow*); (B) internal or primary opening of the fistula at the fascia level of the side of the distal colon, 5 cm from the stoma (*blue arrow*); (C) normal bowel lumen; (D) the enterocutaneous fistula (*yellow arrow*).

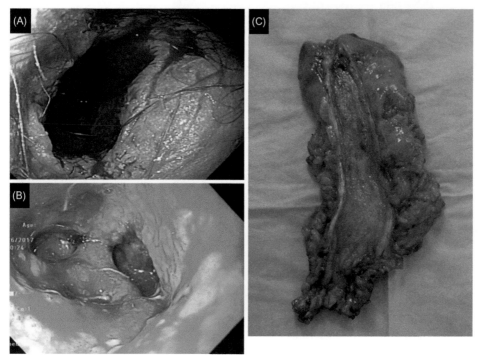

FIGURE 16.17 Persistent perineal sinus after completion proctectomy for Crohn's disease. This complication is common in patients with underlying inflammatory bowel disease, which poses challenge for management. (A and B) A large defect in the skin and pelvis after the surgery; (C) surgical specimen of completion proctectomy.

bowel resection or strictureplasty (Fig. 16.20). Stoma-related complications is classified based on the level of involvement: skin, stoma, and bowel proximal to stoma (Table 16.1). Underlying disease, surgical technique, adhesives, and topical pressure from appliances can all contribute to those complications. At the skin level, there can be peristomal dermatitis, peristomal pressure ulcers, peristomal pyoderma gangrenosum, and ECF (Figs. 16.21 and 16.22). At the

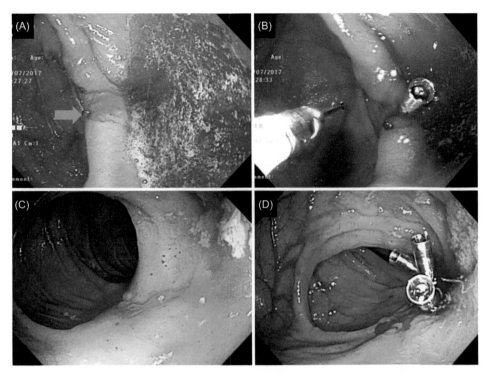

FIGURE 16.18 Bleeding at the anastomosis in two patients with Crohn's disease. (A) Active bleeding at the anastomosis near a dislodged staple (*green arrow*); (B) the bleeding spot was treated with endoclip; (C) active bleeding without obvious lesion at the anastomosis in a separate patient; (D) bleeding was controlled by endoclips × 3.

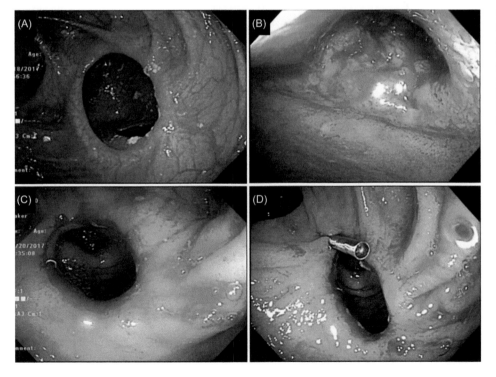

FIGURE 16.19 Bleeding at ileocolonic anastomosis in Crohn's disease. (A) Normal side-to-side ileocolonic anastomosis; (B) diffuse bleeding at the end-to-side anastomosis; (C and D) diffuse bleeding along the whole circumferential anastomosis, with endoscopic clipping.

stoma level, there can be stoma ulcer or granularity, stoma stricture, stoma prolapse, and stoma retraction (Figs. 16.22 and 16.23). Bowel obstruction can result from local anatomical factors [especially the constriction from the fascia (Fig. 16.20A and B) or parastomal hernia (Fig. 16.24)]. The obstruction can be detected by a careful endoscopic examination [10].

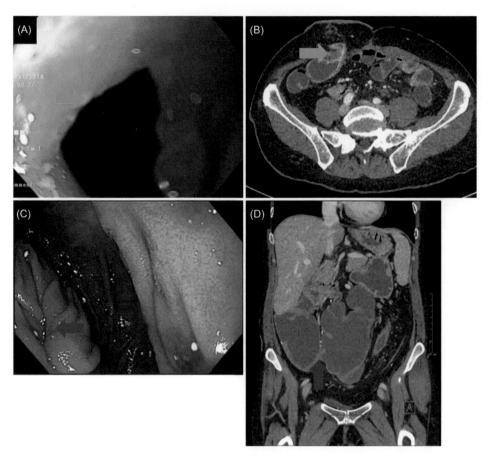

FIGURE 16.20 "Normal" anatomy of ileostomy with or without strictureplasty. There was no active Crohn's disease. (A and B) Extrinsic stricture from pinch of the fascia (*green arrows*); (C and D) strictureplasty close to ilestomy. Outlet of strictureplasty was highlighted (*blue arrows*).

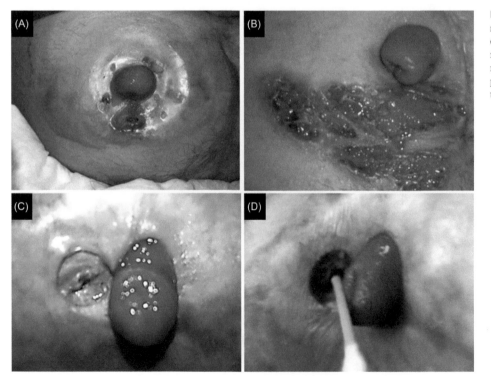

FIGURE 16.21 Peristomal cutaneous lesions in Crohn's disease. (A) Peristomal dermatitis with multiple ulcers; (B and C) peristomal pyoderma gangrenosum; (D) peristoma fistula and abscess with the placement of drainage catheter.

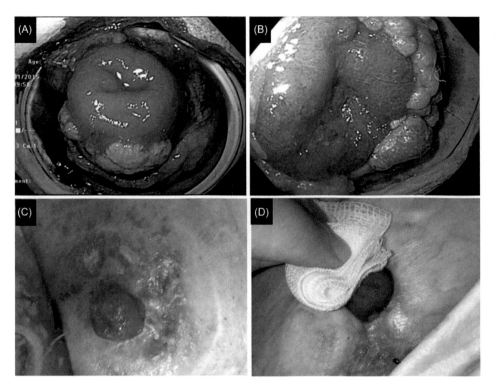

FIGURE 16.22 Common complications at and around stoma. (A) Granular tissue at the stoma; (B) granular tissue with stomal stricture; (C) peristomal dermatitis with ulcers, resulting from pressure and improper use of adhesives; (D) healed peristomal ulcer after hyperbaric oxygen therapy.

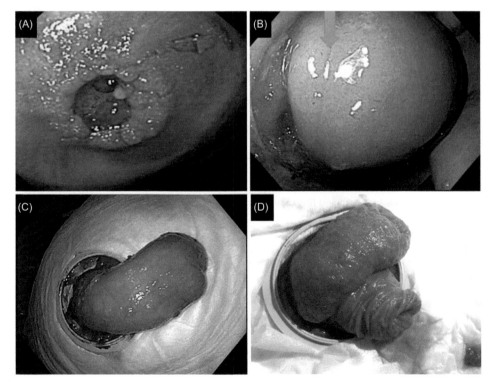

FIGURE 16.23 Common stoma complications. (A) Retracted end ileostomy with granular tissue in the skin; (B) vertical stricture in an ileostomy (*green arrow*); (C) prolapsed end ileostomy; (D) prolapsed loop ileostomy.

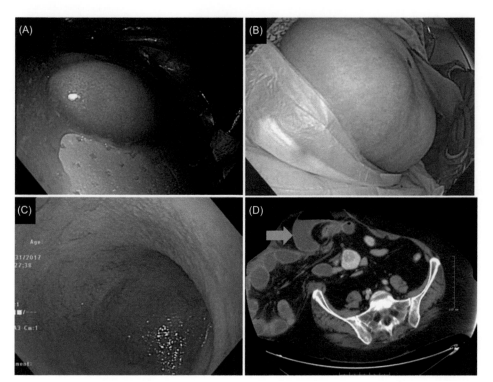

FIGURE 16.24 Parastomal hernia. (A) Ileostomy; (B) large bulging area in right lower abdomen next to the stoma; (C) a loop of small bowel next to the stoma; (D) hernia adjacent to the stoma (*green arrow*).

Summary and recommendations

Postoperative complications in patients undergoing surgery for their CD are common. Endoscopic evaluation should be routinely integrated into the diagnostic strategies. The endoscopist should be familiar with commonly performed surgical procedures in CD and the altered bowel anatomy. The majority of those postoperative complications may be reliably identified and even treated with endoscopy.

References

[1]   Ding Z, Wu XR, Remer EM, Lian L, Stocchi L, Li Y, et al. Association between high visceral fat area and postoperative complications in patients with Crohn's disease following primary surgery. Colorectal Dis 2016;18:163−72.

[2]   Billioud V, Ford AC, Tedesco ED, Colombel JF, Roblin X, Peyrin-Biroulet L. Preoperative use of anti-TNF therapy and postoperative complications in inflammatory bowel diseases: a meta-analysis. J Crohns Colitis 2013;7:853−67.

[3]   Lian L, Stocchi L, Remzi FH, Shen B. Comparison of endoscopic dilation vs surgery for anastomotic stricture in patients with Crohn's disease following ileocolonic resection. Clin Gastroenterol Hepatol 2017;15:1226−31.

[4]   Lan N, Shen B. Endoscopic stricturotomy versus balloon dilation in the treatment of anastomotic strictures in Crohn's disease. Inflamm Bowel Dis 2018;24:897−907.

[5]   Shen B, Kochhar G, Navaneethan U, Liu X, Farraye FA, Gonzalez-Lama Y, et al. Role of interventional inflammatory bowel disease in the era of biologic therapy: a position statement from the Global Interventional IBD Group. Gastrointest Endosc 2019;89:215−37.

[6]   Kochhar GS, Shen B. Use of over-the-scope-clip system to treat ileocolonic transverse staple line leak in patients with Crohn's Disease. Inflamm Bowel Dis 2018;24:666−7.

[7]   Lohsiriwat V. Persistent perineal sinus: incidence, pathogenesis, risk factors, and management. Surg Today 2009;39:189−93.

[8]   Cocorullo G, Tutino R, Falco N, Salamone G, Fontana T, Licari L, et al. Laparoscopic ileocecal resection in acute and chronic presentations of Crohn's disease. A single center experience. G Chir 2017;37:220−3.

[9]   Riss S, Bittermann C, Zandl S, Kristo I, Stift A, Papay P, et al. Short-term complications of wide-lumen stapled anastomosis after ileocolic resection for Crohn's disease: who is at risk? Colorectal Dis 2010;12:e298−303.

[10] Wang X, Shen B. Management of Crohn's disease and complications in patients with ostomies. Inflamm Bowel Dis 2018;24:1167−84.

Chapter 17

Capsule endoscopy in inflammatory bowel disease

Xin-Ying Wang[1] and Side Liu[2]

[1]Department of Gastroenterology, Zhujiang Hospital, Southern Medical University, Guangzhou, P.R. China, [2]Department of Gastroenterology, Nanfang Hospital, Southern Medical University, Guangzhou, P.R. China

Chapter Outline

Abbreviations

CCE	colon capsule endoscopy
CECDAI	the Capsule Endoscopy Crohn's Disease Activity Index
CTE	computed tomographic enterography
GI	gastrointestinal
IBD	inflammatory bowel disease
MRE	magnetic resonance enterography
SBC	small bowel and colon
NSAID	nonsteroidal antiinflammatory drug
SITT	small bowel transit time
VCE	video capsule endoscopy
UC	ulcerative colitis

Introduction

Video capsule endoscopy (VCE) was cleared for the diagnosis of small bowel disease, especially for obscure gastrointestinal (GI) bleeding, by the United States Food and Drug Administration in 2001. Various formats of VCE have been developed, including conventional VCE of the small bowel or colon, and wide-angle panoramic, panenteric, and small bowel and colon (SBC)−VCE [1]. With advances in imaging technology, the already highly sensitive VCE in the detection of small bowel disease still suffers from suboptimal specificity [2].

The parameters of various capsule endoscopies are compared in Table 17.1. VCE provides a noninvasive method to visualize the small intestine via high-quality images in patients with a wide spectrum of disorders such as Crohn's disease (CD), ulcerative colitis (UC), ileal pouch disorders, obscure GI bleeding, polyposis syndromes, celiac disease, small bowel tumors, and other inflammatory disorders.

Atlas of Endoscopy Imaging in Inflammatory Bowel Disease. DOI: https://doi.org/10.1016/B978-0-12-814811-2.00017-7
© 2020 Elsevier Inc. All rights reserved.

Crohn's disease

CD is a chronic inflammatory disease that can involve the entire GI tract, from the mouth to the anus. Small bowel involvement occurs in more than 50% of patients with CD. According to the current guidelines [3], VCE has been recommended as an adjunct endoscopic modality. VCE is indicated for suspected, known, or relapsed CD when ileocolonoscopy and imaging studies are not feasible or conclusive. In addition to diagnosis and differential diagnosis, VCE can also be used in monitoring disease activity and response to treatment as well as evaluating postoperative recurrence of CD [3]. VCE can reliably assess mucosal healing (Fig. 17.1).

Routine small bowel imaging or the use of the patency capsule prior to capsule endoscopy is not recommended in patients with CD [3]. However, in patients with obstructive symptoms or known stenoses, dedicated small bowel cross-sectional imaging modalities such as magnetic resonance enterography (MRE) or computed tomography enterography (CTE) should be performed first. Capsule endoscopy is normally not recommended in patients with chronic abdominal pain or diarrhea.

TABLE 17.1 Capsule endoscopes: small bowel, colon, and patency capsule.

	Manufacturer	Size (mm)	Field of view (degrees)	Image capture	Battery life (h)	Data transmission
PillCam SB	Given Imaging, Yoqneam, Israel	11.4 × 26.3	156	2−6 fps	12	RF
Endocapsule MAJ-I469	Olympus Medical Systems, Tokyo, Japan	1.0 × 26.0	145	2 fps	8	RF
OMOM	Jinshan Science and Technology, Chongqing, China	13.4 × 17.9	140	2 fps	8	RF
MiroCam	IntroMedic, Seoul, Korea	10.8 × 24.5	170	3 fps	12	EFP
CapsoCam SV1	Capsovision Saratoga, United States	11 × 31	360	20 fps	15	Capsule retrieved then USB download
PillCam Patency	Given Imaging, Yoqneam, Israel	11 × 26	n/a	n/a	n/a	n/a
PillCam Colon	Given Imaging, Yoqneam, Israel	11 × 32	172	6 fpm, stomach; 4−35 fps, SB	10	n/a
AKE-1	Ankon Technologies	11.8 × 27.0	140	2 fps	8	RF

EFP, Electrical field propagation; *fps*, frames per second; *fpm*, frames per minute; *RF*, radiofrequency; *SB*, small bowel; *USB*, universal serial bus.

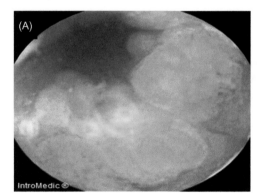 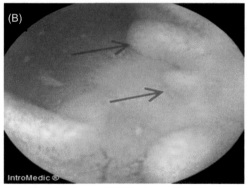

FIGURE 17.1 Documentation of mucosal healing in Crohn's disease after therapy with infliximab. (A) Ulcerated and nodular small bowel mucosa before the medical treatment and (B) resolution of mucosal inflammation (i.e., mucosal healing) with the formation of pseudopolyps after therapy (*blue arrows*).

The common features of CD on VCE include edema, hyperemia, bleeding, exudates, aphthae, erosions, small (≤0.5 cm) and large (>0.5 cm) ulcers, denuded mucosa, and pseudopolyps. The shape of ulcers can be round, linear, circumferential, or stellate (Figs. 17.2 and 17.3). VCE may also detect the primary (i.e., disease associated) or secondary (e.g., anastomotic or drug induced) strictures (Fig. 17.4). The disease can be patchy, segmental, or diffuse. Two scoring systems are used to assess the activity of small intestine mucos, based on the type, location, and severity of

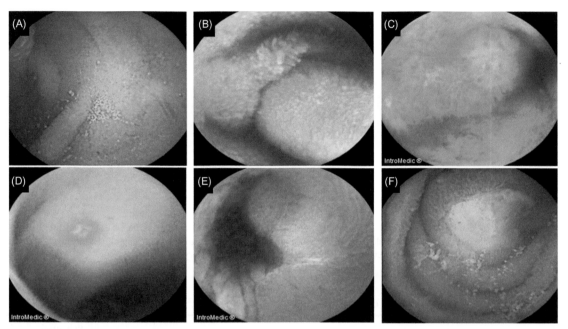

FIGURE 17.2   Patterns of small bowel Crohn's disease on capsule endoscopy. (A) Normal small bowel mucosa as a reference; (B) edema with the villous width is equal to or greater than the villous height; (C) erythema; (D) aphthous ulcer; (E) stellate ulcer with bleeding; and (F) large, clean-based ulcer with exudates.

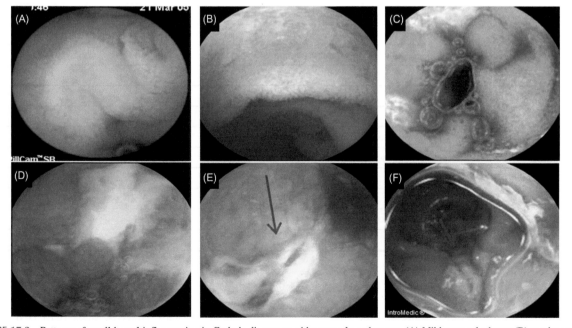

FIGURE 17.3   Patterns of small bowel inflammation in Crohn's disease on video capsule endoscopy. (A) Mild mucosal edema; (B) patchy erythema with granularity of mucosa; (C) mixed edema and ulceration (*black arrow*); and (D−F) longitudinal ulcers with exudates (*arrows*).

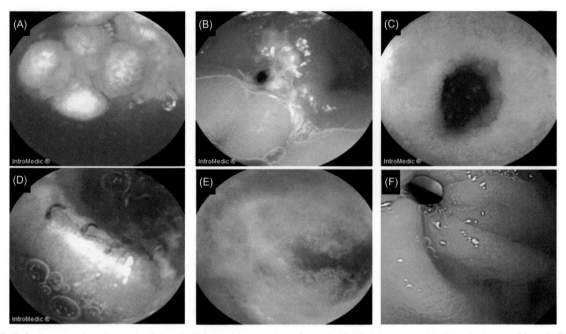

FIGURE 17.4 Patterns of small bowel strictures in IBD on video capsule endoscopy. (A) Nodular mucosa blocking the lumen of bowel; (B) ulcerated stricture with proximal bowel edema; (C) nonulcerated fibrotic stricture; (D) anastomotic stricture with staples; (E) ulcerated stricture with exudates; and (F) retained capsule in the distal ileum viewed with ileocolonoscopy. *IBD*, Inflammatory bowel disease.

TABLE 17.2 Capsule endoscopy Crohn's disease activity index [4] worksheet.[a]

			Score	Proximal	Distal
A	Inflammation	None	0		
		Mild-to-moderate edema/hyperemia/denudation	1		
		Severe edema/hyperemia/denudation	2		
		Bleeding, exudate, aphthae, erosion, small ulcer (≤0.5 cm)	3		
		Moderate ulcer (0.5−2 cm), pseudopolyp	4		
		Large ulcer (>2 cm)	5		
B	Extent of disease	None	0		
		Focal disease (single segment)	1		
		Patchy disease (multiple segments)	2		
		Diffuse disease	3		
C	Stricture[b]	None	0		
		Single passed	1		
		Multiple passed	2		
		Obstruction	3		

[a]*Total score* $= [(A \times B) + C]_{proximal} + [(A \times B) + C]_{distal}$.
[b]*Narrowing.*

small bowel lesions: the Lewis score and the Capsule Endoscopy Crohn's Disease Activity Index (CECDAI). These scoring systems facilitate the assessment of the course of small bowel CD and response to medical therapy. It should be noted that these scoring systems can quantitatively describe the type, distribution, and severity of mucosal lesions, but

TABLE 17.3 Parameters and weightings for the capsule endoscopy scoring index (the Lewis score) [5].

Parameters		Number	Longitudinal extent	Descriptors
First tertile	Villous appearance	Normal—0	Short segment—8	Single—1
		Edematous—1	Long segment—12	Patchy—14
			While tertile—20	Diffuse—17
	Ulcer	None—0	Short segment—5	<1/4—9
		Single—3	Long segment—10	1/4–1/2—12
		Few—5	Whole tertile—15	>1/2—18
		Multiple—10		
Second tertile	Villous appearance	Normal—0	Short segment—8	Single—1
		Edematous—1	Long segment—12	Patchy—14
			While tertile—20	Diffuse—17
	Ulcer	None—0	Short segment—5	<1/4—9
		Single—3	Long segment—10	1/4–1/2—12
		Few—5	Whole tertile—15	>1/2—18
		Multiple—10		
Third tertile	Villous appearance	Normal—0	Short segment—8	Single—1
		Edematous—1	Long segment—12	Patchy—14
			While tertile—20	Diffuse—17
	Ulcer	None—0	Short segment—5	<1/4—9
		Single—3	Long segment—10	1/4–1/2—12
		Few—5	Whole tertile—15	>1/2—18
		Multiple—10		
Stenosis[a]		None—0	Ulcerated—24	Traversed—7
		Single—14	Nonulcerated—2	Nontraversed—10
		Multiple—20		

Lewis score = score of the worst affected tertile [(villous parameter × extent × descriptor) + (ulcer number × extent × size)] + stenosis score
(number × ulcerated × traversed).
Longitudinal extent: Short segment: <10% of the tertile; long segment: 11%–50% of the tertile; whole tertile: >50% of the tertile.
Ulcer number: single: 1; few: 2–7; multiple: ≥8.
Ulcer descriptor (size) is determined by how much of the capsule picture is filled by the largest ulcer.
[a]Rated for the whole study.

they cannot be used as initial diagnostic tools. The CECDAI is the sum of the respective scores for the proximal and distal segments of the small bowel in which the segments are defined by the small bowel transit time (SITT) of the capsule. The total score of each segment, proximal or distal, is the sum of the stricture score and the product of the scores for inflammation and extent of disease for that bowel segment. Thus the CECDAI score is based on the inflammation, extent of disease, and presence of strictures of the proximal or distal segments of the small bowel (Table 17.2) [4].

In 2008 a capsule endoscopy scoring index, also named as the Lewis score, was proposed by Gralnek et al. [5]. The Lewis score is a cumulative scoring system that is based on the presence and distribution of villous edema, ulceration, and stenosis [5]. A score <135 is defined as normal or clinically insignificant mucosal inflammatory change. A score between 135 and 790 is mild, and a score ≥790 is moderate to severe (Table 17.3) [5].

It is difficult for VCE to differentiate CD from nonspecific small lesions or other diseases such as lymphoma, ischemic bowel, small bowel tuberculosis, or drug-induced enteropathy, nonsteroidal antiinflammatory drugs (NSAIDs) in particular (Fig. 17.5). NSAIDs should be discontinued at least 4 weeks prior to small bowel capsule endoscopy in the setting of suspected CD. As a screening tool for CD, capsule endoscopy has high sensitivity and

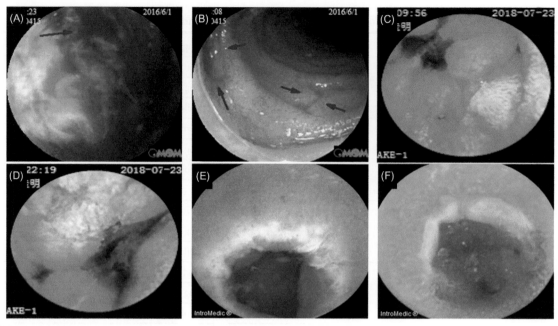

FIGURE 17.5 Small bowel lesions mimicking inflammatory bowel disease. (A) Lymphoma with irregular and bleeding ulcers with exudates (*blue arrows*); (B) lymphoma with multiple small and punched-out ulcers (*blue arrows*); (C and D) Henoch−Schonlein purpura with mucosal nodularity and spontaneous bleeding; and (E and F) nonsteroidal antiinflammatory drug-induced enteropathy with ulcerated diaphragm-like strictures in the small bowel.

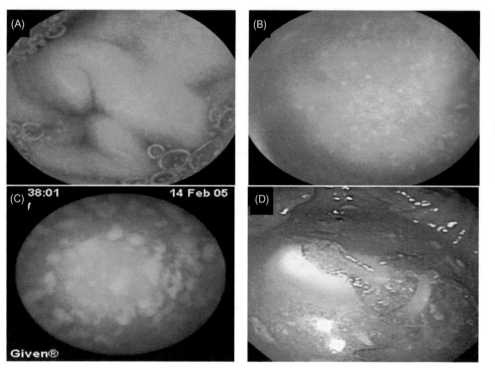

FIGURE 17.6 Capsule endoscopy in UC with or without restorative proctocolectomy. (A and B) Normal small bowel (A) and active colitis with erythema, loss of vascularity, erosions, and exudates (B); (C) loss of vascular pattern and exudates in a patient with UC and a pelvic ileal pouch; and (D) trapped capsule endoscope in a patient with UC and a Kock pouch. *UC*, Ulcerative colitis.

low specificity but is not recommended for determining the initial diagnosis [6]. Rather, the initial diagnosis should be based on clinical features, serum parameters, imaging conventional endoscopy, histology, and sometimes response to medical therapy. Endoscopic mucosal healing has been established as an important target for the treatment of CD as well as UC. In addition to conventional ileocolonoscopy, VCE has been shown to be effective for the measurement with mucosal healing [7,8].

Ulcerative colitis

Colon capsule endoscopy (CCE) and SBC−VCE have been used to evaluate the disease activity in UC. The accuracy of CCE for the assessment of mucosal inflammation in UC seems to be comparable with that of conventional colonoscopy. CCE was better tolerated than the latter [9]. Preoperative assessment of small bowel inflammation with VCE in UC undergoing restorative proctocolectomy and ileal pouch-anal anastomosis has been performed for the prediction of postoperative risk for CD of the pouch, although the value of VCE in this setting is controversial (Fig. 17.6A and B) [10]. There are still no sufficient data to support the use of small bowel VCE or CCE in the diagnosis or surveillance of patients with UC. Nonetheless, CCE is a safe procedure to monitor mucosal healing in UC.

Despite its less invasiveness, CCE has inherited limitations, such as the lack of ability to obtain tissue samples and therapeutic intervention.

Ileal pouch disorders

Restorative proctocolectomy with ileal pouch has become the surgical treatment of choice for UC patients who require colectomy. Various adverse structural, inflammatory, neoplastic, or metabolic sequelae often occur following this reconstructive surgical procedure. Pouchitis, CD of the pouch, and iron deficiency anemia are common complications of pouch surgery, and VCE has been used in clinical trials of pouchitis [11,12]. VCE in combination with conventional esophagogastroduodenoscopy and pouchoscopy has been used for the evaluation of small bowel CD and anemia in patients with ileal pouches [13]. Conventional pouchoscopy with capability of tissue biopsy and therapeutic intervention is preferred. In patients with continent ileostomies undergoing VCE, such as Kock pouches, endoscopic retrieval of the capsule via the nipple valve is needed (Fig. 17.6C and D).

Assessment of small bowel motility in inflammatory bowel disease

VCE has been used to measure SITT in CD and UC patients. For example, SITT was found to be longer in UC patients than non-IBD patients, and SITT was longer in patients with active CD than those with quiescent CD [14].

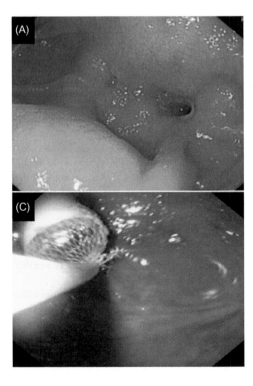
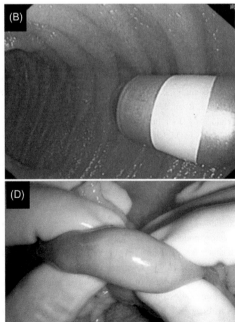

FIGURE 17.7 Capsule retention. (A and B) Stricture at the small bowel and retained capsule endoscope, viewed on enteroscopy; (C) retention and retrieval of capsule in Kock pouch with an endoscopic net device; and (D) retrieval of capsule endoscope with surgical operation. *(C and D) Courtesy Shen B, editor. Interventional inflammatory bowel disease. Cambridge, MA: Elsevier; 2018. p. 33 and Photo courtesy: Dr. Feza Remzi.*

Capsule retention

The most important concern of VCE is the risk of capsule retention. This is particularly true in patients with stricturing CD or those with surgery. The International Conference on Capsule Endoscopy working group defined capsule retention as the persistence of the capsule in the digestive tract for more than 2 weeks, necessitating medical, endoscopic, or surgical intervention (Fig. 17.7) [15]. The risk of capsule retention in patients with suspected CD without obstructive symptoms and without history of small bowel resection or known stenosis is low and comparable to that of obscure GI bleeding [16]. The retention rate ranges from 0.75% to 1.0%. However, Cheifetz et al. [17] reported a retention rate of 13% in patients with established CD. Capsule retention can be managed conservatively, such as potent medical therapy for the component of inflammation. The retained capsule may be retrieved by device-assisted enteroscopy or surgery (Fig. 17.7).

Summary and recommendation

VCE is recommended for suspected inflammatory CD in patients with negative ileocolonoscopy and imaging findings. It is also a promising tool for monitoring disease activity, response to treatment, and postoperative disease recurrence. The Lewis score and the CECDAI are standardized scoring systems for assessing the disease and the response to medical therapy. Capsule retention is the main risk of capsule endoscopy. Imaging examinations such as CTE and MRE should be performed first if patients present with obstructive symptoms or known stenosis.

Disclosure

The authors declare that there is no financial conflict of interest.

References

[1] Leighton JA, Helper DJ, Gralnek IM, Dotan I, Fernandez-Urien I, Lahat A, et al. Comparing diagnostic yield of a novel pan-enteric video capsule endoscope with ileocolonoscopy in patients with active Crohn's disease: a feasibility study. Gastrointest Endosc 2017;85:196−205.e1.

[2] Goldstein JL, Eisen GM, Lewis B, Gralnek IM, Zlotnick S, Fort JG. Video capsule endoscopy to prospectively assess small bowel injury with celecoxib, naproxen plus omeprazole, and placebo. Clin Gastroenterol Hepatol 2005;3:133−41.

[3] Pennazio M, Spada C, Eliakim R, Keuchel M, May A, Mulder CJ, et al. Small-bowel capsule endoscopy and device-assisted enteroscopy for diagnosis and treatment of small-bowel disorders: European Society of Gastrointestinal Endoscopy (ESGE) Clinical Guideline. Endoscopy 2015;47:352−76.

[4] Gal E, Geller A, Fraser G, Levi Z, Niv Y. Assessment and validation of the new capsule endoscopy Crohn's disease activity index (CECDAI). Dig Dis Sci 2008;53:1933−7.

[5] Gralnek IM, Defranchis R, Seidman E, Leighton JA, Legnani P, Lewis BS. Development of a capsule endoscopy scoring index for small bowel mucosal inflammatory change. Aliment Pharmacol Ther 2008;27:146−54.

[6] Shen B. Is it a prime time for small-bowel colon video capsule endoscopy to cover both sides of the ileocecal valve in Crohn's disease? Gastrointest Endosc 2017;85:206−9.

[7] Niv Y. Small-bowel mucosal healing assessment by capsule endoscopy as a predictor of long-term clinical remission in patients with Crohn's disease: a systematic review and meta-analysis. Eur J Gastroenterol Hepatol 2017;29:844−8.

[8] Hall B, Holleran G, Chin JL, Smith S, Ryan B, Mahmud N, et al. A prospective 52 week mucosal healing assessment of small bowel Crohn's disease as detected by capsule endoscopy. J Crohn's Colitis 2014;8:1601−9.

[9] Shi HY, Chan FKL, Higashimori A, Kyaw M, Ching JYL, Chan HCH, et al. A prospective study on second-generation colon capsule endoscopy to detect mucosal lesions and disease activity in ulcerative colitis (with video). Gastrointest Endosc 2017;86:1139−1146.e6.

[10] Murrell Z, Vasiliauskas E, Melmed G, Lo S, Targan S, Fleshner P. Preoperative wireless capsule endoscopy does not predict outcome after ileal pouch-anal anastomosis. Dis Colon Rectum 2010;53:293−300.

[11] Calabrese C, Gionchetti P, Rizzello F, Liguori G, Gabusi V, Tambasco R, et al. Short-term treatment with infliximab in chronic refractory pouchitis and ileitis. Aliment Pharmacol Ther 2008;27:759−64.

[12] Viazis N, Giakoumis M, Koukouratos T, Anastasiou J, Katopodi K, Kechagias G, et al. Long term benefit of one year infliximab administration for the treatment of chronic refractory pouchitis. J Crohn's Colitis 2013;7:e457−60.

[13] Shen B, Remzi FH, Santisi J, Lashner BA, Brzezinski A, Fazio VW. Application of wireless capsule endoscopy for the evaluation of iron deficiency anemia in patients with ileal pouches. J Clin Gastroenterol 2008;42:897−902.

[14] Fischer M, Siva S, Wo JM, Fadda HM. Assessment of small intestinal transit times in ulcerative colitis and Crohn's disease patients with different disease activity using video capsule endoscopy. AAPS PharmSciTech 2017;18:404−9.

[15] Cave D, Legnani P, de Franchis R, Lewis BS. ICCE consensus for capsule retention. Endoscopy 2005;37:1065−7.

[16] Postgate AJ, Burling D, Gupta A, Fitzpatrick A, Fraser C. Safety, reliability and limitations of the given patency capsule in patients at risk of capsule retention: a 3-year technical review. Dig Dis Sci 2008;53:2732−8.

[17] Cheifetz AS, Kornbluth AA, Legnani P, Schmelkin I, Brown A, Lichtiger S, et al. The risk of retention of the capsule endoscope in patients with known or suspected Crohn's disease. Am J Gastroenterol 2006;101:2218−22.

Chapter 18

Enteroscopy in inflammatory bowel disease and inflammatory bowel disease−like conditions

Xiuli Zuo and Tian Ma

Department of Gastroenterology, Qilu Hospital, Shandong University, Jinan, P.R. China

Chapter Outline

Abbreviations

CD Crohn's disease
CMUSE cryptogenic multifocal ulcerous stenosing enteritis
BD Behcet's disease
IBD inflammatory bowel disease
ITB intestinal tuberculosis
MALT mucosa-associated lymphoid tissue
NSAID nonsteroidal antiinflammatory drug
UC ulcerative colitis

Introduction

The small bowel, as the midsection of the gastrointestinal tract between the stomach and colon, is 5−7 m long and divided into the duodenum, jejunum, and ileum. The enteroscopy, which can not only make repeated observation but also take biopsy, is the most accurate visual modality for the detection of the pathology of the entire small bowel, especially the jejunum and ileum. Device-assisted enteroscopy can assess the segment of bowel beyond the conventional ileocolonoscopy or push enteroscopy. The enteroscopic appearance of the normal jejunum and ileum is yellow-orange tubular structure that is ringed by circular Kerckring folds and covered with finger-like villi [1]. In addition, small veins and arteries, occasionally larger vessels, can be visualized (i.e., vascular pattern) in the submucosa of the normal jejunum and ileum. Some variants also appear in the small bowel, such as lymph follicles, which are sporadically distributed in the small bowel and most concentrated in the terminal ileum, especially in children [1]. There is no sharp demarcation of anatomical structure between the jejunum and ileum. However, the ileum has sparser Kerckring folds, sparser and slightly shorter villi as well as more obvious vessels when than that in the jejunum (Fig. 18.1).

In patients with inflammatory bowel disease (IBD), the small bowel involvement is indicative of more aggressive disease with poor prognosis. Small bowel disease is mainly seen in Crohn's disease (CD) and rare in ulcerative colitis (UC) in a form backwash ileitis [2]. For patients suspected of IBD, enteroscopy plays an important role in diagnosis, differential diagnosis, disease monitoring, assessment of treatment response, and even endoscopic treatment. On the

Atlas of Endoscopy Imaging in Inflammatory Bowel Disease. DOI: https://doi.org/10.1016/B978-0-12-814811-2.00018-9
© 2020 Elsevier Inc. All rights reserved.

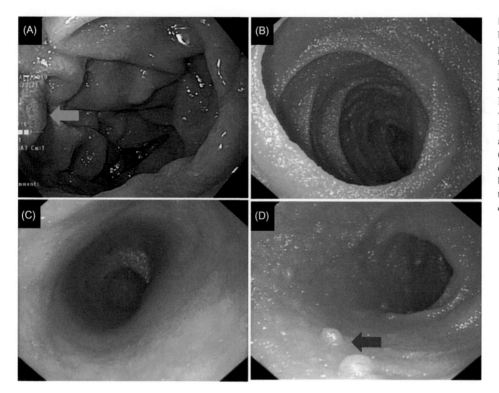

FIGURE 18.1 Normal small bowel: (A) normal duodenum and papilla (*green arrow*) with prominent Kerckring folds; (B) normal jejunum characterized by yellow-orange tubular shape, ringed by Kerckring folds and finger-like villi; (C) normal ileum with sparser Kerckring folds as well as sparser and shorter than the jejunum; and (D) lymphoid follicles sporadically distributed in the entire small bowel, have their highest concentrations in the terminal ileum (*blue arrow*).

other hand, performance of deep enteroscopy or device-assisted enteroscopy requires special equipment and technical skills. In patients with a prior history of bowel surgery, particularly strictureplasty and intestinal bypass surgery, enteroscopy can be challenging, due to altered bowel anatomy.

Crohn's diseases

The enteroscopic appearance of the small bowel lesions of CD has various features. At early stages of disease, aphthous ulcers arise in the involved small bowel segments. With disease progressing, multiple larger and irregular ulcers develop and present with longitudinal arrays, and then merge together to form the classic longitudinal ulcers (also called fissuring ulcers) of CD. The linear or longitudinal ulcers are often prominent along the mesenteric edge. The ulcers of CD are almost always segmentally distributed. Aphthous ulcers, longitudinally arranged ulcers, and fissuring ulcers can be observed during enteroscopy. In patients with CD the characteristic cobblestone appearance in the small bowel is less common than that in the colon (Fig. 18.2). In addition, long-standing inflammation of CD often result in stricture formation of the small bowel, which is called disease-associated or primary stricture (Fig. 18.3). Strictures can also develop at the surgical anastomosis or strictureplasty sites, which are called secondary strictures. The strictures can be short or long with a cutoff of 4-5 cm, inflammatory versus fibrotic versus mixed. Sometimes, those strictures can prevent the passage of enteroscope. Moreover, the primary fistulae can develop after long-standing inflammatory and stricturing disease. In addition, fistulae can develop at the surgical anastomosis or strictureplasty site, in the form of entero-enteric fistula or enterocutaneous fistula (Fig. 18.4). The opening of the fistula may occasionally be identified with a careful enteroscopy. Concurrent strictures are often present distally to the fistula opening.

In addition to diagnosis and differential diagnosis, enteroscopy can also be combined with endoscopic therapy, such as balloon dilation or electroincision for small bowel stricture [3]. It is recommended that preprocedural cross-sectional imaging be obtained before enteroscopy to roadmap extent, degree, and phenotypes of small bowel CD.

Ulcerative colitis

UC has diffused and continuous inflammation that only involves the mucosa. Lesions of UC are almost always limited to the colon, but the distal ileum involvement (backwash ileitis) is also observed in about 20% patients with pancolitis [4].

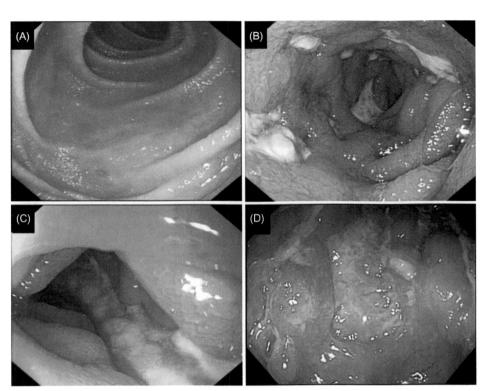

FIGURE 18.2 Patterns of ulcers in small bowel Crohn's disease on single-balloon enteroscopy: (A) multiple aphthous ulcers in the proximal jejunum; (B) multiple irregular ulcers arranged in the axis of ileum before merging into the typical longitudinal ulcers; (C) classic longitudinal ulcer along the mesentery edge in the ileum; and (D) multiple longitudinal ulcers in the terminal ileum.

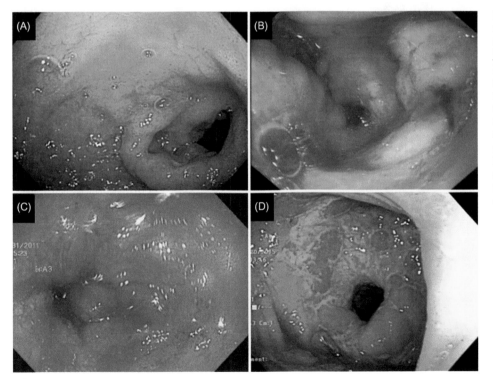

FIGURE 18.3 Patterns of strictures in small bowel Crohn's disease on enteroscopy: (A) mild jejunal stricture with proximal luminal dilation; (B) a distal ileum stricture with an adjacent large ulcer; (C) a tight, nonulcerated stricture resulting from long-term use of antitumor necrosis factor agent; and (D) ulcerated stricture at the outlet of strictureplasty site. Performing balloon-assisted enteroscopy in patients with strictureplasty has been technically difficult.

It should be noted that UC with the distal ileum involvement needs to be differentiated from CD [3]. In this setting, deep enteroscopy as well as ileocolonoscopy can be helpful [5]. The small bowel involvement of UC ausually manifests as diffuse and continuous lesions, and small shallow ulcers also appear occasionally (Fig. 18.5B). Strictures are rare in the small bowel of patients with UC and the cobblestoning of the distal ileum or fistula is hardly seen.

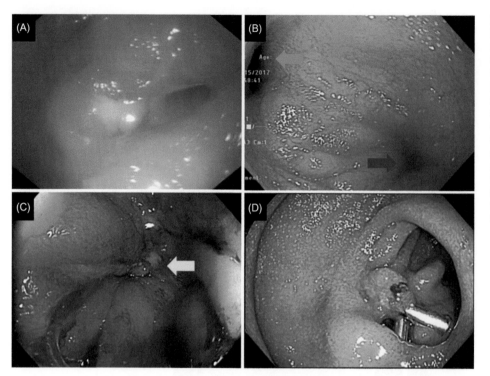

FIGURE 18.4 Patterns of fistula in small bowel CD on enteroscopy: (A) the opening of the fistula in the distal ileum of a patient with CD; (B) primary orifice at the distal ileum (*blue arrow*) of ileosigmoid fistula with current luminal stricture (*green arrow*); (C) primary orifice of enterocutaneous fistula resulting from ileocolonic anastomosis leak; and (D) primary orifice of enterocutaneous fistula from nonhealed jejunostomy for CD. Attempt was made to close the orifice with endoclips. CD, Crohn's disease.

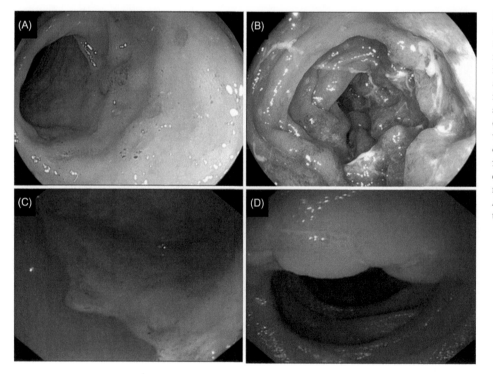

FIGURE 18.5 Disorders with small bowel ulcers: (A) multiple small shallow ulcers with friable mucosa in the distal ileum resulting from NSAID use; (B) diffuse and continuous ileitis (backwash ileitis) with edema, erythema, and small ulcer in ulcerative colitis; (C) giant geographic ulcer with well-defined margin in the junction of the jejunum and ileum resulting from ischemia; and (D) multiple erosions and linear ulcers of jejunum in eosinophilic enteritis. *NSAID*, nonsteroidal antiinflammatory drug.

Non—inflammatory bowel disease small bowel disorders

Not all small bowel ulcers or strictures are CD. Small bowel CD should be differentiated mainly from intestinal tuberculosis (ITB), intestinal Behcet's disease (BD), small bowel lymphoma, and nonsteroidal antiinflammatory drug (NSAID)-associated enteropathy (Table 18.1). The differential diagnosis should be put in clinical context.

TABLE 18.1 The characteristics and differential diagnosis of enteroscopic appearance in inflammatory bowel disease (IBD) and IBD-like diseases.

		Crohn's disease	Intestinal tuberculosis	Intestinal Behcet's disease	NSAIDs-associated enteropathy	UC (backwash ileitis)	Small bowel lymphoma
Ulcer	Aphtha	Common	Less common	Less common	Common	Less common	Less common
	Typical shape	Longitudinal	Transverse	Round	Circular or irregular	Irregular	No
	Number	Multiple	Multiple	Single or multiple	Multiple	Multiple	Single or multiple
	Margin	Relatively clear	Rodent like	Discrete	Relatively clear	Indiscrete	Uncertain
	Distribution	Segmentally distributed; involve more than four segments	Segmentally distributed; involve fewer than four segments	Focally distributed mostly	Usually dispersedly distributed	Diffusedly and continuously distributed	More limited
	Cobblestone appearance	Less common	Rare	Rare	No	No	No
	Fistula	Less common	No	Less common	No	No	No
	Stricture	Common	Common; circular stricture	Less common	Less common; diaphragm-like stenosis	Rare	Common

NSAID, Nonsteroidal antiinflammatory drugs; UC, ulcerative colitis.

The ileocecal area as well as the distal small bowel are the predilection sites for ITB. Like CD, the lesions of ITB are also segmentally distributed but theyusually involve fewer segments. Nodular tubercles are formed first, when the *Mycobacterium tuberculosis* invades the intestinal lymphoid tissue Subsequently tubercles spontaneously become ulcerated and expand along the circular lymph vessels to form the classic transverse or circumferential ulcers with rodent-like margins [6,7]. Circular scars or strictures in the small bowel can also form secondary to fibrosis induced by ulcers healing and tissue remodelling. Since ITB is an infective disease, ulcers in ITB almost exclusively have a necrotic base on endoscopy (Fig. 18.6). Classic histologic features of CD and ITB are also different. For example, granulomas in CD are sparse, small, and noncaseating, while granulomas in ITB are featured with diffuse, large, and central necrosis (Fig. 18.7) [8].

Lesions in intestinal BD almost always appear in the small bowel or ileocecal area, especially the terminal ileum. The distribution of lesions in intestinal BD can be focally single, focally multiple, segmental, or diffuse. Intestinal BD usually manifests as single or multiple large deep round ulcers with a discrete margin and even base (also called punched out ulcers) (Fig. 18.8).

Small bowel lymphoma can be present with ulcers, masses, or diffuse nodularity. Ulcers of small bowel lymphoma have various untypical appearances on enteroscopy. The ulcers are usually large but can be single or multiple, deep or shallow, round or irregular. The adjacent mucosa normally shows no or minimum inflammation. Multiple-piece biopsy, deep biopsy, and immunohistochemistry are helpful for the diagnosis of small bowel lymphoma [9,10]. In some cases, repeated biopsy is required (Fig. 18.9). Enteroscopy has also been used to monitor the response to therapy.

The lesions of NSAID-associated enteropathy arise mainly in the stomach, duodenum, or distal small bowel and are always present with multiple small lesions, which include erosions, red spots, and circular or irregular ulcers. Diaphragm-like stenosis is a representative enteroscopic feature of NSAID-associated enteropathy. Furthermore, diseases of small bowel that should be differentiated from IBD also include eosinophilic enteritis, infectious enteropathy,

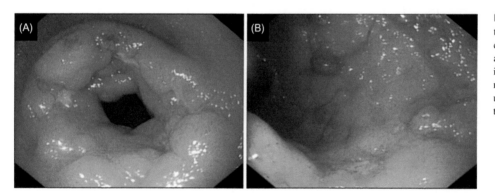

FIGURE 18.6 Enteroscopic features of intestinal tuberculosis: (A) circular ulcer surrounded by edematous mucosa and luminal stenosis in the junction between the jejunum and ileum and (B) small irregular ulcers and nodular tubercles in the terminal ileum.

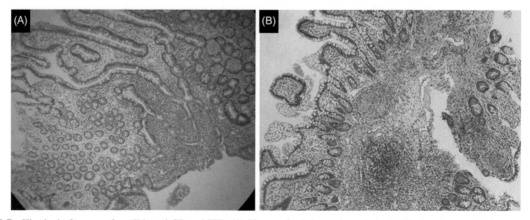

FIGURE 18.7 Histologic features of small bowel CD and ITB: (A) biopsy of aphthous ulcer of the in CD showing distorted small bowel glands between the relatively normal glands and infiltration of mononuclear ulcers in the lamina propria. Noncaseating granulomas are not common in small bowel biopsy and (B) large caseating granulomas characteristic of ITB are only present in less than 30% cases. The biopsy in this particular was taken from nodular tubercles in Fig. 18.5B. CD, Crohn's disease; *ITB*, intestinal tuberculosis.

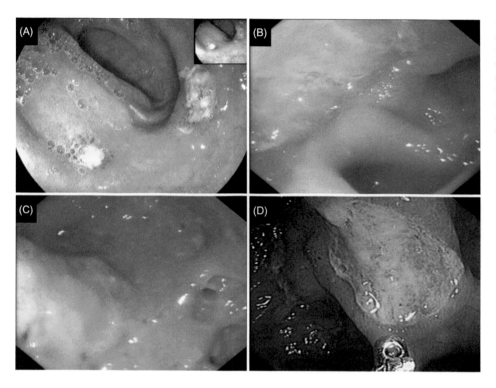

FIGURE 18.8 Intestinal Behcet's disease: (A) deep discrete round ulcers in the terminal ileum; (B and C) multiple deep ulcers with adjacent erythematous mucosa and fistula (*green arrow*) in the terminal ileum; and (D) deep, large, and sharp demarcated ulcer in the ileocecal valve.

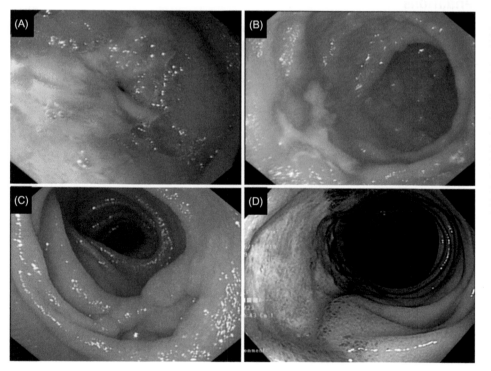

FIGURE 18.9 Intestinal lymphoma: (A) irregular and circumferential ulceration around severe stricture in the jejunum from MALT lymphoma; (B) irregular ulcers and granular change of mucosa of distal ileum from MALT lymphoma; (C) clean-based, punched out, deep ulcer in the proximal jejunum from diffuse large B-cell lymphoma; and (D) large, linear, clean-based, and well demarcated ulcer in the duodenum from NK cell lymphoma. *MALT*, mucosa-associated lymphoid tissue.

ischemic enteropathy, and so on (Fig. 18.5A, C, and D). Cryptogenic multifocal ulcerous stenosing enteritis (CMUSE) is characterized by the presence of multiple circumferential ring-like strictures in the small bowel (Fig. 18.10) [11].

Remarkably, small bowel lesions of IBD do not always present with typical enteroscopic appearance, which makes the diagnosis and differential diagnosis of IBD and IBD-like conditions challenged. Moreover, the diagnostic yield in histopathology of these diseases is low [6]. The final diagnosis is based on a combined assessment of clinical manifestations, laboratory testing, abdominal imaging, endoscopic findings, histopathology of biopsies or surgical specimen, and response to therapy.

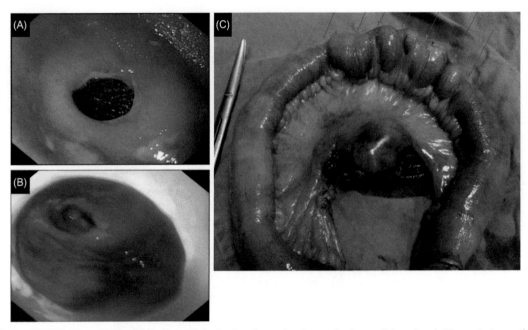

FIGURE 18.10 CMUSE: (A and B) multiple thin-walled, circular ulcerated strictures in the small bowel and (C) surgical resection specimen showing multiple ring-like strictures. *CMUSE*, Cryptogenic multifocal ulcerous stenosing enteritis.

Summary and recommendations

As the small bowel is the frequently involved region and even the only involved area in IBD and IBD-like diseases, deep enteroscopy can be safely and effectively performed in patients suspected of these conditions. The endoscopist should be well acquainted with the enteroscopic characteristics of IBD and IBD-like diseases. The enteroscopic features of IBD and IBD-like conditions are listed in Table 18.1. An accurate diagnosis always depends on a combination of various aspects, especially enteroscopic and histopathological findings.

Disclosure

The authors declared no financial conflict of interest.

References

[1] Keuchel M, Baltes P, Steinbrück I, Matsui U, Hagenmuller F. Normal small bowel. Video J Encycl GI Endosc 2013;1:261−3.

[2] Liverani E, Scaioli E, Digby RJ, Bellanova M, Belluzzi A. How to predict clinical relapse in inflammatory bowel disease patients. World J Gastroenterol 2016;22:1017−33.

[3] Baars JE, Theyventhiran R, Aepli P, Saxena P, Kaffes AJ. Double-balloon enteroscopy-assisted dilatation avoids surgery for small bowel strictures: a systematic review. World J Gastroenterol 2017;23:8073−81.

[4] Ungaro R, Mehandru S, Allen PB, Peyrin-Biroulet L, Colombel JF. Ulcerative colitis. Lancet 2017;389:1756−70.

[5] Bourreille A, Ignjatovic A, Aabakken L, Loftus Jr EV, Eliakim R, Pennazio M, et al. Role of small-bowel endoscopy in the management of patients with inflammatory bowel disease: an international OMED-ECCO consensus. Endoscopy 2009;41:618−37.

[6] Franklin GO, Mohapatra M, Perrillo RP. Colonic tuberculosis diagnosed by colonoscopic biopsy. Gastroenterology 1979;76:362−4.

[7] Zhu QQ, Zhu WR, Wu JT, Chen WX, Wang SA. Comparative study of intestinal tuberculosis and primarysmall intestinal lymphoma. World J Gastroenterol 2014;20:4446−52.

[8] Jung Y, Hwangbo Y, Yoon SM, Koo HS, Shin HD, Shin JE, et al. Predictive factors for differentiating between Crohn's disease and intestinal tuberculosis in Koreans. Am J Gastroenterol 2016;111:1156−64.

[9] Lee JM, Lee KM. Endoscopic diagnosis and differentiation of inflammatory bowel disease. Clin Endosc 2016;49:370−5.

[10] O'Malley DP, Goldstein NS, Banks PM. The recognition and classification of lymphoproliferative disorders of the gut. Hum Pathol 2014;45:899−916.

[11] Chang DK, Kim JJ, Choi H, Eun CS, Han DS, Byeon JS, et al. Double balloon endoscopy in small intestinal Crohn's disease and other inflammatory diseases such as cryptogenic multifocal ulcerous stenosing enteritis (CMUSE). Gastrointest Endosc 2007;66(3 Suppl):S96−8.

Chapter 19

Chromoendoscopy in inflammatory bowel disease

Badar Hasan[1], Peter A. Senada[2], Michael B. Wallace[2], Marietta Iacucci[3] and Bo Shen[4]

[1]Department of Gastroenterology and Hepatology, Cleveland Clinic Florida, Weston, FL, United States, [2]Department of Gastroenterology and Hepatology, Mayo Clinic Florida, Jacksonville, FL, United States, [3]Institute of Translational Medicine, Institute of Immunology and Immunotherapy, NIHR Birmingham Biomedical Research Centre, University Hospitals NHS Foundation Trust, University of Birmingham, Birmingham, United Kingdom, [4]Center for Inflammatory Bowel Diseases, Columbia University Irving Medical Center-New York Presbyterian Hospital, New York, NY, United States

Chapter Outline

Abbreviations

AFI	autofluorescence imaging
BLI	blue-light imaging
CAN	colitis-associated neoplasia
CD	Crohn's disease
CDEIS	Crohn's Disease Endoscopic Index of Severity
CI	confidence interval
CRC	colorectal cancer
DCE	dye chromoendoscopy
CLE	confocal laser endomicroscopy
FACILE	the Frankfurt Advanced Chromoendoscopic IBD LEsions
FICE	flexible spectral imaging color enhancement
HD	high definition
GI	gastrointestinal
IBD	inflammatory bowel disease
IEE	image-enhanced endoscopy
IPAA	ileal pouch-anal anastomosis
ME	magnification endoscopy
MES	the Mayo Endoscopic Score
NBI	narrow-band imaging
NICE	the Narrow-band Imaging International Colorectal Endoscopic (classification)
OCT	optical coherence tomography
OR	odds ratio

Atlas of Endoscopy Imaging in Inflammatory Bowel Disease. DOI: https://doi.org/10.1016/B978-0-12-814811-2.00019-0
© 2020 Elsevier Inc. All rights reserved.

PICaSSO	the Paddington International Virtual Chromoendoscopy
PSC	primary sclerosing cholangitis
SCENIC	the Surveillance for Colorectal Endoscopic Neoplasia Detection and Management in Inflammatory Bowel Disease Patients
SD	standard definition
UC	ulcerative colitis
WLE	white-light endoscopy

Introduction

Endoscopy plays a key role in the diagnosis, differential diagnosis, assessment of response to treatment, and surveillance of colitis-associated neoplasia (CAN). Endoscopy is the most reliable and commonly used modality for the measurement of disease activity. Mucosal healing (MH) has become a treatment goal. MH has been mainly defined with conventional white-light endoscopy (WLE). Therefore the terminology of MH and endoscopic healing has been used interchangeably. The availability of various image-enhanced endoscopy (IEE) technologies may further define MH, one step closer to histology level.

Colorectal cancer (CRC) is one of the leading causes of mortality worldwide with more than 1.8 million new cases diagnosed annually [1]. In the United States, CRC is the second leading cause of cancer death among men and women [2]. Patients with ulcerative colitis (UC) or Crohn's disease (CD) of the colon are at twofold higher risk of the development of CRC than the general population [3,4]. Cancer in patients with UC occurs at a younger age and the incidence increases with time, approaching 18% after 30 years of disease [5,6]. Apart from disease duration, other risk factors for CAN include younger age at diagnosis, increased inflammatory activity, greater extent of chronic inflammation, coexisting primary sclerosing cholangitis (PSC), and a family history of a first-degree relative with CRC [7]. Patients with left-sided or extensive UC and those with CD involving at least one-third of the colon or more than one segment of the colon are the target population for the surveillance program [7]. Colonoscopy plays key role in surveillance of CAN [8]. Guidelines recommend colonoscopy surveillance for CAN 8−10 years after disease diagnosis and continued surveillance every 1−3 years thereafter and the surveillance of UC patients at the time of diagnosis of PSC [4,6].

In the prior era of standard-definition (SD) or high-definition (HD) endoscopy, the practice pattern of endoscopic surveillance of dysplasia in patients with UC or CD of the colon used to be random biopsies in four quadrants every 10 cm of throughout the colon and targeted biopsies of any aberrant, raised or strictured areas with a total of at least 32−33 samples. The random biopsy technique is expensive, time-consuming, and invasive [5,6]. Even with the extensive biopsies, less than 0.1% of the surface area of the colon is surveyed and the detection rate has been low (<2 per 1000 biopsies taken). The multiple random biopsies may still miss dysplastic lesions [4−6,8].

Over the past decade, endoscopic technology has evolved to improve the surveillance of dysplasia in long-standing IBD. HD-WLE provides images with a resolution of more than 1 million pixels, in comparison to SD-WLE with a resolution of 100,000−400,000 pixels. HD-WLE has been instrumental in identifying polyps (especially sessile polyps) that may have been missed by SD-WLE [9]. However, even HD endoscopy is not able to characterize mucosal structure at microscopic level. Efforts, therefore, are being made to develop modalities to guide or even replace standard tissue biopsy. The terminology of optical biopsy has been used [10].

Corresponding to the evolution from SD and HD endoscopy to magnification endoscopy (ME), other imaging techniques have been developed. Various forms of IEE have been developed to characterize mucosal and submucosal features. IEE enhances surface, tone, and contrast. The modalities include ME, dye chromoendoscopy (DCE), dye-less narrow-band imaging (NBI), blue-light imaging (BLI), confocal laser endomicroscopy (CLE) [5], endocytoscopy, and optical coherence tomography (OCT). Applications of CLE, OCT, and ME in inflammatory bowel disease (IBD) are detailed in Chapter 20, Confocal endomicroscopy and other image-enhanced endoscopy in inflammatory bowel disease. The enhanced quality of endoscopic imaging provides better tools to characterize disease activity and CAN. Currently main applications of IEE are the assessment of disease activity and response to treatment, and characterization of CAN.

Disease scoring systems

A variety of instruments have been developed for the measurement of disease activity in CD and UC, for example, the Crohn's Disease Endoscopic Index of Severity (CDEIS) and the Mayo Endoscopic Score (MES). These disease instruments are detailed in Chapter 14, Endoscopic scores in inflammatory bowel disease. These scoring systems were developed based on WLE. Due to the limitation of WLE, currently available disease activity instruments characterize edema, erythema, vascular pattern, bleeding, erosions, and ulcers. None of these instruments included the specific endoscopic features for the definition of MH.

With IEE, attempts have been made to further characterize mucosal (such as scars and drop out of crypts) and superficial vascular structure in IBD, particularly in UC. For example, the Paddington International Virtual Chromoendoscopy (PICaSSO) was proposed to quantify the disease activity in UC [11,12]. The PICaSSO Classification consists of two categories, mucosal architecture (including erosions, crypt abscess, and erosions) and vascular architecture (including vessel dilation and bleeding), accounting extent and distribution of the abnormalities (Figs. 19.1 and 19.2; Table 19.1). The development of IEE-based disease activity scores may help further define MH, endoscopic remission or deep remission.

The main application of IEE has been the surveillance of CAN in patients with UC. The Kudo Classification of the pit pattern of non-IBD colorectal neoplastic lesions was proposed based on ME in 1996. The classification consists of five categories: Type I round pits; Type II stellar or papillary pits; Type III-L large tubular or roundish pits; Type III-S small tubular or roundish pits; Type IV branch-like or gyrus-like pits; and Type V nonstructural pits [13] Kudo pit pattern Types III−V is considered as being dysplastic and Kudo pit pattern Types I−II is considered as being nondysplastic or predictors of dysplasia in patients with long-standing IBD (Figs. 19.3 and 19.4; Table 19.2) [14]. The Kudo Classification was initially designed to depict non-IBD colonic neoplasia. The application of the Kudo Classification in IBD can be challenging, as regenerative hyperplastic villous mucosa with inflammation-associated elongated and irregular pits, making the distinction between inflammatory, hyperplastic, and neoplastic changes difficult.

Other classification systems of colon polyps have been proposed, including the Narrow-band Imaging International Colorectal Endoscopic (NICE) Classification [15,16] and the Showa Classification [17]. The NICE Classification was proposed to differentiate nonneoplastic from neoplastic lesions, based on lesion color, microvascular architecture, and surface pattern [18,19]. It classifies colorectal lesions into three categories: Type I as a hyperplastic lesion, Type II as an adenoma, and Type III as invasive cancer.

In addition to the pit pattern, morphology of mucosal structure and depth of CAN lesions may be categorized by the Paris Classification, which was designed for endoscopic classification of superficial neoplastic lesions in the colon as well as esophagus and stomach (Fig. 19.5; Table 19.3) [20]. The application of the Kudo Classification and Paris Classification are enhanced with the availability and application of various IEE. The Paris Classification was modified in a statement by the Surveillance for Colorectal Endoscopic Neoplasia Detection and Management in Inflammatory

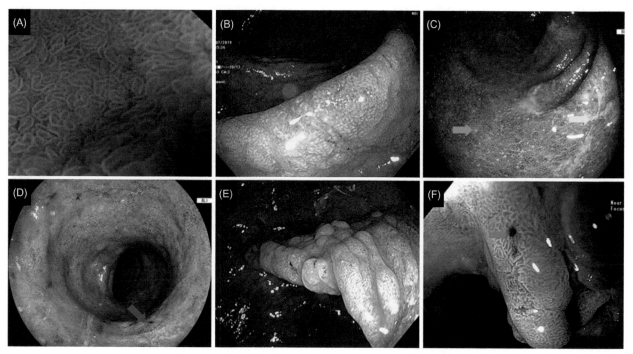

FIGURE 19.1   PICaSSO-mucosal architecture on narrow-band imaging or blue-light imaging: (A) Normal, (B and C) microerosions (*green arrow*), ulcers (*yellow arrow*), (D) mucosal scars (*blue arrow*), (E) mucosal edema, and (F) bleeding (*red arrow*). *PICaSSO*, The Paddington International Virtual Chromoendoscopy.

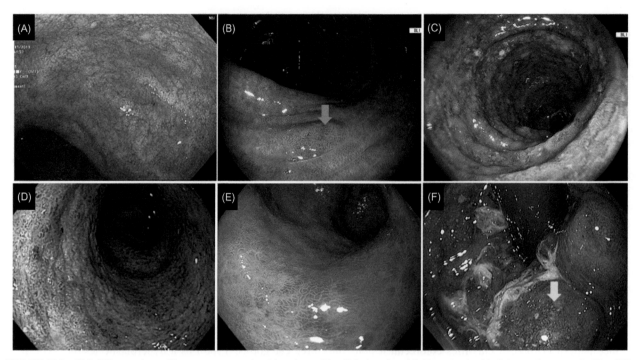

FIGURE 19.2 PICaSSO-vascular architecture on narrow-band imaging or blue-light imaging: (A) Normal subepithelial vascular network, (B) mildly dilated subepithelial vessels (*green arrow*), (C–E) crowded epithelial blood vessels, and (F) edema with crowed subepithelial blood vessels (*blue arrow*). *PICaSSO*, The Paddington International Virtual Chromoendoscopy.

TABLE 19.1 The Paddington International Virtual Chromoendoscopy Score (PICaSSO) in ulcerative colitis [11].

Category	Subcategory	Description
PICaSSO mucosal-architecture	0—No mucosal defect	A: Continuous/regular crypts
		B: Crypts not visible (scar)
		C: Discontinuous and or dilated/elongated crypts
	I—Microerosion or cryptal abscess	1: Discrete
		2: Patchy
		3: Diffuse
	II—Erosions size <5 mm	1: Discrete
		2: Patchy
		3: Diffuse
PICaSSO vascular-architecture	0—Vessels without dilatation	A: Roundish following crypt architecture B: Vessels not visible (scar)
		C: Sparse (deep) vessels without dilatation
	I—Vessels with dilatation	A: Roundish with dilatation
		B: Crowded or tortuous superficial vessels with dilatation
	II—Intramucosal bleeding	A: Roundish with dilatation
		B: Crowded or tortuous superficial vessels with dilatation
	III—Luminal bleeding	A: Roundish with dilatation B: Crowded or tortuous superficial vessels with dilatation

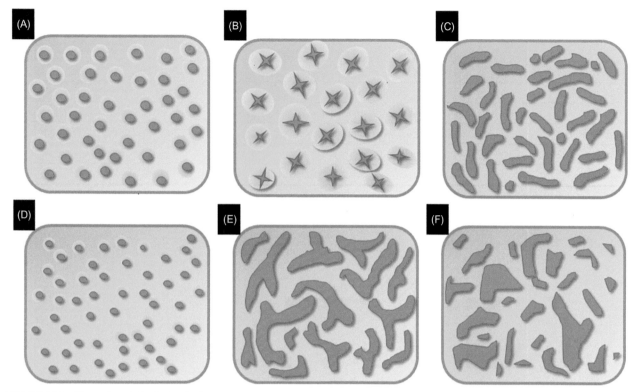

FIGURE 19.3   Illustration of the Kudo pit classification: (A) Type I—round pits (normal pattern), (B) Type II—stellar or papillary pits, (C) Type III-L—large tubular or roundish pits that are larger than normal pit pattern, (D) Type III-S—small tubular or roundish pits that are smaller than normal pit pattern, (E) Type IV—branch-like or gyrus-like pits, and (F) Type V—nonstructural pits.

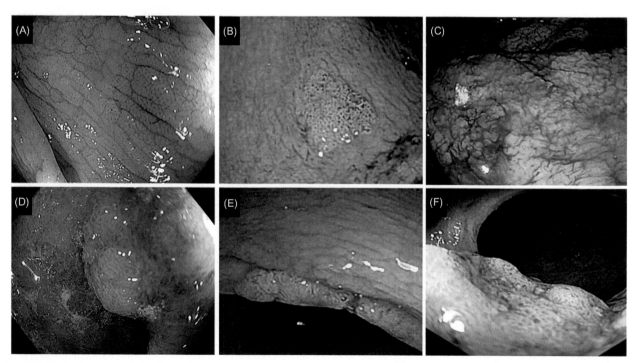

FIGURE 19.4   Illustration of the Kudo pit classification in inflammatory bowel disease with high-definition white-light colonoscopy: (A) Type I—round pits, (B) Type II—stellar or papillary pits, (C) Type III-L—large tubular or roundish pits, (D) Type III-S—small tubular or roundish pits, (E) Type IV—branch-like or gyrus-like pits, and (F) Type V—nonstructural pits.

TABLE 19.2 The Kudo Pit Pattern for colorectal tumorous lesions [13].

	Description	Explanation
Type I	Round pits	Normal pit pattern
Type II	Stellar or papillary pits	
Type III-L	Large tubular or roundish pits	Larger than normal pit pattern
Type III-S	Small tubular or roundish pits	Smaller than normal pit pattern
Type IV	Branch-like or gyrus-like pits	
Type V	Amorphous or nonstructural pattern	

L, Large; S, small.

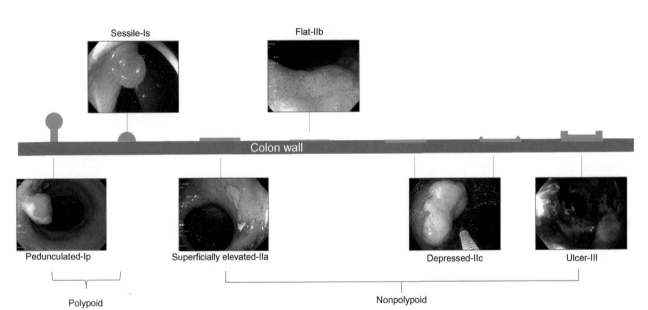

FIGURE 19.5   Superficial neoplastic lesions in inflammation bowel disease according to the Paris Classification.

TABLE 19.3 Adopted from the Paris Classification of neoplastic lesions with "superficial*" morphology (type 0).

	Morphology	Category	Subcategory	Description
Type 0	Polypoid	0-I	0-Ip 0-Is	Pedunculated polypoid Sessile polypoid
	Nonpolypoid and nonexcavated	0-II	0-IIa	Slightly elevated
			0-IIb	Flat
			0-IIc	Slightly depressed without ulcer
	Nonpolypoid with a frank ulcer	0-III		Excavated (ulcer)

"Superficial" is defined the depth of penetration in the digestive wall by the neoplasia is not more than into the submucosa, that is, there is no infiltration of the muscularis propria. P, Pedunculated; S, sessile.

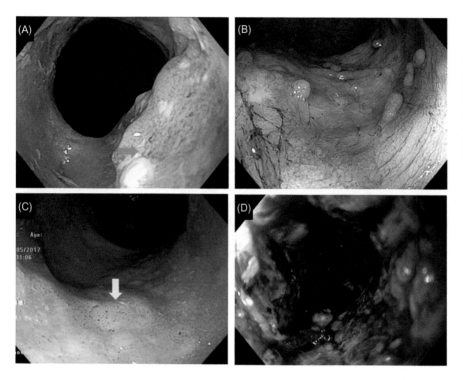

FIGURE 19.6 Components of the SCENIC Classification: (A) Descriptor-ulceration (*green arrow*), (B) distinct border, (C) indistinct border (*yellow arrow*), and (D) strictured ileal pouch-anal anastomosis with poorly defined surface structure. Histology showed adenocarcinoma (D). *SCENIC*, the Surveillance for Colorectal Endoscopic Neoplasia Detection and Management in Inflammatory Bowel Disease Patients.

TABLE 19.4 SCENIC terminology for reporting findings on colonoscopic surveillance of patients with inflammatory bowel disease.

Term	Definition
Visible dysplasia	Dysplasia identified on targeted biopsies from a lesion visualized at colonoscopy
Polypoid	Lesion protruding from the mucosa into the lumen ≥ 2.5 mm
Pedunculated	Lesion attached to the mucosa by a stalk
Sessile	Lesion not attached to the mucosa by a stalk: entire base is contiguous with the mucosa
Nonpolypoid	Lesion with little (<2.5 mm) or no protrusion above the mucosa
Superficial elevated	Lesion with protrusion but <2.5 mm above the lumen (less than the height of the closed cup of a biopsy forceps)
Flat	Lesion without protrusion above the mucosa
Depressed	Lesion with at least a portion depressed below the level of the mucosa
General descriptors	
Ulcerated	Ulceration (fibrinous-appearing base with depth) within the lesion
Border	
Distinct border	Lesion's border is discrete and can be distinguished from surrounding mucosa
Indistinct border	Lesion's border is not discrete and cannot be distinguished from surrounding mucosa
Invisible dysplasia	Dysplasia identified on random (nontargeted) biopsies of colon mucosa without a visible lesion

SCENIC, The Surveillance for Colorectal Endoscopic Neoplasia Detection and Management in Inflammatory Bowel Disease Patients.
Source: Modified from the Paris Classification. The Paris endoscopic classification of superficial neoplastic lesions: esophagus, stomach, and colon: November 30 to December 1, 2002. Gastrointest Endosc. 2003;58:S3—43.

Bowel Disease Patients (SCENIC): International Consensus Recommendations (Fig. 19.6; Table 19.4). The SCENIC consensus statement guides colonoscopic surveillance of CAN which was endorsed by the American Society for Gastrointestinal Endoscopy and American Gastroenterological Association. The endoscopic features of the modified Paris Classification listed in the SCENIC statement help clinicians to make decision on endoscopic resection versus surgical intervention.

Dye chromoendoscopy

Dye-based chromoendoscopy is widely used in the gastrointestinal (GI) disorders, ranging from Barrett's esophagus and early gastric neoplasia to CAN. As compared with ME, OCT, CLE, and confocal endomicroscopy, DCE is available with the conventional endoscope along with staining agents and tubing without the need for special endoscopy equipment. Staining agents for DCE include Lugol's solution, methylene blue, toluidine blue, Congo red, phenol red, and indigo carmine. The most commonly used dyes are 0.03% indigo carmine (a reactive, nonabsorbed dye) and 0.04% methylene blue (an absorbed and contrast dye). Indigo carmine is a deep blue contrast dye that coats mucosal surfaces and persists for a few minutes, highlighting mucosal pits, grooves, and crevices, improving the visualization of mucosal structures and allowing for the delineation of any lesions or areas of inflammation. Methylene blue is absorbed by the small intestinal and colonic epithelial cells. It takes approximately 60 seconds for adequate staining and the dye lasts for approximately 20 minutes. Dysplastic or inflamed cells absorb little or no dye at all, resulting in differential staining which allows for the detection of any lesions and areas of inflammation.

Adequate bowel preparation is important for quality DCE which can be achieved the administration of a clear liquid diet the day before with polyethylene glycol electrolyte solution in a split dose [22]. Sodium phosphate-based bowel preparations should be avoided as they can cause aphthous ulcers (Figs. 19.7 and 19.8) [23].

The role of DCE in the assessment of disease activity of CD or UC has not been well defined. DCE provides more detailed imaging on the mucosal structure and vasculature like virtual chromoendoscopy. To diagnose and assess the severity of IBD the clinician needs to incorporate a constellation of clinical symptoms, lab parameters, and endoscopic findings. One of the earlier trials involving 17 patients with UC suggests that methylene blue–based DCE was effective for assessing the mucosa inflammation in patients with mild ulcerative proctitis. Areas of abnormal staining correlated well areas of inflammation [24]. Another prospective study of 25 patients with UC reported that DCE increased the detection of areas affected by UC [25]. These findings suggest that DCE may be helpful in estimating the extent of inflammation in patients with UC with better characteristics of mucosal details (Figs. 19.9–19.15).

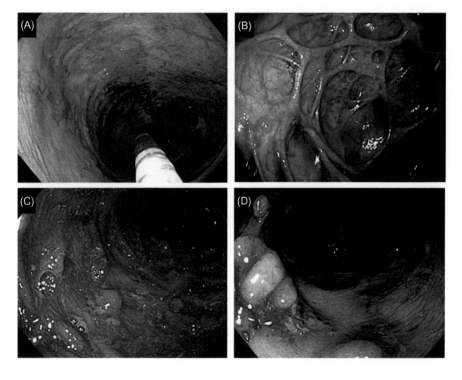

FIGURE 19.7 Bowel preparation and chromoendoscopy: (A and B) Catheter-based dye spray during chromoendoscopy in an ulcerative colitis patient with extensive mucosal scars, (C and D) inadequate bowel preparation with fecal particles which interfered with efficacy of chromoendoscopy.

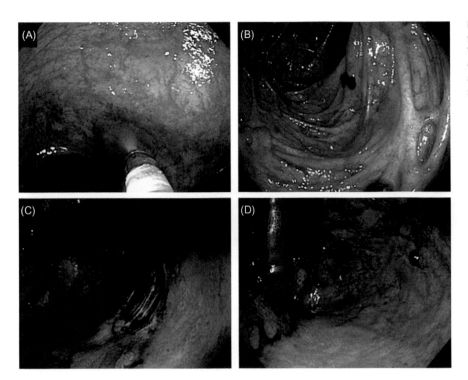

FIGURE 19.8 Dye chromoendoscopy in ulcerative colitis: (A) Spray of methylene blue, (B) the dye spray was too diluted, (C) the dye spray was too concentrated, and (D) retroflex view of the rectum is still needed in patients without active proctitis.

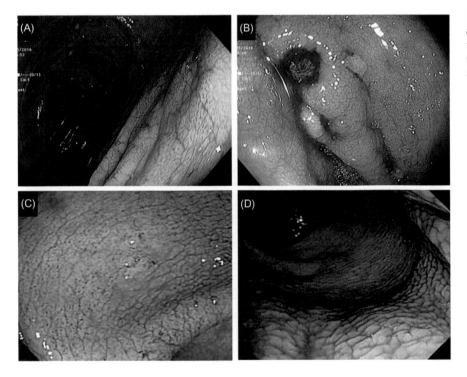

FIGURE 19.9 Dye chromoendoscopy in quiescent ulcerative colitis: (A) Granular mucosa, (B) granular mucosa after biopsy (*green arrow*), (C) punctate pits, and (D) granularity with Kudo Classification-II.

Dye chromoendoscopy may be incidentally performed in patients undergoing the colonoscopy for monitoring activity. The patients may be found to have disease in remission with mucosal scars, nonulcerated strictures, or pseudopolyps. These patients may benefit from the use DCE during the same session to find dysplasia.

Dye chromoendoscopy was shown to yield a better detection rate of CAN than SD or HD-WLE [26−28]. To optimize the quality of surveillance, there should be minimal or absent mucosal inflammation. Iacucci et al. recommended

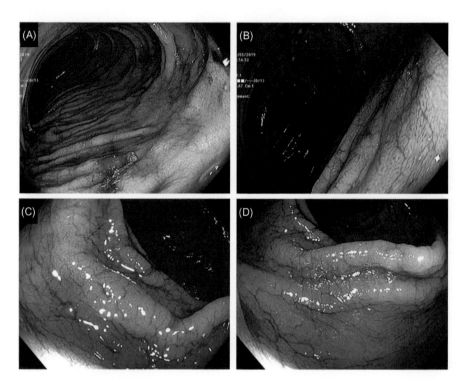

FIGURE 19.10 Dye chromoendoscopy in quiescent ulcerative colitis: (A−D) Granular mucosa with round pit patterns.

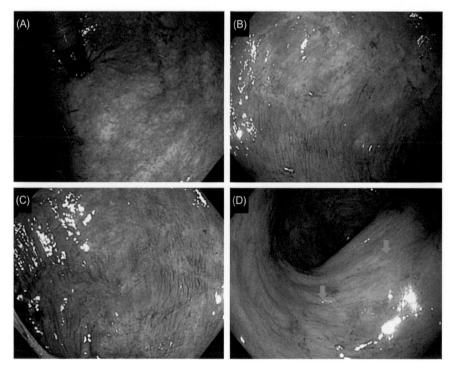

FIGURE 19.11 Dye chromoendoscopy in patients with quiescent ulcerative colitis: (A−C) Mucosal scars and visible mucosal vasculature and (D) hyperplastic lesions with absent staining of methylene blue (*green arrows*). Detailed pit pattern was not well visualized with the high-definition endoscopy.

fecal calprotectin level below 100 μg/g as a marker of minimal disease activity before surveillance colonoscopy [23]. The SCENIC statement has recommended surveillance for CAN be done with HD-WLE and DCE with targeted biopsies in contrast to standard multipiece random biopsy [29]. Earlier studies showed that DCE was more effective to detect CAN than SD-WLE [4−6,21], but not than HD-WLE [7,30], SCENIC metaanalysis combining 8 randomized controlled trials revealed a 1.8-fold increase in the dysplasia detection rate with DCE targeted biopsy as compared to

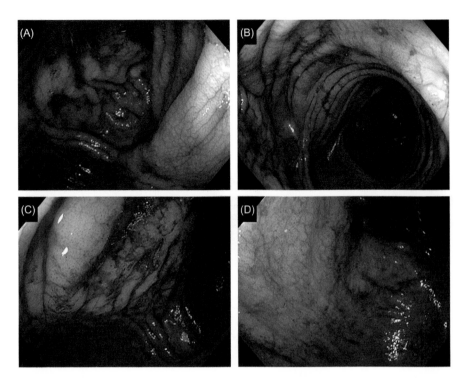

FIGURE 19.12 Dye chromoendoscopy in patients with quiescent ulcerative colitis: (A–D) Kudo type I pit pattern.

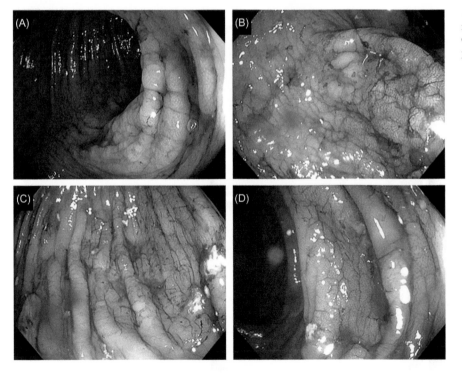

FIGURE 19.13 Dye chromoendoscopy in quiescent ulcerative colitis: (A–D) Granular mucosa with Kudo type I pit pattern.

the WLE random biopsy protocol [4–6]. The SCENIC metaanalysis found the dysplasia detection rate to be 90% on targeted biopsies and only 10% on random biopsies [21,31].

In clinical practice of endoscopy in IBD, DCE is considered the gold standard, while HD-WLE or DCE is preferred over SD-WLE [14]. The purpose of DCE is to enhance mucosal features and guide the targeted biopsy or ablation of aberrant aspects suspected of neoplasia (Figs. 19.16–19.29). In addition to UC and Crohn's colitis, chemoendoscopy has been used for the dysplasia surveillance in patients with ileal pouch-anal anastomosis (IPAA) (Figs. 19.30 and 19.31).

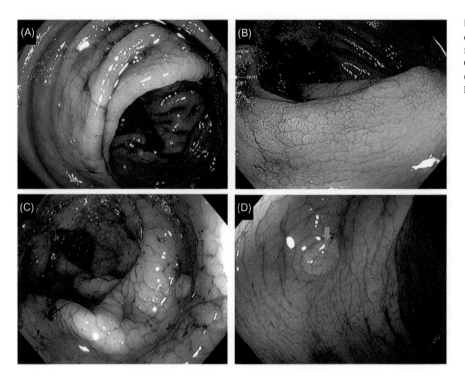

FIGURE 19.14 Dye chromoendoscopy in quiescent ulcerative colitis: (A–C) Granular mucosa and (D) a polypoid lesion (Paris Classification-Is; Kudo Classification-II) (*green arrow*) with a histologic diagnosis of hyperplastic polyp.

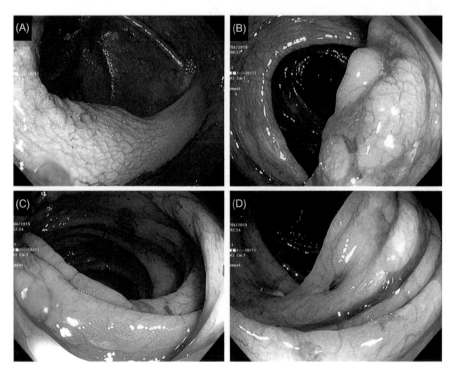

FIGURE 19.15 Mucosal healing on white-light colonoscopy showed mucosal structural changes in quiescent Crohn's colitis on dye chromoendoscopy during routine surveillance colonoscopy: (A–D) Benign pit patterns.

Virtual chromoendoscopy

Conventional colonoscopy can demonstrate intramucosal capillary network of the normal colon. In the inflamed colon mucosa, mucosal edema, granularity, nodularity, fibrosis, and ulceration obscure the view of intramucosal capillary network. Generally speaking, dyes used in DCE are safe to administer, and no adverse events have been reported except

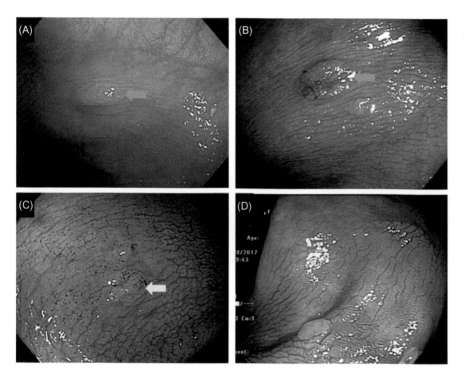

FIGURE 19.16 Dye chromoendoscopy for the assessment of hyperplastic polyps in ulcerative colitis: (A and B) Small sessile hyperplastic polyp on white-light endoscopy (*green arrows*) with staining of methylene blue and (C and D) small superficially elevated hyperplastic lesions, which may (*yellow arrow*) or may not (*red arrow*) be stained by methylene blue.

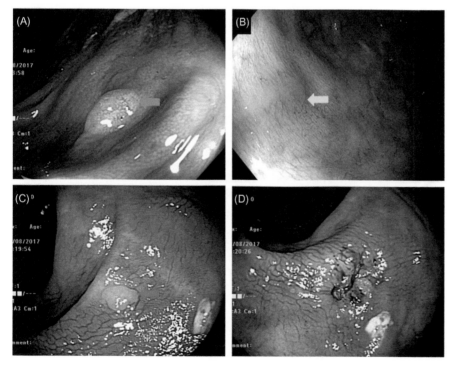

FIGURE 19.17 Dye chromoendoscopy for sessile hyperplastic polyps in ulcerative colitis: (A) Sessile polyp (Paris Classification-Is; Kudo Classification-II) stained with methylene blue (*green arrow*), (B) superficially elevated hyperplastic lesion (Paris Classification-Is; Kudo Classification-unclear) devoid of the dye (*yellow arrow*), and (C and D) sessile polyp (Paris Classification-Is; Kudo Classification-II) devoid of the dye (*red arrow*) which was biopsied (D).

for the one report of methylene blue inducing oxidative DNA damage after chromoendoscopy for the evaluation of Barrett's esophagus [32]. Some patients have dye allergies.

The dye-less, virtual chromoendoscopy, also known as electronic chromoendoscopy, refers to endoscopic imaging technologies with detailed contrast enhancement of the mucosal surface and blood vessels in the GI tract. Regenerative

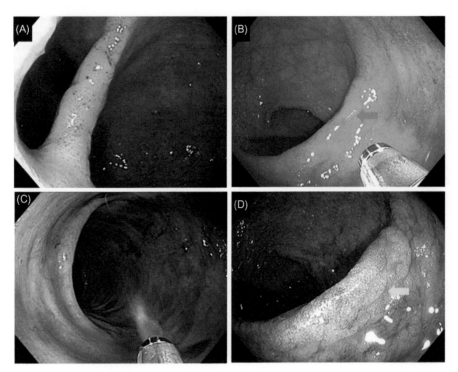

FIGURE 19.18 Dye chromoendoscopy in a patient with quiescent ulcerative colitis and primary sclerosing cholangitis: (A) Patulous ileocecal valve, (B) colon mucosa with loss of vascularity (*green arrow*), (C) spray of methylene blue, and (D) the highlighted area on white-light endoscopy (C) was further defined by staining pattern of dye (*yellow arrow*).

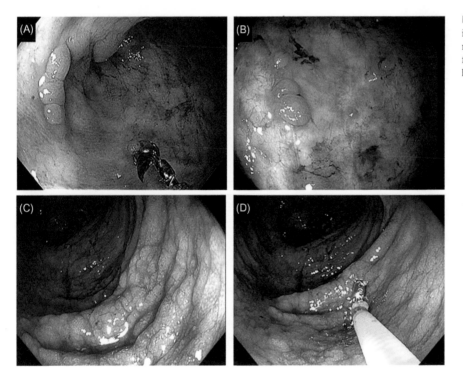

FIGURE 19.19 Dye chromoendoscopy in quiescent ulceration colitis with nodular mucosa: (A–D) Nodular mucosa with surface (Kudo Classification-II) being highlighted by the dye spray.

changes in the mucosa can be misinterpreted as dysplastic lesions. Moreover, chronic inflammation can disrupt the pit pattern, thus limiting the use of the Kudo Classification [14].

Virtual colonoscopy, different from virtual chromocolonoscopy, is specifically referred to as computed tomography colonography. Virtual chromoendoscopy uses either optical or digital technologies. In the former setting, optical lenses are integrated into the endoscope's light source, selectively filtering white light and subsequent resulting in narrow-band

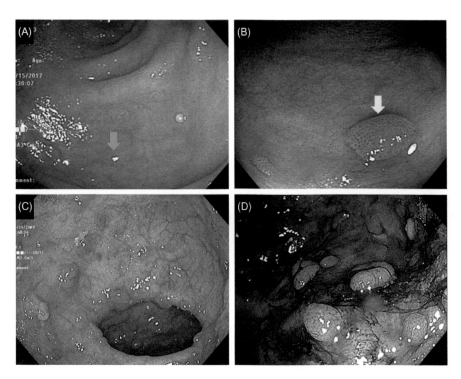

FIGURE 19.20 Dye chromoendoscopy for inflammatory polyps in ulcerative colitis: (A and B) A small sessile polyp with subtle appearance on white-light colonoscopy (*green arrow*) which was highlighted for pit pattern with methylene blue spray (*yellow arrow*) and (C and D) extensive inflammatory polyps with obscure surface on white-light colonoscopy (C), in contrast to the highlighted surface pit pattern on chromoendoscopy (D).

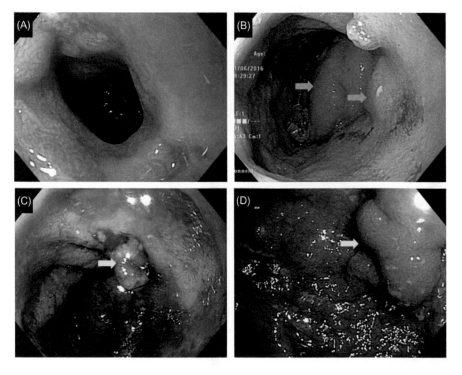

FIGURE 19.21 Chromoendoscopy in ileal pouch prolapse: (A) Ileal pouch-anastomosis with mild stricture, (B) anterior distal pouch prolapse (*green arrows*), and (C and D) prolapsed mucosa was highlighted with methylene blue (*yellow arrows*).

light. In the later setting, chromoendoscopy utilizes digital postprocessing to enhance the real-time images. Virtual chromoendoscopy such as NBI eliminates the need for dye spraying or reinsertion which reduces the procedure time significantly as compared to DCE [21].

Three different dye-less virtual chromoendoscopy systems are commercially available: NBI (Olympus, Tokyo, Japan), flexible spectral imaging color enhancement (FICE; Fujifilm, Tokyo, Japan), and I-SCAN (PENTAX; Tokyo, Japan) [33].

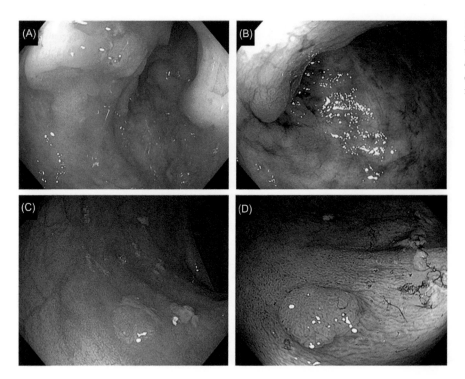

FIGURE 19.22 Dye chromoendoscopy for pseudopolyps in active or quiescent ulcerative colitis: (A−D) Surface features of the polyps were obscure on white-light endoscopy (A and C) but detailed on chromoendoscopy (B and D).

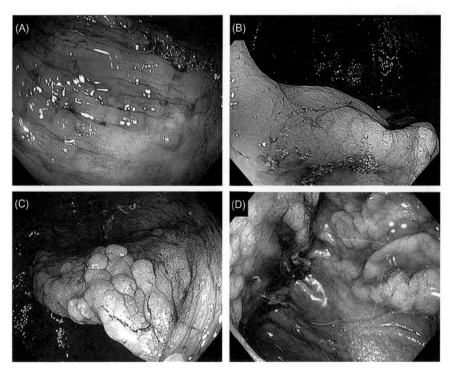

FIGURE 19.23 Dye chromoendoscopy for pseudopolyps in quiescent ulcerative colitis: (A−D) Surface features of the polyps were highlighted on chromoendoscopy.

Narrow-band imaging

Narrow-band imaging is the most commonly used virtual colonoscopy system. The depth of light penetration depends on its wavelength. The longer is the wavelength, the deeper the light penetrates the tissue. NBI is a real-time optical filter placed in front of the white-light source to narrow the wavelength to 30 nm bandwidths in the blue (415 nm) and

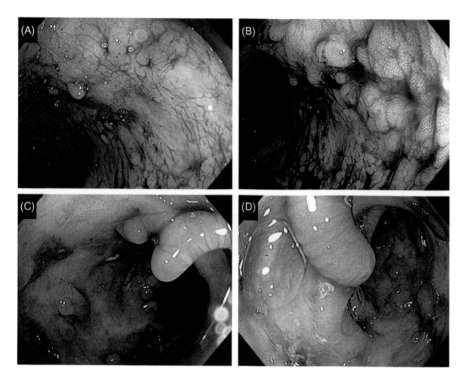

FIGURE 19.24 Dye chromoendoscopy for flat and elongated pseudopolyps in quiescent ulcerative colitis: (A—D) Surface features of the polyps were highlighted on chromoendoscopy.

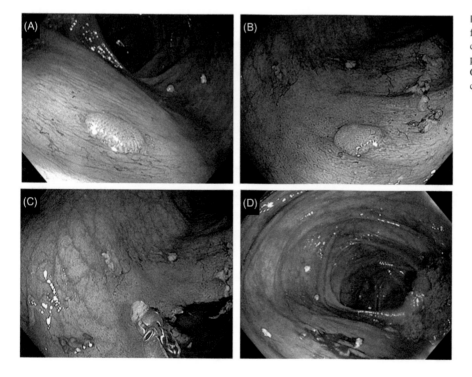

FIGURE 19.25 Dye chromoendoscopy for sessile polyps in quiescent ulcerative colitis: (A—D) Surface features of the polyps (Paris Classification-Is; Kudo Classification-II) were highlighted on chromoendoscopy, which were biopsied.

green (540 nm) region of the spectrum. The short-wave blue light penetrates only superficially, while longer-waved red light penetrates deeper. NBI uses a light source with special red-green and blue filers allowing for bandpass ranges to be narrowed and the relative contribution of blue light to be increased. NBI system can also use white-light illumination in combination with a color charged coupled device-chip. NBI demonstrate enhances pit patterns, mucosal structures, the superficial vasculature of mucosa with a higher contrast than WLE, as NBI has shallower tissue penetration and it is

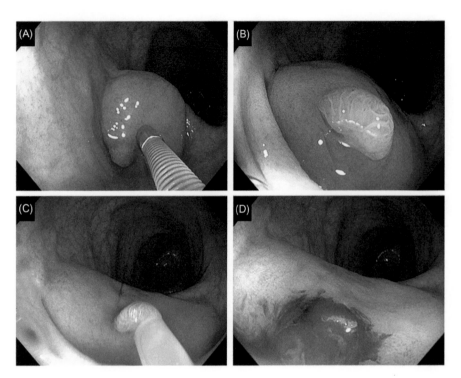

FIGURE 19.26 Dye chromoendoscopy as a guide for colonoscopic polypectomy in quiescent ulcerative colitis: (A−D) The sessile polyp (Paris Classification-Type Is; Kudo Classification-III-L) was highlighted with the dye spray. Endoscopic polypectomy was performed with a saline cushion.

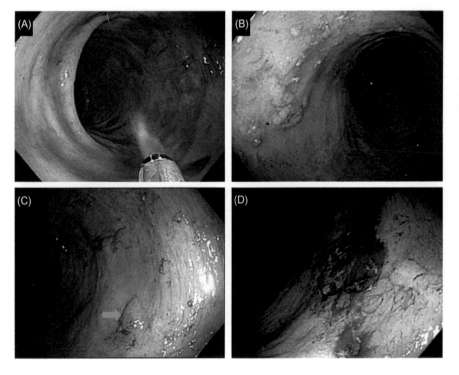

FIGURE 19.27 Dye chromoendoscopy for the surveillance in quiescent ulcerative colitis: (A) Dye spray, (B and C) superficially raised lesions (Paris Classification-Is; Kudo Classification-Type II) devoid of methylene blue stain (*green arrows*), and (D) biopsy taken which showed indefinite for dysplasia.

mostly absorbed by the hemoglobin in the vessels. Various classification systems have been proposed for Barrett's esophagus based on NBI features of the mucosal microstructural pattern (e.g., ridged, villous, circular, irregular, or distorted) and vascular pattern [34].

NBI-based colonoscopy may be useful in grading inflammation in IBD. A pilot study of 14 patients (6 CD and 8 UC) investigated NBI to detect angiogenesis in patients with IBD. Normal-appearing areas on WLE but stronger

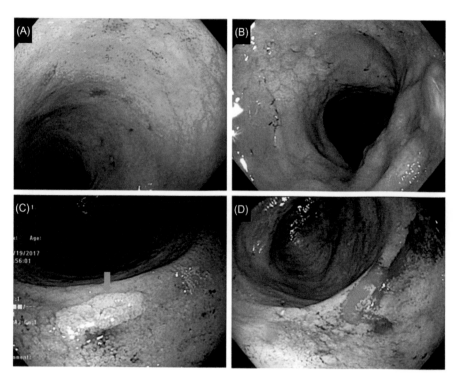

FIGURE 19.28 Dye chromoendoscopy in ulcerative colitis: (A) Loss of mucosal vascular pattern on white-light endoscopy, (B) a slightly raised area of the colon (Paris Classification-IIa) was demonstrated by the dye spray, (C and D) superficially raised lesion (Paris Classification-IIa; Kudo Classification-III-S) (green arrow) which was biopsied, (D) biopsy of B−D lesions showed low-grade dysplasia.

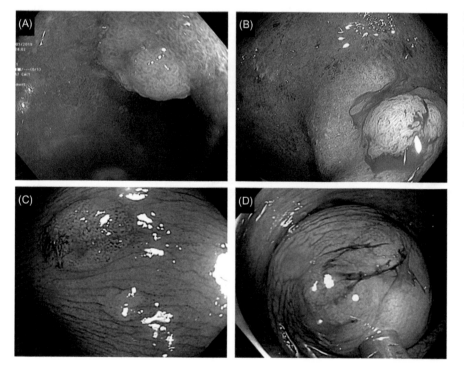

FIGURE 19.29 Dye chromoendoscopy and polypectomy in active or quiescent ulcerative colitis: (A and B) A sessile polyp was removed en bloc polypectomy in mildly active ulcerative colitis, (C and D) a sessile polyp removed with saline-cushion polypectomy in a separate patient. Polyps in both patients showed low-grade dysplasia.

capillary patterns were associated with increased leukocyte infiltration and microvessel density on histology [14]. Kudo et al. assessed mucosal vascular pattern in 157 segments of the large bowel in 30 patients with UC with conventional and NBI colonoscopy and found an association between the obscure mucosal vascular pattern on NBI and acute inflammatory cell infiltrates, goblet cell depletion, and basal plasmacytosis on histology [35]. NBI can provide clear imaging of the microvascular structure and network under the mucosal as well as the pit pattern. NBI depicts (1) whitish round

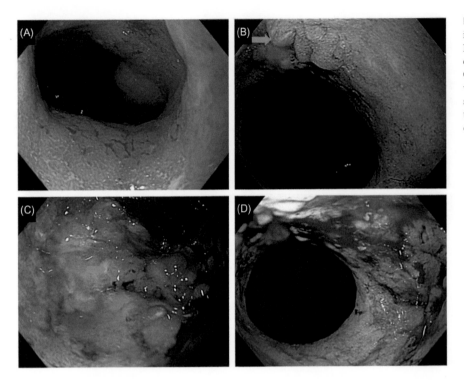

FIGURE 19.30 Dye chromoendoscopy in ileal pouch-anal anastomosis: (A) Normal appearing cuff on white-light endoscopy, (B and C) nodular areas (Paris Classification-IIa; Kudo Classification-II) were demonstrated with methylene blue spray (*green arrows*), and (D) biopsy taken. Final diagnosis is low-grade dysplasia.

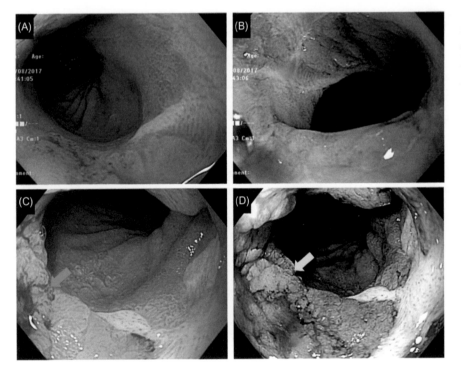

FIGURE 19.31 Dye chromoendoscopy in ileal pouch-anal anastomosis: (A and C) granularity of the anal transition zone (*green arrow*) and (C and D) the lesion was highlighted by the dye spray (*yellow arrow*).

crypts, referred to as crypt opening type; (2) villous structures, referred to as villous type; (3) honeycomb-like microvascular pattern; and (4) irregular microvascular pattern [36]. The NBI patterns were shown to be correlated with histologic findings of crypt distortion, goblet cell depletion, and basal plasmacytosis [25]. The common NBI features in mildly active UC include obscure intramucosal capillary network, reddish granular mucosa; and the features in moderate or severe UC also include mucous exudates, friability, erosion, ulcers, and spontaneous bleeding are observed in the

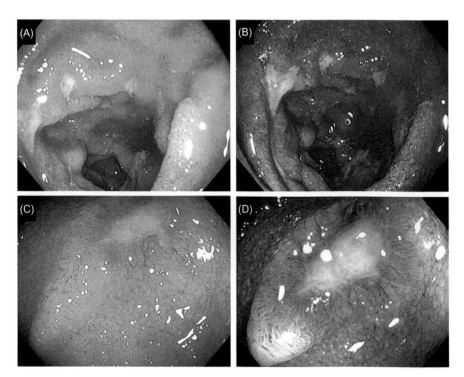

FIGURE 19.32 Narrow-band imaging in the assessment of disease activity in Crohn's disease: (A and B) Edema and ulceration of the terminal ileum on white-light endoscopy and narrow-band imaging and (C and D) clean-based ulcers in the terminal ileum. Adjacent mucosal structure was highlighted better by narrow-band imaging.

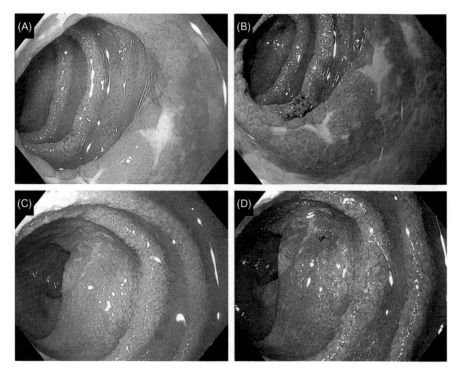

FIGURE 19.33 Narrow-band imaging in the assessment of disease activity of Crohn's disease: (A–D) Multiple small ulcers in the terminal ileum on white-light endoscopy and narrow-band imaging.

moderately or severely active UC. NBI provides high-contrasted, sharp images in the assessment of disease activity in CD (Figs. 19.32–19.35) as well as UC (Figs. 19.36 and 19.37).

The clinical implication of NBI in the surveillance of CAN seems to be unsubstantiated as compared with DCE. Current available studies failed to demonstrate improved detection rate of CAN by NBI when compared with SD- or HD-WLE [31,37,38]. The SCENIC consensus panel did not recommend NBI to replace SD-WLE, HD-WLE, or DCE

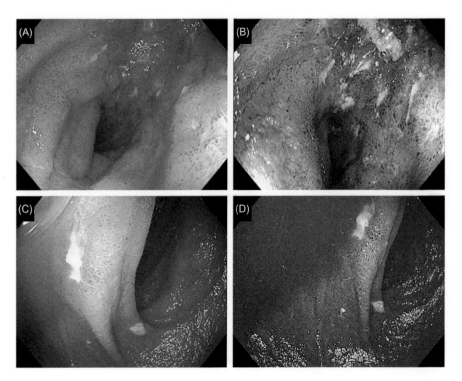

FIGURE 19.34 Narrow-band imaging for the assessment of disease activity in Crohn's disease: (A and B) Mucosal edema, erosions, and loss of vascularity of the terminal ileum which were highlighted with narrow-band imaging and (C and D) small ulcers in the terminal ileum which were highlighted with narrow-band imaging.

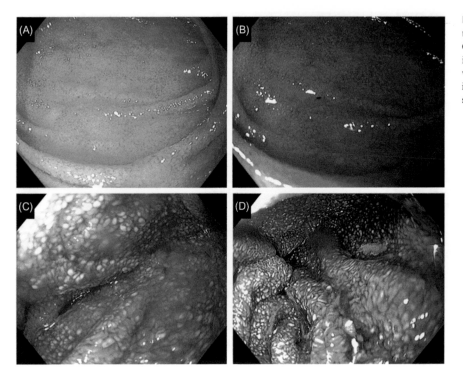

FIGURE 19.35 Narrow-band imaging in the assessment of disease activity in Crohn's disease: (A and B) Normal distal ileum mucosa, (C and D) dilate lacteals on white-light endoscopy and narrow-band imaging in a patient with ileostomy and short gut syndrome from Crohn's disease.

for dysplasia surveillance [21]. A metaanalysis showed that NBI with targeted biopsy for the surveillance of CAN did not yield a sensitivity superior to other endoscopic techniques, including SD-WLE, HD-WLE, or DCE [21]. A randomized clinical trial found that NBI had a numerically higher rate of missing lesions (31.8%) than DCE (13.6%) [39]. The main pitfalls of NBI are its limited depth of penetration and darkfield only by trained eyes. Similar to DCE, the pit

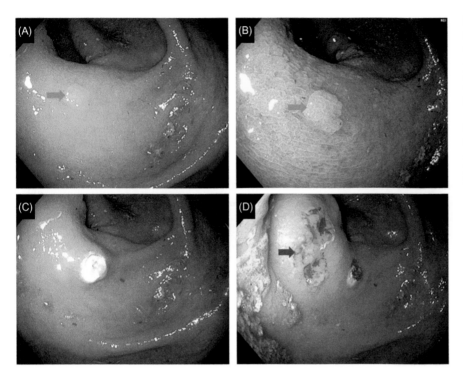

FIGURE 19.36 Narrow-band imaging for the assessment of disease activity and surveillance in quiescent Crohn's colitis: (A and B) A small superficially raised lesion (Paris Classification-Ip; Kudo Classification-II) in the ileocecal valve highlighted by narrow-band imaging (*green arrows*), (C) the lesion was removed with a hot snare, (D) surrounding mucosa of polypectomy site was extensively biopsied for the evaluation of dysplasia as well as underlying chronic inflammatory changes (*below arrow*).

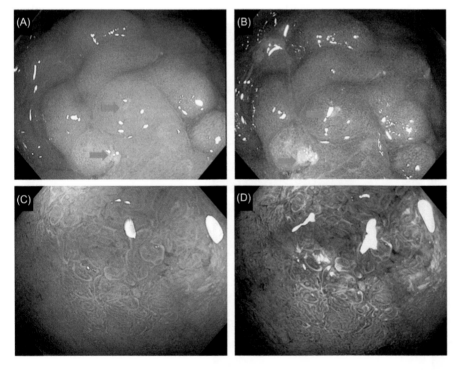

FIGURE 19.37 Narrow-band imaging for the assessment of disease activity in ulcerative colitis: (A and B) Mucosal edema and small erosions (*green arrows*), (C and D) mucosal vascular patterns with intramucosal capillaries, and distorted crypts on white-light endoscopy and narrow-band imaging.

pattern for dysplasia on NBI interferes with chronic inflammation in IBD [40]. It should be noted that the findings of NBI of the earlier literature are largely based on SD-WLE. Since then NBI has been used in conjunction with DCE, HD, or ME for surveillance of CAN [41]. Nonetheless, the detection rate of CAN was not found to be different between HD-NBI and HD-DCE.

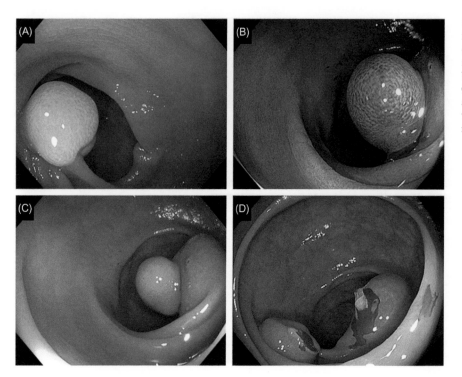

FIGURE 19.38 Narrow-band imaging for the characterization of colon polyps: (A) A pedunculated polyp; (B) characterization of the surface structure of the polyp by narrow-band imaging (Paris Classification-Ip; Kudo Classification-II) and (C and D) endoscopic polypectomy with a saline cushion. The histologic diagnosis was tubular adenoma.

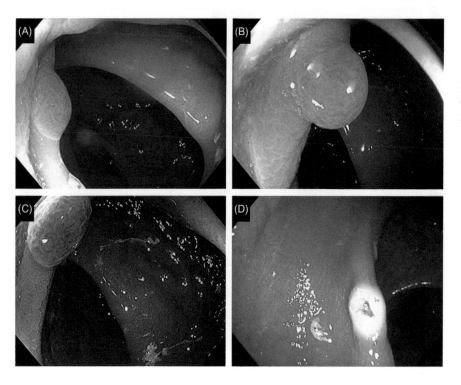

FIGURE 19.39 Narrow-band imaging for the surveillance of neoplasia in quiescent Crohn's colitis: (A–C) A small sessile polyp at the distal rectum with surface structure highlighted by narrow-band imaging (Paris Classification-Is; Kudo Classification-II), (D) polypectomy was performed with a hot snare.

In clinical practice, NBI can still be used to identify subtle mucosal changes suspected of dysplasia on WLE (Figs. 19.38–19.47). NBI may be applied for the surveillance of dysplasia in IPAA (Figs. 19.48 and 19.49). NBI may even be used to assess perianal lesions for dysplasia or infection (Fig. 19.50).

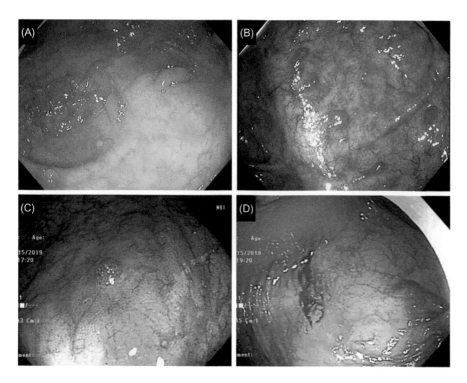

FIGURE 19.40 Narrow-band imaging for the surveillance of neoplasia in ulcerative colitis: (A) Inflammatory pseudopolyps on white-light endoscopy, (B and C) the polyps highlighted with narrow-band imaging (Paris Classification-Is; Kudo Classification-II), and (D) targeted biopsy taken.

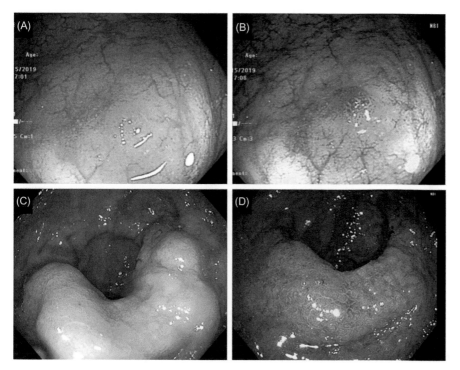

FIGURE 19.41 Narrow-band imaging in a patient with ulcerative colitis and primary sclerosing cholangitis: (A and B) A small sessile hyperplastic polyp (Paris Classification-Is; Kudo Classification-II) on white-light endoscopy and narrow-band imaging and (C–D) rectal varices presenting with a mass lesion. Intramucosal capillaries with irregular, tortuous structure were highlighted with narrow-band imaging. Narrow-band imaging provides contrasted images better than white-light endoscopy.

I-SCAN and FISE

Similar to NBI, the I-SCAN system has also been shown to improve the prediction of inflammatory activity and disease extent in IBD on HD-WLE endoscopy [42]. I-SCAN can identify subtle histologic abnormalities underlying the endoscopically healed mucosa in UC [43]. The new I-SCAN-optical enhancement combines optical and a digital

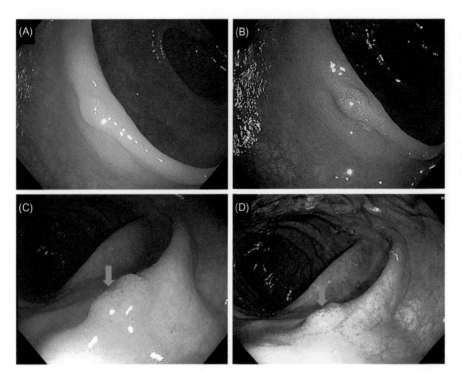

FIGURE 19.42 Narrow-band imaging for the surveillance of neoplasia in quiescent ulcerative colitis: (A and B) Small sessile hyperplastic polyp (Paris Classification-Is; Kudo Classification-II) on white-light endoscopy and narrow-band imaging and (C and D) Small sessile dysplastic polyp with central indentation (Paris Classification-Is; Kudo Classification-II) on white-light endoscopy and narrow-band imaging (*green arrows*). The histologic diagnosis is low-grade dysplasia.

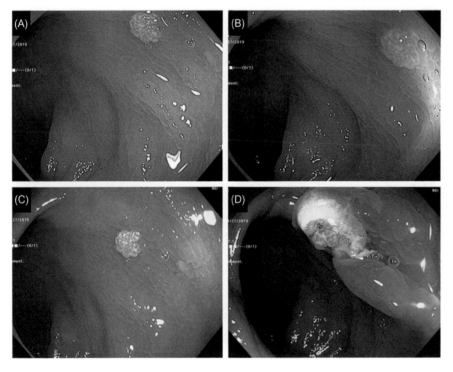

FIGURE 19.43 Narrow-band imaging for the surveillance of neoplasia in ulcerative colitis: (A and B) Small sessile hyperplastic polyp on white-light endoscopy, (C) the surface structure of the polyp (Paris Classification-Is; Kudo Classification-II) highlighted by narrow-band endoscopy, and (D) polypectomy was performed with a hot snare.

enhancement chromoendoscopy into a single system. It has three modes to enhance contrast (to digitally add blue color to relatively dark areas), surface (to modify luminance intensity), and tone. I-SCAN has been used to assess the disease activity in CD (Fig. 19.51) or UC (Figs. 19.52–19.54). The features on I-SCAN may further define MH.

I-SCAN may be used for the surveillance of CAN (Figs. 19.55 and 19.56). There are scant data on the application of flexible spectral imaging color enhancement (FISE) in IBD.

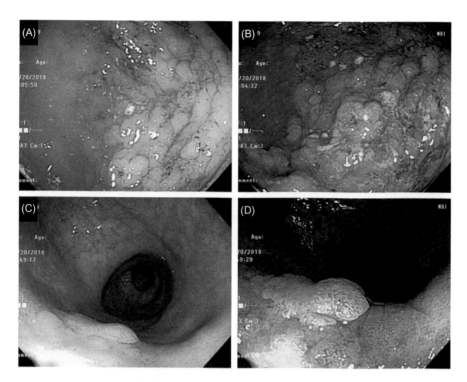

FIGURE 19.44 Narrow-band imaging for the surveillance of neoplasia in ulcerative colitis: (A and B) Diffuse nodular mucosa in the hepatic flexure on white-light endoscopy and narrow-band imaging. Biopsy showed inflammatory polyps and (C and D) sessile polypoid lesions on white-light endoscopy and narrow-band imaging (Paris Classification-Is; Kudo Classification-III-S). The histologic diagnosis was low-grade dysplasia.

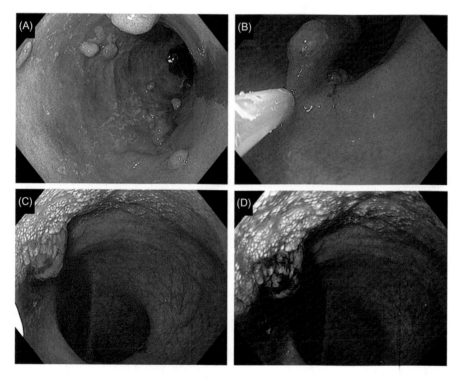

FIGURE 19.45 Narrow-band imaging for the surveillance of neoplasia in ulcerative colitis: (A–C) Inflammatory polyps (Paris Classification-Ip; Kudo Classification-II) before and after polypectomy on white-light endoscopy and (D) postpolypectomy inspection with narrow-band imaging.

Blue-light imaging

Blue-light imaging is another form of optical chromoendoscopy. The excitation of blue light is able to enhance mucosal surface contrast, as short-wave used and only penetrates superficially into the tissue with less scattering. Currently available ELUXEO 7000 (Fujifilm, Tokyo, Japan) is equipped with high-intensity light-emitting diodes excitation and

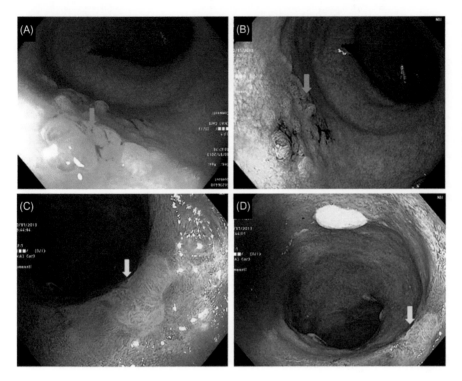

FIGURE 19.46 Narrow-band imaging for the surveillance of neoplasia in ulcerative colitis: (A and B) Superficially raised lesions (Paris Classification-IIa; Kudo Classification-III-S) on white-light endoscopy and narrow-band imaging (*green arrows*) and (C and D) flat lesions (Paris Classification-IIb; Kudo Classification-IIIS) on narrow-band imaging (*yellow arrows*). The final histologic diagnosis for both types of lesions is low-grade dysplasia.

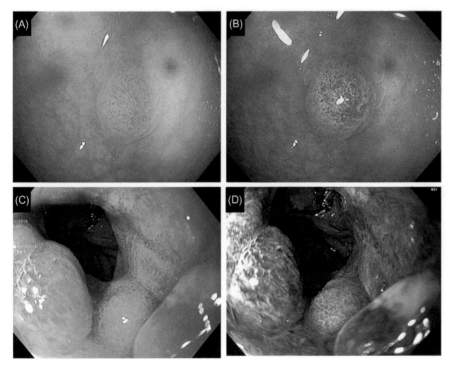

FIGURE 19.47 Narrow-band imaging for the assessment of colitis-associated neoplasia in two patients with ulcerative colitis: (A and B) A sessile hyperplastic polyp in the colon (Paris Classification-Is; Kudo Classification-II) on white-light endoscopy and narrow-band imaging and (C and D) nodular lesions in the rectal cuff (Paris Classification Is; Kudo Classification-III-S) in a patient with ileal pouch-anal anastomosis on white-light endoscopy and narrow-band imaging. The surface structure of the polypoid/nodular lesions is demonstrated better with narrow-band imaging. Biopsy showed inflammatory polyps.

megapixel complementary metal−oxide−semiconductor technology, which enable full HD display. The system uses two types of lasers with a wavelength of 410 nm (for blue light laser) and 450 nm (for white-light laser). BLI provides surface relief or surface rendering, demarcation line, vascular pattern, and delineation of lesions with brighter images than NBI. BLI was evaluated in Barrett's esophagus [44−47] and colon polyps and adenoma [48,49]. In the distinction between small neoplastic and nonneoplastic polyps (<10 mm), the accuracy of nonmagnifying BLI was 95.2%,

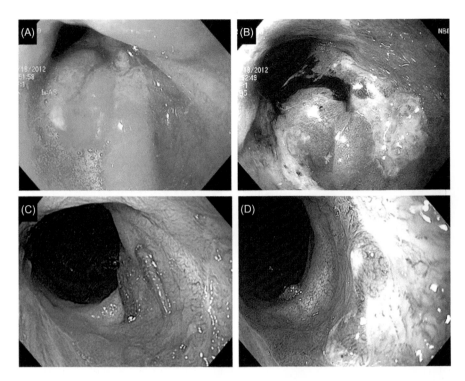

FIGURE 19.48 Narrow-band imaging in the surveillance endoscopy in ileal pouch-anal anastomosis. The patients had surgery for refractory ulcerative colitis: (A and B) Nodular cuff mucosa with superficial mucosa structure highlighted better on narrow-band imaging (Paris Classification-Is; Kudo Classification-I) and (C and D) anal transition zone between squamous and columnar epithelia on white-light endoscopy and narrow-band imaging.

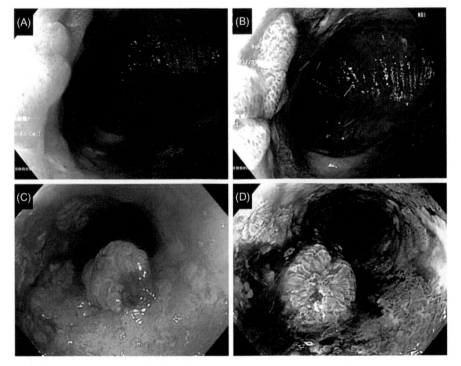

FIGURE 19.49 Narrow-band imaging in patients with ulcerative colitis and ileal pouch-anal anastomoses: (A and B) Raised lesions in the cuff with surface structures highlighted by narrow-band imaging (Paris Classification-Is; Kudo Classification-II). The biopsy showed indefinite for dysplasia and (C and D) inflammatory polyp in the pouch body with surface structure highlighted by narrow-band imaging (Paris Classification-Ip; Kudo Classification-II).

as compared with 83.2% of WLE [50]. A multicenter randomized controlled study showed more adenomas detected by BLI than WLE [51]. A separate randomized trial found a lower missing rate of colon adenoma by BLI than WLE (1.6% vs 10.0%) [49]. The clinical utility of BLI in IBD is currently under investigation. BLI has been used for the assessment of disease activity of IBD (Figs. 19.57−19.61). BLI can potentially be used for the surveillance of CAN (Fig. 19.62).

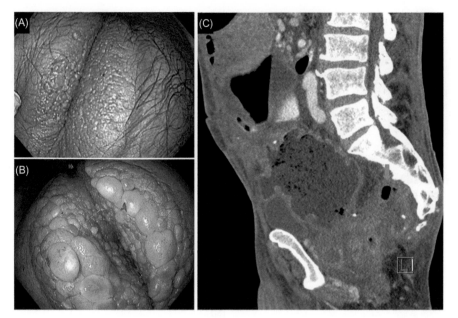

FIGURE 19.50 Narrow-band imaging for the evaluation of perianal lesions inflammatory bowel disease: (A) Perianal lesions in a patient with chronic pouchitis, persistent diarrhea, and pelvic abscess and (B) the perianal lesions were highlighted by narrow-band imaging. The biopsy of the lesion showed condyloma accumulate and (C) presacral abscess in the patient with ileal pouch-anal anastomosis (*green arrow*).

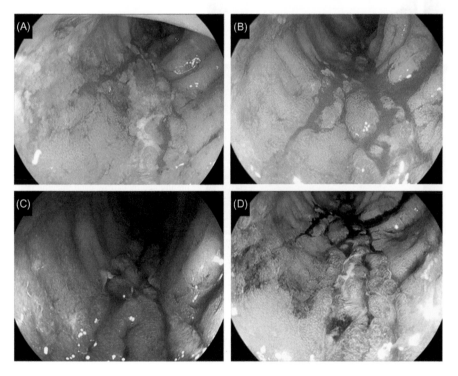

FIGURE 19.51 Virtual electronic chromoendoscopy I-SCAN assessment of moderately active Crohn's disease: (A) White-light endoscopy, (B) I-SCAN mode 1, (C) I-SCAN mode 2, and (D) I-SCAN mode 3 optical enhancement. Virtual chromoendoscopy defined better size of serpiginous ulcers as well as surface affected by disease and ulcers as Simple Endoscopic Scoring system = 6. *Courtesy Dr. Marietta Iacucci of University of Birmingham.*

Autofluorescence

Like NBI, the push-button autofluorescence imaging (AFI) makes its routine application in clinical practice amenable. The main principle of AFI is based on the demonstration of natural fluorescence emitted by endogenous molecules such as elastin, collagen, flavins, and porphyrins. The captured signals from the mucosa excited by a short-wave light source are transformed into red, green, or blue colors by a video processor and displayed as distinct color images in real time [40]. With blue light, normal GI mucosa emits green fluorescence. In contrast, thickened mucosa to inflammation or neoplasia blocks the blue light, converting the fluorescence signal to purple color [52].

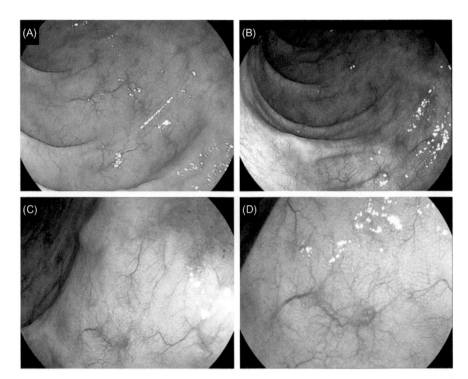

FIGURE 19.52 Virtual electronic chromoendoscopy I-SCAN assessment of disease activity in ulcerative colitis: (A) mucosal healing with high-definition I-SCAN mode 1-Mayo Endoscopic Score of 0, UCEIS = 0 and (B−D) I-SCAN mode 2 and optical enhancement mode 3, respectively, defined mucosal and vascular architecture as PICaSSO endoscopic findings of scars and spars vessels. *PICaSSO:* The Paddington International Virtual Chromoendoscopy. *UCEIS:* The Ulcerative Colitis Endoscopic Index of Severity *Courtesy Dr. Marietta Iacucci of University of Birmingham.*

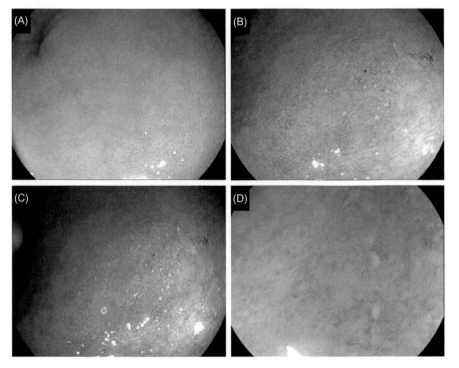

FIGURE 19.53 Virtual electronic chromoendoscopy I-SCAN assessment of mild colon inflammation: (A) I-SCAN mode 1. The colon mucosa appeared normal, (B and C) I-SCAN mode 2 and I-SCAN mode 3 optical enhancement-defined better the mild inflammatory activity of the mucosa with multiple erosions, and (D) magnification characterized the small erosions and vessels architecture. *Courtesy Dr. Marietta Iacucci of University of Birmingham.*

The application of AFI in GI diseases has been the evaluation of reflux esophagitis, chronic atrophic gastritis, Barrett's esophagus, and early gastric cancer [50,53,54] The data on the use of AFI in IBD are emerging. An AFI-histology correlation study showed that the stronger color purple on AFI corresponded to the degree of histologic inflammation. The diagnostic accuracy of AFI in predicting active inflammation of UC was 92% with an excellent interobserver agreement [52]. The evaluation of color components of AFI showed that green color was most often

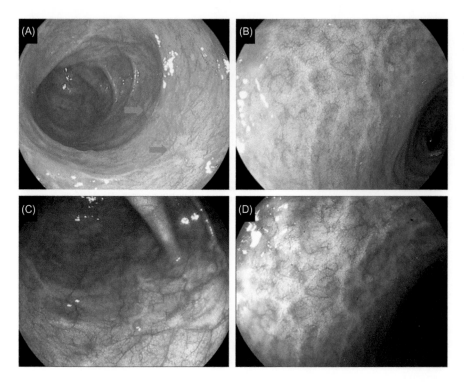

FIGURE 19.54 Virtual electronic chromoendoscopy I-SCAN for the assessment of mucosal healing in ulcerative colitis: (A) Mucosal healing with scars (*green arrows*) on white-light endoscopy and (B–D) the area was highlighted by I-SCAN mode 1, 2, and 3, respectively. *Courtesy Dr. Marietta Iacucci of University of Birmingham.*

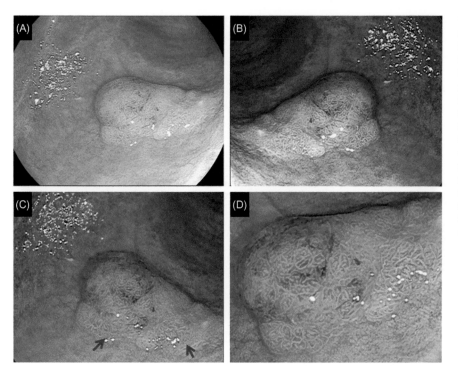

FIGURE 19.55 Virtual electronic chromoendoscopy I-SCAN for the surveillance in ulcerative colitis: (A) High-definition I-SCAN mode 1 showed a polypoid dysplastic lesion (Paris Classification-Is) in a patient with long-standing ulcerative colitis, (B and C) I-SCAN modes 2 and 3 optical enhancement characterized better the mucosal and vascular pattern (Kudo Classification-III-L) as well as the border of the lesion as irregular, (D) I-SCAN mode 3 optical enhancement with zoom-defined borders and mucosa surrounded the lesion. The final diagnosis was sessile serrated adenoma with low-grade dysplasia. *Courtesy Dr. Marietta Iacucci of University of Birmingham.*

(and inversely) related to sites of mucosal inflammation and the MES. Mucosal inflammation with a thickened mucosal layer makes penetration of autofluorescence into the submucosa difficult, leading to a reduced green color signal. The intensity of green color signal may be useful to measure the severity of inflammation of UC [53].

A combined use of AFI and HD-WLE improves the detection rate of CAN, as compared with WLE in patients with IBD [21,55]. A metaanalysis found that AFI was associated with a higher likelihood of detecting dysplasia than WLE

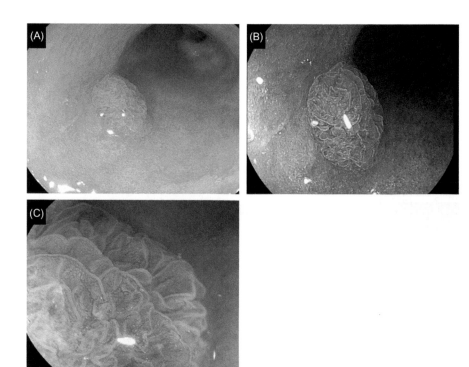

FIGURE 19.56 Virtual electronic chromoendoscopy I-SCAN for the surveillance in a patient with long-standing ulcerative colitis: (A) High definition I-SCAN mode 1 showed a polypoid sessile dysplastic lesion on Paris Classification 1s; (B and C) I-SCAN mode 2 and optical enhancement mode 3 with magnification defined the mucosal pit pattern as villous Kudo Classification-IV pit pattern and regular border without inflammation. Final diagnosis was low-grade dysplasia. *Courtesy Dr. Marietta Iacucci of University of Birmingham.*

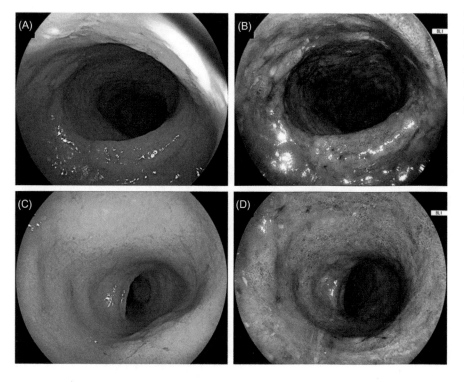

FIGURE 19.57 Blue-light imaging for the evaluation of colitis. (A–D) Mild colitis with mucosal granularity and loss of vascularity on white light (A and C), which were more dramatic on blue light (B and D). *Courtesy Drs. Peter A. Senada and Michael B. Wallace of Mayo Clinic Florida.*

with an odds ratio (OR) of 3.055 per-dysplastic lesion analysis and OR of 2.502 analysis per-patient analysis [49]. A recent randomized controlled trial 210 patients undergoing surveillance colonoscopy for long-standing UC as compared with AFI with DCE involving showed that the mean number of detected dysplastic lesions per patient was 0.13 and 0.37 for AFI and DCE, respectively. It appears that DCE may be more effective for the surveillance of CRC than AFI [56].

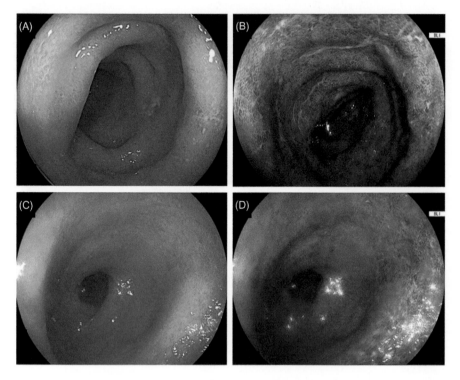

FIGURE 19.58 Blue-light imaging for the evaluation of colitis. (A—D) Mild colitis with mucosal erythema and loss of vascularity on white light (A and C), which were more obvious on blue light (B and D). *Courtesy Drs. Peter A. Senada and Michael B. Wallace of Mayo Clinic Florida.*

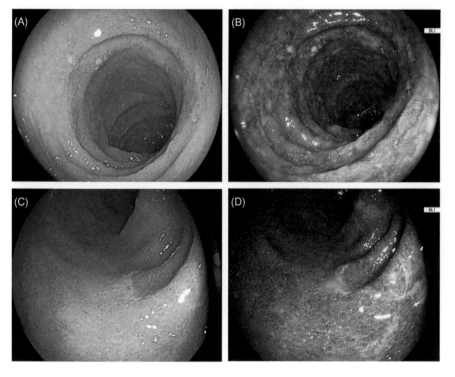

FIGURE 19.59 Blue-light imaging for the evaluation of ileitis. (A—D) Diffuse mild-to-moderate with erythema and erosions on white light (A and C), which were more obvious on blue light (B and D). *Courtesy Drs. Peter A. Senada and Michael B. Wallace of Mayo Clinic Florida.*

Multimode imaging

WLE and various IEE modalities have been compared for the detection of CAN. A metaanalysis of 15 studies evaluated IEE (including DCE, NBI, and AFI) with targeted biopsy versus WLE with random biopsy. The detection rate of dysplasia was 17.3% in the IEE-targeted biopsy group and 0.33% in the WLE-random biopsy group [49]. A separate

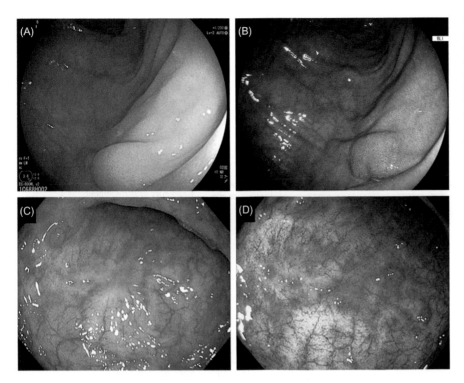

FIGURE 19.60 Blue-light imaging and narrow-band imaging in ulcerative colitis: (A and B) A polypoid lesion in the colon on white light (A) and blue light (B)and (C and D) mucosal scars and pseudopolyps in a separate patient with ulcerative colitis on white light (C) and narrow-band imaging (D). *Photo A and B—Courtesy Drs. Peter A. Senada and Michael B. Wallace of Mayo Clinic Florida.*

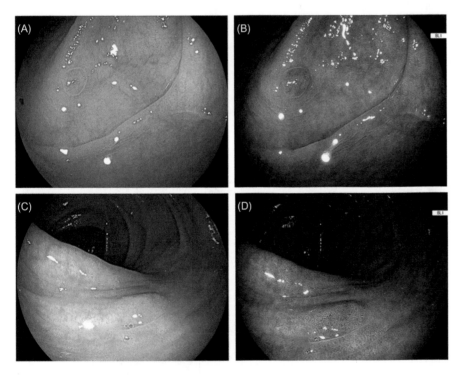

FIGURE 19.61 Blue-light imaging for the evaluation of colon lesions: (A and B) A small sessile polyp on white light (A) and blue light (B)and (C and D) a slightly raised lesion with mild pit pattern change in the colon on white light (C) and blue light (D). *Courtesy Drs. Peter A. Senada and Michael B. Wallace of Mayo Clinic Florida.*

network metaanalysis of 27 studies of IEE (i.e., DCE, NBI, I-SCAN, FISE, and AFI) versus WLE confirmed a higher likelihood of detecting dysplasia by IEE than WLE (19.3% vs 8.5%, OR = 2.036) with an incremental yield of 10.8%. A subanalysis showed that DCE (OR = 2.605) and AFI (OR = 3.055) had a higher likelihood of detecting dysplasia than WLE. However, NBI (OR = 0.650), I-SCAN (OR = 1.096), and FISE (OR = 1.118) were not superior to WLE

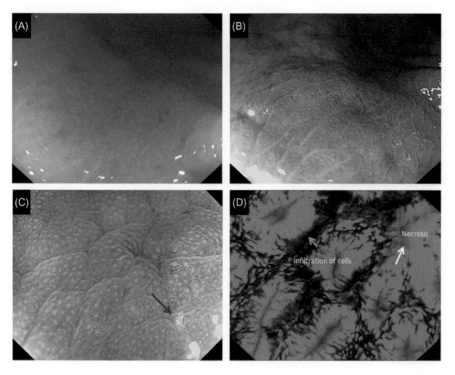

FIGURE 19.62 Multimode imaging for the assessment of mild inflammation versus mucosal healing in ulcerative colitis before discontinuation of biological therapy: (A) Colon mucosa on high-definition white-light endoscopy, (B and C) small tiny microerosion on narrow-band imaging with magnification (*blue arrow*), (D) crypts architecture distortion with some areas of crypts necrosis on endocytoscope (*white arrow*) and increase cell infiltration in the lamina propria (*green arrow*). *Courtesy Dr. Marietta Iacucci of University of Birmingham.*

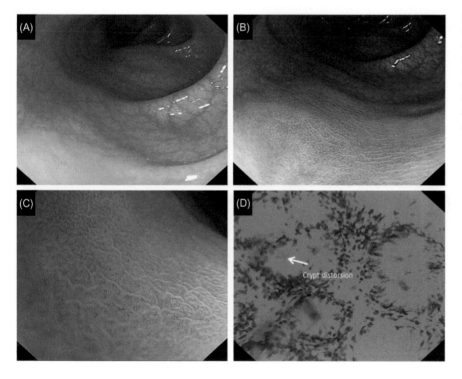

FIGURE 19.63 Multimode imaging for the assessment of mucosal healing in ulcerative colitis: (A) White-light colonoscopy, (B and C) narrow-band imaging with (C) and without (B) optical magnification, and (D) endocytoscopy showing crypt distortion (*arrow*). *Courtesy Dr. Marietta Iacucci of University of Birmingham.*

[49]. Despite the higher detection rate of DCE with a smaller number of biopsy than WLE, its application in daily clinical practice has to overcome technical hurdles.

Attempts have been made to enhance the detection rate of CAN with combined use of various endoscopy modalities. For example, chromoendoscopy is being used in conjunction with other IEE such as ME, CLE, and confocal

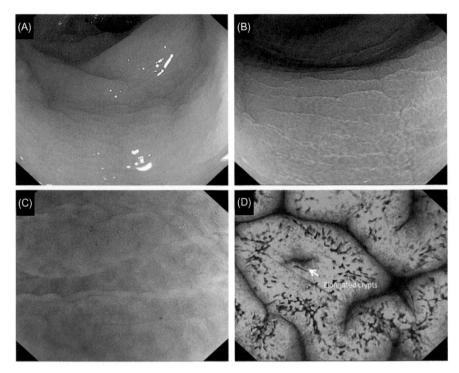

FIGURE 19.64 Multimode imaging for the assessment of mucosal healing in ulcerative colitis: (A) White-light endoscopy, (B and C) elongated and distorted crypts on narrow-banding imaging with (C) or without (B) magnification, and (D) the elongated and distorted crypt structure confirmed by endocytoscope (*arrow*). *Courtesy Dr. Marietta Iacucci of University of Birmingham.*

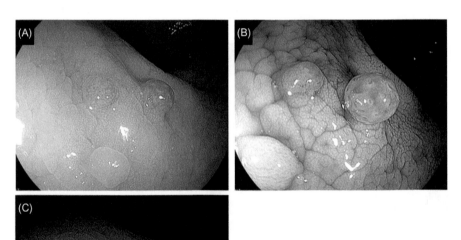

FIGURE 19.65 Multimode imaging of vascular hyperplastic polyps incidentally found in a 28-year-old male patient with infectious colitis. The surrounding mucosa was edematous: (A) White-light endoscopy, (B) dye chromoendoscopy, and (C) narrow-band imaging. *Courtesy Dr. Bolin Yang of Jiangsu Province Hospital of Chinese Medicine and Affiliated Hospital of Nanjing University of Chinese Medicine.*

endomicroscopy to assess the disease activity (Figs. 19.62–19.64) or CAN (Figs. 19.65–19.73). The multimode imaging can be used for dysplasia surveillance in patients with IPAA (Figs. 19.74–19.76).

Tri-mode endoscopy with HD-WLE, AFI, and NBI has been reported for surveillance of CAN [56,57]. In a prospective study of 52 dysplastic and 255 nondysplastic lesions, the overall sensitivity for real-time prediction of dysplasia was 76.9% [95% confidence interval (CI) 46.2–95.0] for endoscopic trimodal imaging, and 81.6% (95% CI 65.7–92.3)

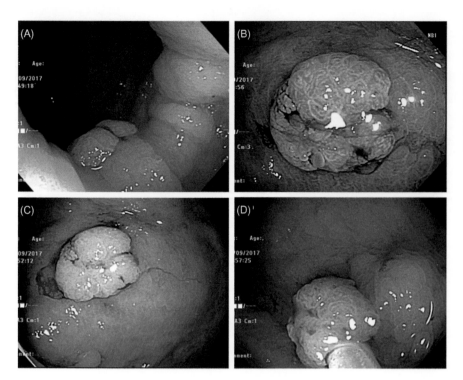

FIGURE 19.66 Multimode imaging for the assessment of polypoid lesions in ulcerative colitis. The combined assessment yielded a Paris Classification-Is; Kudo Classification-IIIs lesion: (A) White-light endoscopy, (B) narrow-band imaging, (C) dye chromoendoscopy, and (D) endoscopic polypectomy was performed with a histologic diagnosis of adenoma with low-grade dysplasia.

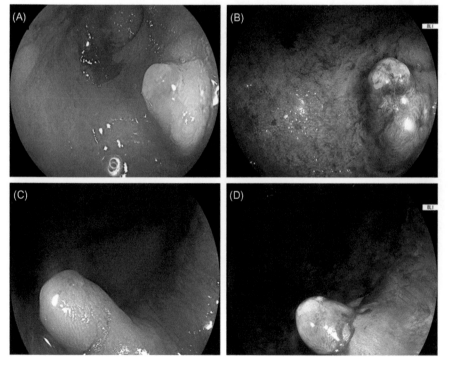

FIGURE 19.67 Blue-light imaging for the evaluation of polypoid lesions in uclerative colitis: (A and B) An inflammatory polyp of the colon with a white cap on white light (A) and blue light (B); (C and D) a polypoid lesion at the dentate line on white-light (C) and blue-light (D) imaging. *Courtesy Drs. Peter A. Senada and Michael B. Wallace of Mayo Clinic Jacksonville.*

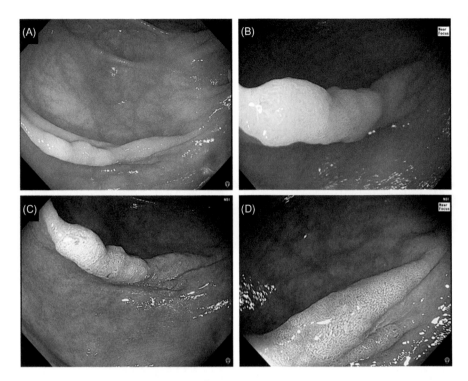

FIGURE 19.68 Multimode imaging for the surveillance in a patient with long-standing ulcerative colitis: (A and B) High-definition and magnification white-light endoscopy showed a flat lesion (Paris Classification-IIa) and (C and D) narrow-band imaging with or without magnification characterized (Kudo Classification-III-L) and borders as regular without inflammation. The final diagnosis was low-grade dysplasia. *Courtesy Dr. Marietta Iacucci of University of Birmingham.*

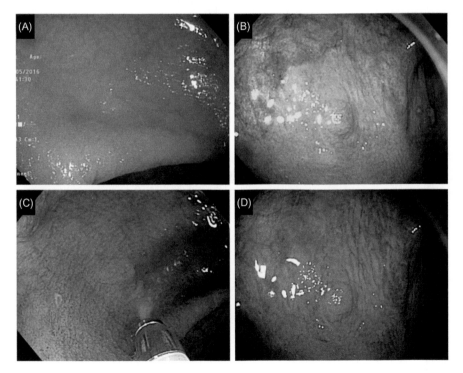

FIGURE 19.69 Multimode imaging for the surveillance of neoplasia in ulcerative colitis. There was a small sessile polyp (Paris Classification-Is; Unclear Kudo Classification) in the transverse colon: (A) White-light endoscopy, (B) narrow-band imaging, and (C and D) dye chromoendoscopy. *Courtesy Dr. Marietta Iacucci of University of Birmingham.*

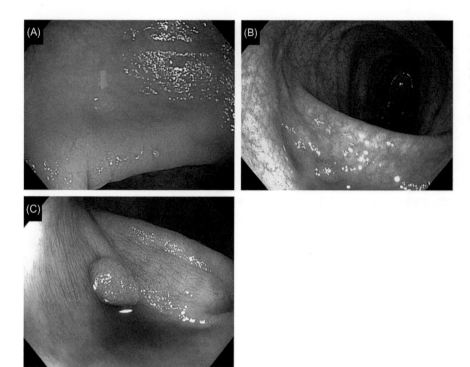

FIGURE 19.70 Multimode imaging for the surveillance of neoplasia in quiescent ulcerative colitis. There was a small sessile polyp (Paris Classification-Is; Kudo Classification-IIIS) in the ascending colon: (A) White-light endoscopy (*green arrow*), (B) narrow-band imaging, and (C) dye chromoendoscopy.

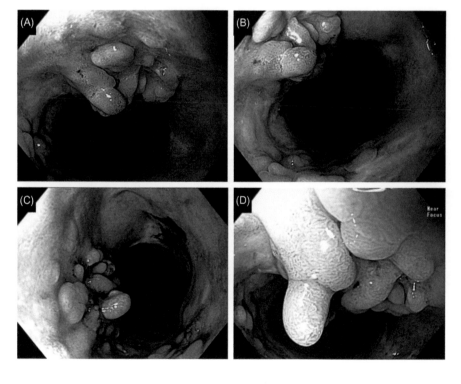

FIGURE 19.71 Multimode imaging for the surveillance of neoplasia in ulcerative colitis. Inflammatory polyps (Paris Classification-Ip; Kudo Classification-II) were found in the distal rectum: (A and B) Narrow-band imaging and (C and D) dye chromoendoscopy.

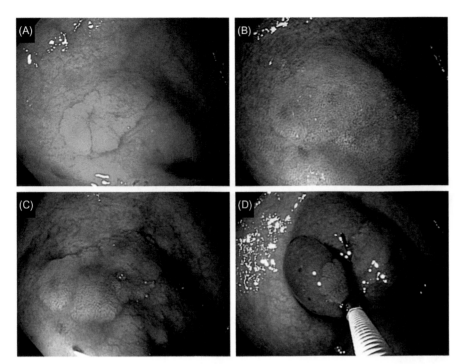

FIGURE 19.72 Multimode imaging of flat lesion in ulcerative colitis: (A) A large flat lesion with distinct border (Paris Classification-IIb) in the ascending colon on white-light endoscopy, (B and C) the lesion was highlighted with narrow-band imaging (B) and dye chromoendoscopy (C) (Kudo Classification-II), and (D) endoscopic mucosal resection of the lesion.

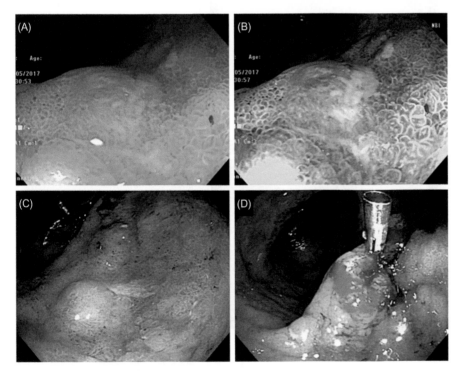

FIGURE 19.73 Multimode imaging for the management of dysplastic lesions in ulcerative colitis. This was a Paris Classification-IIa and Kudo Classification-III-S lesion with an indistinct border: (A) White-light endoscopy, (B) narrow-band imaging, (C) dye chromoendoscopy, and (D) tissue biopsy of the lesion resulting in excessive bleeding which was controlled by the deployment of an endoclips. The histologic diagnosis was low-grade dysplasia.

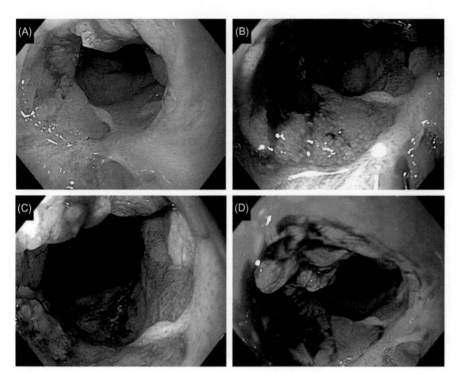

FIGURE 19.74 Multimode imaging for the surveillance in a patient with ileal pouch-anal anastomosis: (A) Granular and nodular mucosa at the anal transition zone on white-light endoscopy; (B and C) the area was surveyed with dye chromoendoscopy and (D) narrow-band imaging was also used. The histologic diagnosis was hyperplastic lesion.

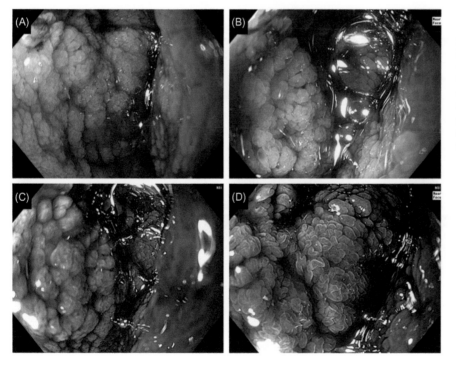

FIGURE 19.75 Multimode imaging for the assessment of neoplasia in the ileal pouch-anal anastomosis. Diffuse dysplastic lesion was located in the distal ileal pouch (Paris Classification-Ip; Kudo Classification-III-S): (A) Dye chromoendoscopy, (B) dye chromoendoscopy in a magnification mode, (C) narrow-band imaging, and (D) narrow-band imaging in a magnification mode. The histologic diagnosis was high-grade dysplasia.

for DCE. The overall negative predictive value for endoscopic trimodal imaging was 96.9% (95% CI 92.0−98.8) and 94.7% (90.2−97.2) for DCE [57]. The Frankfurt Advanced Chromoendoscopic IBD LEsions (FACILE) Classification was developed based on multimode endoscopy to distinguish neoplastic from nonneoplastic lesions. This classification was based on four predictors of dysplasia: morphology (nonpolypoid lesions), irregular vessel architecture, irregular

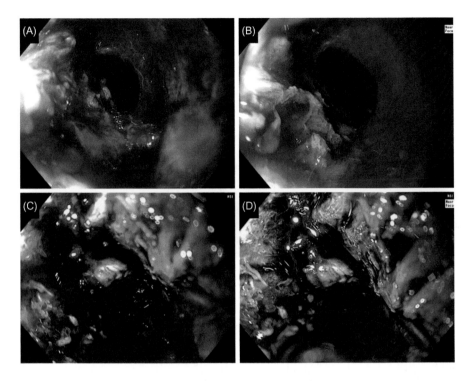

FIGURE 19.76 Multimode imaging for the surveillance of neoplasia in the ileal pouch-anal anastomosis. The patient was found to have invasive cancer in the cuff and anastomosis. The tissue structure was damaged and the mucosal surface was almost obliterated: (A) Dye chromoendoscopy, (B) dye chromoendoscopy in a magnification mode, (C) narrow-band imaging, and (D) narrow-band imaging in a magnification mode.

surface pattern, and signs of inflammation within the lesions. The diagnostic accuracy of the FACILE Classification was estimated to be 85% [58].

Summary and recommendations

Advances in IEE have delineated mucosal and submucosal structures. The current main applications of IEE are the assessment of disease activity and surveillance for CAN. Patients with a long-standing history of IBD are at a higher risk of developing CRC as compared to the general population. An effective endoscopic surveillance program remains the cornerstone to identify dysplastic lesions. DCE with targeted biopsy is the surveillance tool of choice. Once a lesion is identified, it should be characterized and evaluated for resectability using the modified Paris classification. Other IEE modalities such as NBI, BLL, and AFI appear promising, but more studies are needed to evaluate and to further define their role in disease activity monitoring dysplasia surveillance.

References

[1]   World Cancer Research Fund. Colorectal cancer statistics. World Cancer Research Fund. <https://www.wcrf.org/dietandcancer/cancer-trends/colorectal-cancer-statistics>; August 22, 2018.

[2]   CDC. Colorectal (colon) cancer. CDC. <https://www.cdc.gov/cancer/colorectal/>; July 31, 2019.

[3]   Neumann H. New endoscopic approaches in IBD. World J Gastroenterol 2011;17:63—8.

[4]   Goran L, Negreanu L, Negreanu AM. Role of new endoscopic techniques in inflammatory bowel disease management: has the change come? World J Gastroenterol 2017;23:4324—9.

[5]   Iannone A, Ruospo M, Palmer SC, Principi M, Barone M, Di Leo A, et al. Systematic review with network meta-analysis: endoscopic techniques for dysplasia surveillance in inflammatory bowel disease. Aliment Pharmacol Ther 2019;50:858—71.

[6]   Feuerstein JD, Rakowsky S, Sattler L, Yadav A, Foromera J, Grossberg L, et al. Meta-analysis of dye-based chromoendoscopy compared with standard- and high-definition white-light endoscopy in patients with inflammatory bowel disease at increased risk of colon cancer. Gastrointest Endosc 2019;90:186—95.

[7]   Rubin DT, Ananthakrishnan AN, Siegel CA, Sauer BG, Long MD. ACG clinical guideline—ulcerative colitis in adults. Am J Gastroenterol 2019;114:384—413.

[8]   Galanopoulos M, Tsoukali E, Gkeros F, Vraka M, Karampekos G, Matzaris GJ. Screening and surveillance methods for dysplasia in inflammatory bowel disease patients: where do we stand? World J Gastrointest Endosc 2018;10:250—8.

[9]   Sharma P, Gupta N, Kuipers EJ, Repici A, Wallace M. Advanced imaging in colonoscopy and its impact on quality. Gastrointest Endosc 2014;79:28—36.

[10]   Liu J, Dlugosz A, Neumann H. Beyond white light endoscopy: the role of optical biopsy in inflammatory bowel disease. World J Gastroenterol 2013;19:7544−51.

[11]   Iacucci M, Daperno M, Lazarev M, Arsenascu R, Tontini GE, Akinola O, et al. Development and reliability of the new endoscopic virtual chromoendoscopy score: the PICaSSO (Paddington International Virtual ChromoendoScopy ScOre) in ulcerative colitis. Gastrointest Endosc 2017;86:1118−27.

[12]   Trivedi PJ, Kiesslich R, Hodson J, Bhala N, Boulton RA, Cooney R, et al. The Paddington International Virtual Chromoendoscopy Score in ulcerative colitis exhibits very good inter-rater agreement after computerized module training: a multicenter study across academic and community practice (with video). Gastrointest Endosc 2018;88:95−106.

[13]   Kudo S, Tamura S, Nakajima T, Yamano H, Kusaka H, Watanabe H. Diagnosis of colorectal tumorous lesions by magnifying endoscopy. Gastrointest Endosc 1996;44:8−14.

[14]   Danese S, Fiorino G, Angelucci E, Vetrano S, Pagano N, Rando G, et al. Narrow-band imaging endoscopy to assess mucosal angiogenesis in inflammatory bowel disease: a pilot study. World J Gastroenterol 2010;16:2396−400.

[15]   Hewett DG, Kaltenbach T, Sano Y, Tanaka S, Saunders BP, Ponchon T, et al. Validation of a simple classification system for endoscopic diagnosis of small colorectal polyps using narrow-band imaging. Gastroenterology 2012;143:599−607.

[16]   Hayashi N, Tanaka S, Hewett DG, Kaltenbach TR, Sano Y, Ponchon T, et al. Endoscopic prediction of deep submucosal invasive carcinoma: validation of the narrow-band imaging international colorectal endoscopic (NICE) classification. Gastrointest Endosc 2013;78:625−32.

[17]   Wada Y, Kudo SE, Kashida H, Ikehara N, Inoue H, Yamamura F, et al. Diagnosis of colorectal lesions with the magnifying narrow-band imaging system. Gastrointest Endosc 2009;70:522−31.

[18]   Puig I, Kaltenbach T. Optical diagnosis for colorectal polyps: a useful technique now or in the future? Gut Liver 2018;12:385−92.

[19]   Patrun J, Okreša L, Iveković H, Rustemović N. Diagnostic accuracy of NICE Classification system for optical recognition of predictive morphology of colorectal polyps. Gastroenterol Res Pract 2018;2018:1−10.

[20]   The Paris endoscopic classification of superficial neoplastic lesions: esophagus, stomach, and colon: November 30 to December 1, 2002. Gastrointest Endosc 2003;58:S3−43.

[21]   Lv X-H, Wang B-L, Cao G-W. Narrow band imaging for surveillance in inflammatory bowel disease: a systematic review and meta-analysis. J Clin Gastroenterol 2019;53:607−15.

[22]   Hassan C, East J, Radaelli F, Spada C, Benamouzig R, Bisschops R, et al. Bowel preparation for colonoscopy: European Society of Gastrointestinal Endoscopy (ESGE) Guideline—update 2019. Endoscopy 2019;51:775−94.

[23]   Iacucci M, Cannatelli R, Tontini GE, Panaccione R, Danese S, Fiorino G, et al. Improving the quality of surveillance colonoscopy in inflammatory bowel disease. Lancet Gastroenterol Hepatol 2019;4:971−83.

[24]   Baldi F, di Febo G, Biasco G, Gizzi G, Ferrarini F, Milazzo G, et al. Methylene blue dye spraying method in patients with ulcerative proctitis: a comparative study with morphological findings and functional capacity of the rectal epithelium. Endoscopy 1979;11:179−84.

[25]   Ibarra-Palomino J, Barreto-Zúñiga R, Elizondo-Rivera J, Bobadilla-Díaz J, Villegas-Jiménez A. Application of chromoendoscopy to evaluate the severity and interobserver variation in chronic non-specific ulcerative colitis. Rev Gastroenterol Mex 2002;67:236−40.

[26]   Picco MF, Pasha S, Leighton JA, Bruining D, Loftus Jr EV, Thomas CS, et al. Procedure time and the determination of polypoid abnormalities with experience: implementation of a chromoendoscopy program for surveillance colonoscopy for ulcerative colitis. Inflamm Bowel Dis 2013;19:1913−20.

[27]   Subramanian V, Mannath J, Ragunath K, Hawkey CJ. Meta-analysis: the diagnostic yield of chromoendoscopy for detecting dysplasia in patients with colonic inflammatory bowel disease. Aliment Pharmacol Ther 2011;33:304−12.

[28]   Soetikno R, Subramanian V, Kaltenbach T, Rouse RV, Sanduleanu S, Suzuki N, et al. The detection of nonpolypoid (flat and depressed) colorectal neoplasms in patients with inflammatory bowel disease. Gastroenterology 2013;144:1349−52.

[29]   Laine L, Kaltenbach T, Barkun A, McQuaid KR, Subramanian V, Soetikno R, et al. SCENIC international consensus statement on surveillance and management of dysplasia in inflammatory bowel disease. Gastroenterology 2015;148:639−51.

[30]   Iannone A, Ruospo M, Wong G, Principi M, Barone M, Strippoli GFM, et al. Chromoendoscopy for surveillance in ulcerative colitis and Crohn's disease: a systematic review of randomized trials. Clin Gastroenterol Hepatol 2017;15:1684−97.

[31]   van den Broek FJ, Fockens P, van Eeden S, Stokkers PC, Ponsioen CY, Reitsma JB, et al. Narrow-band imaging versus high-definition endoscopy for the diagnosis of neoplasia in ulcerative colitis. Endoscopy 2011;43:108−15.

[32]   Olliver JR, Wild CP, Sahay P, Dexter S, Hardie LJ. Chromoendoscopy with methylene blue and associated DNA damage in Barrett's oesophagus. Lancet 2003;362(9381):373−4.

[33]   Iacucci M, Kaplan GG, Panaccione R, Akinola O, Lethebe BC, Lowerison M, et al. A randomized trial comparing high definition colonoscopy alone with high definition dye spraying and electronic virtual chromoendoscopy for detection of colonic neoplastic lesions during IBD surveillance colonoscopy. Am J Gastroenterol 2018;113:225−34.

[34]   Sharma P, Bansal A, Mathur S, Wani S, Cherian R, McGregor D, et al. The utility of a novel narrow band imaging endoscopy system in patients with Barrett's esophagus. Gastrointest Endosc 2006;64:167−75.

[35]   Kudo T, Matsumoto T, Esaki M, Yao T, Iida M. Mucosal vascular pattern in ulcerative colitis: observations using narrow band imaging colonoscopy with special reference to histologic inflammation. Int J Colorectal Dis 2009;24:495−501.

[36]   Esaki M, Kubokura N, Kudo T, Matsumoto T. Endoscopic findings under narrow band imaging colonoscopy in ulcerative colitis. Dig Endosc 2011;23(Suppl. 1):140−2.

[37]   Dekker E, van den Broek FJ, Reitsma JB, Hardwick JC, Offerhaus GJ, van Deventer SJ, et al. Narrow-band imaging compared with conventional colonoscopy for the detection of dysplasia in patients with longstanding ulcerative colitis. Endoscopy 2007;39:216−21.

[38] Ignjatovic A, East JE, Subramanian V, Suzuki N, Guenther T, Palmer N, et al. Narrow band imaging for detection of dysplasia in colitis: a randomized controlled trial. Am J Gastroenterol 2012;107:885−90.

[39] Pellisé M, López-Cerón M, Rodríguez de Miguel C, Jimeno M, Zabalza M, Ricart E, et al. Narrow-band imaging as an alternative to chromoendoscopy for the detection of dysplasia in long-standing inflammatory bowel disease: a prospective, randomized, crossover study. Gastrointest Endosc 2011;74:840−8.

[40] Kiesslich R, Neurath MF. Advanced endoscopy imaging in inflammatory bowel diseases. Gastrointest Endosc 2017;85:496−508.

[41] Kawasaki K, Nakamura S, Esaki M, Kurahara K, Eizuka M, Nuki Y, et al. Clinical usefulness of magnifying colonoscopy for the diagnosis of ulcerative colitis-associated neoplasia. Dig Endosc 2019;31(Suppl. 1):36−42.

[42] Neumann H, Vieth M, Günther C, Neufert C, Kiesslich R, Grauer M, et al. Virtual chromoendoscopy for prediction of severity and disease extent in patients with inflammatory bowel disease: a randomized controlled study. Inflamm Bowel Dis 2013;19:1−42.

[43] Iacucci M, Fort Gasia M, Hassan C, Panaccione R, Kaplan GG, Ghosh S, et al. Complete mucosal healing defined by endoscopic Mayo subscore still demonstrates abnormalities by novel high definition colonoscopy and refined histological gradings. Endoscopy 2015;47:726−34.

[44] Yoshida N, Yagi N, Inada Y, Kugai M, Okayama T, Kamada K, et al. Ability of a novel blue laser imaging system for the diagnosis of colorectal polyps. Dig Endosc 2014;26:250−8.

[45] Yoshida N, Hisabe T, Hirose R, Ogiso K, Inada Y, Konishi H, et al. Improvement in the visibility of colorectal polyps by using blue laser imaging (with video). Gastrointest Endosc 2015;82:542−9.

[46] de Groof AJ, Swager AF, Pouw RE, Weusten BLAM, Schoon EJ, Bisschops R, et al. Blue-light imaging has an additional value to white-light endoscopy in visualization of early Barrett's neoplasia: an international multicenter cohort study. Gastrointest Endosc 2019;89:749−58.

[47] Shimoda R, Sakata Y, Fujise T, Yamanouchi K, Tsuruoka N, Hara M, et al. The adenoma miss rate of blue-laser imaging vs. white-light imaging during colonoscopy: a randomized tandem trial. Endoscopy 2017;49:186−90.

[48] Ang TL, Li JW, Wong YJ, Tan YJ, Fock KM, Tan MTK, et al. A prospective randomized study of colonoscopy using blue laser imaging and white light imaging in detection and differentiation of colonic polyps. Endosc Int Open 2019;7:E1207−13.

[49] Imperatore N, Castiglione F, Testa A, De Palma GD, Caporaso N, Cassese G, et al. Augmented endoscopy for surveillance of colonic inflammatory bowel disease: systematic review with network meta-analysis. J Crohns Colitis 2019;13:714−24.

[50] Osada T, Arakawa A, Sakamoto N, Ueyama H, Shibuya T, Ogihara T, et al. Autofluorescence imaging endoscopy for identification and assessment of inflammatory ulcerative colitis. World J Gastroenterol 2011;17:5110−16.

[51] Ikematsu H, Sakamoto T, Togashi K, Yoshida N, Hisabe T, Kiriyama S, et al. Detectability of colorectal neoplastic lesions using a novel endoscopic system with blue laser imaging: a multicenter randomized controlled trial. Gastrointest Endosc 2017;86:386−94.

[52] Moriichi K, Fujiya M, Okumura T. The efficacy of autofluorescence imaging in the diagnosis of colorectal diseases. Clin J Gastroenterol 2016;9:175−83.

[53] ASGE Technology Committee, Song L-MWK, Banerjee S, Desilets D, Diehl DL, Farraye FA, et al. Autofluorescence imaging. Gastrointest Endosc 2011;73:647−50.

[54] Borovicka J, Fischer J, Neuweiler J, Netzer P, Gschossmann J, Ehmann T, et al. Autofluorescence endoscopy in surveillance of Barrett's esophagus: a multicenter randomized trial on diagnostic efficacy. Endoscopy 2006;38:867−72.

[55] van den Broek FJC, Fockens P, van Eeden S, Hardwick JC, Stokkers PC, Dekker E. Endoscopic tri-modal imaging for surveillance in ulcerative colitis: randomised comparison of high-resolution endoscopy and autofluorescence imaging for neoplasia detection; and evaluation of narrow-band imaging for classification of lesions. Gut 2008;57:1083−9.

[56] Vleugels JLA, Rutter MD, Ragunath K, Rees CJ, Ponsioen CY, Lahiff C, et al. Chromoendoscopy versus autofluorescence imaging for neoplasia detection in patients with longstanding ulcerative colitis (FIND-UC): an international, multicentre, randomised controlled trial. Lancet Gastroenterol Hepatol 2018;3:305−16.

[57] Vleugels JLA, Rutter MD, Ragunath K, Rees CJ, Ponsioen CY, Lahiff C, et al. Diagnostic accuracy of endoscopic trimodal imaging and chromoendoscopy for lesion characterization in ulcerative colitis. J Crohns Colitis 2018;12:1438−47.

[58] Iacucci M, McQuaid K, Gui X, Iwao Y, Lethebe BC, Lowerison M, et al. A multimodal (FACILE) classification for optical diagnosis of inflammatory bowel disease associated neoplasia. Endoscopy 2019;51:133−41.

Chapter 20

Confocal endomicroscopy and other image-enhanced endoscopy in inflammatory bowel disease

Charles A. Lavender[1], Xiuli Zuo[2], Marietta Iacucci[3], Bo Shen[4] and Julia J. Liu[1]

[1]Division of Gastroenterology, University of Arkansas for Medical Sciences, Little Rock, AR, United States, [2]Department of Gastroenterology, Qilu Hospital, Shandong University, Jinan, P.R. China, [3]Institute of Translational Medicine, Institute of Immunology and Immunotherapy, NIHR Birmingham Biomedical Research Centre, University Hospitals NHS Foundation Trust, University of Birmingham, Brimingham, United Kingdom, [4]Center for Inflammatory Bowel Diseases, Columbia University Irving Medical Center-New York Presbyterian Hospital, New York, NY, United States

Chapter Outline

Abbreviations

CAN colitis-associated neoplasia
CD Crohn's disease
CLE confocal laser endomicroscopy
DCE dye chromoendoscopy
eCLE endoscopy-based confocal laser endomicroscopy
GI gastrointestinal
HD high-definition
IBD inflammatory bowel disease
IEE image-enhanced endoscopy
ME magnification endoscopy
MH mucosal healing
NBI narrow-band imaging
OCT optical coherence tomographyp
CLE confocal laser endomicroscopy
PSC primary sclerosing cholangitisTNF tumor necrosis factor
UC ulcerative colitis
WLE white-light endoscopy

Atlas of Endoscopy Imaging in Inflammatory Bowel Disease. DOI: https://doi.org/10.1016/B978-0-12-814811-2.00020-7
© 2020 Elsevier Inc. All rights reserved.

Introduction

Advances in imaging technology have made in vivo endoscopic inspection of mucosal and submucosal microstructures possible, to the level of spatial resolution and depth beyond standard white-light endoscopy (WLE). The image-enhanced endoscopy (IEE) includes confocal laser endomicroscopy (CLE), endocytoscopy, magnification endoscopy (ME), optical coherence tomography (OCT), and ultrasound elastography. The current main applications of these IEE techniques in inflammatory bowel disease (IBD) are the assessment of disease activity, surveillance of colitis-associated neoplasia (CAN), and transmural disease process. Accurate measurement of disease activity in IBD is imperative for the treatment and prognosis. In addition to clinical, laboratory, and radiographic evaluation, most clinicians prefer direct visualization of disease activity. Mucosal healing (MH) has been set as one of the primary targets for medical therapy. While the criteria and application of histologic MH are evolving, WLE-based endoscopic MH is increasingly used in clinical trials and clinical practice. However, deep tissue healing consists of clinical, endoscopic, histologic, and biological aspects; and from mucosal to transmural. IEE positions well in between WLE and histology. In contrast to tedious and invasive tissue sampling in histologic evaluation, IEE such as CLE and endocytoscopy can provide in vivo virtual ultramagnification images as well as the areas for targeted biopsy. Furthermore, high-resolution imaging makes surveillance of CAN more accurate.

Dye-based chromoendoscopy and virtual chromoendoscopy are discussed in Chapter 19, Chromoendoscopy in inflammatory bowel disease.

Confocal laser endomicroscopy

CLE is a novel endoscopic imaging modality that allows real-time visualization of subcellular architecture details of the gastrointestinal (GI) tract. This imaging technique is designed to provide noninvasive, 3D optical biopsies of the luminal surface allowing instantaneous virtual histologic diagnosis. CLE delivers an excitation wavelength of 488 nm and confocal image data are collected at different scan rates depending on the system used. Confocal images can be collected simultaneously with endoscopic images. It can be performed using a probe-based system (probe-based confocal laser endomicroscopy by Cellvizio, Mauna Kea Technologies, Paris, France) passed through the channel of a standard endoscope or with a miniaturized confocal microscope integrated into the distal tip of an endoscope [endoscopy-based confocal laser endomicroscopy (eCLE) by Pentax, Fort Wayne, NJ]. Both the systems are approved by the Food and Drug Administration of the United States; currently, the probe-based confocal laser endomicroscopy (pCLE) is the only modality in clinical use as eCLE is no longer commercially available (Fig. 20.1).

To illuminate tissue structures, CLE is performed with the aid of contrast agents such as intravenous fluorescein (2.5 or 5 mL of 10% solution) or topical contrast agents, including acriflavine hydrochloride and cresyl violet (both not available in the United States). The CLE scope is gently placed on the mucosal surface and scans the area to create an optical biopsy of the tissue in the horizontal plane. In the eCLE system, a scan of different depths can be performed to generate a collection of images termed z-stacks.

Common clinical indications for CLE include surveillance of dysplasia in Barrett's esophagus and identification and characterization of colorectal polyps. In patients with IBD, CLE has been used for both structural and functional assessment of the intestinal epithelium. Specifically, CLE was used to evaluate the severity of intestinal inflammation in the diagnosis of IBD, and mucosal barrier functional status to determine the probability of disease relapse and guide the selection of biological therapy. Here, we will highlight the role of CLE and in vivo imaging in advancing our understanding of the pathophysiology of IBD, thereby improving the clinical management of patients.

Structural assessment of the intestine

The ability to accurately gauge the severity of mucosal inflammation and detect microscopic inflammation in IBD is highly relevant in the care of patients as endoscopic MH has emerged as an important goal of therapy. MH is associated with more favorable clinical outcomes, including decreased disease activity, reduced rate of hospitalization or surgical resections, and increased rate of sustained clinical remission [1,2]. However, conventional WLE is not a modality with adequate sensitivity to assess mucosal inflammation; therefore the gold standard remains histopathological assessment of intestinal biopsy samples.

CLE provides a real-time evaluation of mucosal inflammation and can detect microscopic inflammation in areas with a normal macroscopic appearance. When examined by CLE, inflamed mucosa in IBD patients has multiple abnormalities, including irregular and tortuous crypts with irregular and wider lumens in addition to fluorescein leakage into the interstitial space [3]. CLE has been used to develop classification systems for inflammatory activity in patients with ulcerative colitis (UC). Kiesslich et al. [4] reported a three-grade classification system for inflammation consisting of

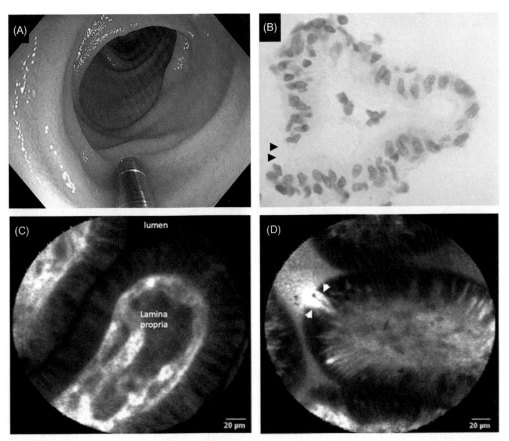

FIGURE 20.1 Whit-light endoscopy, pCLE, and light microscopy of the terminal ileum and epithelial gaps. (A) Endoscopic appearance of the mucosa of pCLE imaged area; (B) light microscopy of the mucosa of Crohn's disease imaged with pCLE (D). The nuclei were stained red with nuclear fast red, while goblet cells stained blue with Alcian blue and epithelial gaps (*black arrows*). (C) pCLE showing the normal terminal ileum in the normal; (D) pCLE showing an increased epithelial gap in Crohn's disease (*white arrows*). pCLE, Probe-based confocal laser endomicroscopy. *Courtesy Liu JJ, et al. Gastrointest Endosc 2011;73:1174−80 with permission from Elsevier.*

TABLE 20.1 Classification of features on confocal laser endomicroscopy to grade inflammatory activity [4].

Grade of inflammation	Crypt architecture	Cellular infiltration	Vascular architecture
None	Regular luminal openings and distribution of crypts covered by a homogeneous layer of epithelial cells, including goblet cells	Absent	Normal hexagonal, honeycomb appearance that presents a network of capillaries outlining the stroma surrounding the luminal openings of the crypts
Mild to moderate	Differences in shape, size, and distribution of crypts; increased distance between crypts, focal crypt destruction	Present; <50% of crypts involved	Mild to moderate increase of capillaries, dilated, and distorted capillaries
Severe	Unequivocal crypt destruction	Present; >50% of crypts involved	Marked increase of dilated and distorted capillaries; leakage of fluorescein

classification of crypt architecture, microvascular alterations, and cellular infiltration (Table 20.1; Figs. 20.1 and 20.2). Crypt architecture was defined using another CLE-based system by Li et al. (Table 20.2; Figs. 20.3−20.6). Fluorescein leakage was shown to have a significant correlation with histologic findings in UC patients, with more than half of the patients with normal-appearing mucosa on conventional WLE revealed acute inflammation on histopathology [3]. Another study prospectively demonstrated that a composite score based on fluorescein leakage and crypt diameter was

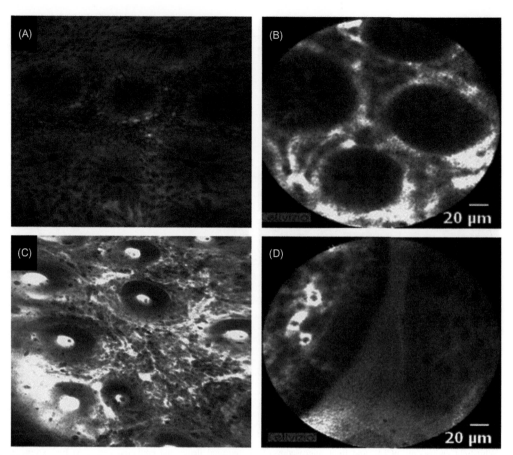

FIGURE 20.2   Evaluation of noninflamed and inflamed colon mucosa. (A) Noninflamed mucosa on integrated CLE. Colonic crypts, surrounded by dark-appearing goblet cells, became evident; (B) noninflamed mucosa IBD visualized with probe-based CLE. Confocal imaging displays colonic crypts and microvessels in the lamina propria; (C) inflamed mucosa in IBD on integrated CLE. Colonic crypts were variously shaped and irregular. Fluorescein sodium leaked into the lamina propria, highlighting damage of minute mucosal microvessels. (D) Mucosal gaps appearing as white incisions of the mucosal surface on probe-based CLE. These gaps are predictive of an acute flare of the disease within the next 12 months. *CLE*, Confocal laser endomicroscopy; *IBD*, inflammatory bowel disease. *Courtesy Neumann H, Kiesslich R. Gastrointest Endosc Clin N Am 2013;23:695−705.*

TABLE 20.2 Classification of crypt architecture on confocal laser endomicroscopy in ulcerative colitis [3].

Crypt architecture on confocal laser endomicroscopy	Description
A	Regular arrangement and size of crypts
B	Irregular arrangements of crypts, enlarged spaces between crypts
C	Dilation of crypt openings, more irregular arrangement of crypts, and enlarged spaces between crypts as compared to type B
D	Crypt destruction and/or crypt abscess

predictive of disease flare in UCin a 12-month follow-up [5]. In Crohn's disease (CD), another system to grade inflammatory activity known as the CD Endomicroscopic Activity Score has been proposed to differentiate active versus quiescent disease and includes parameters such as crypt number, colonic crypt distortion, and microerosions, augmented vascularization, quantification of goblet cells, and increased cellular infiltration [6]. Overall, CLE appears to be a sensitive tool to instantly assess inflammatory activity in IBD patients and reliably detect microscopic inflammation in macroscopically noninflamed mucosa.

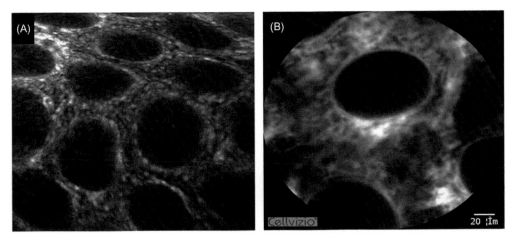

FIGURE 20.3 Grade A in the CLE classification of ulcerative colitis. Normal-sized crypts are regularly arranged. The lumen of crypts presents as dark spots without fluorescein leakage. (A) Crypts of Grade A observed by endoscope-based CLE; (B) few crypts can be observed by probe-based CLE. *CLE*, Confocal laser endomicroscopy. *Courtesy Dr. Xiuli Zuo of Shandong University Qilu Hospital.*

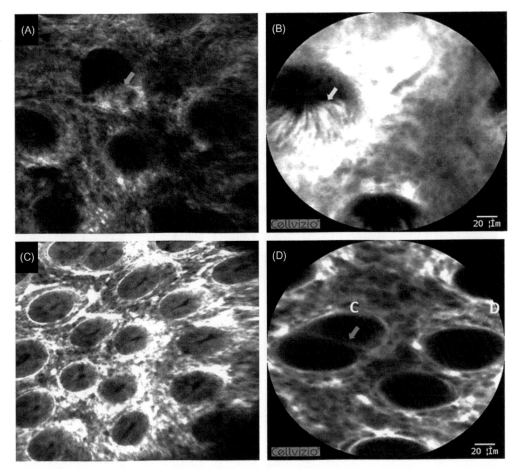

FIGURE 20.4 Grade B in the CLE classification of ulcerative colitis. Crypts are normal in size but irregularly arranged. The lumen of crypts is free of fluorescein into spaces between epithelial cells. Sporadic crypt fusion can be seen. (A) eCLE: fluorescein leakage in the spaces between epithelial cells (*green arrow*); (B) pCLE: fluorescein leakage in the spaces between epithelial cells (*yellow arrow*); (C) crypt fusion on eCLE (*blue arrow*); (D) crypt fusion (*red arrow*) on pCLE. *eCLE*, Endoscope-based confocal laser endomicroscopy; *pCLE*, probe-based confocal laser endomicroscopy. *Courtesy Dr. Xiuli Zuo of Shandong University Qilu Hospital.*

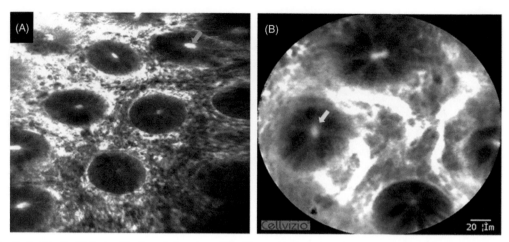

FIGURE 20.5 Grade C in the CLE classification of ulcerative colitis. Fluorescein leaks into the lumen of crypts shown as bright spots. The epithelium of crypts is intact. (A) eCLE: fluorescein leakage in the lumen of crypts (*green arrow*); (B) pCLE: fluorescein leakage in the lumen of crypts (*yellow arrow*). *eCLE*, Endoscope-based confocal laser endomicroscopy; *pCLE*, probe-based confocal laser endomicroscopy. *Courtesy Dr. Xiuli Zuo of Shandong University Qilu Hospital.*

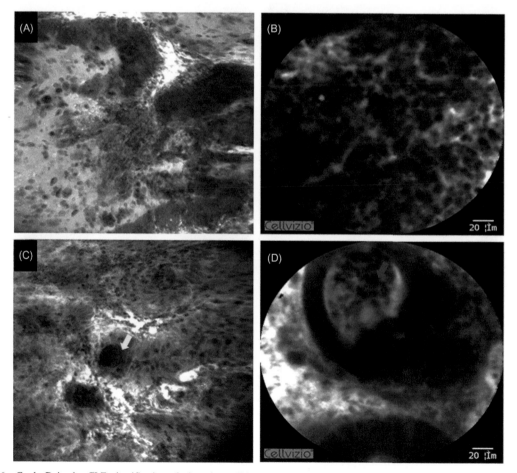

FIGURE 20.6 Grade D in the CLE classification of ulcerative colitis. Most crypts are replaced by diffuse necrosis, and remaining crypts are destroyed. Crypt abscesses are seen. (A) Destroyed crypts (*green arrow*) and diffuse necrosis on eCLE; (B) diffuse necrosis on eCLE; (C) crypt abscess (*yellow arrow*) on eCLE; (D) crypt abscess (*blue arrow*) on pCLE. *eCLE*, Endoscope-based confocal laser endomicroscopy; *pCLE*, probe-based confocal laser endomicroscopy. *Courtesy Dr. Xiuli Zuo of Shandong University Qilu Hospital.*

Functional assessment of the mucosal barrier

Impaired mucosal barrier function resulting in loss of tolerance to igut microbes and ongoing inflammation plays a crucial role in the pathogenesis of IBD [7]. CLE allows for the visualization and quantification of epithelial gaps, which are extrusion zones created with the shedding of intestinal epithelial cells [7]. Patients with CD or UC were found to have significantly increased numbers of epithelial gaps compared to non-IBD controls [8,9]. The increased cell extrusion could be demonstrated using a mouse model of IBD, the interleukin-10 knockout or IL-10 knock-out mouse. A quantitative assessment of epithelial cells and gaps resulting from intestinal epithelial cell extrusion could be validated against conventional multiphoton confocal microscopy and light microscopy in the IL-10 knock-out mice [10]. Besides, gap density measured via CLE correlated with quantitative analysis of immunohistochemical staining of mucosal biopsy samples of intestinal epithelial cells destined to undergo extrusion. Another more descriptive classification system of mucosal barrier dysfunction is the 3-grade Watson grading system of cell shedding, contrast leakage, and microerosions [8]. Watson Grade I is normal, physiological cell shedding in the absence of fluorescein leakage or microerosions resulting from multiple cells being shed from one area. Grade II is marked by the loss of barrier function with fluorescein leakage into the lumen. Grade III is defined by the presence of microerosions caused by multiple intestinal epithelial cells extruded from a single area [8]. In addition, CLE was used to identify intramucosal bacteria in the colon and terminal ileum; and IBD patients had a significantly higher quantity and area of involvement of intramucosal bacteria [9].

The clinical application of mucosal barrier function assessment using CLE was first shown in the prediction of disease relapse. Using the Watson Grade (increased contrast leakage and microerosions of Grade II or III) or increased gap density on CLE as markers of mucosal barrier dysfunction, both were shown to predict disease relapse in both adult and pediatric IBD patients [8,10,11]. In addition to providing a marker of barrier dysfunction that is predictive of disease relapse, CLE was recently shown to hold the promising potential to refine personalized biologic treatment algorithms. Using molecular imaging techniques, CLE was shown to yield a clinical response rate of 94% to antitumor necrosis factor (TNF) antibody [12]. Using epithelial gap density as a measure of mucosal barrier dysfunction, personalized selection of biologic agents based on barrier dysfunction criteria enabled a clinical response rate of 90% to be achieved in IBD patients prospectively in a single centered setting. This translated into an 80% reduction in IBD-related hospitalization over a year [13].

Surveillance of dysplasia

IBD confers an increased risk for the development of colitis-associated dysplasia or cancer. CLE may play an important role in dysplasia surveillance in patients with UC or extensive CD by providing an in vivo histologic view of concerning lesions. Using CLE, a classification system was developed to predict dysplasia based on vessel and crypt architecture (Table 20.3) that was prospectively shown to detect neoplastic changes (intraepithelial neoplasia and/or cancer) with a high (99.2%) accuracy compared to histopathologic diagnosis [8]. A metaanalysis of 15 studies of CLE for dysplasia detection showed it could distinguish neoplasia in IBD patients for surveillance with excellent sensitivity (93%) and specificity (90%) [9]. Another study in patients with UC showed that CLE enabled differentiation of dysplasia-associated lesions and adenoma-like masses with a 97% of accuracy and strong agreement between endomicroscopy and histopathology [15].

TABLE 20.3 Confocal pattern classification to predict colorectal pathology [14].

Grading	Vessel architecture	Crypt architecture
Normal	Hexagonal, honeycomb appearance that presents a network of capillaries outlining the stroma surrounding the luminal openings of the crypts	Regular luminal openings and distribution of the crypts covered by a homogeneous layer of epithelial cells, including goblet cells
Regeneration	Hexagonal, honeycomb appearance with no or mild increase in the number of capillaries	Star-shaped luminal crypt openings or focal aggregation of regular-shaped crypts with a regular or reduced amount of goblet cells
Neoplasia	Dilated and distorted vessels with elevated leakage; irregular architecture with little or no orientation to adjunct tissue	Ridged-lined irregular epithelial layer with loss of crypts and goblet cells; irregular cell architecture with little or no mucin

The current standard of practice for random four-quadrant biopsies every 10 cm is limited as it only has moderate sensitivity; and neoplasia may be missed in up to a third of standard colonoscopies [16]. In conjunction with a red-flag technique such as chromoendoscopy, CLE enables targeted biopsies to be collected to increase diagnostic yield. CLE can be used to define the histology of macroscopically concerning lesions during surveillance colonoscopy, providing an immediate microscopic view of the lesion.. One randomized controlled trial using this approach demonstrated markedly increased detection of dysplastic lesions compared to standard endoscopy while requiring fewer biopsies. CLE was shown to be predictive of neoplasia with 97.8% accuracy [4].

Primary sclerosing cholangitis (PSC)-associated IBD, namely, PSC-IBD, represents a specific phenotype of IBD which right-sided colitis that appeared to carry a particularly high risk of dysplasia [17]. In patients with PSC-IBD, CLE was shown to be effective in predicting dysplasia and may be a useful tool for surveillance in addition to high-definition (HD) WLE [14]. Overall, by providing real-time optical biopsies during surveillance colonoscopy, CLE appears to have a promising role in providing targeted biopsieswith a improved sensitivity and specificity in the detection of neoplasia.

Other potential applications in inflammatory bowel disease

CLE has been used for structural evaluations and diagnosis of extraintestinal manifestations and/or complications in IBD patients such as differentiating benign from malignant biliary strictures in PSC [14]. Another useful application of CLE in IBD patients is to identify the presence of other mucosal lesions such as celiac disease. The features on CLE in celiac disease include a reduction in the number of folds, total villous atrophy, and mosaic pattern, and increased number of lymphocytes [18].

Endocytoscopy

Similar to CLE, endocytoscopy enables endoscopists to visualize microvessels, epithelial crypts, and infiltration of inflammatory cells. Therefore the technology has been used in the assessment of disease activity in UC [19] and dysplastic lesions in the GI tract [20]. Endocytoscope is equipped with a standard WLE mode and ultramagnifying mode. Endocytoscopy can be combined with dye-based (e.g., methylene blue) chromoendoscopy (DCE) or narrow-band imaging (NBI). Endocytoscopy has even been combined with computer-aided artificial intelligence to quantify mucosal inflammation in UC responding to histology (Figs. 20.7–20.9) [21].

Magnification endoscopy

The newer generation of endoscopy is equipped with dual-focus or magnification function and NBI. ME includes an adjustable focusing system that can enlarge the image from 1.5 × to 150 × with a resolution of 850,000 or higher pixel density, as compared with 100,000- to 200,000-pixel densities of conventional endoscopy. However, the current model of HD endoscopy can yield 850,000 to 2 million pixels. The newest version of ME can discriminate objects with 10−71 μm in diameter. In translucent mucosal epithelium, the light beam can reach the subepithelial vascular network.

There are scant published data in the use of plain ME in the assessment of disease activity in IBD. ME may be useful in the assessment of subtle mucosal diseases in the small bowel or colon and provision of targeted biopsy. These disease conditions include celiac disease, protein-losing enteropathy, and acute rejection of small bowel transplantation. ME may be powerful to evaluate villous abnormalities [22]. The ability of ME to delineate subepithelial vasculature as well as detailed mucosal structure makes it particularly useful in the assessment of disease activity in IBD (Figs. 20.10−20.12) or radiation-associated bowel injury (Fig. 20.13). In patients with UC the endoscopic appearance on magnifying NBI was shown to correlate with histologic disease activity, possibly making ME a more reliable tool to assess the true disease extent [23] and histologic MH [24].

The current main applications of ME in IBD are to assess flat and depressed dysplastic lesions [25−27], dysplasia in UC [28], and distinction between neoplastic versus nonneoplastic polyps, in a combination with dye chromoendoscopy (DCE) or NBI [29]. The imaging resolution of the old versions of ME used in the abovementioned studies more than a decade ago is comparable to that of current HD-WLE; and both have been used for the assessment of disease activity and CAN (see next). The application of ME in IBD has evolved into the use of a combined ME with DCE (with or without magnification) or NBI (with or without magnification) [30−32]. Magnifying dye-based chromoendoscopy is valuable for observation of surface structure, while magnifying NBI provides detailed information in surface and vessel pattern. On magnifying chromoendoscopy, the pit density of the neoplastic lesions was shown to be greater

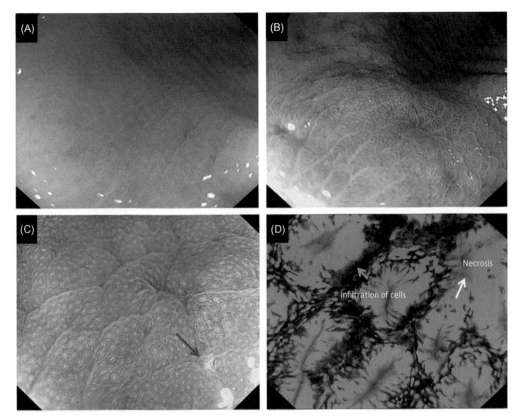

FIGURE 20.7 Narrow-band imaging with and without magnification and endocytoscopy to assess mild inflammation versus mucosal healing in an ulcerative colitis patient before discontinuation of biological therapy. (A–C) High-definition and narrow-band imaging with (C) and without (B) optical magnification showing a tiny microerosion of colonic mucosa (*blue arrow*); (D) endocytoscope confirming crypts architecture distortion with some areas of crypts necrosis (*white arrow*) and increase cell infiltration in the lamina propria (*green arrow*).

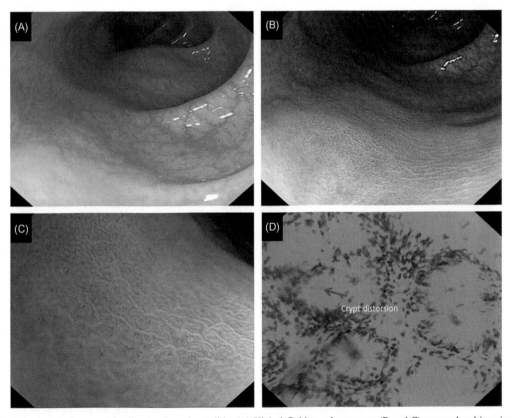

FIGURE 20.8 Assessment of mucosal healing in ulcerative colitis. (A) High-definition colonoscopy; (B and C) narrow-band imaging with (C) and without (B) optical magnification; (D) endocytoscopy showing crypt distortion. *Courtesy Dr. Marietta Iacucci of University of Birmingham.*

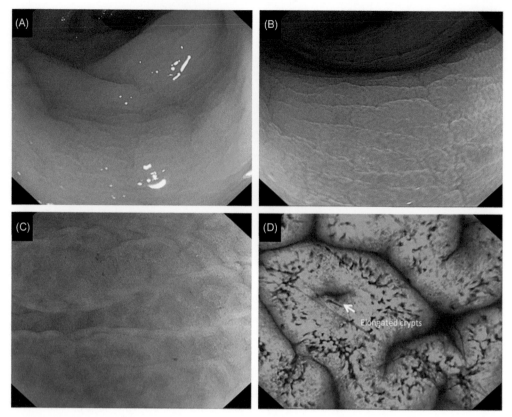

FIGURE 20.9   Assessment of mucosal healing in ulcerative colitis. (A–C) Narrow-band imaging with magnification characterized crypts architecture as elongated and distorted crypts in ulcerative colitis; (D) confirmation by endocytoscope (*arrow*). *Courtesy Dr. Marietta Iacucci of University of Birmingham.*

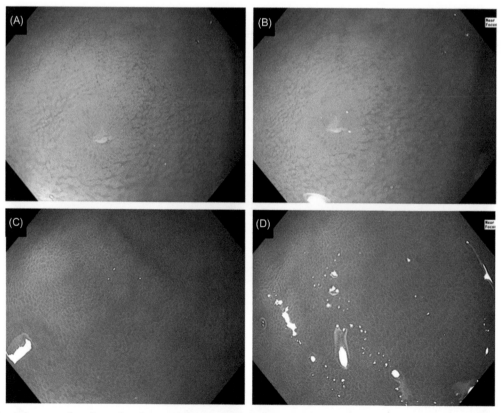

FIGURE 20.10   Assessment of erosion and erythema with high-definition and magnification endoscopy. (A and C) Erosion (A) and erythema (C) on high-definition endoscopy; (B and D) the same erosion (B) and erythema (D) on magnification endoscopy. In this case the resolution of images does not appear to be different between the two endoscopy modalities.

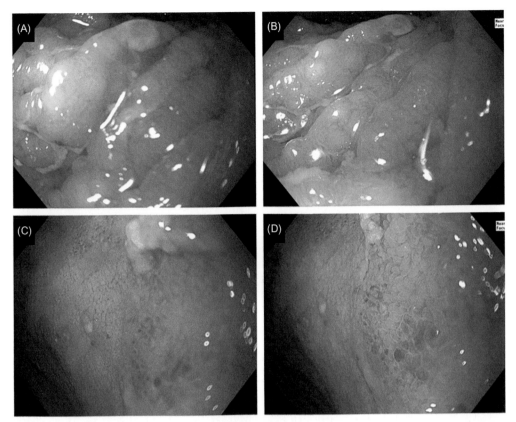

FIGURE 20.11 Assessment of mucosal edema and erythema with high-definition and magnification endoscopy. (A and C) Mucosal edema (A) and erythema (C) on high-definition endoscopy; (B and D) the same mucosal edema (B) and erythema (D) on magnification endoscopy. In this case, detailed subepithelial vasculature and mucosal structure are shown better on magnification endoscopy.

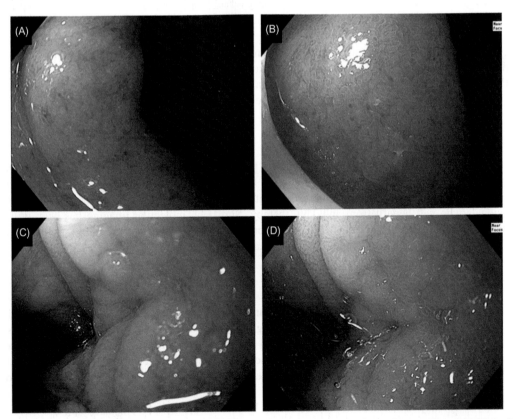

FIGURE 20.12 Assessment of mucosal erythema and edema with high-definition and magnification endoscopy. (A and C) Mucosal erythema (A) and edema (C) on high-definition endoscopy; (B and D) the same mucosal erythema (B) and edema (D) on magnification endoscopy. Notice that pit patterns are better demonstrated on magnification endoscopy.

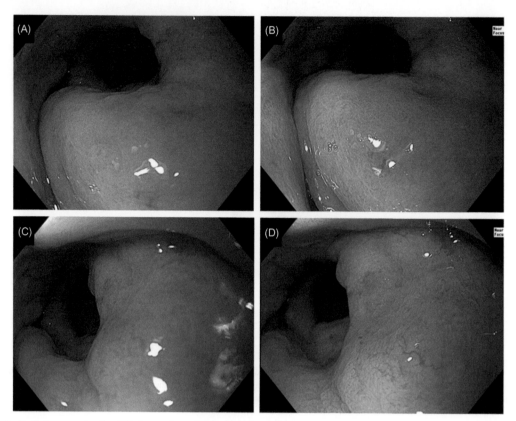

FIGURE 20.13 Radiation-associated cuffitis in a patient with ileal pouch—anal anastomosis for hereditary colon cancer. (A and C) Telangiectasia in the cuff on high-definition white-light endoscopy; (B and D) telangiectasia visualized better with a magnification mode.

than that of the nonneoplastic lesions and pit margins were more often irregular in the neoplastic lesions than in the nonneoplastic lesions [33]. There are limited published data on the current version of ME with zooming capability with or without DCE or NBI in IBD (Figs. 20.14−20.19). It can be challenging to scan the large area of the colon with the zooming lens equipped in ME. Practically, ME may only be applied in targeted areas.

Molecular imaging

Molecular imaging is developed for the assessment of patient's response to targeted biological therapy such as anti-TNF and formulating a tailored treatment plan in IBD. The membrane-bound TNF can be evaluated with endoscopy-based molecular imaging with a fluorescent anti-TNF antibody. This endoscopy practice may help one to identify individuals who will benefit from anti-TNF therapy. Similarly, molecular imaging has been utilized to assess the response to vedolizumab by assessing integrin expression [34]. Molecular imaging is further discussed in Chapter 21, Molecular imaging in inflammatory bowel disease.

Optical coherence tomography

OCT uses the backscattering of light to obtain cross-sectional images of tissue. OCT can be performed via a catheter-based probe on endoscopy. OCT enhances endoscopic imaging of the superficial layers of the esophagus, stomach, bile ducts, pancreatic duct, and colon. Depth of penetration and resolution varies between equipment. The newer generation of OCT is able to identify and characterize anatomic structures of the GI tract such as crypts and glands. The main advantage of OCT over endoscopic ultrasound is its higher spatial resolution. The major disadvantage of OCT over endoscopic ultrasound is its shallower depth of penetration.

Most OCT studies were performed in the assessment of Barrett's esophagus [35,36]. The application of OCT in IBD is limited to the assessment of the distinction between CD and UC due to its ability to penetrate deeper layers of the bowel wall [37,38]. Shen et al. used a prototype of OCT to study transmural diseases in CT and use ex vivo [39] and

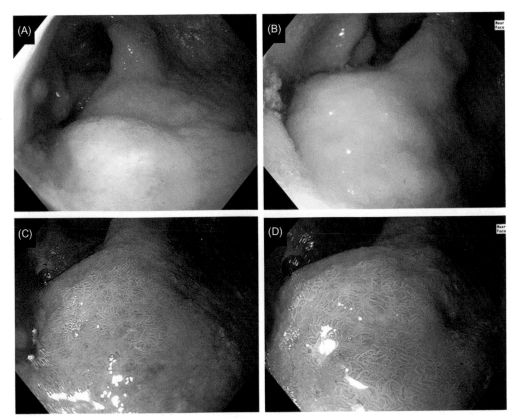

FIGURE 20.14 Assessment of mild mucosal inflammation with loss of vascular pattern of the ileal pouch. (A) High-definition white-light endoscopy; (B) white-light endoscopy in a magnification mode; (C) narrow-banding imagingin a conventional mode; (D) narrow-band imaging in a magnification mode. Narrow-band imaging in the magnification mode provided the most detailed mucosal features.

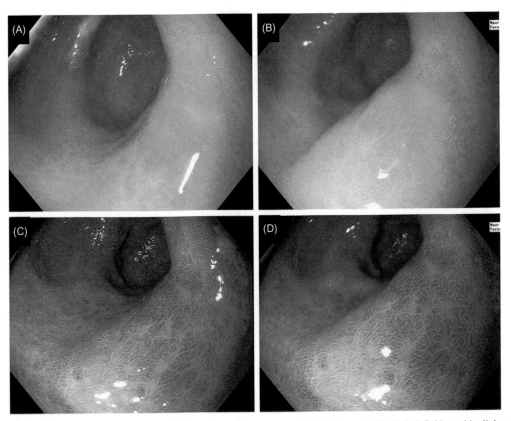

FIGURE 20.15 Assessment of mild inflammation with loss of vascular pattern of the ileal pouch. (A) High-definition white-light endoscopy; (B) white-light endoscopy in a magnification mode; (C) narrow-banding imaging in a conventional mode; (D) narrow-band imaging in magnification mode. Narrow-band imaging in the magnification mode provided the most detailed mucosal features.

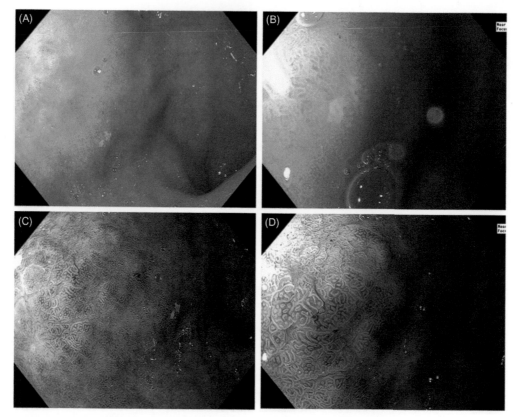

FIGURE 20.16 Assessment of mucosal inflammation with erosions of the ileal pouch. (A) High-definition white-light endoscopy; (B) white-light endoscopy in a magnification mode; (C) narrow-banding imaging in a conventional mode; (D) narrow-band imaging in magnification mode. Narrow-band imaging in the magnification mode provided the most detailed mucosal features.

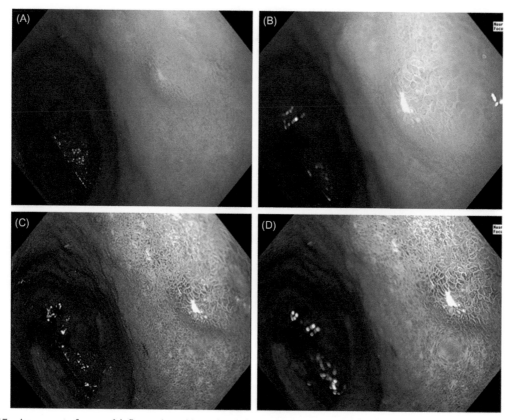

FIGURE 20.17 Assessment of mucosal inflammation with erosions of the ileal pouch. (A) High-definition white-light endoscopy; (B) white-light endoscopy in a magnification mode; (C) narrow-banding imaging in a conventional mode; (D) narrow-band imaging in magnification mode. Narrow-band imaging in the magnification mode provided the most detailed mucosal features.

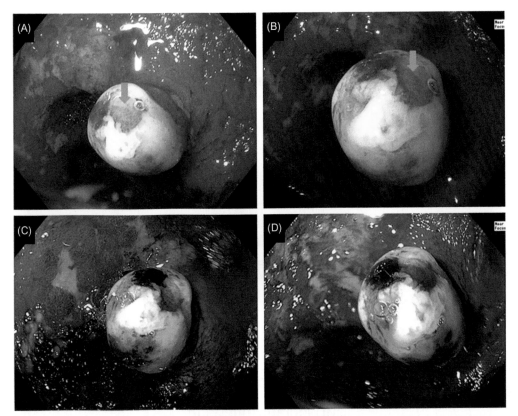

FIGURE 20.18 Assessment of inflammatory polyps. (A) High-definition white-light endoscopy; (B) white-light endoscopy in a magnification mode showing more detailed mucosal structure than white-light endoscopy (*green arrow*); (C) narrow-banding imaging in a conventional mode; (D) narrow-band imaging in magnification mode. Narrow-band imaging in the magnification mode did not appear to provide more detailed mucosal features.

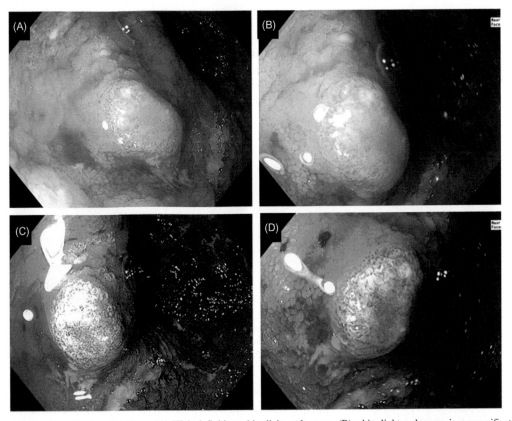

FIGURE 20.19 Assessment of polypoid lesions. (A) High-definition white-light endoscopy; (B) white-light endoscopy in a magnification mode; (C) narrow-banding imaging in a conventional mode; (D) narrow-band imaging in magnification mode. Narrow-band imaging in the magnification mode provided more detailed mucosal features. Histologic diagnosis was indefinite for dysplasia.

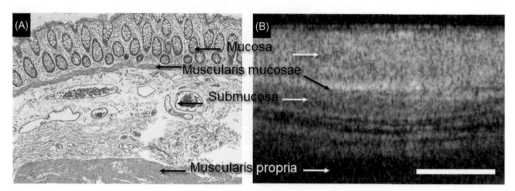

FIGURE 20.20   Ex vivo evaluation of normal colon anatomy. (A) Surgically resected colon specimen showing layered structures, as labeled on histology; (B) corresponding features on optical coherence tomography. *Courtesy Shen B, Zuccaro G, Gramlich TL,et al. Clin Gastroenterol Hepatol 2004;2:754−60.*

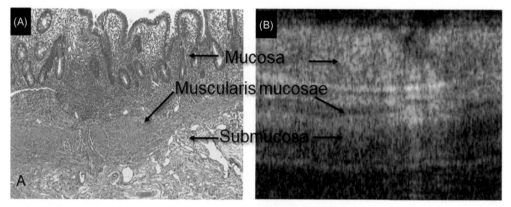

FIGURE 20.21   Ex vivo evaluation of ulcerative colitis. (A) Despite inflammation in the lamina propria, muscularity mucosae (with hyperplasia), and superficial submucosa, the layered structure was preserved on histology; (B) the layered structure of the bowel wall was still preserved on optical coherence tomography. *Courtesy Shen B, Zuccaro G, Gramlich TL, et al. Clin Gastroenterol Hepatol 2004;2:754−60.*

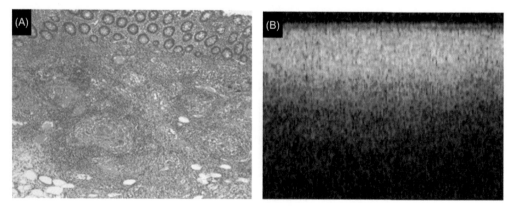

FIGURE 20.22   Ex vivo evaluation of Crohn's disease of the terminal ileum. (A) Disruption of the layered structure with inflammatory infiltration in the lamina propria, muscularis mucosae, submucosa, and beyond; (B) obliteration of the layered structure on optical coherence tomography. *Courtesy Shen B, Zuccaro G, Gramlich TL, et al. Clin Gastroenterol Hepatol 2004;2:754−60.*

in vivo via colonoscopy [40]. T features on OCT were shown to be helpful for the distinction between CD and UC (Figs. 20.20−20.25). The main limitations of the OCT are its depth of penetration (around 1 mm), which is shorter than whole thickness of the bowel wall, and lower spatial resolution than CLE, creating a barrier for being used in routine surveillance of CAN.

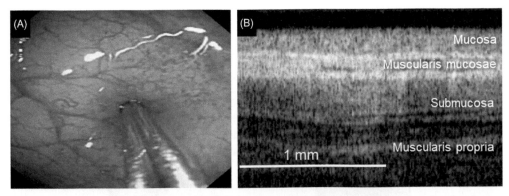

FIGURE 20.23 In vivo evaluation of normal colon structure. (A) Probe-based optical coherence tomography in action via colonoscopy; (B) Layered bowel wall structure on optical coherence tomography. *Courtesy Shen B, Zuccaro G Jr, Gramlich TL, et al. Clin Gastroenterol Hepatol 2004;2:1080−87.*

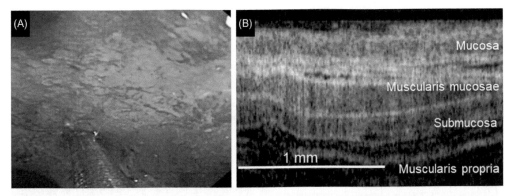

FIGURE 20.24 In vivo evaluation of ulcerative colitis. (A) The through-the-scope optical coherence tomography in action via colonoscopy; (B) preserved layered structure on optical coherence tomography. *Courtesy Shen B, Zuccaro G Jr, Gramlich TL, et al. Clin Gastroenterol Hepatol 2004;2:1080−87.*

FIGURE 20.25 In vivo evaluation of Crohn's colitis. (A) Probe-based optical coherence tomography in action via colonoscopy. Diffuse mucosal inflammation mimicked ulcerative colitis; (B) disrupted layered structure of the colon wall on optical coherence tomography, indicative of Crohn's disease. *Courtesy Shen B, Zuccaro G Jr, Gramlich TL, et al. Clin Gastroenterol Hepatol 2004;2:1080−87.*

Elastography

Ex vivo optical coherence elastography in a murine model is used to assess biomechanical properties of the colonby the application of elastic waves [41]. On the other hand, ultrasound elastography with strain or shear-wave imaging has been described. Trans-abdominal real-time shear-wave ultrasound elastography estimates the speed of the wave and quantifies tissue stiffness. The technology has been investigated for the assessment of bowel wall fibrosis and the

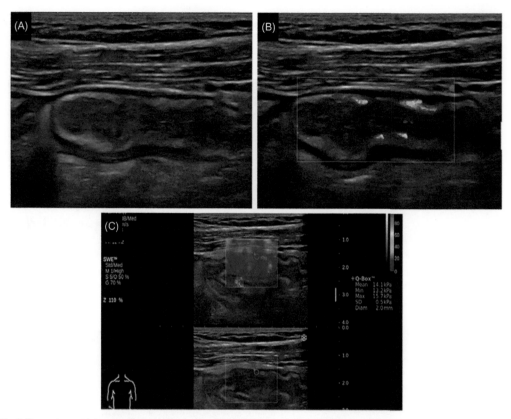

FIGURE 20.26 Inflammatory strictures in Crohn's disease on ultrasound elastography. (A) Thickened bowel wall in B-mode imaging; (B) signals in color imaging mode; (C) predominant blue signals (indicative of minimum stiffness) in elastography mode.*Courtesy Dr. Ren Mao, The First Affiliated Hospital of Sun Yat-Sen University*

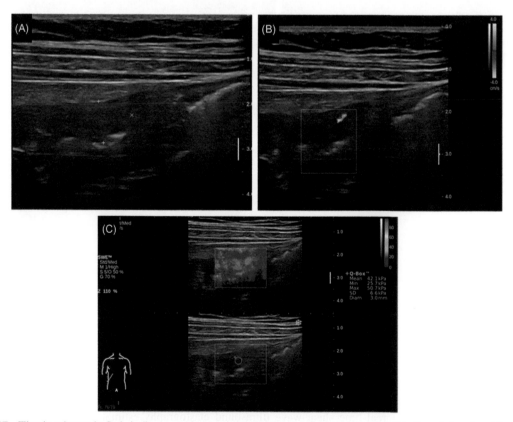

FIGURE 20.27 Fibrotic strictures in Crohn's disease on ultrasound elastography. (A) Thickened bowel wall in B-mode imaging; (B) signals in color imaging mode; (C) predominant green signals (indicative of moderate stiffness) in elastography mode. *Courtesy Dr. Ren Mao of the First Affiliated Hospital of Sun Yat-Sen University*

distinction between inflammatory and fibrostenotic strictures in CD. Bowel wall elasticity estimates were color-coded to generate a quantitative shear-wave images (Figs. 20.26 and 20.27) [42,43]. We expect that probe-based endoscopy ultrasound elastography may provide a more accurate measurement of bowel fibrosis than computed tomography or magnetic resonance imaging. Ultrasound assessment of transmural disease in CD is detailed in Chapter 35, Transmural imaging in inflammatory bowel disease.

Summary and recommendations

Advanced endoscopic techniques such as CLE and endocytoscopy allow for real-time in vivo microscopic evaluation of surface of the GI tract. In IBD patients, advanced endoscopy provides structural assessment of the epithelium for grading of mucosal inflammation, targeted surveillance for neoplasia, and diagnosis of other mucosal lesions. Moreover, CLE can be performed to evaluate the mucosal barrier function status, qualitatively using the Watson grading system or semiquantitatively using epithelial cell gap density. Both markers of epithelial barrier dysfunction were found to be predictive of disease relapse. CLE, endocytoscopy, and combined ME, chromoendoscopy, or NBI are clinically applicable for the more accurate assessment of disease activity, MH or tissue healing, and CAN. The endoscopic imaging techniques need to be improved to assess deeper layers of the bowel for transmural inflammation and fibrosis.

Disclosure

JJL receives grant support from Mauna Kea Technologies. CAL, ZL, MI, and BS declare no financial conflict of interest.

References

[1] Neurath M, Travis S. Mucosal healing in inflammatory bowel diseases: a systematic review. Gut 2012;61:1619−35.

[2] Baert F, Moortgat L, Van Assche G, Caenepeel P, Vergauwe P, De Vos M, et al. Mucosal healing predicts sustained clinical remission in patients with early-stage Crohn's disease. Gastroenterology 2010;138:463−8.

[3] Li CQ, Xie XJ, Yu T, Gu XM, Zuo XL, Zhou CJ, et al. Classification of inflammation activity in ulcerative colitis by confocal laser endomicroscopy. Am J Gastroenterol 2010;105:1391−6.

[4] Kiesslich R, Goetz M, Lammersdorf K, Schneider C, Burg J, Stolte M, et al. Chromoscopy-guided endomicroscopy increases the diagnostic yield of intraepithelial neoplasia in ulcerative colitis. Gastroenterology 2007;123:874−82.

[5] Buda A, Hatem G, Neumann H, D'Incà R, Mescoli C, Piselli P, et al. Confocal laser endomicroscopy for prediction of disease relapse in ulcerative colitis: a pilot study. J Crohns Colitis 2014;8:301−11.

[6] Neumann H, Vieth M, Atreya R, Grauer M, Siebler J, Bernatik T, et al. Assessment of Crohn's disease activity by confocal laser endomicroscopy. Inflamm Bowel Dis 2012;18:2261−9.

[7] Kiesslich R, Goetz M, Angus EM, Hu Q, Guan Y, Potten C, et al. Identification of epithelial gaps in the human small and large intestine by confocal endomicroscopy. Gastroenterology 2007;113:1769−78.

[8] Kiesslich R, Duckworth CA, Moussata D, Gloeckner A, Lim LG, Goetz M, et al. Local barrier dysfunction identified by confocal laser endomicroscopy predicts relapse in inflammatory bowel disease. Gut 2012;61:1146−53.

[9] Moussata D, Goetz M, Gloeckner A, Kerner M, Campbell B, Hoffman A, et al. Confocal laser endomicroscopy is a new imaging modality for recognition of intramucosal bacteria in inflammatory bowel disease in vivo. Gut 2011;60:26−33.

[10] Turcotte JF, Wong K, Mah SJ, Dieleman LA, Kao D, Kroeker K, et al. Increased epithelial gaps in the small intestine are predictive of hospitalization and surgery in patients with inflammatory bowel disease. Clin Transl Gastroenterol 2012;3:e19.

[11] Shavrov A, Kharitonova AY, Claggett B, Claggett B, Morozov DA, Brown DK, et al. A pilot study of the predictive value of probe-based confocal laser endomicroscopy for relapse in pediatric inflammatory bowel disease patients. J Pediatr Gastroenterol Nutr 2016;6:873−8.

[12] Atreya R, Neumann H, Neufert C, Waldner MJ, Billmeier U, Zopf Y, et al. In vivo imaging using fluorescent antibodies to tumor necrosis factor predicts therapeutic response in Crohn's disease. Nat Med 2014;20:313−18.

[13] Liu JJ, Rosson TB, Xie JJ, Harris ZP, McBride RG, Siegel E, et al. Personalized inflammatory bowel disease care reduced hospitalizations. Dig Dis Sci 2019;64:1809−14.

[14] Slivka A, Gan I, Jamidar P, Costamagna G, Cesaro P, Giovannini M, et al. Validation of the diagnostic accuracy of probe-based confocal laser endomicroscopy for the characterization of indeterminate biliary strictures: results of a prospective multicenter international study. Gastrointest Endosc 2015;81:282−90.

[15] Hurlstone D, Thomson M, Brown S, Tiffin N, Cross SS, Hunter MD. Confocal endomicroscopy in ulcerative colitis: differentiating dysplasia-associated lesional mass and adenoma-like mass. Clin Gastroenterol Hepatol 2007;5:1235−41.

[16] Rex D, Cutler C, Lemmel G, Rahmani EY, Clark DW, Helper DJ, et al. Colonoscopic miss rates of adenomas determined by back-to-back colonoscopies. Gastroenterology 1997;112:24−8.

[17] Dlugosz A, Barakat A, Björkström NK, Öst Å. Diagnostic yield of endomicroscopy for dysplasia in primary sclerosing cholangitis associated inflammatory bowel disease: a feasibility study. Endosc Int Open 2016;4:e901−11.

[18] Trovato C, Sonzogni A, Ravizza D, Fiori G, Rossi M, Tamayo D, et al. Celiac disease: in vivo diagnosis by confocal endomicroscopy. Gastrointest Endosc 2007;65:1096−9.

[19] Nakazato Y, Naganuma M, Sugimoto S, Bessho R, Arai M, Kiyohara H, et al. Endocytoscopy can be used to assess histological healing in ulcerative colitis. Endoscopy 2017;49:560−3.

[20] Mori Y, Kudo SE, Ogawa Y, Wakamura K, Kudo T, Misawa M, et al. Diagnosis of sessile serrated adenomas/polyps using endocytoscopy (with videos). Dig Endosc 2016;28(Suppl. 1):43−8.

[21] Maeda Y, Kudo SE, Mori Y, Misawa M, Ogata N, Sasanuma S, et al. Fully automated diagnostic system with artificial intelligence using endocytoscopy to identify the presence of histologic inflammation associated with ulcerative colitis (with video). Gastrointest Endosc 2019;89:408−15.

[22] Cammarota G, Martino A, Pirozzi GA, Cianci R, Cremonini F, Zuccalà G, et al. Direct visualization of intestinal villi by high-resolution magnifying upper endoscopy: a validation study. Gastrointest Endosc 2004;60:732−8.

[23] Hurlstone DP, Sanders DS, McAlindon ME, Thomson M, Cross SS. High-magnification chromoscopic colonoscopy in ulcerative colitis: a valid tool for in vivo optical biopsy and assessment of disease extent. Endoscopy 2006;38:1213−17.

[24] Sasanuma S, Ohtsuka K, Kudo SE, Ogata N, Maeda Y, Misawa M, et al. Narrow band imaging efficiency in evaluation of mucosal healing/relapse of ulcerative colitis. Endosc Int Open 2018;6:E518−23.

[25] Hurlstone DP, Cross SS, Adam I, Shorthouse AJ, Brown S, Sanders DS, et al. A prospective clinicopathological and endoscopic evaluation of flat and depressed colorectal lesions in the United Kingdom. Am J Gastroenterol 2003;98:2543−9.

[26] Hurlstone DP, Brown S, Cross SS. The role of flat and depressed colorectal lesions in colorectal carcinogenesis: new insights from clinicopathological findings in high-magnification chromoscopic colonoscopy. Histopathology 2003;43:413−26.

[27] Hurlstone DP, Cross SS, Drew K, Adam I, Shorthouse AJ, Brown S, et al. An evaluation of colorectal endoscopic mucosal resection using high-magnification chromoscopic colonoscopy: a prospective study of 1000 colonoscopies. Endoscopy 2004;36:491−8.

[28] Sada M, Igarashi M, Yoshizawa S, Kobayashi K, Katsumata T, Saigenji K, et al. Dye spraying and magnifying endoscopy for dysplasia and cancer surveillance in ulcerative colitis. Dis Colon Rectum 2004;47:1816−23.

[29] Togashi K, Konishi F, Ishizuka T, Sato T, Senba S, Kanazawa K. Efficacy of magnifying endoscopy in the differential diagnosis of neoplastic and non-neoplastic polyps of the large bowel. Dis Colon Rectum 1999;42:1602−8.

[30] Hiyama S, Iijima H, Shinzaki S, Mukai A, Inoue T, Shiraishi E, et al. Narrow band imaging with magnifying endoscopy for Peyer's patches in patients with inflammatory bowel disease. Digestion 2013;87:269−80.

[31] Nishiyama S, Oka S, Tanaka S, Sagami S, Hayashi R, Ueno Y, et al. Clinical usefulness of narrow band imaging magnifying colonoscopy for assessing ulcerative colitis-associated cancer/dysplasia. Endosc Int Open 2016;4:E1183−7.

[32] Kawasaki K, Nakamura S, Esaki M, Kurahara K, Eizuka M, Nuki Y, et al. Clinical usefulness of magnifying colonoscopy for the diagnosis of ulcerative colitis-associated neoplasia. Dig Endosc 2019;31(Suppl. 1):36−42.

[33] Nishiyama S, Oka S, Tanaka S, Hayashi N, Hayashi R, Nagai K, et al. Is it possible to discriminate between neoplastic and nonneoplastic lesions in ulcerative colitis by magnifying colonoscopy? Inflamm Bowel Dis 2014;20:508−13.

[34] Iacucci M, Furfaro F, Matsumoto T, Uraoka T, Smith S, Ghosh S, et al. Advanced endoscopic techniques in the assessment of inflammatory bowel disease: new technology, new era. Gut 2019;61:562−72 https://doi.org/10.1136/gutjnl-2017-315235.

[35] Kohli DR, Schubert ML, Zfass AM, Shah TU. Performance characteristics of optical coherence tomography in assessment of Barrett's esophagus and esophageal cancer: systematic review. Dis Esophagus 2017;30:1−8.

[36] Konda VJ, Koons A, Siddiqui UD, Xiao SY, Turner JR, Waxman I. Optical biopsy approaches in Barrett's esophagus with next-generation optical coherence tomography. Gastrointest Endosc 2014;80:516−17.

[37] Consolo P, Strangio G, Luigiano C, Giacobbe G, Pallio S, Familiari L. Optical coherence tomography in inflammatory bowel disease: prospective evaluation of 35 patients. Dis Colon Rectum 2008;51:1374−80.

[38] Familiari L, Strangio G, Consolo P, Luigiano C, Bonica M, Barresi G, et al. Optical coherence tomography evaluation of ulcerative colitis: the patterns and the comparison with histology. Am J Gastroenterol 2006;101:2833−40.

[39] Shen B, Zuccaro G, Gramlich TL, Gladkova N, Lashner BA, Delaney CP, et al. Ex vivo histology-correlated optical coherence tomography in the detection of transmural inflammation in Crohn's disease. Clin Gastroenterol Hepatol 2004;2:754−60.

[40] Shen B, Zuccaro Jr G, Gramlich TL, Gladkova N, Trolli P, Kareta M, et al. In vivo colonoscopic optical coherence tomography for transmural inflammation in inflammatory bowel disease. Clin Gastroenterol Hepatol 2004;2:1080−7.

[41] Nair A, Liu CH, Singh M, Das S, Le T, Du Y, et al. Assessing colitis ex vivo using optical coherence elastography in a murine model. Quant Imaging Med Surg 2019;9:1429−40.

[42] Chen YJ, Mao R, Li XH, Cao QH, Chen ZH, Liu BX, et al. Real-time shear wave ultrasound elastography differentiates fibrotic from inflammatory strictures in patients with Crohn's disease. Inflamm Bowel Dis 2018;24:2183−90.

[43] Giannetti A, Matergi M, Biscontri M, Tedone F, Falconi L, Franci L. Real-time elastography in Crohn's disease: feasibility in daily clinical practice. J Ultrasound 2017;20:147−55.

Chapter 21

Molecular imaging in inflammatory bowel disease

Timo Rath, Markus F. Neurath and Raja Atreya
The Ludwig Demling Endoscopy Center of Excellence, Erlangen University Hospital, Erlangen, Germany

Chapter Outline

Abbreviations

CAC colitis-associated cancer
CD Crohn's disease
CLE confocal laser endomicroscopy
CT computed tomography
FITC fluorescein isothiocyanate
IBD inflammatory bowel disease
SSA sessile serrated adenoma
SPECT single-photon emission computed tomography
TNF tumor necrosis factor
VDZ vedolizumab

Introduction

Molecular imaging is based on the principle of utilizing fluorescently labeled probes with high specificity toward defined molecular targets and their subsequent detection and visualization with endoscopic devices (Table 21.2), thereby enabling visualization of single molecules or receptors [1]. Generally speaking, different type of probes can be utilized for molecular imaging such as antibodies, enzymes, affibodies, lectins, or peptides, all of which have certain advantages and disadvantages, as summarized in Table 21.1. The ideal probe for molecular imaging should exhibit high affinity toward the molecular target, rapid binding kinetics with sufficient tissue penetration, and at the same time low immunogenicity.

Confocal laser endomicroscopy

Confocal laser endomicroscopes (CLE) are among the most commonly used devices for visualization of the labeled molecular targets with microscopic resolution. CLE is based on the emission of low-power laser light into the tissue and the subsequent detection of the fluorescent signal, which is, prior to the recapture, directed through a pinhole, so that only light from the same focal plane is detected, thereby enabling high lateral resolution and microscopic imaging at 1000-fold magnification. Even without molecular probes, CLE has been shown to allow for precise in vivo

TABLE 21.1 Characteristics of various probes used for molecular imaging.

	Antibodies	Peptides	Enzymes	Affibodies	Lectins
Advantages	High specificity High affinity Clinically approved	High affinity Low toxicity Low costs Good tissue penetration	High specificity High stability High signal-to-background ratio	High specificity High affinity Small size	High specificity pH stability Low costs Low toxicity
Disadvantages	Immunogenicity High costs Low tissue penetration	Unknown binding site Formulation is complex	High costs Complex formulation	Complex formulation	Large size

Source: Adapted from Atreya R, Goetz M. Molecular imaging in gastroenterology. Nat Rev Gastroenterol Hepatol 2013;10:704–12; Sturm MB, Wang TD. Emerging optical methods for surveillance of Barrett's oesophagus. Gut 2015;64:1816–23 [11].

TABLE 21.2 Endoscopic devices used for molecular imaging.

Wide-field detection devices	Autofluorescence imaging	
	Multimodal colonoscope CF-Y0012 (Olympus)	Collection of wide-field images of white light, fluorescence, and reflectance
	Fiber-optic colonoscope FC-38 LV (Pentax)	Collection of wide-field images from probes labeled with Cy5
Narrow field of view devices	Probe-based CLE (Mauna Kea, France)	
	Endoscope-based CLE (Pentax Medical, discontinued)	
	Handheld confocal endomicroscopy probe (FIVE 1, Optiscan, Australia)	

CLE, confocal laser endomicroscopy.

assessment of the degree of histological inflammation in inflammatory bowel disease (IBD) and for discrimination between Crohn's disease (CD) and ulcerative colitis by visualization of microscopic features, which are conventionally used in standard histopathology to differentiate between the two disease entities. Furthermore, CLE is the only endoscopic device to date that is capable of visualizing disintegrity of the epithelial barrier in IBD patients, and this finding is a strong predictor for the occurrence of a flare within the next 12 months, as shown in several studies. A summary of CLE based classification systems for the diagnosis and prediction of relapse in IBD can be found in Table 21.3.

Through the integration of specific labeling of single subcellular targets and their endoscopic visualization with high-resolution imaging devices, molecular imaging has been successfully evaluated in mucosal inflammation and cancer development in both, mice and humans.

Molecular imaging for grading intestinal inflammation and prediction of response to medical therapy

In one of the first studies on mucosal imaging for grading intestinal inflammation, induction of intestinal cell apoptosis by the anti-tumor necrosis factor (TNF) antibody infliximab was visualized using single-photon emission computed tomography and ^{99m}Tc-annexin V as an apoptosis marker (Fig. 21.1). Interestingly, colonic uptake of ^{99m}Tc-annexin V significantly increased not only in two models of murine colitis upon administration of anti-TNF antibodies but also in patients with active CD responding to infliximab treatment [2]. Hence, this study not only successfully demonstrated molecular imaging of intestinal inflammation but also for the first time that the induction of T-cell apoptosis under anti-TNF treatment can be visualized and that the later correlates with short-term clinical response. A phase-II clinical trial

TABLE 21.3 CLE based classifications for diagnosis and prediction of relapse in IBD.

	Parameters on CLE	Diagnostic performance and clinical relevance
Chang-Qing Score [12]	A: No inflammation, *regular arrangement, and size of crypts* B: Chronic inflammation, *irregular crypt arrangement, and enlarged space between crypts* C: Acute inflammation, *enlarged crypt openings, crypt more enlarged, and irregularly arranged compared to B* D: Acute inflammation, *severe crypt destruction, crypt abscesses*	Sensitivity, specificity, and accuracy for prediction of relapse by CLE of 64%, 89%, and 74%
Watson Score [13]	I: Normal, *cell shedding confined to single cells per shedding site* II: Structural defect, *cell shedding confined to single cells per shedding Site + Fluorescein leak into the lumen*III: Functional defect, *microerosions with exposure of the lamina propria to the lumen in any field + Fluorescein leak into the lumen*	Sensitivity, specificity, and accuracy for the prediction of flare for Watson grades II/III of 62.5%, 91.2%, and 79%
CD Endomicroscopic Activity Score [14]	I: Quiescent disease II: Active disease *Distinction based on increase/decrease of crypt number/crypt lumen, tortuosity, microerosions, vascularity, goblet cells, cellular infiltrate*	
IBD differentiation based on Endomicroscopic Assessment Score [15]	Ulcerative colitis: *Severe, widespread crypt distortion, decreased crypt density, frankly irregular surface* Crohn's disease: *Discontinuous inflammation, focal cryptitis, discontinuous crypt architectural abnormality*	Diagnosis of Crohn's disease by CLE: sensitivity 90%, specificity 97.4%, positive predictive value 97.3%, negative predictive value 90.5% Diagnosis of ulcerative colitis by CLE: sensitivity 97.4%, specificity 90%, positive predictive value 90.5%, negative predictive value 97.3%

CD, Crohn's disease; *CLE*, confocal laser endomicroscopy; *IBD*, inflammatory bowel disease.

on 25 patients with CD made use of good manufacturing practice−conformed fluorescein isothiocyanate (FITC)-labeled adalimumab as a molecular probe to visualize membranous TNF (mTNF) during live endoscopy with CLE (Fig. 21.2). Importantly, this study proved not only that real-time molecular imaging of single mTNF$^+$ cells is safe and feasible, but also that the amount of mTNF$^+$ cells is a significant predictor of the success of subsequent anti-TNF therapy (Fig. 21.3) [3]. Hence, this landmark clinical trial was the first to establish an algorithm in which endoscopic molecular imaging of mTNF$^+$ cells can be utilized to make an a priori predication about the success of subsequent anti-TNF therapy and was therefore the first study to open the possibility of personalized medicine in IBD through molecular imaging. That the same principle might hold true also for other biological therapies, which was recently shown in a case series on patients with CD. Herein, molecular imaging of the integrin $\alpha4\beta7$ was achieved via labeling with FITC-labeled vedolizumab (VDZ) and its subsequent visualization with CLE. Further, while CD patients with sustained clinical and endoscopic remission toward subsequent VDZ therapy revealed pericryptal $\alpha4\beta7^+$ cells in the mucosa, no $\alpha4\beta7^+$ cells were observed in CD patients with CD not responding to VDZ [4] (Fig 21.4).

Molecular imaging in colitis-associated neoplasia and sporadic colonic cancer

First studies have also shown the potential of molecular imaging for visualization of the development of colitis-associated cancer (CAC). Using gGlu-HMRG as an enzymatically activatable probe which fluoresces in the presence of γ-glutamyltranspeptidase, an enzyme that is specifically associated with cancer, in a mouse model of CAC, fluorescence endoscopy allowed identification of cancerous or dysplastic lesions 5 minutes after application with clear discrimination against inflammatory changes and a higher target to background ratio than white-light endoscopy (Figs. 21.5 and 21.6)

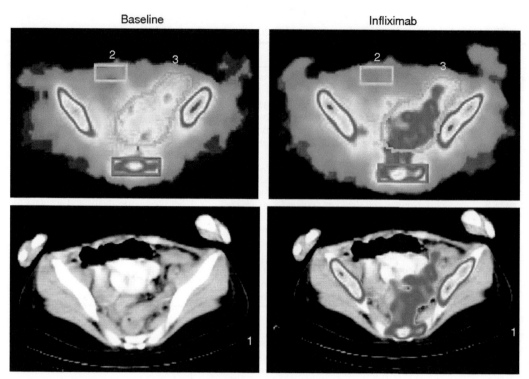

FIGURE 21.1 Molecular imaging of intestinal T-cell apoptosis with SPECT. ^{99m}Technetium-annexin as an apoptosis marker is significantly increased in patients with active Crohn's disease responding to infliximab treatment. Upper row: SPECT images in pseudocolors, Lower row: CT image with superimposed SPECT image. *CT*, Computed tomography; *SPECT*, single-photon emission computed tomography. *Reproduced from Van den Brande JM, Koehler TC, Zelinkova Z, et al. Prediction of antitumour necrosis factor clinical efficacy by real-time visualisation of apoptosis in patients with Crohn's disease. Gut 2007;56:509–17, with permission by BMJ publishing group.*

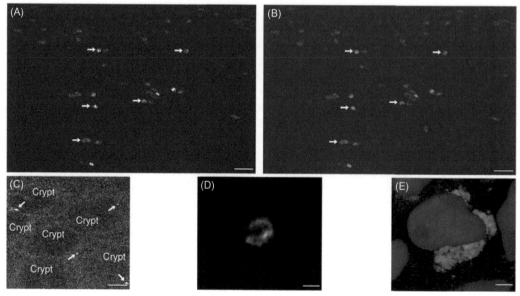

FIGURE 21.2 In vivo molecular imaging of intestinal mTNF$^+$ mucosal cells in Crohn's disease. Upon topical spraying of FITC-labeled adalimumab onto inflamed mucosa, single mTNF expressing (mTNF$^+$) cells can be visualized by CLE (*arrows*, A). Contrast enhancement of the image shown in A (*arrows*, B). Single pericryptal mTNF$^+$ cells (*arrows*, C) with a membranous fluorescence pattern (D) in digitally postprocessed and magnified confocal in vivo images (C and D). Immunohistochemical visualization of membranous expression of mTNF after staining with FITC-adalimumab (E). *CLE*, Confocal laser endomicroscopy; *FITC*, fluorescein isothiocyanate; *mTNF*, membranous tumor necrosis factor. *Reproduced from Atreya R, Neumann H, Neufert C, et al. In vivo imaging using fluorescent antibodies to tumor necrosis factor predicts therapeutic response in Crohn's disease. Nat Med 2014;20:313–8.*

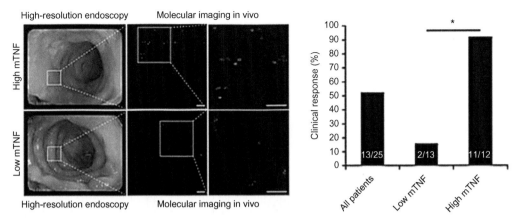

FIGURE 21.3 **Clinical significance of molecular imaging of mTNF expression in CD.** In a phase-II clinical trial on 25 patients, patients with CD exhibited either high (upper panel) or low numbers of mTNF$^+$ cells (lower panel) during molecular imaging, despite a similar endoscopic degree of inflammation (upper and lower left panel). Upon subsequent anti-TNF therapy with adalimumab, a significantly higher percentage of patients with high levels of mTNF$^+$ cells prior to therapy achieved clinical response compared to patients with low numbers of mTNF$^+$ expressing cells (right panel). *CD*, Crohn's disease; *mTNF*, membranous tumor necrosis factor. *Reproduced from Atreya R, Neumann H, Neufert C, et al. In vivo imaging using fluorescent antibodies to tumor necrosis factor predicts therapeutic response in Crohn's disease. Nat Med 2014;20:313–8.*

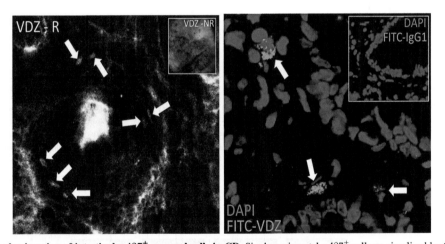

FIGURE 21.4 **Molecular imaging of intestinal α4β7$^+$ mucosal cells in CD.** Single pericryptal α4β7$^+$ cells as visualized by CLE using a handheld probe after topical administration of FITC−VDZ in a patient responding to subsequent VDZ therapy (VDZ-R, left picture). Immunohistochemistry of α4β7$^+$ expressing cells in the lamina propria of a CD patient responding to VDZ (right picture) (*white arrows*). By contrast, no α4β7$^+$ cells were observed during ex vivo CLE in CD patients not responding to VDZ (VDZ-NR, left picture). *CD*, Crohn's disease; *CLE*, confocal laser endomicroscopy; *FITC*, fluorescein isothiocyanate; *VDZ*, vedolizumab. *Reproduced from Rath T, Bojarski C, Neurath MF, et al. Molecular imaging of mucosal alpha4beta7 integrin expression with the fluorescent anti-adhesion antibody vedolizumab in Crohn's disease. Gastrointest Endosc 2017;86:406−8, with permission of Elsevier.*

[5]. Other murine studies have utilized cathepsin-activatable near-infrared fluorescent probes to visualize and discriminate dysplastic areas within the inflamed colon [6,7]. Clearly, these studies open up the possibility of identifying CAC with high sensitivity and specificity through molecular imaging and first in-human clinical trials are eagerly awaited (Figs. 21.8 and 21.9). It is indeed possible in humans in vivo has been shown for the sequence of sporadic colorectal carcinogenesis. Here, various studies have successfully utilized molecular imaging for the visualization of not only cancer cells and colorectal cancer but also for the in vivo detection of otherwise hardly visible polyps or sessile serrated adenomas using specific fluorescently labeled peptides topically [8,9] or systemically (Fig. 21.7) [10]. Hence, these results indicate that molecular imaging holds the potential to facilitate or even enable the demarcation of dysplastic or cancerous tissue especially under circumstances in which conventional endoscopy is currently limited such as accompanying inflammation, lesions at difficult anatomical positions or flat and subtle lesions.

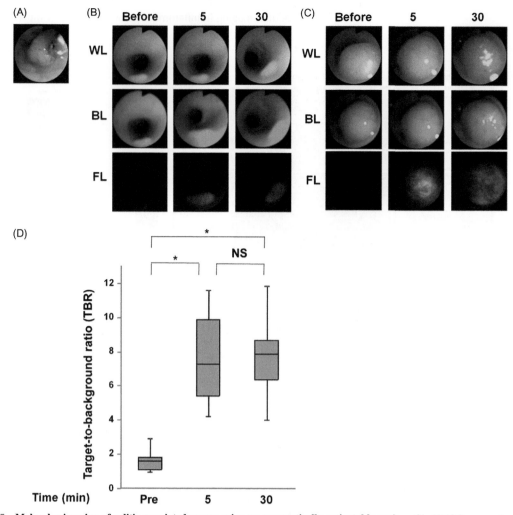

FIGURE 21.5 **Molecular imaging of colitis-associated cancer using an enzymatically activatable probe.** gGlu-HMRG as a probe that fluoresces in the presence of the cancer-associated enzyme GGT can visualize tumors in a model of colitis-associated cancer 5 and 30 min after topical administration. *BL*, blue light; *FL*, fluorescence endoscopy; *GGT*, γ-glutamyltranspeptidase; *WL*, white-light endoscopy. *Reproduced by Mitsunaga M, Kosaka N, Choyke PL, et al. Fluorescence endoscopic detection of murine colitis-associated colon cancer by topically applied enzymatically rapid-activatable probe. Gut 2013;62:1179—86, with permission of BMJ publishing group.*

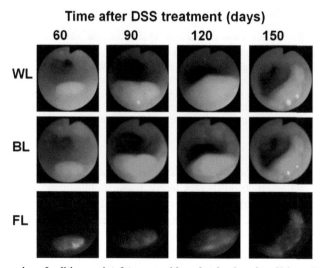

FIGURE 21.6 **Monitoring of progression of colitis-associated tumors with molecular imaging.** Using gGlu-HMRG as a fluorescent probe it is possible to observe tumor progression over time by repeated fluorescence colonoscopy. *BL*, Blue excitation light; *FL*, fluorescence endoscopy *WL*, white-light endoscopy. *Reproduced by Mitsunaga M, Kosaka N, Choyke PL, et al. Fluorescence endoscopic detection of murine colitis-associated colon cancer by topically applied enzymatically rapid-activatable probe. Gut 2013;62:1179—86, with permission of BMJ publishing group.*

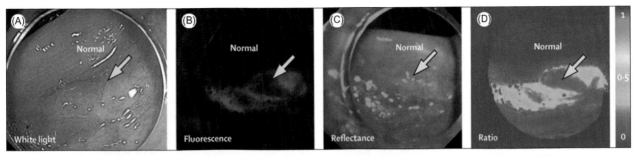

FIGURE 21.7 **Detection of SSA using a FITC-labeled peptide.** (A) On white-light endoscopy, the SSA is barely visible. (B) Fluorescence imaging after staining with a FITC-labeled peptide shows an increased intensity and high contrast from the SSA (*arrow*). (C) Reflectance and coregistered fluorescence images are combined in a ratio to correct for differences in a distance over the image field of view. (D) The ratio image shows increased signal derived from the SSA (*arrow*) with clear delineation against a background (*red line*). *FITC*, Fluorescein isothiocyanate; *SSA*, sessile serrated adenomas. *Reproduced from Joshi BP, Dai Z, Gao Z, et al. Detection of sessile serrated adenomas in the proximal colon using wide-field fluorescence endoscopy. Gastroenterology 2017;152:1002−13 and Lee JH, Wang TD. Molecular endoscopy for targeted imaging in the digestive tract. Lancet Gastroenterol Hepatol 2016;1:147−55 [16], with permission of Elsevier.*

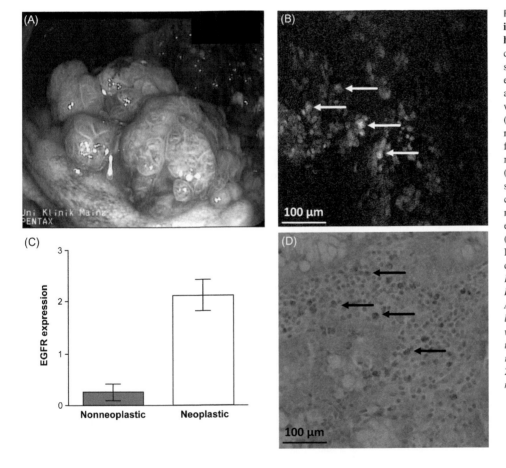

FIGURE 21.8 **Molecular imaging of colorectal cancer in humans.** (A and B) Individual cells from a neoplastic lesion as shown in (A) exhibit EGFR expression as visualized by CLE after incubation of intestinal tissue with FITC-labeled anti-EGFR (B) (*white arrows*). Mean specific fluorescence of human tissue samples for the normal non-neoplastic mucosa and neoplastic mucosa ($P < 0.002$, C). Bars indicate standard errors. Histopathology confirmed malignancy, and immunohistochemistry showed EGFR expression in multiple tumor cells (*black arrows*, D). *CLE*, Confocal laser endomicroscopy; *EGFR*, epidermal growth factor receptor; *FITC*, fluorescein isothiocyanate. *Reproduced from Goetz M, Ziebart A, Foersch S, et al. In vivo molecular imaging of colorectal cancer with confocal endomicroscopy by targeting epidermal growth factor receptor. Gastroenterology 2010;138:435−46 [17], with permission of Elsevier.*

Summary and recommendations

Molecular imaging is a rapidly evolving field that makes visualization of single receptors or molecules possible. It detects molecular changes rather than relying on morphologic appearances. This approach can visualize otherwise subtle or even invisible lesions. It may have a role in the prediction of response of biological therapies used in IBD as well as in cancer. Molecular imaging may find its role in the accurate detection of colitis-associated neoplasia. We hope the approach can be incorporated into clinical routine.

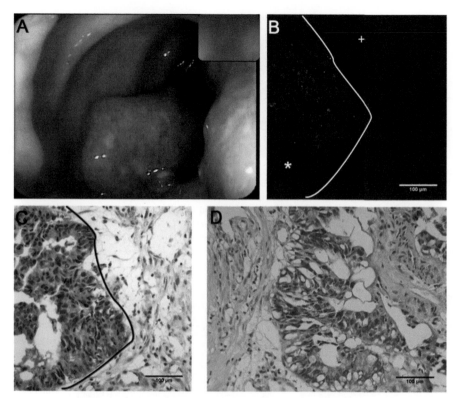

FIGURE 21.9 **In vivo molecular imaging of colorectal cancer.** By staining EGFR with an Alex Fluor 488–labeled anti-EGFR antibody and subsequent visualization with CLE, colorectal cancer as well the transition zone between cancer (*asterisk*) and normal tissue (*cross*) can be clearly visualized (A and B). Transition zone on EGFR immunohistochemistry (C) and in hematoxylin and eosin staining (D). *CLE*, Confocal laser endomicroscopy; *EGFR*, epidermal growth factor receptor. *Reproduced from Liu J, Zuo X, Li C, et al. In vivo molecular imaging of epidermal growth factor receptor in patients with colorectal neoplasia using confocal laser endomicroscopy. Cancer Lett 2013;330:200–7 [18], with permission of Elsevier.*

References

[1]   Atreya R, Goetz M. Molecular imaging in gastroenterology. Nat Rev Gastroenterol Hepatol 2013;10:704–12.

[2]   Van den Brande JM, Koehler TC, Zelinkova, Z, Bennink RJ, tel Vende AA, te Cate FJW. Prediction of antitumour necrosis factor clinical efficacy by real-time visualisation of apoptosis in patients with Crohn's disease. Gut 2007;56:509–17.

[3]   Atreya R, Neumann H, Neufert C, Walner MJ, Billmeier U, Zopf Y. In vivo imaging using fluorescent antibodies to tumor necrosis factor predicts therapeutic response in Crohn's disease. Nat Med 2014;20:313–18.

[4]   Rath T, Bojarski C, Neurath MF, Atreya R. Molecular imaging of mucosal alpha4beta7 integrin expression with the fluorescent anti-adhesion antibody vedolizumab in Crohn's disease. Gastrointest Endosc 2017;86:406–8.

[5]   Mitsunaga M, Kosaka N, Choyke PL, Young MR, Dextras CR, Saud SM. Fluorescence endoscopic detection of murine colitis-associated colon cancer by topically applied enzymatically rapid-activatable probe. Gut 2013;62:1179–86.

[6]   Ding S, Blue RE, Moorefield E, Yuan H, Lund PK. Ex vivo and in vivo noninvasive imaging of epidermal growth factor receptor inhibition on colon tumorigenesis using activatable near-infrared fluorescent probes. Mol Imaging 2017;16 1536012117729044.

[7]   Gounaris E, Martin J, Ishihara Y, Khan MM, Lee G, Sinh P. Fluorescence endoscopy of cathepsin activity discriminates dysplasia from colitis. Inflamm Bowel Dis 2013;19:1339–45.

[8]   Joshi BP, Dai Z, Gao Z, Lee JH, Ghimire N, Chen J. Detection of sessile serrated adenomas in the proximal colon using wide-field fluorescence endoscopy. Gastroenterology 2017;152:1002–13.

[9]   Joshi BP, Pant A, Duan X, Prahbu A, Wamsteker EJ, Kwon RS. Multimodal video colonoscope for targeted wide-field detection of nonpolypoid colorectal neoplasia. Gastroenterology 2016;150:1084–6.

[10] Burggraaf J, Kamerling IM, Gordon PB, Schrier L L, de Kam ML, Kales AJ. Detection of colorectal polyps in humans using an intravenously administered fluorescent peptide targeted against c-Met. Nat Med 2015;21:955–61.

[11] Sturm MB, Wang TD. Emerging optical methods for surveillance of Barrett's oesophagus. Gut 2015;64:1816–23.

[12] Li C-Q, Liu J, Ji R, Li Z, Xie XJ, Li YQ. Use of confocal laser endomicroscopy to predict relapse of ulcerative colitis. BMC Gastroenterol 2014;14:45.

[13] Kiesslich R, Duckworth CA, Moussata D, et al. Local barrier dysfunction identified by confocal laser endomicroscopy predicts relapse in inflammatory bowel disease. Gut 2012;61:1146−53.

[14] Neumann H, Vieth M, Atreya R, Grauer M, Seibler J, Bernatik T. Assessment of Crohn's disease activity by confocal laser endomicroscopy. Inflamm Bowel Dis 2012;18:2261−9.

[15] Tontini GE, Mudter J, Vieth M, Atreya R, Günther C, Zopf Y. Confocal laser endomicroscopy for the differential diagnosis of ulcerative colitis and Crohn's disease: a pilot study. Endoscopy 2015;47:437−43.

[16] Lee JH, Wang TD. Molecular endoscopy for targeted imaging in the digestive tract. Lancet Gastroenterol Hepatol 2016;1:147−55.

[17] Goetz M, Ziebart A, Foersch S, Vieth M, Waldner MJ, Delaney P. In vivo molecular imaging of colorectal cancer with confocal endomicroscopy by targeting epidermal growth factor receptor. Gastroenterology 2010;138:435−46.

[18] Liu J, Zuo X, Li C, Yu T, Gu X, Zhou C. In vivo molecular imaging of epidermal growth factor receptor in patients with colorectal neoplasia using confocal laser endomicroscopy. Cancer Lett 2013;330:200−7.

Chapter 22

Inflammatory bowel disease−like conditions: radiation injury of the gut

Bo Shen

Center for Inflammatory Bowel Diseases, Columbia University Irving Medical Center-New York Presbyterian Hospital, New York, NY, United States

Chapter Outline

Abbreviations

AVM arteriovenous malformation
GI gastrointestinal
IBD inflammatory bowel disease
UC ulcerative colitis

Introduction

Gastrointestinal (GI) tract is vulnerable to radiation injury, consisting acute and chronic forms. The acute injury exerts cytotoxic effect on the fast dividing intestinal epithelium, particularly intestinal stem cells, with subsequent changes in cell death. Downregulation of Bcl-2 plays a central role. The disruption of intestinal epithelial barrier may lead to bacterial translocation [1,2]. Acute radiation injury is usually reversible [3]. Chronic radiation injury is characterized by epithelial atrophy, fibrosis, and ischemia resulting from obliterative arteritis and stem cell death [4,5]. The latter one is more common than the previous.

Risk factors consist of the elements related to patients (such as previous abdominal or pelvic surgery, low body weight, advanced age, female gender, tobacco use, diabetes, and cardiovascular disease) and therapies (such as dosage and interval). Since the majority of patients with chronic disease and a portion of ulcerative colitis (UC) have to undergo various operations in the lifetime, surgery-associated adhesion may result in fixed bowel loops, posing a risk for radiation injury. The group of authors reported that prior pelvic radiation for colitis-associated cancer increased the risk for pouch failure after restorative proctocolectomy [6].

Common histologic features in acute and chronic radiation injury include epithelial injury, crypt distortion, Paneth cell metaplasia, intramucosal hemorrhage, dilated, tortuous capillaries, hyalinization, and fibrosis of the lamina propria and submucosal area, and microthrombi (Fig. 22.1) [7]. The injury can be superficial or transmural, resulting in specific inflammation and fibrosis. Radiation enteritis, colitis, or proctitis can mimic inflammatory bowel disease (IBD) in clinical, endoscopic, and radiographic presentations [6].

Clinical history makes differential diagnosis between radiation injury to the GI tract and IBD straightforward, despite having shared endoscopic features between the disease entities. However, clinical scenarios of IBD, IBD-associated

Atlas of Endoscopy Imaging in Inflammatory Bowel Disease. DOI: https://doi.org/10.1016/B978-0-12-814811-2.00022-0
© 2020 Elsevier Inc. All rights reserved.

cancer, IBD patients undergoing radiation therapy for non-GI malignancy, and IBD versus radiation—associated anastomotic lesions, may overlap. Although the injury from external beam radiation accounts for the majority of patients, brachytherapy can also cause radiation injury to the bowel [8—11].

Radiation enteritis

Endoscopic features of radiation enteritis include mucosal pallor, telangiectasia, friability, spontaneous bleeding, and superficial or deep ulcers (Figs. 22.2—22.4, 22.5A, and 22.6A). Telangiectasia, similar to arteriovenous malformation (AVM), petechiae can be focal, patchy, and diffuse with continuous or segmental distribution pattern (Fig. 22.6A). The ulcers can be multiple, circumferential, or longitudinal. Patient with chronic radiation enteritis may develop "lead pipe" of the bowel lumen (Fig. 22.6D), inflammatory polyps, masses, strictures (Fig. 22.4B), and fistulas. Patients with small bowel resection and anastomosis who undergoing radiation therapy may carry a high risk for anastomotic strictures.

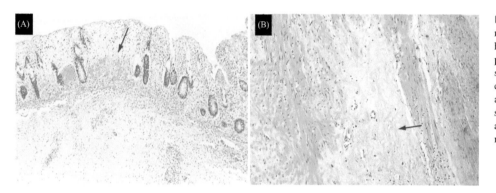

FIGURE 22.1 Acute and chronic radiation injury to the rectum on histology. (A) Edema of the lamina propria and submucosa; intramucosal hemorrhage (*arrow*), drop of crypts, and Paneth cell metaplasia and (B) dilated, tortuous submucosal capillaries with hyalinization and fibrosis (*arrow*), and microthrombi.

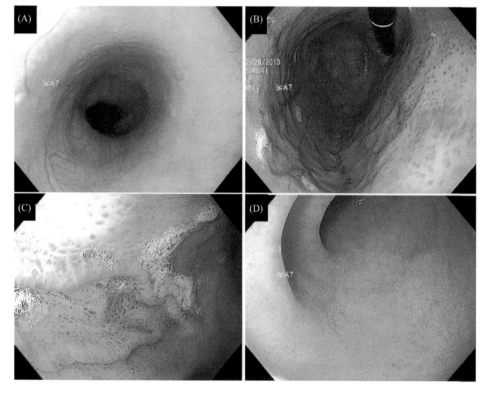

FIGURE 22.2 Radiation injury to the upper GI tract. (A) Normal esophagus. Squamous epithelia of the esophagus is more resistant to radiation injury than the columnar epithelia in other parts of GI tract; (B and C) radiation diffuse petechiae in the entire stomach; and (D) duodenopathy from radiation with flat mucosa and "lead pipe" appearance of the lumen. *GI*, Gastrointestinal.

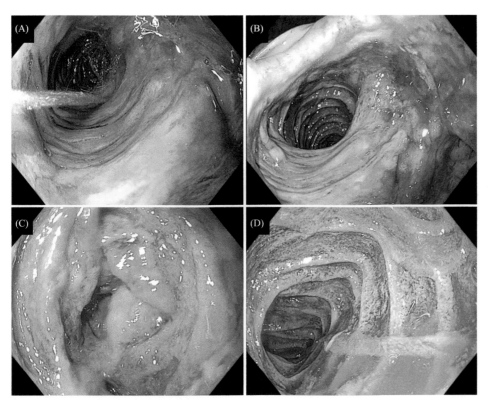

FIGURE 22.3 Acute radiation enteritis in a patient with Whipple procedure for pancreas cancer. (A−D) Diffuse mucosaa edema, erythema, and exudates of the duodenum and jejunum.

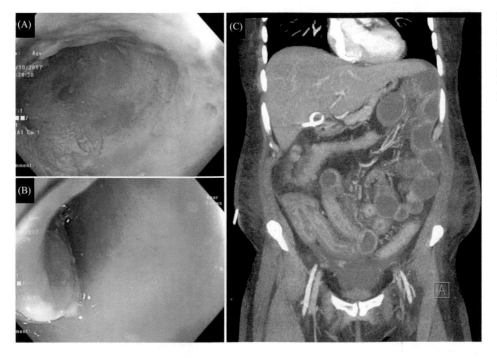

FIGURE 22.4 Radiation enteritis. (A) Erythematous mucosa with spontaneous bleeding, (B) luminal stricture with an inflammatory polyp, and (C) diffuse small bowel wall thickening with segmental luminal dilations.

Histologic findings of chronic radiation enteritis include drop of crypts, diffuse collagen deposition with mucosal and serosal thickening, acute and chronic inflammatory cell infiltration, vascular sclerosis, and obliterative vasculitis (Fig. 22.1). The physiological consequences can be altered intestinal transit, reduced bile acid absorption, increased intestinal permeability, small intestine bacterial overgrowth, and lactose malabsorption.

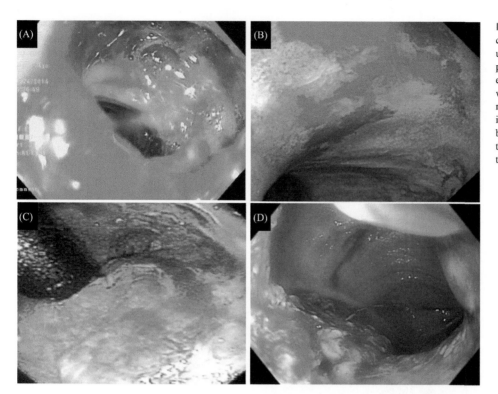

FIGURE 22.5 Radiation enteritis, colitis, proctitis, and pouchitis with ulcers and bleeding in different patients. (A) Segment of radiation enteritis in the proximal jejunum with profuse bleeding, (B) spontaneous bleeding with friable mucosa in the colon, (C) inflammatory bleeding lesions in the distal rectum; and (D) bleeding lesions in the distal ileal pouch.

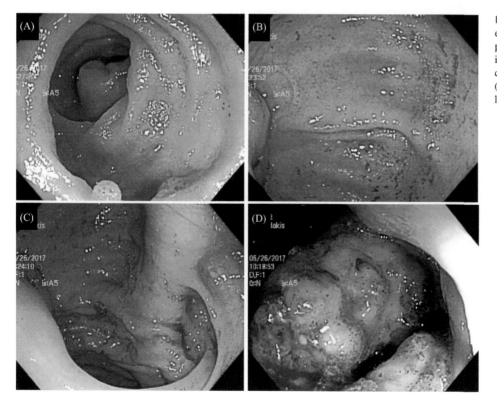

FIGURE 22.6 Range of radiation enteritis and colitis in the same patient. (A) Petechiae-type lesions in the distal ileum, (B and C) petechiae type lesions in the colon, and (D) deep ulcerated necrotic mass lesion in the distal rectum.

Capsule endoscopy may have a role in the diagnosis of radiation enteritis [12,13]. However, radiation enteritis is often associated with intestinal stricture, which put the patient at risk for capsule retention. A patency capsule should be used before video capsule endoscopy. Push enteroscopy [14] or device-assisted enteroscopy may be used for observation, tissue biopsy, or possible therapy (such as stricture dilation).

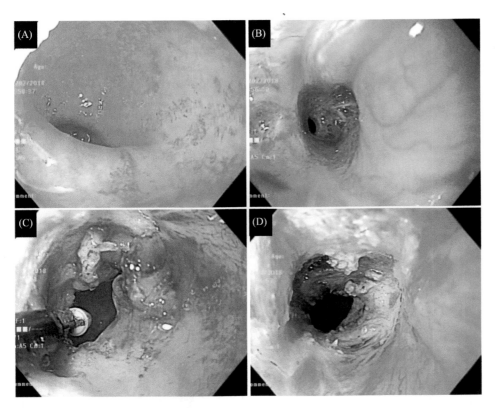

FIGURE 22.7 Radiation proctitis and anorectal stricture. (A) Telangiectasia of the rectal mucosa. The patient presented with anemia, without visible hematochezia and (B−D) tight anorectal stricture, which was treated with insulated-tip endoscopic stricturotomy in a separate patient.

Radiation colitis and proctitis

Radiation injury to the lower GI tract results from the therapy for cancers of the anorectum, anus, cervix, uterus, prostate, testes, and urinary bladder, with the rectum and sigmoid colon being mostly affected. Due to the lack of predominant neutrophil or mononuclear cell infiltration, radiation proctitis has been termed radiation proctopathy. Chronic radiation proctitis presents five phenotypes: (1) inflammation-predominant form (e.g., edema, mucosal pallor and ulcer) (Fig. 22.5D), (2) bleeding-predominant form (e.g., friability, spontaneous bleeding, and telangiectasia or AVM−like lesions) (Figs. 22.5B and C, 22.6B and C, 22.7A, and 22.8), (3) mixed form (having both bleeding-predominant and inflammation-predominant forms' features) (Figs. 22.9−22.11), (4) stricturing form (Figs. 22.7B−D, 22.12, and 22.13A and B), and (5) penetrating form with either sinus (Fig. 22.13C and D) or fistula (Figs. 22.14 and 22.15). Understanding and recognition of these phenotypes are important for the management and differential diagnosis between sole radiation injury and radiation injury to the bowel in patients with underlying IBD (see next).

The endoscopic changes are usually continuous in distribution but can be patchy in intensity [15]. The mucosa of the large bowel often is friable even with spontaneous bleeding. Large and deep ulcers may exist. In severe cases, radiation can result in strictures and fistulas.

Radiation gastrointestinal injury in inflammatory bowel disease-associated cancer

Chronic IBD can lead to colitis-associated neoplasia. Some of the colitis associated may be treated with radiation therapy with or without surgery. In addition, IBD patients on long-term purine analog therapy may have an increased risk for lymphoma, for which radiation therapy is a part of multidisciplinary treatment. With radiation therapy, IBD patients with rectal cancer carried a higher risk for small bowel obstruction than non-IBD patients with rectal cancer, although long-term toxicities in the two groups are comparable [16].

UC patients with prior radiation therapy for colitis-associated cancer were found to have an increased risk for pouch failure, should they undergo restorative proctocolectomy and ileal pouch−anal anastomosis (IPAA) [6]. It is not clear whether the development of radiation-associated pouchitis or cuffitis contributes to pouch failure.

Radiation injuries to the gut in patients with IBD-associated cancer also include strictures or fistula. Radiation injury can lead it or contribute to anastomotic strictures, anastomotic fistulas, or anastomotic sinuses. Strictures (Figs. 22.7B−D, 22.12, and 22.13A and B) fistulas (Figs. 22.14 and 22.15), and anastomotic sinuses (Fig. 22.13C and

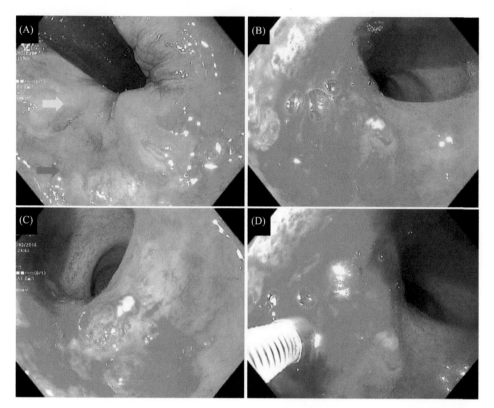

FIGURE 22.8 Radiation proctitis with predominantly bleeding in a patient with prostate cancer. (A) Retroflex view demonstrating telangiectasia of the rectal columnar mucosa (*green arrow*) and sparing squamous epithelia in the anal canal (*yellow arrow*); (B and C) telangiectasia, friability, and spontaneous bleeding of the rectum; and (D) endoscopic spray of 50% glucose to treat bleeding.

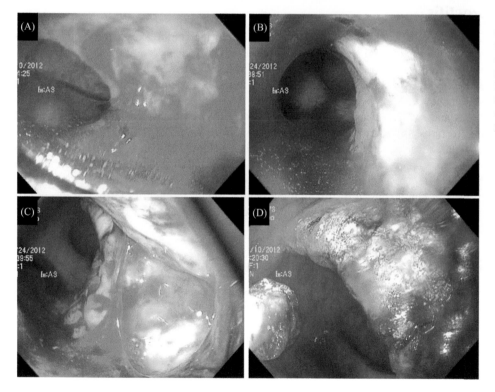

FIGURE 22.9 Mixed type radiation proctitis with both inflammation and telangiectasia bleeding. (A) Diffuse bleeding treated with spray of 50% glucose, (B and C) large ulcer at the anterior wall of the rectum with active bleeding, and (D) treatment of with argon plasma coagulation.

D) can occur in any noncancer, nonradiation-exposed IBD patients. Despite the fact that the distinction between radiation-induced versus surgical ischemia-associated anastomotic strictures, fistulas, and sinuses can be difficult, radiation injuries are characterized by friability and spontaneous bleeding of mucosal or ulcer surface.

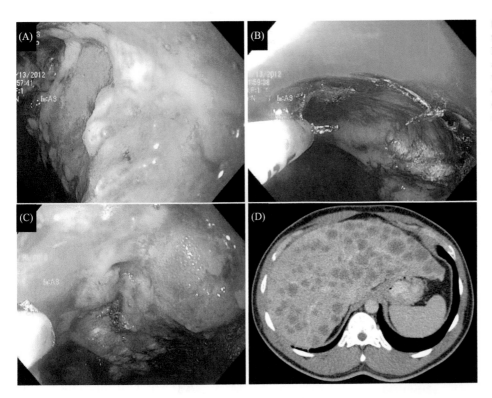

FIGURE 22.10 Radiation proctitis in a patient prostate cancer with large ulcers. (A) Large ulcerated area at the anterior wall of the rectum, (B and C) endoscopic therapy of bleeding with argon plasma coagulation, and (D) liver metastasis of prostate cancer of the patient.

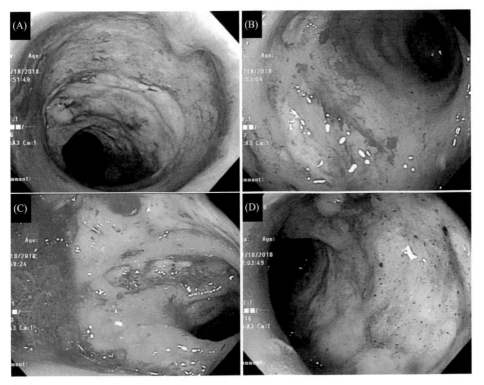

FIGURE 22.11 Patterns of severe inflammatory/bleeding mixed type of radiation sigmoiditis and proctitis with various shapes of ulcers and bleeding. (A) Large circumferential ulcers in the sigmoid colon, (B) linear ulcers arranged in a circumferential fashion of the sigmoid colon, (C) longitudinal ulcers in the sigmoid colon, and (D) large ulcer at the anterior wall of the distal rectum.

Inflammatory bowel disease patients undergoing radiation for extraintestinal cancer

In patients with concurrent IBD and prostate cancer, radiation therapy did not have an adverse impact on the disease course of IBD, as compared with those without radiation therapy [17]. A less number of patients with IBD and prostate

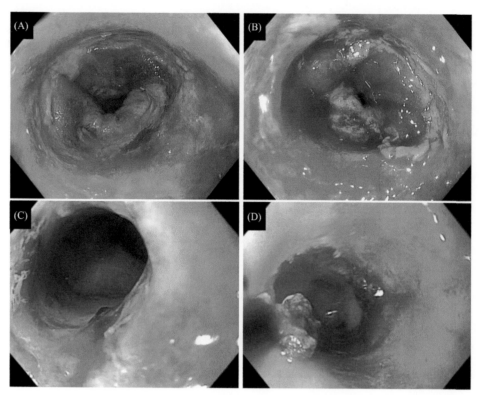

FIGURE 22.12 Radiation-induced colorectal anastomosis strictures in two patients with a history of rectal cancer. (A and B) Inflammatory stricture with friable colon mucosa at both sides of anastomosis and (C and D) anastomotic stricture treated with endoscopic stricturotomy.

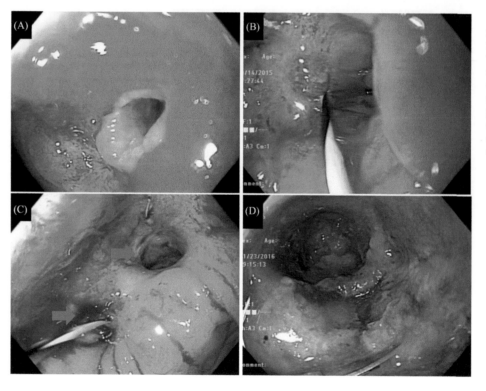

FIGURE 22.13 Radiation injury in patients with the bowel anastomosis after bowel resection. (A) Tight, ulcerated radiation-induced colorectal anastomosis stricture; (B) tight inflammatory stricture at the colorectal anastomosis; (C) two presacral sinuses from colorectal anastomosis leak (*green arrows*); and (D) large sinus cavity from colorectal anastomosis leak.

cancer or cervical/urine cancer developed exacerbation of diarrhea or toxicity after radiation therapy [18,19]. It is clear however that post-IPAA external beam radiation can cause pouchitis or cuffitis [20,21]. Radiation-associated pouchitis often presents with telangiectasia lesions and mucosal edema (Fig. 22.16). The similar lesions can be seen in the rectal

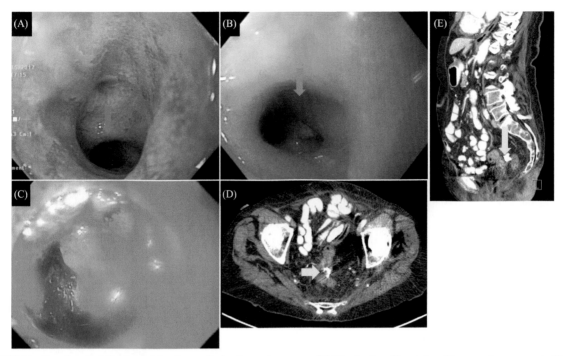

FIGURE 22.14   Rectal vaginal fistula resulting from radiation for uterine cancer. (A) The fistula orifice at the mid rectum (*green arrow*); (B) the fistula opening at the proximal vagina (*red arrow*); (C) the fistula was treated with over-the-scope-clip at the rectum side; and (D and E) the metal clip on computed tomography (*yellow arrow*).

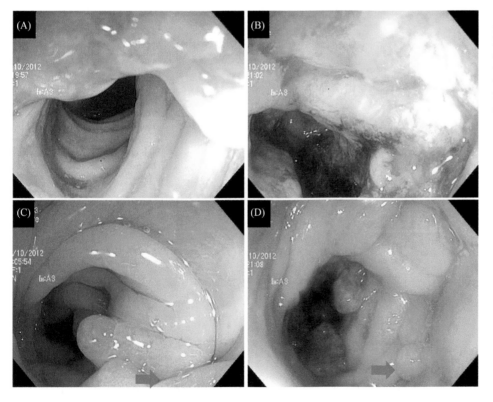

FIGURE 22.15 Radiation-induced colon—bladder fistula. (A) Inflammatory polyp at the sigmoid colon, (B) large ulcerated area at the anterior wall of the distal rectum, and (C and D) colon—bladder fistula with an opening at the distal sigmoid colon (*green arrows*).

cuff, that is, radiation cuffitis (Fig. 22.17). Therefore external pelvic radiation therapy may be avoided or administered at a reduced total dose, if oncological benefits of the do not exist. Of note, brachytherapy may also lead to injury to the adjacent ileal pouch or rectum (Fig. 22.16) [9−12].

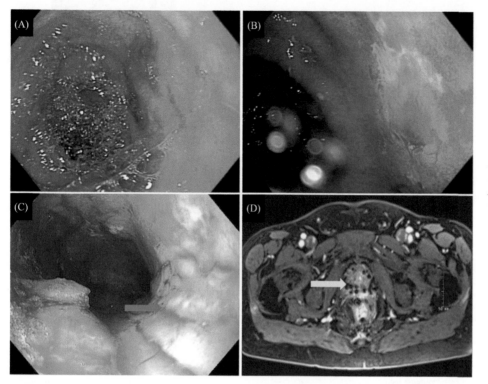

FIGURE 22.16 Radiation injury to the ileal pouch. (A and B) Distal pouchitis from external beam radiation with ulcers, friability, and hypervascularity in the distal pouch (B). Proximal pouch was normal (A); (C and D) anterior wall pouch inflammation from brachytherapy (*green arrow*). Radiation beads in the adjacent prostate (*yellow arrow*). *Photo courtesy Shen B. Male issues of the ileal pouch. In: Shen B, editor. Pouchitis and ileal pouch disorders. Elsevier; 2019. p. 513−4.*

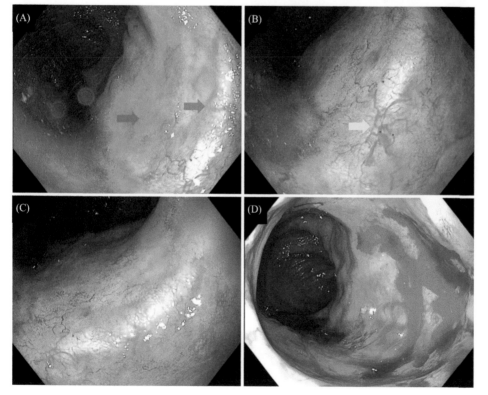

FIGURE 22.17 Radiation distal pouchitis and cuffitis in ileal pouch−anal anastomosis for ulcerative colitis. (A−C) Telangiectasia in the distal pouch body (*green arrow*) and cuff (*yellow arrow*), the pouch−anal anastomosis dividing the pouch body and cuff is highlighted with a red arrow and (D) excessive bleeding after tissue biopsy.

TABLE 22.1 Comparison of endoscopic features between radiation-induced gut injury and inflammatory bowel disease.

Characteristics	Radiation enteritis/colitis/proctitis	Crohn's disease	Ulcerative colitis
Shared endoscopic features	Edema, erythema, friability, ulcer, granularity, exudates, strictures	Edema, erythema, friability, ulcer, granularity, exudates, strictures	Edema, erythema, friability, ulcer, granularity, exudates, strictures
Disease-"specific" endoscopic features	Small or large bowel "bleeding" strictures, spontaneous bleeding with or without superficial ulceration, telangiectasia-type lesions; large carpet-like ulcers or necrotic mass lesions in severe cases	Linear longitudinal ulcers, skip lesions, fistula-associated abscess, perianal lesions	Sharp demarcation of diseased and nondiseased segments of distal large bowel; inflammatory polyps

Summary and recommendations

Both external beam and brachytherapy can induce radiation injury to the GI tract. The pattern of injury ranges from inflammation, bleeding lesions to strictures and fistulas. The GI radiation injury can mimic that of IBD in endoscopy and histology. More importantly, the radiation to IBD-associated colorectal cancer or extraintestinal cancer may impact the disease course, diagnosis, and differential diagnosis of IBD in those with or without bowel surgery. Therefore differential diagnosis should cover several aspects, including distinctions among radiation-induced bowel injury, IBD, radiation for IBD-associated colorectal cancer, and radiation for extraintestinal cancer in IBD patients. Endoscopy along with histology provides valuable information in the diagnosis and differential diagnosis (Table 22.1).

References

[1] Langley RE, Bump EA, Quartuccio SG, Medeiros D, Braunhut SJ. Radiation-induced apoptosis in microvascular endothelial cells. Br J Cancer 1997;75:666—72.

[2] Guzman-Stein G, Bonsack M, Liberty J, Delaney JP. Abdominal radiation causes bacterial translocation. J Surg Res 1989;46:104—7.

[3] Schultheiss TE, Lee WR, Hunt MA, Hanlon AL, Peter RS, Hanks GE. Late GI and GU complications in the treatment of prostate cancer. Int J Radiat Oncol Biol Phys 1997;37:3—11.

[4] Denham JW, Hauer-Jensen M. The radiotherapeutic injury—a complex 'wound'. Radiother Oncol 2002;63:129—45.

[5] Denton A, Forbes A, Andreyev J, Maher EJ. Non surgical interventions for late radiation proctitis in patients who have received radical radiotherapy to the pelvis. Cochrane Database Syst Rev 2002;(1):CD003455.

[6] Wu XR, Kiran RP, Remzi FH, Katz S, Mukewar S, Shen B. Preoperative pelvic radiation increases the risk for ileal pouch failure in patients with colitis-associated colorectal cancer. J Crohns Colitis 2013;7:e419—26.

[7] Wu XR, Liu XL, Katz S, Shen B. Pathogenesis, diagnosis, and management of ulcerative proctitis, chronic radiation proctopathy, and diversion proctitis. Inflamm Bowel Dis 2015;21:703—15.

[8] Zeitlin SI, Sherman J, Raboy A, Lederman G, Albert P. High dose combination radiotherapy for the treatment of localized prostate cancer. J Urol 1998;160:91—5.

[9] Sudha SP, Kadambari D. Efficacy and safety of argon plasma coagulation in the management of extensive chronic radiation proctitis after pelvic radiotherapy for cervical carcinoma. Int J Colorectal Dis 2017;32:1285—8.

[10] Kishan AU, Kupelian PA. Late rectal toxicity after low-dose-rate brachytherapy: incidence, predictors, and management of side effects. Brachytherapy 2015;14:148—59.

[11] Cherian S, Kittel JA, Reddy CA, Kolar MD, Ulchaker J, Angermeier K, et al. Safety and efficacy of iodine-125 permanent prostate brachytherapy in patients with J-pouch anastomosis after total colectomy for ulcerative colitis. Pract Radiat Oncol 2015;5:e437—42.

[12] Kim HM, Kim YJ, Kim HJ, Park SW, Bang S, Song SY. A pilot study of capsule endoscopy for the diagnosis of radiation enteritis. Hepato-gastroenterology 2011;58:459—64.

[13] Kopelman Y, Groissman G, Fireman Z. Radiation enteritis diagnosed by capsule endoscopy. Gastrointest Endosc 2007;66:599 discussion.

[14] Tian C, Mehta P, Shen B. Endoscopic therapy of bleeding from radiation enteritis with hypertonic glucose spray. ACG Case Rep J 2014;1:181.

[15] O'Brien PC, Hamilton CS, Denham JW, Gourlay R, Franklin CI. Spontaneous improvement in late rectal mucosal changes after radiotherapy for prostate cancer. Int J Radiat Oncol Biol Phys 2004;58:75—80.

[16] Mudgway R, Bryant AK, Heide ES, Riviere P, O'Hare C, Rose BS, et al. A matched case-control analysis of clinical outcomes for inflammatory bowel disease patients with rectal cancer treated with pelvic radiation therapy. Int J Radiat Oncol Biol Phys 2019; [Epub ahead of print].

[17] Feagins LA, Kim J, Chandrakumaran A, Gandle C, Naik KH, Cipher DJ, et al. Rates of adverse IBD-related outcomes for patients with IBD and concomitant prostate cancer treated with radiation therapy. Inflamm Bowel Dis 2019; [Epub ahead of print].

[18] Gestaut MM, Swanson GP. Long term clinical toxicity of radiation therapy in prostate cancer patients with Inflammatory Bowel Disease. Rep Pract Oncol Radiother 2017;22:77–82.

[19] White EC, Murphy JD, Chang DT, Koong AC. Low toxicity in inflammatory bowel disease patients treated with abdominal and pelvic radiation therapy. Am J Clin Oncol 2015;38:564–9.

[20] Kulkarni G, Liu X, Shen B. Pouchitis associated with pelvic radiation for prostate cancer. ACG Case Rep J 2016;3:e129.

[21] Kani HT, Shen B. Male issues of the ileal pouch. Inflamm Bowel Dis 2015;21:716–22.

Chapter 23

Superimposed infections in inflammatory bowel diseases

Geeta Kulkarni[1] and Bo Shen[2]

[1]Center for Inflammatory Bowel Diseases, Digestive Disease and Surgery Institute, Cleveland Clinic, Cleveland, OH, United States, [2]Center for Inflammatory Bowel Diseases, Columbia University Irving Medical Center-New York Presbyterian Hospital, New York, NY, United States

Chapter Outline

Abbreviations

CD Crohn's disease
CDI *Clostridioides difficile* infection
CJI *Campylobacter jejuni* infection
CMV cytomegalovirus
EBV Epstein−Barr virus
H & E hematoxylin and eosin
HPV human papillomavirus
HSV herpes simplex virus
GI gastrointestinal
IBD inflammatory bowel disease
ICV ileocecal valve
ITB intestinal tuberculosis
TB tuberculosis
TNF tumor necrosis factor
UC ulcerative colitis

Introduction

Crohn's disease (CD) and ulcerative colitis (UC) are the two most common forms of immune-mediated chronic inflammatory bowel diseases (IBD). The disease course of IBD is characterized by relapsing or persistent inflammation. Disease flare-up may reflect intrinsic natural disease course or triggered by superimposed or opportunistic infections.

Atlas of Endoscopy Imaging in Inflammatory Bowel Disease. DOI: https://doi.org/10.1016/B978-0-12-814811-2.00023-2
© 2020 Elsevier Inc. All rights reserved.

Gut microbiota plays an important role in the pathogenesis and disease progression of IBD mainly in the form of dysbiosis or alteration in gut commensal bacteria or fungi. On the other hand, dysbiosis may promote superimposed infections either by opportunistic or acute extrinsic pathogens. Furthermore, immunomodulator and biological therapies help to subside inflammation and attain remission in IBD. However, these therapies could impair the immune system and make IBD patients prone to superimposed infections. Patients' general health and nutritional status, and underlying disease, along with the use of immunosuppressive medications predispose them to infections of opportunistic or extraneous pathogens. Commonly used immunosuppressive agents include corticosteroids, immunomodulators (such as purine analogs and methotrexate), anti−tumor necrosis factor (anti-TNF), antiintegrin, and antiinterleukin agents, and Janus kinase inhibitors. The severity of the complications secondary to infections in IBD patients depends upon geographic location, age, nutritional status, comorbidities, complexity of IBD, degree of immunosuppression. Overall, a multidisciplinary approach is often required to manage superimposed infections in IBD, consisting of IBD specialists, infectious disease specialists, pathologists, and colorectal surgeons. The managing clinician should remain vigilant to infectious etiology of disease flare-up. In addition to the evaluation of clinical presentation, imaging, and laboratory tests, endoscopy with the ability of aspiration of luminal contents and tissue biopsy plays a key role in the identification of pathogens and diagnosis. Some superimposed infections may present with characteristic endoscopic features. The identification of infectious causes of disease flare-up or refractoriness to conventional immunosuppressive therapy makes proper medical treatment possible, which alter the disease course, in some case, avoid unnecessary escalation of immunosuppressive therapy or surgery.

Superimposed infections in IBD harbor a wide spectrum of etiologies including bacterial, viral, fungal, or parasitic infections (Table 23.1) [1]. These infections are more likely limited to the gut but also could manifest systemic or extraintestinal complications, leading to significant morbidities [2].

Bacterial infections

Infections of *Clostridium* (also known as *Clostridioides*) *difficile*, *Campylobacter* sp. (spp.), and *Mycobacterium tuberculosis* are frequently encountered in IBD patients. Other agents include *Campylobacter*, *Salmonella*, *Shigella*, and *Escherichia coli* O157:H7, *Yersinia*, *Aeromonas* species, and *Plesiomonas* [3−5]. These bacterial infections may contribute to pathogenesis or exacerbation of IBD. Endoscopic features of the bacterial infections may be different from that in IBD patients. It is critical to distinguish superimposed bacterial infections in IBDfrom flare-up of IBD. However, the distinction is sometimes difficult. Stool culture or even gene-based multipathogen panel should be obtained. In addition, flexible sigmoidoscopy or ileocolonoscopy may be performed.

Clostridioides difficile

C. difficile, a Gram-positive, anaerobic, spore-forming bacillus, is a common pathogen identified during flare-ups of CD, UC [3], pouchitis [6,7], or enteritis (in patients with ileostomies) (Figs. 23.1−23.3) [8,9]. *C. difficile* infection

TABLE 23.1 Common microbiological agents associated with superimposed infection in inflammatory bowel disease.

Etiology	Pathogens
Bacterial	*Clostridium difficile*
	Clostridium perfringens
	Campylobacter jejuni
	Enterobacteriaceae
	Mycobacterium tuberculosis
Viral	Cytomegalovirus
	Epstein−Barr virus
	Herpes simplex virus
	Human papilloma virus
Fungal	*Candida albicans*
	Histoplasma capsulatum
Parasitic	*Cryptosporidium parvum*
	Entamoeba histolytica

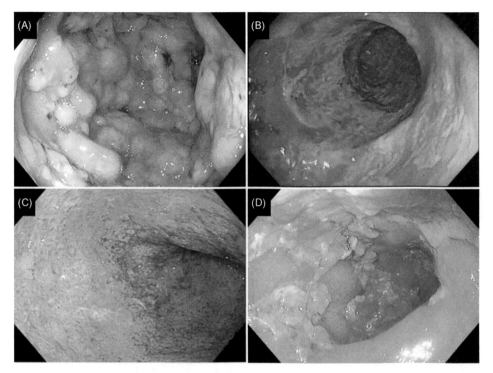

FIGURE 23.1 *Clostridium difficile* in non-IBD and IBD patients. (A) Mucosal edema and pseudomembrane in a non-IBD patient with chronic renal failure; (B–D) *C. difficile* infection in three patients with underlying ulcerative colitis. (B) Ulcers covered with while plaques mimicking pseudomembrane in the descending colon; (C) diffuse erythema and edema of the sigmoid colon; (D) and linear ulcers in the descending colon. *IBD*, Inflammatory bowel disease.

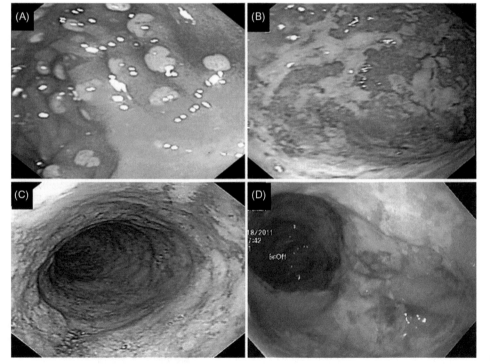

FIGURE 23.2 Pouchitis in patients with or without *Clostridium difficile* infection. (A) Edematous and nodular mucosa with pseudomembranes of the afferent limb and pouch body in a *C. difficile*–negative patient; (B and C) diffuse pseudomembranes in an ileal pouch patient with *C. difficile* infection; and (D) larges ulcers in the pouch body, and cuff in a patient with *C. difficile* infection.

(CDI) was reported that nearly 10% of the IBD patients would develop CDI and approximately 40% of them did not have prior exposure to antibiotics [10]. In addition, the incidence of CDI may be gradually increasing in IBD patients [11]. Concerns exist on relapse or refractoriness of CDI after medical therapy [12].

C. *difficile* can exist in the gut as asymptomatic colonization in favorable conditions the pathogen may induce mild-to-severe diarrhea, or even toxic megacolon. Interestingly, clinical symptoms of acute IBD flare and CDI can plausibly

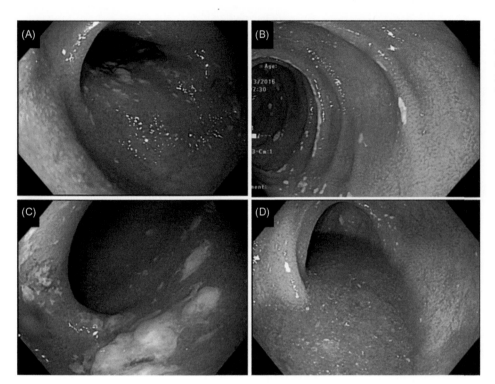

FIGURE 23.3 *Clostridium diffi-cile*—associated enteritis in ileost-omy patient. (A—D) Diffuse edema and various forms of ulcers in the distal ileum. The lumen of the distal ileum is also dilated.

overlap; therefore the existence of both in any severity prompts precise diagnosis and effective management to avoid the adverse outcomes [13]. As severity of clinical presentations and laboratory tests do not necessarily correlate, endo-scopic assessment of mucosal inflammation is often required. However, endoscopic features of CDI in IBD can range from completely normal mucosa to severe or fulminant colitis or enteritis. Classic pseudomembranes of CDI in non-IBD patients is often absent on ileoscopy or colonoscopy in IBD patients (Fig. 23.1) [2]. On the other hand, pseudo-membrane may be present in IBD without CDI (Fig. 23.2). Notably, CDI is not limited to the colon. In patients with colectomy or ileostomy, CDI can be presented as pouchitis (Fig. 23.2) [4,5] or enteritis (Fig. 23.3) [6].

Campylobacter spp

Campylobacter is a nonspore forming, Gram-negative, helical-shaped bacillus. *Campylobacter jejuni* infection (CJI) is the most commonly found foodborne infections in the United States. Uncontrolled CJI can lead to Guillain—Barré syn-drome and reactive arthritis. In addition, CJI is one of the enteropathogenic infections in UC patients, which can mimic or exacerbate UC [14]. Patients with CJI are often present with fever and leukocytosis. The diagnosis of CJI is mainly based on stool tests. Colonoscopy findings of CJI in IBD patients appear to be nonspecific. Patients with CJI and UC may have worse clinical outcomes than UC patients without CJI. However, it is not clear whether the eradication of this superimposed infection with oral erythromycin or azithromycin can lead to improved outcomes of underlying IBD.

Mycobacterium spp

Globally, there is a great concern over morbidity and mortality from tuberculosis (TB). *M. tuberculosis* is considered as a pathogenic bacterium responsible for causing active TB or latent TB infection. Overlap of endoscopic, histologic, and imaging features exists between CD and intestinal tuberculosis (ITB) (Chapter 25). Both granulomatous diseases have a predilection toward the involvement of the ileocecal valve (ICV) and terminal ileum[15]. It was reported that the gut could be the sixth most common site of extrapulmonary spread in patients with latent TB [16]. Clinically ITB could lead to stricture, fistula, intestinal perforation, abscess, and ascites [17].

The distinction between CD and ITB is critical, as clinical management is different. Endoscopy and histology along with microbiological workup play a key role in the distinction [18—20]. Their distinguishing features are listed in Table 25-2. CD is characterized by longitudinal ulcers along the edge of the mesentery and deformed or strictured ICV, while ITB is featured with circumferential ulcers and patent or nodular ICV [12,14].

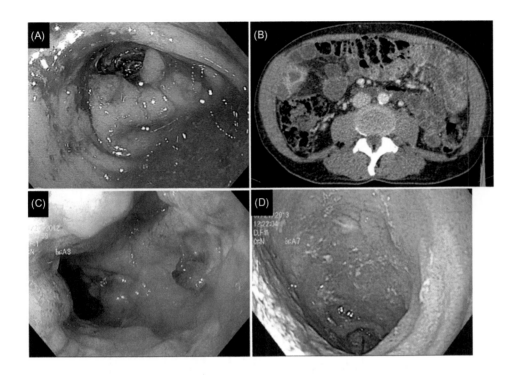

FIGURE 23.4 *Salmonella* and *Aeromonas* infections in two patients with IBD. (A and B) *Salmonella* enteritis in CD with nodular and friable mucosa of the terminal ileum on ileocolonoscopy (A) and mucosal and near-transmural enhancement of the terminal ileum on CT enterography (B); (C and D) *Aeromonas* infection in the ileal pouch/cuff with ulcers, nodularity, and edema of the cuff (C) and pouch body (D). *CD*, Crohn's disease; *IBD*, Inflammatory bowel disease.

On the other hand, CD and latent TB may coexist, suggesting that latent TB may be a part of etiopathogenesis of CD. This notion has been supported by the fact that *Mycobacteria* are isolated from intestinal mucosal specimens in IBD patients as well as non-IBD immunocompromised patients [21]. A recent study from South Korea showed a higher incidence of TB in IBD patients treated with TNF inhibitors than the general population regardless of their latent TB status at baseline [22]. Superimposed ITB infection can potentially lead to IBD flare.

Other enteric bacterial pathogens

Enteric gut flora including Enterobacteriaceae is implicated in the pathogenesis and disease flare-up of IBD. Superimposed infections by *Salmonella*, *Enterococcus faecalis* [23], enterovirulent *E. coli* [24,25], adherent invasive *E. coli* [26], *Aeromonas* spp. [5], and *Clostridium perfringens* [27] may trigger disease exacerbation of IBD (Fig. 23.4).

Viral infection

Viral infections in IBD patients result from *cytomegalovirus* (CMV), *Epstein—Barr virus* (EBV), less commonly from herpes simplex virus (HSV) and human papillomavirus (HPV). These viral infections can occur in patients with CD, UC, or ileal pouches. Endoscopy with tissue biopsy in junction with serology is the most reliable modalities for the diagnosis. The viral infection may be a trigger for the flare-up of underlying IBD and its treatment may be beneficial However, the presence of the virus in tissue or body fluid specimens may represent innocent bystander in some patients.

Cytomegalovirus

CMV, a member of the Herpesviridae family, can cause mild-to-severe enteritis or colitis mainly in immunocompromised hosts. It is also a common viral pathogen identified in IBD patients. CMV serology and histology have been routinely checked during disease relapse or refractory course of IBD. CMV enteritis or colitis is characterized by the presence of various forms and ulcers on endoscopy [28] (Figs. 23.5—23.7).

Classic endoscopic features include well-demarcated, deep, punched-out ulcers with or without coated white membrane mainly at the right colon and less often in other parts of the large bowel in non-IBD patients [29,30], whereas endoscopic manifestations of CMV infection in IBD patients may also show a mucosal defect, well-demarcated ulcers, irregularly shaped ulcers, punched-out deep ulcers, longitudinal ulcers in the small and large colon or ileal pouch

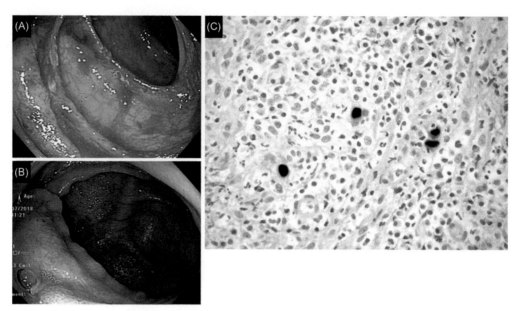

FIGURE 23.5 CMV colitis in immune-compromised non-IBD patients (with liver transplant). (A) Superficial ulcers along the folds of ascending colon; (B) large clean-based ulcer in the ileocecal valve with a raised edge; and (C) virus-loaded mononuclear cells in tissue biopsy from the ulcer base in immunohistochemistry. *CMV*, Cytomegalovirus.

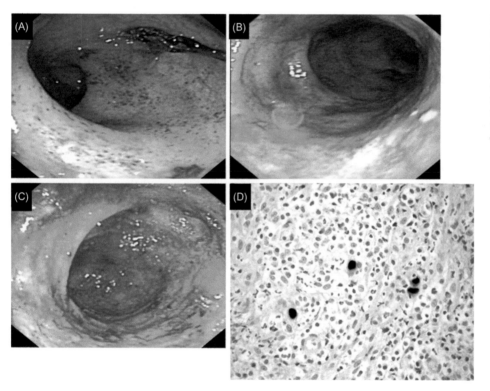

FIGURE 23.6 CMV infection superimposed on underlying ulcerative colitis. (A and B) Diffuse colitis from the rectum (A) to proximal colon (B and C), with edema, erythema, and friable mucosa; (D) immunohistochemistry showed CMV-infected cells with nuclear stains. *CMV*, Cytomegalovirus.

patients [31,32], along with mucosal edema and erythema, cobblestone-like changes [33−35]. Chronic CMV infection may cause pseudotumor-like appearance [36] (Figs. 23.5−23.7). Heavy-load CMV infection may contribute to fulminant colitis or toxic megacolon in UC patients [37,38]. It is interesting that coinfection of *C. difficile* and CMV [39], and coinfection of CMV and EBV [40] can occur in IBD patients, further exacerbating disease course. Despite the endoscopic features, the diagnosis of CMV colitis is based on positive histopathology by hematoxylin and eosin (H&E) stain or immunohistochemistry, and/or positive qPCR for CMV DNA in colonic mucosal tissues [37].

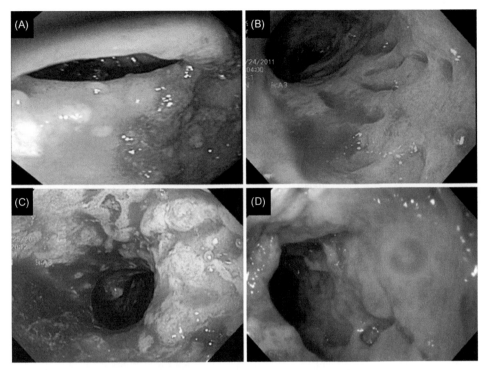

FIGURE 23.7 CMV infection in patients with underlying IBD. (A) Concurrent CMV infection with mucosal nodularity and edema in the ascending colon in a patient with ulcerative colitis; (B–D) CMV-associated pouchitis and cuffitis characterized by multiple punched-out ulcers with mucosal edema in the pouch body (B and C) and rectal cuff (D) in a patient with ileal pouch—anal anastomosis. *CMV*, Cytomegalovirus; *IBD*, inflammatory bowel disease.

Superimposed infection of CMV in IBD patients usually requires antiviral therapy with intravenous ganciclovir or oral valganciclovir [41], which may be beneficial for controlling active IBD [42], especially in those with heavy viral load [34].

Epstein–Barr virus

EBV, another member of the Herpesviridae family, has an association with diverse clinical and endoscopic manifestations in IBD patients [43]. Both acute infection (infectious mononucleosis) and activation of latent EBV infection can occur. Chronic EBV infection can cause mucosal inflammation and even inflammatory mass in the colon or small bowel, mimicking IBD [44]. Correlation between superimposed EBV infection and disease course and outcome of IBD has been demonstrated [45].

Chronic EBV infection may lead to lymphoproliferative diseases in IBD patients [46–50], especially in those with a long-term use of purine analogs and possibly anti-TNF agents. The endoscopic and histologic distinction between superimposed EBV in the gut in IBD and sole IBD flare-up, and between gastrointestinal (GI) lymphoma and CD, can be difficult. Common endoscopic manifestations include diffuse erythematous, thickened, or nodular mucosa, multiple, small, deep ulcers, or pseudotumors, in the small bowel or large bowel (Figs. 23.8 and 23.9). Endoscopic features of EBV-associated lymphoma are discussed in Chapter 33, Inflammatory bowel disease-like conditions: gastrointestinal lymphoma and other neoplasms.

Herpes simplex virus

HSV infection is usually mild and self-limiting in immunocompetent individuals. Approximately 90% of the general population is seropositive for HSV by the middle age [51]. Although superimposed infection with HSV is not common in IBD, immunosuppressive therapies and immunocompromised state in those patients are still at risk. The chronically epithelial cells (particularly squamous epithelia) in IBD patients could easily favor HSV penetration across the epithelial barrier and thus allowing subsequent IBD flare. Clinical presentations of HSV infection vary from odynophagia, dysphagia, nausea, vomiting, fatigue, fever, and abdominal pain to with bloody diarrhea [52]. Since HSV tends to affect squamous cells, the common locations of GI infection are the esophagus, anorectum, and perianal skin (Figs. 23.10 and 23.11), less commonly in the small or large bowel [53]. HSV infection can cause erythema multiforme involving skin or oral cavity [54]. Therefore oral inspection of IBD patients during physical examination or upper and lower GI tract

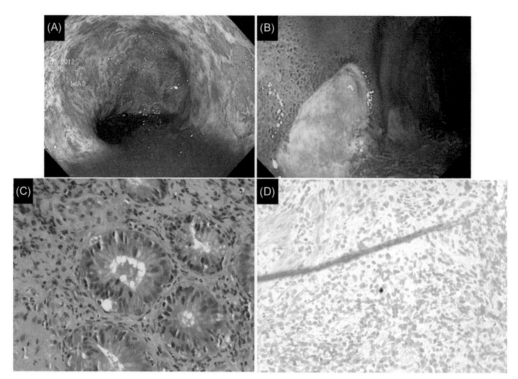

FIGURE 23.8 EBV infection in a patient with ileal pouch–anal anastomosis for ulcerative colitis who presented with fever and increased bowel frequency. (A) diffuse pouchitis with erythema and exudates; (B) large, discrete, clean-based ulcers in the afferent limb; (C) hematoxylin and eosin stain showed acute and chronic inflammation of pouch mucosa; and (D) chromogen in situ hybridization showed a virus-infected mononuclear cell. *EBV*, Epstein–Barr virus.

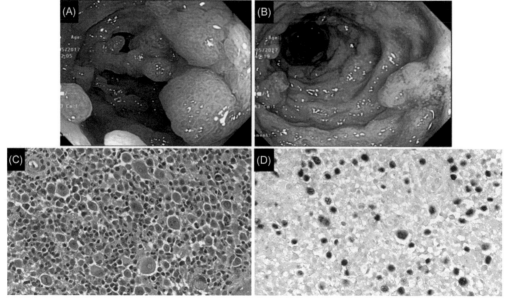

FIGURE 23.9 EBV infection in a patient with ulcerative colitis in the absence of the use of immunosuppressive agents. (A and B) Mucosal edema, nodularity, and pseudopolyps at the ascending colon; (C) marked lymphoplasmacytic inflammation on hematoxylin and eosin stain; and (D) EBER-positive neoplastic cells EBER with in situ hybridization. *EBER*, Epstein–Barr virus–encoded RNA; *EBV*, Epstein–Barr virus. *Histology photo courtesy Claudiu Cotta, MD, of Cleveland Clinic.*

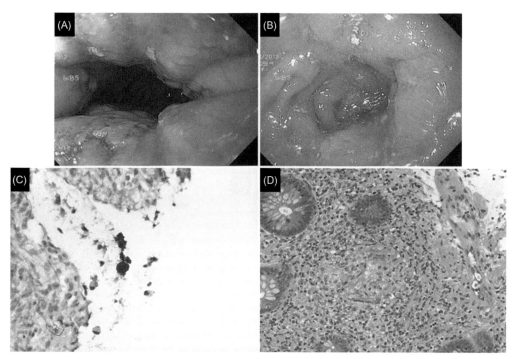

FIGURE 23.10 HSV-2 infection in a patient with ileal pouch—anal anastomosis. (A) Anal fissures and edema of the anal canal; (B) pouchitis with diffuse mucosal edema and linear ulcers; (C) immunohistochemistry demonstrated virus-infected epithelial cells from anal canal biopsy; and (D) non-caseating granulomas in the lamina propria from anal transitional zone biopsy. *HSV*, Herpes simplex virus.

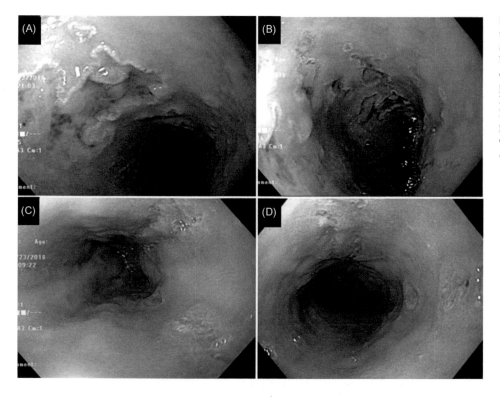

FIGURE 23.11 HSV-2 esophagitis versus Crohn's esophagitis. (A and B) HSV esophagitis with "rat-bite" ulcers in an immune-compromised bone marrow—transplant patient. (C and D) Esophageal Crohn's disease with linear erythema and ulceration in a patient treated with a combination of adalimumab and vedolizumab. *HSV*, Herpes simplex virus.

endoscopy to assess HSV-related vesicular lesions should be performed. IBD patients under severe immunosuppression, who present with dysphagia, odynophagia, or sharp perianal pain, with or without skin blisters should undergo endoscopic examination. Coinfection of HSV and CMV is associated with poor disease outcomes, such as the need for colectomy in IBD patients [55].

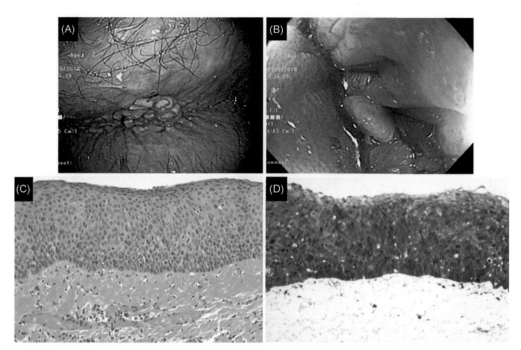

FIGURE 23.12 HPV infection in a patient with Crohn's disease on long-term azathioprine. (A) Perianal fistulas with setons in place and perianal dermatitis; (B) nodularity of the anal canal; (C) the lower rectal biopsies showed full-thickness squamous dysplasia with hyperchromatic, overlapping nuclei with numerous mitotic figures and dyskeratotic keratinocytes; and (D) P16 immunohistochemistry confirmed diagnosis of high-grade squamous intraepithelial lesion with full-thickness, diffuse, block positivity in the nuclei and cytoplasm, the pattern of high-risk HPV DNA integration. *HPV*, Human papilloma virus. *Histology photo courtesy Eric Willis, MD of Cleveland Clinic.*

The endoscopic manifestations of esophageal HSV infection include severe inflammation with deep, punched-out ulcers with white raised edges and friability of surrounding tissue [43]. Sometimes it is difficult to distinguish HSV associated esophagitis and CD of the esophagus (Fig. 23.11). Tissue biopsy and cytology are often needed. Patients with active HSV infection should avoid the use of corticosteroids or other immunosuppressive medications until the infection is completely resolved.

Human papillomavirus

Patients with IBD carry a higher risk for colorectal cancer than the general population. While colitis-associated neoplasia occurs in patients with long-stand colitis, anorectal cancer can develop in patients with anal and perianal CD [56–58]. HPV infection has contributed to the etiopathogenesis of anorectal cancer in CD. Routine colonoscopy or flexible sigmoidoscopy examination should include careful inspection of distal rectum, anal canal, and perianal skin. Inflammatory lesion should be biopsied for HPV, HSV, sexually transmitted diseases, and cancer (Fig. 23.12).

Endoscopic features of IBD-like sexual transmitted diseases are discussed in Chapter 25.

Fungal infection

Fungi are commensal flora in the GI tract. Dysbiosis, that is, alterations in commensal microbiota, including fungi, is implicated for the pathogenesis of IBD [59]. Common etiological agents of opportunistic fungal infections in IBD are *Candida* spp. and *Histoplasma* spp. [60]. including *Candia* spp. and less frequent, but potentially encountered infection of *Histoplasma* spp. These fungi, same as other previously described pathogens, invade inflamed intestinal mucosa in IBD patients. Invasive fungal infection is pathological, bearing clinical significance.

Candida spp

Candia spp. is the most dominant fungal genera in the GI tract [61,62]. *Candida albicans,* the most common species adapted to the GI tract, is prone to causing superimposed infection [63]. Colonization with *C. albicans* has been

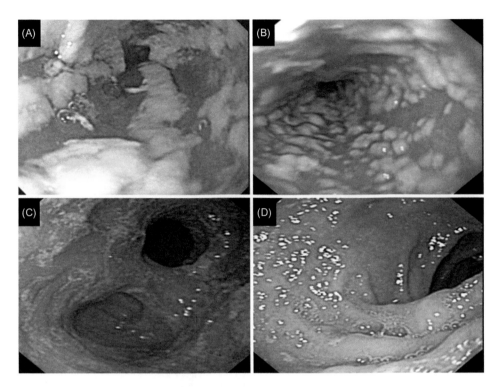

FIGURE 23.13 *Candida* esophagitis after antibiotic use in a patient with chronic pouchitis. (A and B) Diffuse esophagitis with white plaques and (C and D) mild diffuse pouchitis with mucosal edema, small erosions, and exudates.

demonstrated in CD and UC [53]. In addition, oral or esophageal candidiasis is seen in IBD patients, particularly in those on corticosteroids [64]. A study from Japan demonstrated that the prevalence of *Candida* spp. was greater in CD than UC or healthy controls [65]. Treatment of UC flare-up with antifungal agent fluconazole exhibited a reduced colon inflammation as well as clinical symptoms [66].

Infection from *Candida* spp. can promote acute and chronic inflammation [67], hypothetically worsening IBD. Common risk factors for *candidiasis* are the use of corticosteroids and antibiotics. Endoscopic features of esophageal candidiasis typically are white plaque-like lesions with surrounding erythematous mucosa and underlying friable mucosa (Fig. 23.13).

Histoplasma capsulatum

Histoplasma capsulatum is a dimorphic fungus, which can lead to the second most common fungal infections after candidiasis in immunocompromised IBD patients [68]. It is isolated from endemic areas, such as river valleys of Ohio, Missouri, and Mississippi in the United States. Histoplasmosis is caused by inhaling airborne spores. The spores are residing in the soil, which normally get contaminated from fecal dropping of bats or birds infected with *Histoplasma* spp. Histoplasmosis can be diagnosed by fungal culture of blood or detecting serum antibody or antigen of *Histoplasma* in the urine. Histology or Gomori's methenamine silver staining of a tissue biopsy can also be used for the detection of *Histoplasma* spp.

Disseminated histoplasmosis can affect the GI tract, from the mouth to anus. Patients with IBD with mucosal or transmural inflammation, especially those on immunosuppressive agents, may be prone to injury from disseminated histoplasmosis [69,70]. Superimposed infection with *H. capsulatum* and its treatment can affect the disease course of CD [71,72]. The endoscopic manifestations of *Histoplasma* infection in immunocompetent patients showed isolated to diffuse mucosal nodularity, central erosions, ulcers polypoid lesions, in the terminal ileum, colon [73], or ileal pouch (Fig. 23.14).

Parasite infections

The most frequently reported parasitic superimposed GI infections in IBD in literature resulted from *Entamoeba histolytica* followed by *Cryptosporidium* spp. The diagnosis should be made promptly in immunocompromised patients, particularly in those on anti-TNF therapy.

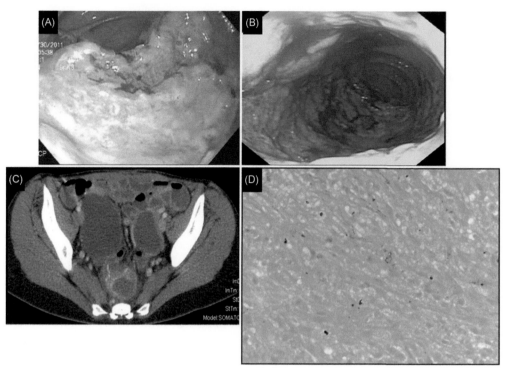

FIGURE 23.14 *Histoplasmosis capsulatum* infection in a patient on anti—tumor necrosis factor for Crohn's disease of the pouch. (A and B) A large, "dirty"-looking ulcerated area in the pouch body; (C) thickened pouch body wall on CT enterography; and (D) silver stain showed spores in the tissue biopsy.

Entamoeba histolytica

The genus *Entamoeba* comprises an anaerobic parasite named *E. histolytica*, which can cause amebic dysentery or diarrhea. The infection is most prevalent in tropical and developing countries. Almost 50 million people are infected with *Entamoeba* [74]. This is the foremost frequently reported superimposed infection in IBD patients. Most of the non-IBD patients can be asymptomatic, but plausibly IBD patients may exhibit GI symptoms due to their immunocompromised status. A Turkish study reported that *Entamoeba* was more prevalent in IBD patients than the general population and more frequent in UC patients than CD patients [75]. It has been difficult to distinguish between IBD-related colitis versus combined amebiasis and IBD, due to the overlap between symptomatic and endoscopic features of both conditions. Therefore the microscopic diagnosis of *E. histolytica* by trichome staining may be more reliable [66,76].

Cryptosporidium spp

Cryptosporidium is a protozoan parasite of domestic and wild animals in tropical countries. The feces from animals normally contaminate food or water resources. Cryptosporidiosis transmission occurs via the fecal—oral route through ingesting invasive oocysts form of cryptosporidium. Relapse of IBD in patients at risk should be assessed for cryptosporidiosis [77]. It is speculated that superimposed cryptosporidium infection in IBD may become even more intense, in those on anti-TNF therapy. The relapse in cryptosporidiosis can also trigger relapses in IBD and hence should be managed judiciously [78]. The cryptosporidium antigen can be detected in stool samples of infected patients. The endoscopic manifestations in non-IBD patients with cryptosporidiosis include nonspecific ileitis with focal mucosal congestion [79], and pseudopolyps with inflammation in the small and large bowel.

Summary and recommendations

Superimposed infections from bacterial, viral, fungal, or parasitic agents are common in patients with IBD, particularly in those on long-term, immunosuppressive agents. Systemic presentations, such as fever, chills, night sweats, weight loss, and leukocytosis, should trigger a full scale of evaluation. While microbiology workup of tissue or blood fluid specimens are the standard care, the endoscopic evaluation may demonstrate characteristic features, such as white plaque for

Candida esophagitis and punched-out ulcers for CMV colitis. In addition, tissue and luminal specimens for microbiology workup can be obtained through the biopsy. Nonetheless, endoscopic features of combined IBD and superimposed infection, IBD alone, and infectious esophagitis, gastritis, enteritis, or colitis, may largely overlap.

Acknowledgment

Dr. Bo Shen is supported by the Ed and Joey Story Endowed Chair.

Disclosures

The authors declare no financial conflicts of interest.

References

[1] Kucharzik T, Maaser C. Infections and chronic inflammatory bowel disease. Viszeralmedizin 2014;30:326−32.

[2] Antonelli E, Baldoni M, Giovenali P, et al. Intestinal superinfections in patients with inflammatory bowel diseases. J Crohns Colitis 2012;6:154−9.

[3] Lin WC, Chang CW, Chen MJ, Chu CH, Shih SC, Hsu TC, et al. Challenges in the diagnosis of ulcerative colitis with concomitant bacterial infections and chronic infectious colitis. PLoS One 2017;12:e0189377.

[4] Hanada Y, Khanna S, Loftus Jr EV, Raffals LE, Pardi DS. Non-*Clostridium difficile* bacterial infections are rare in patients with flares of inflammatory bowel disease. Clin Gastroenterol Hepatol 2018;16:528−33.

[5] Lobatón T, Hoffman I, Vermeire S, Ferrante M, Verhaegen J, Van Assche G. Aeromonas species: an opportunistic enteropathogen in patients with inflammatory bowel diseases? A single center cohort study. Inflamm Bowel Dis 2015;21:71−8.

[6] Seril DN, Ashburn JH, Lian L, Shen B. Risk factors and management of refractory or recurrent *Clostridium difficile* infection in ileal pouch patients. Inflamm Bowel Dis 2014;20:2226−33.

[7] Li Y, Qian J, Queener E, Shen B. Risk factors and outcome of PCR-detected *Clostridium difficile* infection in ileal pouch patients. Inflamm Bowel Dis 2013;19:397−403.

[8] Kochhar G, Edge P, Blomme C, Wu XR, Lopez R, Ashburn J, et al. *Clostridium difficile* enteropathy is associated with a higher risk for acute kidney injury in patients with an ileostomy—a case-control study. Inflamm Bowel Dis 2018;24:402−9.

[9] Navaneethan U, Venkatesh PG, Shen B. *Clostridium difficile* infection and inflammatory bowel disease: understanding the evolving relationship. World J Gastroenterol 2010;16:4892−904.

[10] Bosca-Watts M, Tosca J, Anton R, Mora M, Minguez M, Mora F. Pathogenesis of Crohn's disease: bug or no bug. World J Gastrointest Pathophysiol 2015;6:1−12.

[11] Binion D. *Clostridium difficile* infection and inflammatory bowel disease. Gastroenterol Hepatol (NY) 2016;12:334−7.

[12] DePestel D, Aronoff D. Epidemiology of *Clostridium difficile* infection. J Pharm Pract 2013;26:464−75.

[13] Sokol H, Lalande V, Landman C, Bourrier A, Nion-Larmurier I, Rajca S, et al. *Clostridium difficile* infection in acute flares of inflammatory bowel disease: a prospective study. Dig Liver Dis 2017;49:643−6.

[14] Arora Z, Mukewar S, Wu X, Shen B. Risk factors and clinical implication of superimposed *Campylobacter jejuni* infection in patients with underlying ulcerative colitis. Gastroenterol Rep (Oxf) 2016;4:287−92.

[15] Venkatesh PGK, Navaneethan U. Mimickers of intestinal tuberculosis: could this be Crohn's disease? An unsolved enigma. Saudi J Gastroenterol 2011;17:95−6.

[16] Zhou D, Quyang Q, Xiong M, Zhang Y. Crohn's disease with positive Ziehl-Neelsen stain: three case reports. Niger J Clin Pract 2018;10:1387−90.

[17] Singh H, Mandavdhare H, Sharma V. All that fistulises is not Crohn's disease: multiple entero-enteric fistulae in intestinal tuberculosis. Pol Przegl Chir 2019;91:35−7.

[18] Kedia S, Das P, Madhusudhan KS, Dattagupta S, Sharma R, Sahni P, et al. Differentiating Crohn's disease from intestinal tuberculosis. World J Gastroenterol 2019;25:418−32.

[19] Sato R, Nagai H, Matsui H, Yamane A, Kawashima M, Higa K, et al. Ten cases of intestinal tuberculosis which were initially misdiagnosed as inflammatory bowel disease. Intern Med 2019;58:2003−8.

[20] Li Y, Qian JM. The challenge of inflammatory bowel disease diagnosis in Asia. Inflamm Intest Dis 2017;1:159−64.

[21] Graham DY, Markesich DC, Yoshimura HH. Mycobacteria and inflammatory bowel disease. Results of culture. Gastroenterology 1987;92:438−42.

[22] Kang J, Jeong DH, Han M, Yang SK, Byeon JS, Ye BD, et al. Incidence of active tuberculosis within one year after tumor necrosis factor inhibitor treatment according to latent tuberculosis infection status in patients with inflammatory bowel disease. J Korean Med Sci 2018;33:e292.

[23] Zhou Y, Chen H, He H, Du Y, Hu J, Li Y, et al. Increased *Enterococcus faecalis* infection is associated with clinically active Crohn disease. Medicine (Baltimore) 2016;95:e5019.

[24] Curová K, Kmetová M, Sabol M, Gombosová L, Lazúrová I, Siegfried L. Enterovirulent *E. coli* in inflammatory and noninflammatory bowel diseases. Folia Microbiol (Praha) 2009;54:81−6.

[25] Rhodes JM. The role of *Escherichia coli* in inflammatory bowel disease. Gut 2007;56:610−12.

[26] Martinez-Medina M, Garcia-Gil LJ. *Escherichia coli* in chronic inflammatory bowel diseases: an update on adherent invasive *Escherichia coli* pathogenicity. World J Gastrointest Pathophysiol 2014;5:213−27.

[27] Banaszkiewicz A, Kądzielska J, Gawrońska A, Pituch H, Obuch-Woszczatyński P, Albrecht P, et al. Enterotoxigenic *Clostridium perfringens* infection and pediatric patients with inflammatory bowel disease. J Crohns Colitis 2014;8:276−81.

[28] Kang EA, Yoon H, Seo AY, Shin CM, Im JP, Park YS, et al. Characteristics of cytomegalovirus enterocolitis in patients with or without inflammatory bowel diseases. Scand J Gastroenterol 2018;53:453−8.

[29] Le PH, Kuo CJ, Wu RC, Hsu JT, Su MY, Lin CJ, et al. Pancolitis associated with higher mortality risk of cytomegalovirus colitis in patients without inflammatory bowel disease. Ther Clin Risk Manag 2018;14:1445−51.

[30] Kim CH, Bahng S, Kang KJ, Ku BH, Jo YC, Kim JY, et al. Cytomegalovirus colitis in patients without inflammatory bowel disease: a single center study. Scand J Gastroenterol 2010;45:1295−301.

[31] Mabvuure NT, Maclean L, Oien K, Gaya D. Cytomegalovirus pouchitis in a patient with Crohn's disease. BMJ Case Rep 2014;2014:1−4.

[32] He X, Bennett AE, Lian L, Shen B. Recurrent cytomegalovirus infection in ileal pouch-anal anastomosis for ulcerative colitis. Inflamm Bowel Dis 2010;16:903−4.

[33] Yang H, Zhou W, Lv H, Wu D, Feng Y, Shu H, et al. The association between CMV viremia or endoscopic features and histopathological characteristics of CMV colitis in patients with underlying ulcerative colitis. Inflamm Bowel Dis 2017;23:814−21.

[34] Hirayama Y, Ando T, Hirooka Y, Watanabe O, Miyahara R, Nakamura M, et al. Characteristic endoscopic findings and risk factors for cytomegalovirus-associated colitis in patients with active ulcerative colitis. World J Gastrointest Endosc 2016;8:301−9.

[35] Inflammatory Bowel Disease Group, Chinese Society of Gastroenterology, Chinese Medical Association. Evidence-based consensus on opportunistic infections in inflammatory bowel disease (republication). Intest Res 2018;16:178−93.

[36] Vegunta AS, Dasar SK, Joshi SK, Rao RV. Spontaneous partial vanishing cytomegalovirus pseudotumour of colon in an immunocompetent patient. J Clin Diagn Res 2015;9:TD07−9.

[37] Inoue K, Wakabayashi N, Fukumoto K, et al. Toxic megacolon associated with cytomegalovirus infection in a patient with steroid-naive ulcerative colitis. Intern Med 2012;51:2739−43.

[38] Jones A, McCurdy JD, Loftus Jr EV, Bruining DH, Enders FT, et al. Effects of antiviral therapy for patients with inflammatory bowel disease and a positive intestinal biopsy for cytomegalovirus. Clin Gastroenterol Hepatol 2015;13:949−55.

[39] McCurdy JD, Enders FT, Khanna S, Bruining DH, Jones A, Killian JM, et al. Increased rates of *Clostridium difficile* infection and poor outcomes in patients with IBD with cytomegalovirus. Inflamm Bowel Dis 2016;22:2688−93.

[40] Matsumoto H, Kimura Y, Murao T, Osawa M, Akiyama T, Mannoji K, et al. Severe colitis associated with both Epstein-Barr virus and cytomegalovirus reactivation in a patient with severe aplastic anemia. Case Rep Gastroenterol 2014;8:240−4.

[41] Subramanian V, Finlayson C, Harrison T, Rice P, Pollok R. Primary cytomegalovirus infectious colitis complicating Crohn's disease successfully treated with oral valganciclovir. J Crohns Colitis 2010;4:199−202.

[42] Wang Y, Aggarwal P, Liu X, Lu H, Lian L, Wu X, et al. Antiviral treatment for colonic cytomegalovirus infection in ulcerative colitis patients significantly improved their surgery free survival. J Clin Gastroenterol 2018;52:e27−31.

[43] Wu S, He C, Tang TY, Li YQ. A review on co-existent Epstein-Barr virus-induced complications in inflammatory bowel disease. Eur J Gastroenterol Hepatol 2019;31:1085−91.

[44] Osman M, Al Salihi M, Abu Sitta E, Al Hadidi S. A rare case of Epstein-Barr virus mucocutaneous ulcer of the colon. BMJ Case Rep 2017, July 6, 2017. Epub.

[45] Li X, Chen N, You P, Peng T, Chen G, Wang J, et al. The status of Epstein-Barr virus infection in intestinal mucosa of Chinese patients with inflammatory bowel disease. Digestion 2019;99:126−32.

[46] Goetgebuer RL, van der Woude CJ, de Ridder L, Doukas M, de Vries AC. Clinical and endoscopic complications of Epstein-Barr virus in inflammatory bowel disease: an illustrative case series. Int J Colorectal Dis 2019;34:923−6.

[47] Akamatsu T, Watanabe N, Chiba T. Epstein-Barr virus-associated lymphoma developed shortly after immunosuppressive treatment for ulcerative colitis. Clin Gastroenterol Hepatol 2007;5:521.

[48] Schwartz LK, Kim MK, Coleman M, Lichtiger S, Chadburn A, Scherl E. Case report: lymphoma arising in an ileal pouch anal anastomosis after immunomodulatory therapy for inflammatory bowel disease. Clin Gastroenterol Hepatol 2006;4:1030−4.

[49] Juffermans NP, Jager A, Kersten MJ, van Oers MH, Hommes DW. [Epstein-Barr virus-related lymphomas in patients with inflammatory bowel disease]. Ned Tijdschr Geneeskd 2005;149:1859−63.

[50] N'guyen Y, Andreoletti L, Patey M, Lecoq-Lafon C, Cornillet P, Léon A, et al. Fatal Epstein-Barr virus primo infection in a 25-year-old man treated with azathioprine for Crohn's disease. J Clin Microbiol 2009;47:1252−4.

[51] Schunter MO, Walles T, Fritz P, Meyding-Lamadé U, Thon KP, Fellermann K, et al. Herpes simplex virus colitis complicating ulcerative colitis: a case report and brief review on superinfections. J Crohns Colitis 2007;1:41−6.

[52] Jafri H, Kalina DR, Aziz T, Serrano PE, Haider S. Herpes simplex virus colitis in a patient with newly diagnosed Crohn's disease. Case Rep Med 2018;2018:1−3.

[53] Lee BH, Um WH, Jeon SR, Kim HG, Lee TH, Kim WJ, et al. Herpes simplex virus duodenitis accompanying Crohn's disease. Korean J Gastroenterol 2013;62:292−5.

[54] Georgesen C, Huang J, Avarbock A, Harp J, Magro C. Orofacial granulomatosis and erythema multiforme in an adolescent with Crohn's disease. Pediatr Dermatol 2018;35:e294−7.

[55] Hosomi S, Watanabe K, Nishida Y, Yamagami H, Yukawa T, Otani K, et al. Combined infection of human herpes viruses: a risk factor for subsequent colectomy in ulcerative colitis. Inflamm Bowel Dis 2018;24:1307—15.

[56] Ky A, Sohn N, Weinstein MA, Korelitz BI. Carcinoma arising in anorectal fistulas of Crohn's disease. Dis Colon Rectum 1998;41:992—6.

[57] Biancone L, Orlando A, Kohn A, Colombo E, Sostegni R, Angelucci E, et al. Infliximab and newly diagnosed neoplasia in Crohn's disease: a multicentre matched pair study. Gut 2006;55:228—33.

[58] Vuitton L, Jacquin E, Parmentier AL, Crochet E, Fein F, Dupont-Gossart AC, et al. High prevalence of anal canal high-risk human papillomavirus infection in patients with Crohn's disease. Clin Gastroenterol Hepatol 2018;16:1768—76.

[59] Li J, Chen D, Yu B, He J, Zheng P, Mao X, et al. Fungi in gastrointestinal tracts of human and mice: from community to functions. Microb Ecol 2018;75:821—9.

[60] Navaneethan U, Shen B. Secondary pouchitis: those with identifiable etiopathogenetic or triggering factors. Am J Gastroenterol 2010;105:51—64.

[61] Gouba N, Drancourt M. Digestive tract mycobiota: a source of infection. Med Mal Infect 2015;45:9—16.

[62] Mukherjee PK, Sendid B, Hoarau G, Colombel JF, Poulain D, Ghannoum A. Mycobiota in gastrointestinal diseases. Nat Rev Gastroenterol Hepatol 2015;12:77—87.

[63] Salehi F, Esmaeili M, Mohammadi R. Isolation of *Candida* species from gastroesophageal lesions among pediatrics in Isfahan, Iran: identification and antifungal susceptibility testing of clinical isolates by E-test. Adv Biomed Res, 6. 2017. p. 103.

[64] Lisciandrano D, Ranzi T, Carrassi A, Sardella A, Campanini MC, Velio P, et al. Prevalence of oral lesions in inflammatory bowel disease. Am J Gastroenterol 1996;91:7—10.

[65] Imai T, Inoue R, Kawada Y, Morita Y, Inatomi O, Nishida A, et al. Characterization of fungal dysbiosis in Japanese patients with inflammatory bowel disease. J Gastroenterol 2019;54:149—59.

[66] Zwolinska-Wcislo M, Brzozowski T, Budak A, Kwiecien S, Sliwowski Z, Drozdowicz D, et al. Effect of *Candida* colonization on human ulcerative colitis and the healing of inflammatory changes of the colon in the experimental model of colitis ulcerosa. J Physiol Pharmacol 2009;60:107—18.

[67] Kumamoto CA. Inflammation and gastrointestinal *Candida* colonization. Curr Opin Microbiol 2011;14:386—91.

[68] Stamatiades GA, Ioannou P, Petrikkos G, Tsioutis C. Fungal infections in patients with inflammatory bowel disease: a systematic review. Mycoses 2018;61:366—76.

[69] Dotson JL, Crandall W, Mousa H, Honegger JR, Denson L, Samson C, et al. Presentation and outcome of histoplasmosis in pediatric inflammatory bowel disease patients treated with antitumor necrosis factor alpha therapy: a case series. Inflamm Bowel Dis 2011;17:56—61.

[70] Lee JH, Slifman NR, Gershon SK, Edwards ET, Schwieterman WD, Siegel JN, et al. Life-threatening histoplasmosis complicating immunotherapy with tumor necrosis factor alpha antagonists infliximab and etanercept. Arthritis Rheum 2002;46:2565—70.

[71] Anderson B, Sweetser S. Ileocolonic histoplasmosis complicating Crohn's disease. Clin Gastroenterol Hepatol 2017;15:e135—6.

[72] Galandiuk S, Davis BR. Infliximab-induced disseminated histoplasmosis in a patient with Crohn's disease. Nat Clin Pract Gastroenterol Hepatol 2008;5:283—7.

[73] Zhu LL, Wang J, Wang ZJ, Wang YP, Yang JL. Intestinal histoplasmosis in immunocompetent adults. World J Gastroenterol 2016;22:4027—33.

[74] Kataria H, Seth A, Attri AK, Singh Punia RP. Ameboma of colon simulating colonic adenocarcinoma. Int J Appl Basic Med Res 2018;8:42—4.

[75] Ustun S, Dagci H, Aksoy U, Guruz Y, Ersoz G. Prevalence of amebiasis in inflammatory bowel disease in Turkey. World J Gastroenterol 2003;9:1834—5.

[76] Fleming R, Cooper CJ, Ramirez-Vega R, Huerta-Alardin A, Boman D, Zuckerman MJ. Clinical manifestations and endoscopic findings of amebic colitis in a United States-Mexico border city: a case series. BMC Res Notes 2015;8:781.

[77] Manthey MW, Ross AB, Soergel KH. Cryptosporidiosis and inflammatory bowel disease. Experience from the Milwaukee outbreak. Dig Dis Sci 1997;42:1580—6.

[78] Vadlamudi N, Maclin J, Dimmitt RA, Thame KA. Cryptosporidial infection in children with inflammatory bowel disease. J Crohns Colitis 2013;7:e337—43.

[79] Ogata S, Suganuma T, Okada C, Inoue K, Kinoshita A, Sato K. A case of sporadic intestinal cryptosporidiosis diagnosed by endoscopic biopsy. Acta Med Okayama 2009;63:287—91.

Chapter 24

Inflammatory bowel disease—associated digestive disorders

Bo Shen

Center for Inflammatory Bowel Diseases, Columbia University Irving Medical Center-New York Presbyterian Hospital, New York, NY, United States

Chapter Outline

Abbreviations

AIH autoimmune hepatitis
AIP autoimmune pancreatitis
CD Crohn's disease
EoE eosinophilic esophagitis
GAVE gastric antral vascular ectasia
GERD gastroesophageal reflux disorder
GI gastrointestinal
IBD inflammatory bowel disease
IBS irritable bowel syndrome
ICV ileocecal valve
IPAA ileal pouch—anal anastomosis
NSAIDs nonsteroidal anti-inflammatory drugs
PSC primary sclerosing cholangitis
UC ulcerative colitis

Introduction

Inflammatory bowel disease (IBD), that is, Crohn's disease (CD) and ulcerative colitis (UC), can affect any organ of the digestive disease system. The involvement of gastrointestinal (GI) tract, liver, gall bladder, or pancreas in IBD may manifest as a part of the disease process [such as primary sclerosing cholangitis (PSC)], associated autoimmune disorders [such as autoimmune hepatitis (AIH) and autoimmune pancreatitis (AIP)], or adverse consequences of IBD-associated bowel obstruction, IBD medications, or IBD surgery. Complications of these associated digestive system disorders may in return affect the GI tract, such as esophageal or gastric varices from portal hypertension in patients with concurrent IBD and PSC. Finally, functional bowel diseases may overlap with the disease course of IBD.

Atlas of Endoscopy Imaging in Inflammatory Bowel Disease. DOI: https://doi.org/10.1016/B978-0-12-814811-2.00024-4
© 2020 Elsevier Inc. All rights reserved.

Esophagus and stomach

While gastroesophageal reflux disorder (GERD) is common in general population, reflux esophagitis with erosions is common in patients with partial bowel obstruction from CD, small intestinal bacterial overgrowth, and IBD-associated surgery (such as strictureplasty) or surgical anastomosis strictures(Fig. 24.1). There is an association between CD, eosinophilic esophagitis (EoE), and lymphocytic esophagitis, particularly in the pediatric population [1−3]. In histology, CD of the esophagus is featured with chronic active inflammation, and rarely granulomas [4]. The histologic hallmark of EoE is the increased number of eosinophils in the esophageal epithelium with at least 15 eosinophils per high-power field [5]. Lymphocytic esophagitis is characterized by the prominence of peripapillary intraepithelial lymphocytes without remarked granulocytosis. True esophageal CD is rare in adult patients. However, CD-associated esophagitis should be distinguished from reflux esophagitis, EoE, and lymphocytic esophagitis (Fig. 24.2). Classic histologic features of esophageal CD include the infiltration of mononuclear cells with clusters and occasional granuloma formation. CD patients with severe malnutrition and iron deficiency anemia may present with esophageal strictures (i.e., Plummer−Vinson syndrome) [6]. Long-term use of antibiotics or corticosteroids, particularly in the presence of partial gastric outlet obstruction from small bowel obstruction in IBD, may predispose the patient to the development of *Candida* esophagitis (Fig. 24.3). Esophageal varices can occur in patients with IBD. The patients often have concurrent PSC, porto-mesenteric vein thrombosis, or non-alcoholic fatty liver disease−associated cirrhosis (Fig. 24.4).

Various forms of gastritis may be found in patients with CD or UC, ranging from nonspecific gastritis, nonsteroidal antiinflammatory drug (NSAID)−associated gastritis, *Helicobacter pylori*−associated gastritis [7], and proton pump inhibitor−associated nodular gastric mucosa or gastric pseudopolyps (Fig. 24.5) [8]. Nonspecific antral gastritis that is not associated with the use of NSAIDs or *H. pylori* infection is common in patients with CD. The association between *H. pylori* infection and CD is not settled [7,9].

Isolated pyloric stenosis can occur in patients with CD [10,11]. Pyloric stenosis in this setting typically shows no ulceration or inflammation in the stenosis site, gastric antrum or body, or duodenum (Figs. 24.6−24.8). These patients may or may not have CD in the small bowel or colon. The pattern may share a similar pathogenesis of immune-mediated achalasia. Another pattern in CD may be observed in the ileocecal valve (ICV) (see next). Pyloric stenosis in CD typically responds poorly to medical therapy or endoscopic balloon dilation (Fig. 24.6). These patients may be treated with endoscopic stricturotomy with a needle knife or insulated-tip knife (Fig. 24.7). Topical injection of botulinum toxin A (Fig. 24.6B and D) and gastric peroral endoscopic myotomy may be attempted (Fig. 24.8). Some patients may even require pyloroplasty surgery (Fig. 24.9). Peroral endoscopic gastrostomy tube has been used for a supplement of enteral nutrition, relief of symptoms of gastric outlet obstruction, or decompression of the stomach (Fig. 24.10).

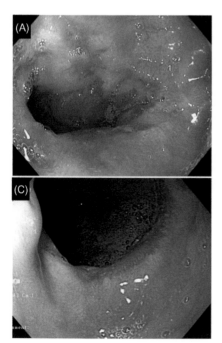
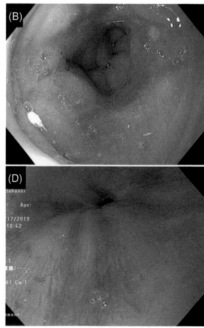

FIGURE 24.1 Erosive esophagitis in patients with Crohn's disease. Patients with stricturing Crohn's disease in the intestine leading to intermittent bowel obstruction are at risk for reflux-associated esophagitis: (A−D) linear or circumferential erosions in the distal esophagus.

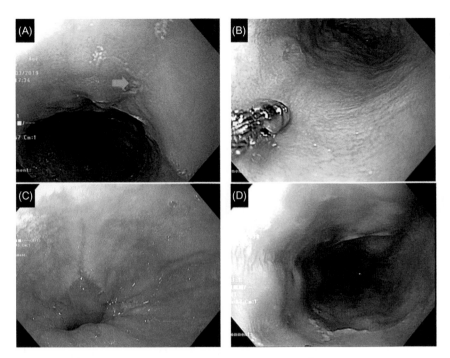

FIGURE 24.2 Esophageal Crohn's disease versus reflux esophagitis in Crohn's disease: (A and B) esophageal Crohn's disease characterized by ulcers with a raised age (*green arrows*). Biopsy showed infiltration of lymphocytes and histiocytes and (C and D) linear or circumferential ulcers in reflux esophagitis in two patients with Crohn's disease. Differential diagnosis needs histologic evaluation.

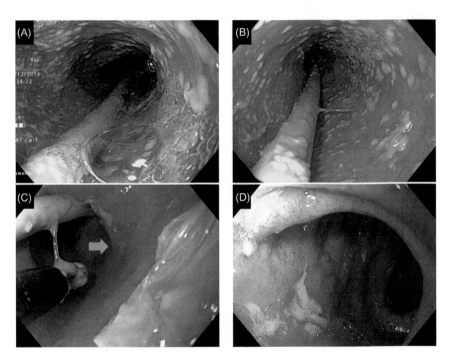

FIGURE 24.3 Candida esophagitis and esophageal strictures related to long-time antibiotic use in a patient with antibiotic-dependent pouchitis and current use of budesonide: (A—C) white plaques and films with a mild stricture (*green arrow*) in the esophagus. The patient was also on tube feeding due to dysphagia and (D) pouchitis with ulcers in the body of the ileal pouch.

Gastric lesions may present with gastric antral vascular ectasia (GAVE) or portal hypertensive gastropathy in PSC patients with portal hypertension and liver cirrhosis. Gastric varices may develop in IBD patients with portal hypertension from concurrent PSC, primary biliary cirrhosis, AIH-associated cirrhosis, porto-mesenteric vein thrombosis, or even nonalcoholic fatty liver disease (Figs. 24.11—24.13).

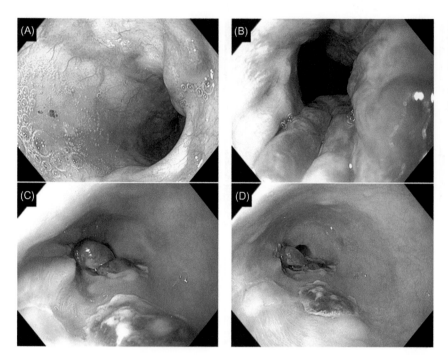

FIGURE 24.4 Esophageal varices in two patients with concurrent primary sclerosing cholangitis and ulcerative colitis: (A and B) prominent varices in the esophagus in patient 1 and (C and D) status post of banding ligation of esophageal varices resulting in ulceration in patient 2.

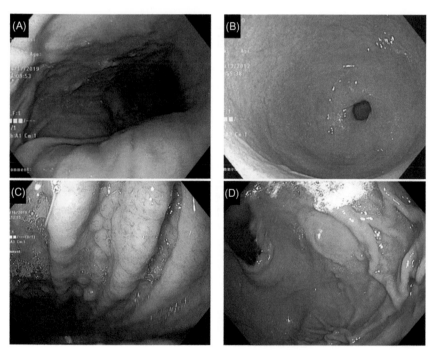

FIGURE 24.5 Various forms of gastritis and gastropathy in inflammatory bowel disease: (A) gastritis with linear erosions in a patient with Crohn's disease; (B) atrophic gastritis with loss of vascularity in a patient with primary sclerosing cholangitis and ulcerative colitis; and (C and D) granular mucosa and fundic gland polyps in two patients with Crohn's disease with a long-term use of proton pump inhibitors.

Small intestine

IBD may be associated with celiac disease, GI lymphoma, and IgG-associated enteropathy (Chapter 12: Inflammatory bowel disease—associated neoplasia, and Chapter 26: Inflammatory bowel disease—like conditions: other immune-mediated gastrointestinal disorders). IBD, particularly UC, may be associated with PSC. Nonspecific duodenitis

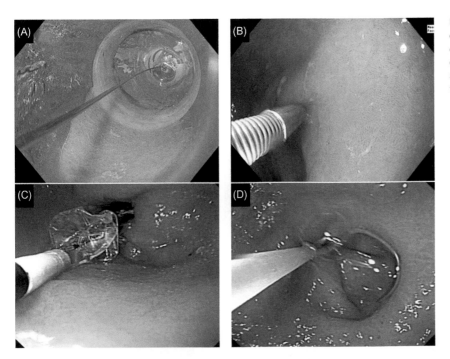

FIGURE 24.6 Pyloric stenosis in Crohn's disease treated with balloon dilation and topical injection of botulinum toxin: (A and C) endoscopic balloon dilation in two patients and (B and D) subsequent intralesional injection of botulinum toxin.

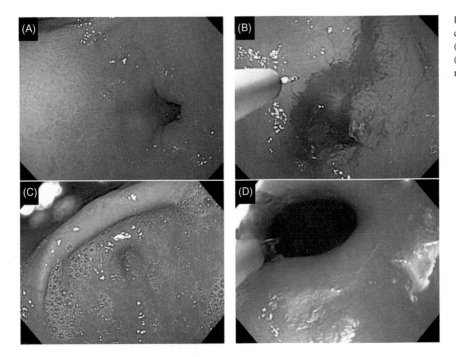

FIGURE 24.7 Pyloric stenosis in Crohn's disease treated with endoscopic eletroincision: (A and C) pyloric stenoses in two patients and (B and D) endoscopic electroincision with needle knife.

(i.e., the absence of noncaseating granulomas or chronic duodenitis with alterations of tissue structures) may be present on both CD and UC (Fig. 24.14A). IBD patients with underlying PSC or other liver disease or porto-mesenteric thrombosis may present with portal hypertensive duodenopathy (Fig. 24.15).

PSC-associated endoscopic or histologic inflammation in the duodenum, especially in the area around the papilla, can be encountered. Duodenitis may present with erythema, loss of folds, exudates, and small erosions. The etiology

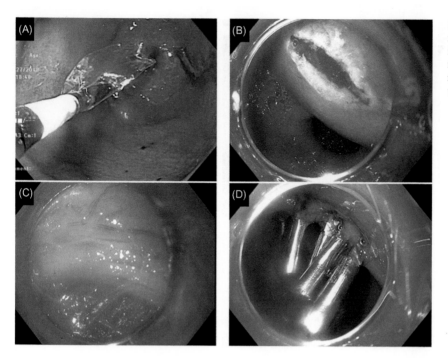

FIGURE 24.8 Gastric peroral endoscopic myotomy for the treatment of pyloric stenosis in Crohn's disease: (A) initial attempted was made to dilate pyloric stenosis with a poor response and (B–D) myotomy was performed with tunnel incision and clipping.

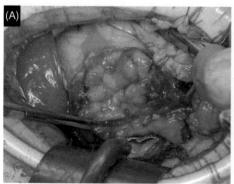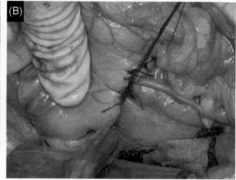

FIGURE 24.9 Pyloric stenosis in Crohn's disease treated with surgical strictureplasty: (A) pyloric stenosis and (B) completion of stricture-plasty. *Photo courtesy Dr. Tracy L. Hull of Cleveland Clinic.*

and pathogenesis of PSC-associated duodenitis are not clear. Speculated contributing factors for duodenitis or duodeno-pathy in patients with concurrent IBD and PSC include portal hypertension, contents of bile acids, IgG4, autoimmunity, immune suppressive–associated cytomegalovirus infection, and liver transplantation–associated graft-versus-host disease. Side-view duodenoscopy may find a retracted or embedded duodenum papilla, which may be more common in patients with PSC than those without (Fig. 24.16) [12]. In patients with PSC, varices may be found in the duodenum and peristomal area [13], as well as the esophagus, stomach, or large bowel (see later).

Surgical strictureplasty has been routinely performed in CD patients with strictured small intestine (Fig. 24.17A and B). Occasionally, gastric bypass surgery with gastrojejunostomy (Fig. 24.17C and D; see Fig. 24.18) or Whipple's procedure (i.e., pancreaticoduodenectomy with the resection of the head of the pancreas, duodenum, gall bladder, and common bile duct) (Fig. 24.19) is performed in patients with duodenum CD. These patients are assumed to have small intestinal bacterial overgrowth due to the presence of the blind loop or change in small bowel anatomy or motility. Food residues or bezoars may be retained in the bowel segment (Fig. 24.17).

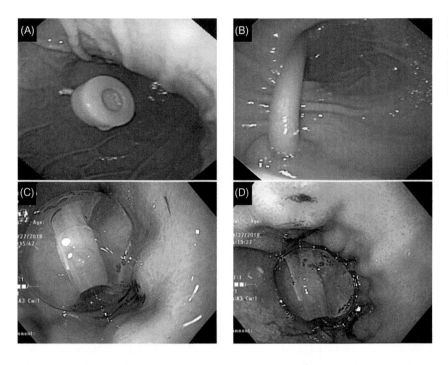

FIGURE 24.10 Pyloric stenosis treated with ventilating peroral endoscopic gastrostomy: (A) older version of bumper viewed from the stomach; (B) dislodged bumper; and (C and D) balloon-type bumper viewed from the stomach.

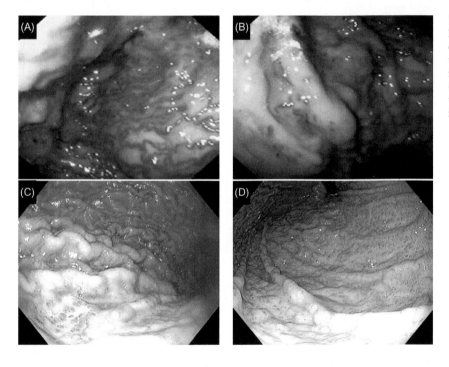

FIGURE 24.11 Gastric varices and portal hypertensive gastropathy in two patients with concurrent primary sclerosing cholangitis and ulcerative colitis: (A and B) prominent varices in the proximal gastric body in patient 1 and (C and D) portal hypertensive gastropathy with diffuse discrete erythema predominantly in the proximal stomach.

Colon

Patients with CD may present with isolated ICV stenosis without bowel inflammation on both sides of the valve. The ICV stenosis in this setting typically has normal-appearing mucosa with no ulcers or inflammation. The isolated ICV stenosis in CD may share the similar etiopathogenesis of achalasia of the distal esophageal sphincter and isolated

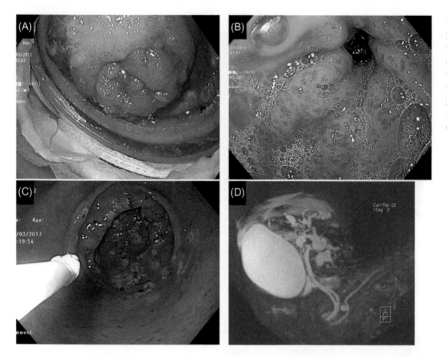

FIGURE 24.12 GAVE in a patient with primary sclerosing cholangitis, ulcerative colitis, colitis-associated cancer, failed ileal pouch-anal anastomosis: (A) end ileostomy created for the failed ileal pouch; (B) diffuse red spots in the gastric antrum from GAVE; (C) argon plasma coagulation for the treatment of GAVE; and (D) dilated intra- and extrahepatic bile ducts. *GAVE*, Gastric antral vascular ectasia.

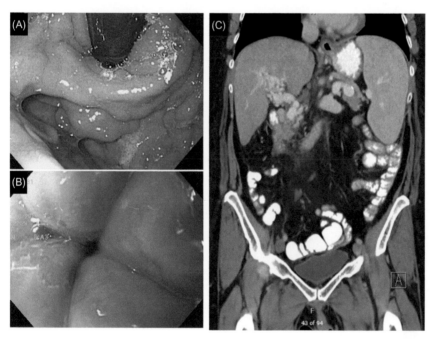

FIGURE 24.13 Gastric and esophageal varices resulting from portal and mesenteric vein thrombosis with cavernous formation in a patient with ulcerative colitis, restorative proctocolectomy, and ileal pouch−anal anastomosis: (A) gastric varices in the proximal stomach on a retroflex view of endoscopy; (B) esophageal varices; and (C) cavernous formation of portal and mesenteric vein thrombosis (*green arrow*).

pyloric stenosis in CD (see earlier). Patients commonly present with symptoms from a partial small bowel obstruction which poorly respond to conventional medical therapy. Endoscopic balloon dilation, endoscopic stricturotomy (i.e., endoscopic valvectomy) (Figs. 24.20 and 24.21) [10,14], or surgical resection are eventually needed.

Diverticulosis can coexist with IBD. However, IBD hardly presents with diverticular bleeding or diverticulitis. Diverticular colitis may mimic CD or ischemic colitis, due to its segmental distribution and presence of structural

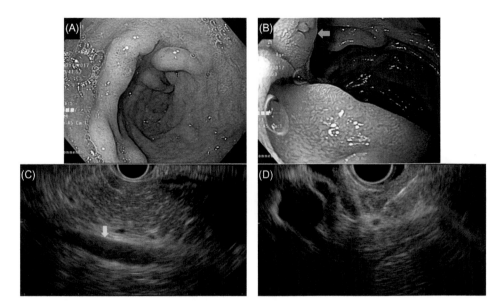

FIGURE 24.14 Duodenitis and cyst of the pancreas in a patient with ulcerative colitis: (A) duodenum erosions with dilated lumen; (B) bile flowing out from the papilla (*green arrow*); (C) liver parenchyma and portal vein (*yellow arrow*); and (D) small benign cyst (*blue arrow*) in the body of the pancreas.

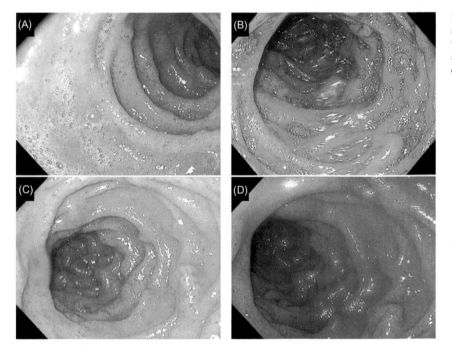

FIGURE 24.15 Portal hypertension—associated duodenopathy in a patient with concurrent ulcerative colitis and primary sclerosing cholangitis: (A—D) diffuse punctate erythema and friable mucosa in the duodenum.

changes on histology. Colonoscopy plays a key role in the differential diagnosis of IBD and diverticular disease. While colonoscopy is usually contraindicated in patients with suspected active diverticulitis with or without abscess, the procedure with a minimum air insufflation may incidentally be performed for the differential diagnosis of CD, ischemia, or GI lymphoma. Colonoscopic features of diverticulitis include the presence of excessive mucopurulent exudates, inflammation in the diverticulum, and inflammatory polypoid lesions (Fig. 24.22). Diverticular colitis is featured with the presence of inflammation in the mucosa around the orifice of diverticular while sparing the diverticulum per se (Fig. 24.23C and D). In contrast, patients with concurrent diverticulosis and

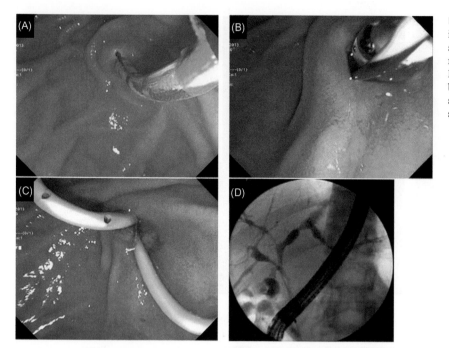

FIGURE 24.16 Primary sclerosing cholangitis in a patient with ulcerative colitis: (A−C) sphincterotomy, cannulation, and stent placement of the papilla to drain the bile duct. Notice that duodenum mucosa was normal, but the papilla was retracted and (D) and strictured intra- and extrahepatic ducts were shown on radiographic imaging.

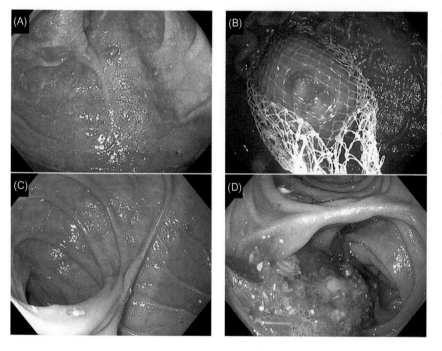

FIGURE 24.17 Small bowel bacterial overgrowth or bezoars in strictureplasty and gastric bypass surgery: (A and B) Finney strictureplasty for distal ileum stricture. Retained fecal bezoar in the lumen of strictureplasty site was removed by the endoscopic net and (C and D) retained food residues in Roux-en-Y gastric bypass site. The surgery was performed for prior operative bowel injury in the patient with restorative proctocolectomy.

active IBD typically show the presence of inflammation in the diverticulum and surrounding colon mucosa (Fig. 24.23A and B).

Diarrhea with frequent, loose bowel movements is common presentations in patients with IBD. These patients are not immune to functional bowel disorders [15]. Coexisting IBD and diarrhea-predominant irritable bowel syndrome

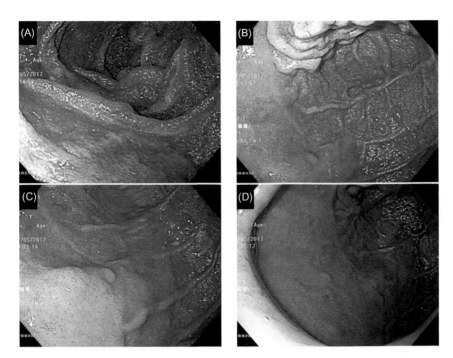

FIGURE 24.18 Billroth-II surgery for the treatment duodenum Crohn's disease: (A) normal efferent limb and (B–D) Crohn's disease in remission with patent anastomosis and normal efferent and afferent limbs.

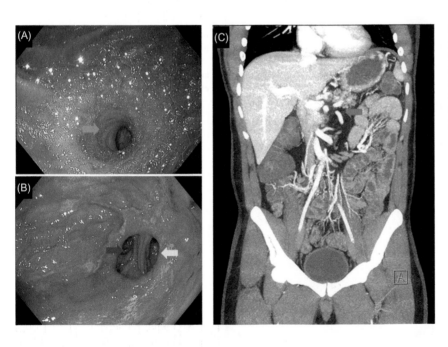

FIGURE 24.19 Duodenum Crohn's disease treated with Whipple's procedure: (A) orifice of the common bile duct at the stump of the afferent limb of the gastrojejunostomy (*green arrow*); (B) afferent (*yellow arrow*) and (*blue arrow*) limbs of the gastrojejunostomy; and (C) the site of Whipple's procedure on computed tomography (*red arrow*).

(IBS) are common, which may partially explain the poor correlation between patients' subjective symptomatology and objective endoscopic, radiographic, laboratory findings of inflammation, and other structural abnormalities. Whether IBD is a contributing factor for the development of IBS is not clear. The past few years have witnessed a growing interest in other functional bowel disorders in IBD, including constipation-predominant IBS, intestinal pseudoobstruction, colonic inertia or general GI dysmotility, and dyssynergic defecation [16,17].

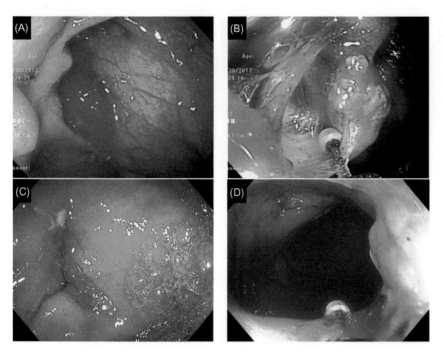

FIGURE 24.20 Isolated ileocecal valve stenosis in Crohn's disease treated with endoscopic valvectomy: (A and C) stenoses of the ileocecal valve with no or minimum inflammation on the epithelia in two patients and (B and D) endoscopic valvectomy with an insulated-tip knife.

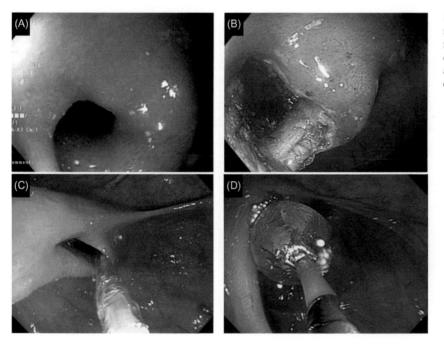

FIGURE 24.21 Isolated ileocecal valve stenosis in Crohn's disease: (A) nonulcerated stenosis at the ileocecal valve; (B) status post endoscopic stricturotomy/valvectomy; and (C and D) ileocecal valve stenosis treated with endoscopic balloon dilation.

Various structural and functional disorders can occur after restorative proctocolectomy and ileal pouch—anal anastomosis (IPAA) or continent ileostomy. Dyssynergic defecation with or without pouch outlet obstruction [18] and floppy pouch complex [19] can develop (Fig. 24.24). Dyssynergic evacuation is often associated with distal pouchitis or anterior cuffitis, which should be distinguished from classic microbiota-associated diffuse pouchitis and classic

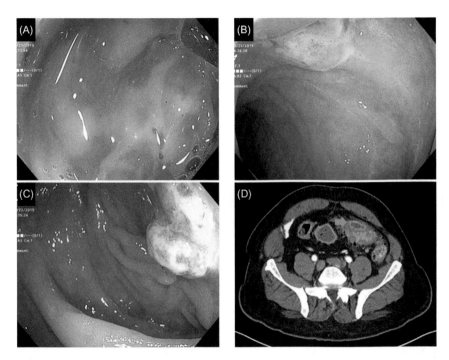

FIGURE 24.22 Recurrent diverticulitis in the sigmoid colon with abscess: (A) muco-purulent exudates in the lumen of the sigmoid colon; (B and C) diverticulitis with the orifice being covered with inflammatory polyps between ulcerative colitis with diverticulosis and diverticular colitis; and (D) abscess along the sigmoid colon (*green arrow*). Differential diagnosis should include diffuse ulcerative colitis with diverticuli and diverticular colitis. Notice that mucosal inflammation also involves diverticuli (*green arrow*).

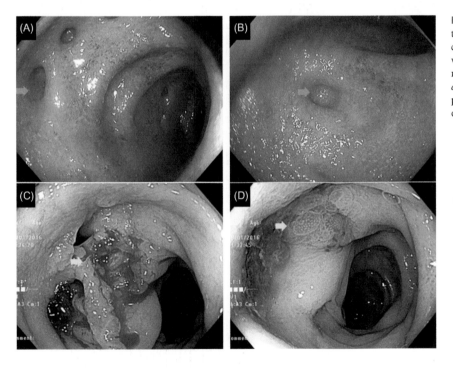

FIGURE 24.23 Distinction between ulcerative colitis with diverticulosis and diverticular colitis: (A and B) diffuse ulcerative colitis with diverticuli. Notice that mucosal inflammation also involves diverticuli (*green arrows*) and (C and D) diverticular colitis with predominantly involves the area around the orifice of diverticuli (*yellow arrows*).

circumferential cuffitis, a form of remnant UC (Figs. 24.25 and 24.26). The etiology and pathogenesis of floppy pouch complex are not clear. Risk factors for floppy pouch complex include a low body weight [20]. Other speculated factors include thing bowel wall of the pouch body, or decreased support from surgically created adhesions after laparoscopic IPAA or peripouch fat. Intestinal pseudoobstruction may occur in patients with IPAA or continent ileostomy with an

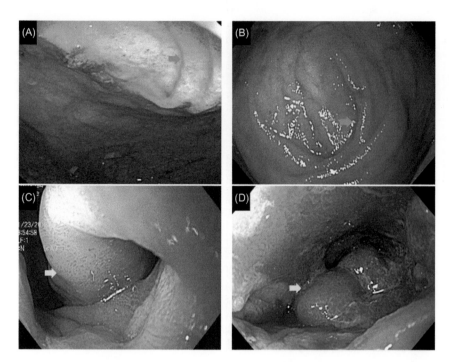

FIGURE 24.24 Phenotypes of flappy pouch complex associated with dyssynergic defecation: (A and B) afferent limb syndrome in a J pouch (A) and an S pouch (B) with a sharp angulation between pouch body and afferent limb (*green arrows*) and (C and D) distal pouch prolapse partially blocking the outlet (*yellow arrows*).

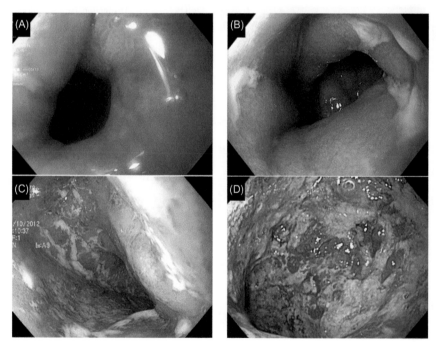

FIGURE 24.25 Distal pouchitis from dyssynergic defecation versus classic diffuse pouchitis: (A and B) tight, spastic anus (A) with paradoxical contractions on manometry in a patient with distal pouch ulcers from excessive straining (B); (C and D) classic pouchitis with diffuse distribution of inflammation in the entire pouch body.

unknown etiology, the patients present with megapouch with or without diffuse dilation of the small bowel (Figs. 24.27 and 24.28) [21].

Inflammatory or neoplastic polyps may occur in patients with long-standing enteritis, colitis, proctitis, pouchitis, or cuffitis. Diagnostic and therapeutic polypectomy may be performed. In patients with concurrent portal hypertension due

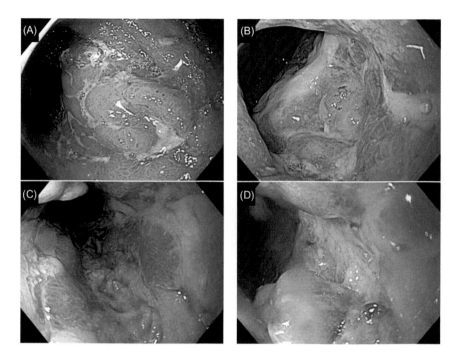

FIGURE 24.26 Anterior cuffitis from dyssynergic defecation versus classic circumferential cuffitis: (A and B) cuffitis at the anterior wall only at 4—5 o'clock from excessive straining and (C and D) classic circumferential cuffitis representing a remnant form of ulcerative colitis.

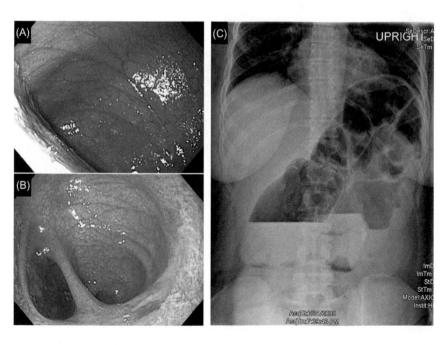

FIGURE 24.27 Intestinal pseudoobstruction in a patient with a J pouch for refractory ulcerative colitis: (A) dilated efferent limb; (B) dilated lumen of the J pouch; and (C) diffuse dilation of the small bowel and the pouch body.

to PSC or other liver or vascular conditions, polypectomy may carryer a high risk of bleeding than those without (Fig. 24.29).

Rectum and rectal cuff

Prolapse, hemorrhoids, and skin tags can develop in patients with IBD. Prominent veins in the healthy and patients with IBD should be differentiated from that seen in those with portal hypertension conditions, such as PSC

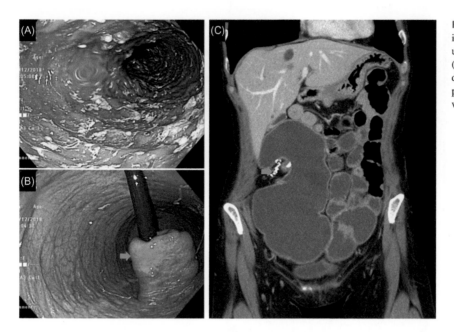

FIGURE 24.28 Intestinal pseudoobstruction in a patient with a Kock pouch for refractory ulcerative colitis: (A) dilated efferent limb; (B) dilated lumen of the Kock pouch; and (C) diffuse dilation of the small bowel and the pouch body. The nipple valve is highlighted with green arrows.

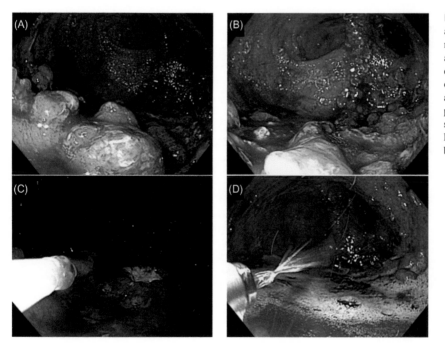

FIGURE 24.29 Endoscopic polypectomy in a patient with primary sclerosing cholangitis, restorative proctocolectomy, and ileal pouch—anal anastomosis for ulcerative colitis: (A and B) chronic pouchitis with the extensive formation of pseudopolyps in the distal pouch and cuff and (C and D) endoscopic polypectomy was performed for the concern of dysplasia in the setting of primary sclerosing cholangitis. Resulted polypectomy bleeding was controlled by spraying of 50% glucose.

(Figs. 24.30 and 24.31). Patients with concurrent active UC and varices may have mucosal inflammation on top of the dilated veins (Fig. 24.30C and D). Varices even can occur in the rectal cuff in PSC patients with IPAA (Fig. 24.32).

Pilonidal cyst occurs in the area close to the tailbone, which can be mistaken as presacral sinus from surgical ileorectal, colorectal, or IPAA leak (Fig. 24.33).

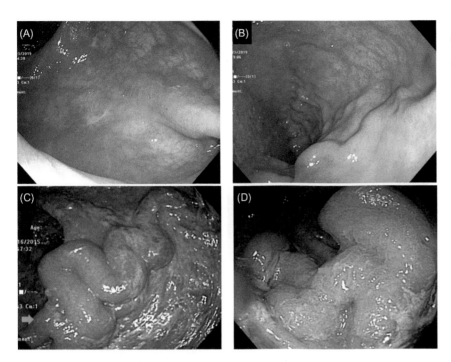

FIGURE 24.30 Prominent rectal veins versus rectal varices in the healthy and diseased: (A) prominent rectal veins in a healthy individual; (B) prominent rectal veins in a patient with quiescent ulcerative colitis; and (C and D) rectal varices with small erosions (*green arrows*) in a patient with concurrent ulcerative colitis and primary sclerosing cholangitis.

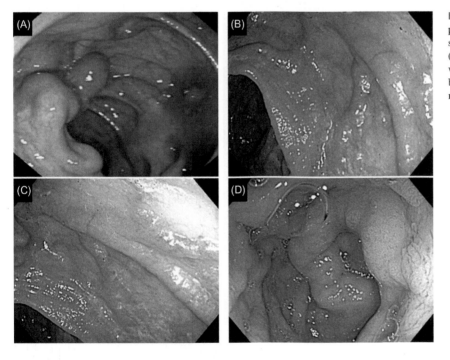

FIGURE 24.31 Varices in the large bowel in patients with ulcerative colitis and primary sclerosing cholangitis: (A) large cecal varices; (B and C) rectal varices; and (D) rectal varices with overlying granular mucosa, which should be distinguished from bowel inflammation or neoplasm.

Liver

The association between UC and PSC has been well established. The presence of PSC is shown to be risk factors for chronic pouchitis and enteritis in the afferent limb of the ileal pouch in UC patients with restorative proctocolectomy [22—24]. The presence of portal hypertension with or without thrombocytopenia predisposes the patients to the

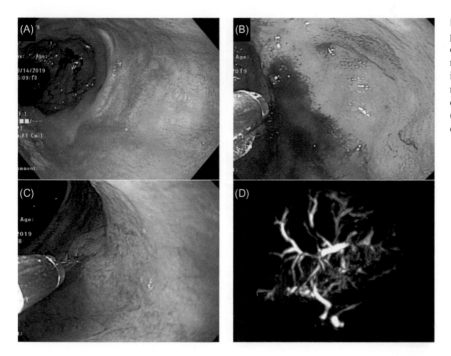

FIGURE 24.32 Varices in the rectal in a patient with concurrent primary sclerosing cholangitis and ulcerative colitis: (A) prominent veins in the rectal cuff; (B and C) bleeding from surveillance endoscopy for dysplasia resulting in excessive bleeding, which was controlled by spraying of 50% glucose; and (D) intrahepatic bile duct dilations and strictures on magnetic resonance cholangiopancreatography.

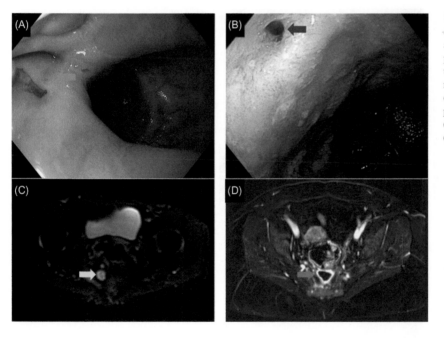

FIGURE 24.33 Pilonidal cyst versus presacral sinus in ileal pouch−anal anastomosis: (A and C) opening of the pilonidal cyst at the posterior wall of the rectal cuff (*green arrow*); the cyst located at the presacral space on magnetic resonance imaging (*yellow arrow*) and (B and D) presaral sinus from chronic pouch−anal anastomosis leak on endoscopy (*blue arrow*) and magnetic resonance imaging (*red arrow*).

development of endoscopic procedure (e.g., biopsy, polypectomy, and stricture dilation)−associated bleeding. Endoscopic retrograde cholangiopancreatography, magnetic resonance cholangiopancreatography, and endoscopic ultrasonography have been routinely used to assess the strictures and dilations of bile ducts and mass lesions in PSC (Figs. 24.34−24.37).

In addition to PSC, patients with IBD may carry a higher risk for primary biliary cirrhosis, AIH, overlap syndrome (Figs. 24.38 and 24.39), nonalcoholic fatty liver disease, hepatotoxicity from IBD medications (particularly purine

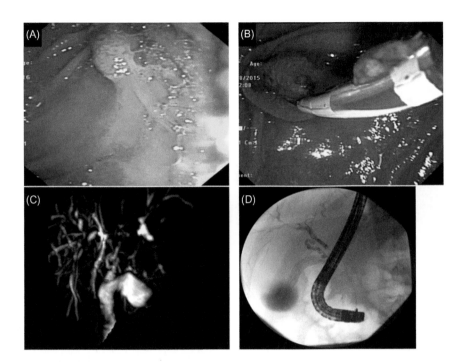

FIGURE 24.34 Primary sclerosing cholangitis coexisting with ulcerative colitis: (A and B) cannulation of the mildly inflamed duodenum papilla and (C and D) dilated and strictured intra- and extrahepatic bile ducts on magnetic resonance cholangiopancreatography and endoscopic retrograde cholangiopancreatography.

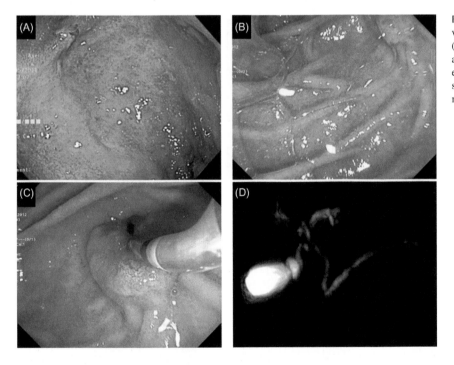

FIGURE 24.35 Concurrent ulcerative colitis, vasculitis, and primary sclerosing cholangitis: (A) diffuse ulcerative colitis with moderate activity; (B) normal duodenum mucosa; (C) endoscopic papillotomy; and (D) dilated and strictured intra- and extrahepatic bile duct on magnetic resonance cholangiopancreatography.

analogs, methotrexate, and antitumor necrosis factor agents), and cholelithiasis (Fig. 24.40) than the general population [25−28].

Portal vein or mesenteric vein thromboses are common in IBD, even in patients with quiescent disease [29,30]. Among patients with CD, mesenteric vein thrombosis was found to be associated with bowel stenosis and intestinal

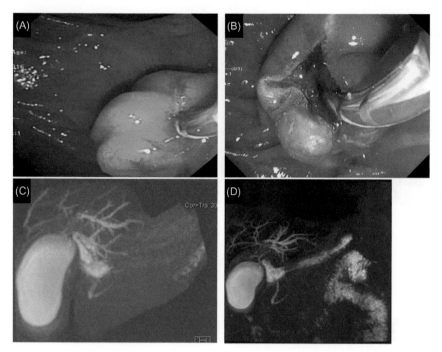

FIGURE 24.36 Primary sclerosing cholangitis in an ulcerative patient with restorative procto-colectomy and ileal pouch—anal anastomosis: (A and B) endoscopic cannulation of the papilla. Notice that the papilla was prominent, which can occur in patients with primary sclerosing cholangitis or IgG4-associated cholangiopathy and (C and D) dilated intrahepatic bile ducts with the intact gall bladder on magnetic resonance cholangiopancreatography.

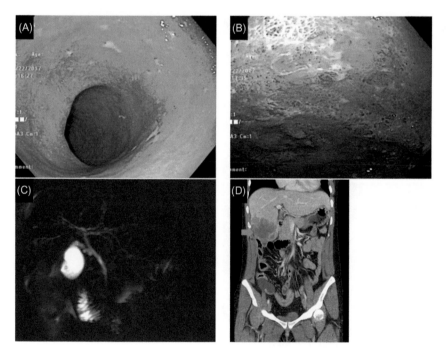

FIGURE 24.37 Primary sclerosing cholangitis—associated cholangiocarcinoma in a 32-year-old female patient with ulcerative colitis, restorative proctocolectomy, and ileal pouch—anal anastomosis: (A) extensive enteritis of the afferent limb with loss of vascularity and erosions; (B) diffuse pouchitis with erythema and loss of vascular pattern; (C) dilated with intra- and extrahepatic bile ducts; and (D) the large malignant mass at the right lobe of the liver (*green arrow*).

surgery [21]. UC or CD patients with porto-mesenteric vein thrombosis have more aggressive form of colitis and poorer prognosis [31]. UC patients undergoing colectomy are at risk for porto-mesenteric vein thrombosis and postoperative complications [32]. Patients with IPAA who had porto-mesenteric vein thrombosis were found to carry a higher risk for ischemic pouchitis [33] or chronic pouchitis (Figs. 24.41 and 24.42).

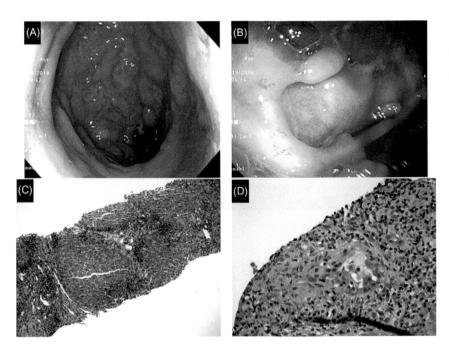

FIGURE 24.38 Concurrent ulcerative colitis and primary sclerosing cholangitis and autoimmune hepatitis overlap: (A and B) diffuse ulcerative colitis with ulcers and pseudopolyps; (C) moderate to severe hepatitis; and (D) destructive cholangitis. *Histology photo courtesy Dr. Xiuli Liu of University of Florida College of Medicine.*

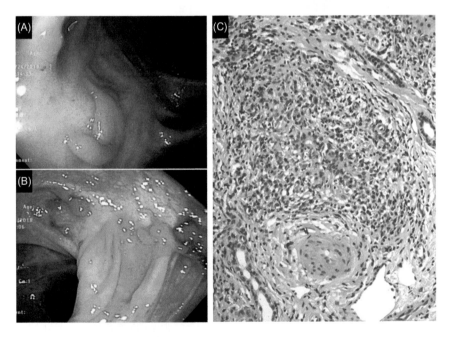

FIGURE 24.39 Concurrent ulcerative colitis and primary biliary cirrhosis and autoimmune hepatitis overlap: (A and B) ulcerative colitis in remission with pseudopolyps; and (C) injured duct in the center of that cluster of inflammatory cells from primary biliary cirrhosis and interface activity, lobular inflammation, and plasma cells infiltration from autoimmune hepatitis. *Histology photo courtesy Dr. Xiuli Liu of University of Florida College of Medicine.*

Chronic porto-mesenteric vein thrombosis in IBD is associated with esophageal, gastric, small or large bowel, and peristomal varices (Fig. 24.13).

Pancreas

Patients with IBD may present with various pancreas disorders, including acute and chronic pancreatitis, drug-induced pancreatitis, idiopathic pancreas, and AIP [34]. AIP is increasingly recognized in IBD, particularly for type 2 AIP in UC patients. There are two forms of AIP, with type 1 AIP being a part of a systemic IgG4-positive disease and meeting

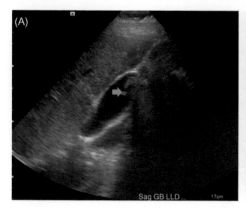
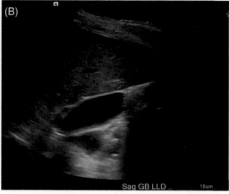

FIGURE 24.40 Cholelithiasis in a patient with Crohn's disease patient: (A and B) Gall bladder stones (*green arrow*) with a thickened wall.

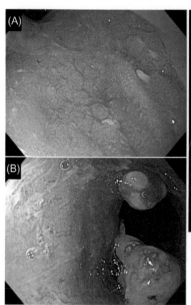
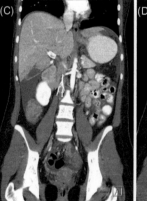
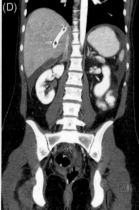

FIGURE 24.41 Extensive deep vein thrombosis involving the veins in the low extremities and portal vein in a patient with decompensated primary sclerosing cholangitis, ulcerative colitis, restorative proctocolectomy, and ileal pouch—anal anastomosis: (A and B) diffuse pouchitis with granular mucosa and pseudopolyps; (C) the filter was placed in the inferior vena cava (*green arrow*); and (D) transjugular intrahepatic portosystemic shunt stent in the liver.

the HISORt criteria [35,36], and type 2 AIP being idiopathic duct-centric pancreatitis with granulocytic lesions and absent IgG4-positive plasma cells and systemic involvement (Fig. 24.43). There is possible association between AIP-, PSC-, and IgG4-associated cholangitis. Duodenitis may be present in patients with concurrent pancreas disease (Fig. 24.14).

Summary and recommendations

In addition to classic disease features in the GI tract, IBD patients may present with intra- as well as extraintestinal manifestations. These disease conditions in the digestive systems may share some of pathogenetic pathways (such as PSC with UC) and presentations (such as varices in porto-mesenteric vein thrombosis and PSC) with IBD. The intraintestinal manifestations in IBD may result from IBD disease process per se, IBD medications, superimposed infections, IBD surgery, or IBD-associated complications. The involvement of IBD in the hepato-biliary-pancreatic system is beyond PSC. The evaluation of these disease conditions is accessible to endoscopy along with other transmural imaging. Investigations are needed for isolated pyloric stenosis or ICV stenosis in CD patients using esophageal achalasia as a control.

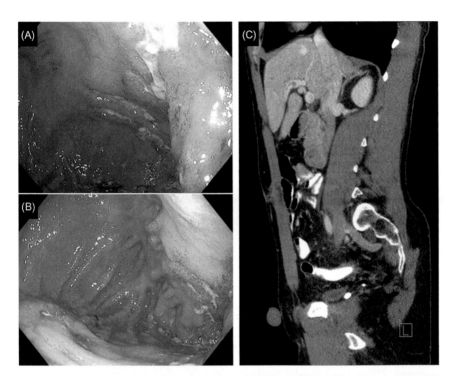

FIGURE 24.42 Portal vein thrombosis is considered a risk factor for ischemic pouchitis: (A and B) ischemic pouchitis with asymmetric ulcers and inflammation limited to the afferent limb part of the pouch body and (C) blood clot in the portal vein (*green arrow*).

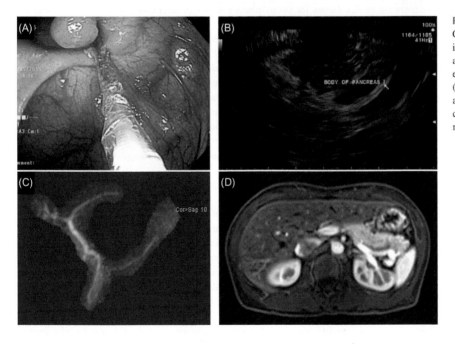

FIGURE 24.43 Concurrent ileal stricturing Crohn's disease and autoimmune pancreatitis in a 34-year-old female patient: (A) deformed and stricture ileocecal valve from Crohn's disease undergoing endoscopic dilation and (B–D) pancreatitis dominantly at the body and tail of pancreas with heterogenic parenchyma on endoscopic ultrasonography (B) and magnetic resonance imaging (C and D).

References

[1] Taft TH, Mutlu EA. The potential role of vedolizumab in concomitant eosinophilic esophagitis and Crohn's disease. Clin Gastroenterol Hepatol 2018;16:1840–1.

[2] Nguyen AD, Dunbar KB. How to approach lymphocytic esophagitis. Curr Gastroenterol Rep 2017;19:24.

[3] Haque S, Genta RM. Lymphocytic oesophagitis: clinicopathological aspects of an emerging condition. Gut 2012;61:1108–14.

[4] Decker GA, Loftus Jr EV, Pasha TM, Tremaine WJ, Sandborn WJ. Crohn's disease of the esophagus: clinical features and outcomes. Inflamm Bowel Dis 2001;7:113–19.

[5] Furuta GT, Katzka DA. Eosinophilic esophagitis. N Engl J Med 2015;373:1640–8.

[6] Park JM, Kim KO, Park CS, Jang BI. A case of Plummer-Vinson syndrome associated with Crohn's disease. Korean J Gastroenterol 2014;63:244–7.

[7] Lahat A, Kopylov U, Neuman S, Levhar N, Yablecovitch D, Avidan B, et al. on behalf of the Israeli IBD research Network (IIRN). *Helicobacter pylori* prevalence and clinical significance in patients with quiescent Crohn's disease. BMC Gastroenterol 2017;17:27 [E journal].

[8] Petrolla AA, Katz JA, Xin W. The clinical significance of focal enhanced gastritis in adults with isolated ileitis of the terminal ileum. J Gastroenterol 2008;43:524–30.

[9] Shah A, Talley NJ, Walker M, Koloski N, Morrison M, Burger D, et al. Is there a link between *H. pylori* and the epidemiology of Crohn's disease? Dig Dis Sci 2017;62:2472–80.

[10] Raghu Subramanian C, Triadafilopoulos G. The gates of hell: Crohn's disease isolated to the pylorus and ileo-cecal valve. Dig Dis Sci 2014;59:1108–11.

[11] Nakamura H, Yanai H, Miura O, Minamisono Y, Mitani N, Higaki S, et al. Pyloric stenosis due to Crohn's disease. J Gastroenterol 1998;33:739–42.

[12] Parlak E, Ciçek, Dişibeyaz S, Köksal AS, Sahin B. An endoscopic finding in patients with primary sclerosing cholangitis: retraction of the main duodenal papilla into the duodenum wall. Gastrointest Endosc 2007;65:532–6.

[13] Wiesner RH, LaRusso NF, Dozois RR, Beaver SJ. Peristomal varices after proctocolectomy in patients with primary sclerosing cholangitis. Gastroenterology 1986;90:316–22.

[14] Yang Y, Lyu W, Shen B. Endoscopic valvectomy of ileocecal valve stricture resulting in resolution of ileitis in Crohn's disease. Gastrointest Endosc 2018;88:195–6.

[15] Colombel JF, Shin A, Gibson PR. AGA clinical practice update on functional gastrointestinal symptoms in patients with inflammatory bowel disease: expert review. Clin Gastroenterol Hepatol 2019;17:380–90.

[16] Barros LL, Farias AQ, Rezaie A. Gastrointestinal motility and absorptive disorders in patients with inflammatory bowel diseases: prevalence, diagnosis and treatment. World J Gastroenterol 2019;25:4414–26.

[17] Abdalla SM, Kalra G, Moshiree B. Motility evaluation in the patient with inflammatory bowel disease. Gastrointest Endosc Clin N Am 2016;26:719–38.

[18] Khanna R, Li Y, Schroeder T, Brzezinski A, Lashner BA, Kiran RP, et al. Manometric evaluation of evacuatory difficulty (dyschezia) in ileal pouch patients. Inflamm Bowel Dis 2013;19:569–75.

[19] Khan F, Hull TL, Shen B. Diagnosis and management of floppy pouch complex. Gastroenterol Rep (Oxf) 2018;6:246–56.

[20] Freeha K, Gao XH, Hull TL, Shen B. Characterization of risk factors for floppy pouch complex in ulcerative colitis. Int J Colorectal Dis 2019;34:1061–7.

[21] Shashi P, Shen B. Characterization of megapouch in patients with restorative proctocolectomy. Surg Endosc 2019;33:2293–303.

[22] Shen B. Pouchitis: what every gastroenterologist needs to know. Clin Gastroenterol Hepatol. 2013;11:1538–49.

[23] Navaneethan U, Venkatesh PG, Bennett AE, Patel V, Hammel J, Kiran RP, et al. Impact of budesonide on liver function tests and gut inflammation in patients with primary sclerosing cholangitis and ileal pouch anal anastomosis. J Crohns Colitis 2012;6:536–42.

[24] Shen B, Bennett AE, Navaneethan U, Lian L, Shao Z, Kiran RP, et al. Primary sclerosing cholangitis is associated with endoscopic and histologic inflammation of the distal afferent limb in patients with ileal pouch-anal anastomosis. Inflamm Bowel Dis 2011;17:1890–900.

[25] Navaneethan U, Shen B. Hepatopancreatobiliary manifestations and complications associated with inflammatory bowel disease. Inflamm Bowel Dis 2010;16:1598–619.

[26] Chen CH, Lin CL, Kao CH. Association between inflammatory bowel disease and cholelithiasis: a nationwide population-based cohort study. Int J Environ Res Public Health 2018;15(3) https://doi.org/0.3390/ijerph15030513 pii: E513.

[27] Zou ZY, Shen B, Fan JG. Systematic review with meta-analysis: epidemiology of nonalcoholic fatty liver disease in patients with inflammatory bowel disease. Inflamm Bowel Dis 2019;25:1764–72.

[28] Silva J, Brito BS, Silva INN, Nóbrega VG, da Silva MCSM, Gomes HDN, et al. Frequency of hepatobiliary manifestations and concomitant liver disease in inflammatory bowel disease patients. Biomed Res Int 2019;2019:7604939 [E journal].

[29] Violi NV, Schoepfer AM, Fournier N, Guiu B, Bize P, Denys A. Prevalence and clinical importance of mesenteric venous thrombosis in the Swiss Inflammatory Bowel Disease Cohort Swiss Inflammatory Bowel Disease Cohort Study Group AJR Am J Roentgenol 2014;203:62–9.

[30] Landman C, Nahon S, Cosnes J, Bouhnik Y, Brixi-Benmansour H, Bouguen G, et al. Groupe d'Etude Thérapeutique des Affections Inflammatoires du Tube Digestif. Portomesenteric vein thrombosis in patients with inflammatory bowel disease. Inflamm Bowel Dis 2013;19:582–9.

[31] Arora Z, Wu X, Navaneethan U, Shen B. Non-surgical porto-mesenteric vein thrombosis is associated with worse long-term outcomes in inflammatory bowel diseases. Gastroenterol Rep (Oxf) 2016;4:210–15.

[32] Fichera A, Cicchiello LA, Mendelson DS, Greenstein AJ, Heimann TM. Superior mesenteric vein thrombosis after colectomy for inflammatory bowel disease: a not uncommon cause of postoperative acute abdominal pain. Dis Colon Rectum 2003;46:643–8.

[33] Shen B, Plesec TP, Remer E, Kiran P, Remzi FH, Lopez R, et al. Asymmetric endoscopic inflammation of the ileal pouch: a sign of ischemic pouchitis? Inflamm Bowel Dis 2010;16:836–46.

[34] Ramos LR, Sachar DB, DiMaio CJ, Colombel JF, Torres J. Inflammatory bowel disease and pancreatitis: a review. J Crohns Colitis 2016;10:95–104.

[35] Chari ST, Takahashi N, Levy MJ, Smyrk TC, Clain JE, Pearson RK, et al. A diagnostic strategy to distinguish autoimmune pancreatitis from pancreatic cancer. Clin Gastroenterol Hepatol 2009;7:1097–103.

[36] Chari ST. Diagnosis of autoimmune pancreatitis using its five cardinal features: introducing the Mayo Clinic's HISORt criteria. J Gastroenterol 2007;42(Suppl. 18):39–41.

Chapter 25

Inflammatory bowel disease–like conditions: infectious

Danfeng Lan and Yinglei Miao

Department of Gastroenterology, The First Affiliated Hospital of Kunming Medical University, Yunnan Institute of Digestive Disease, Kunming, P.R. China

Chapter Outline

Abbreviations

AIDS	acquired immune deficiency syndrome
CD	Crohn's disease
CMV	*Cytomegalovirus*
GI	gastrointestinal
HIV	human immunodeficiency virus
IBD	inflammatory bowel disease
ICV	ileocecal valve
UC	ulcerative colitis

Introduction

Infectious bowel diseases are more common in developing countries than industrialized countries. On the other hand, the incidence and prevalence of inflammatory bowel disease (IBD) in the former are increasing [1]. Superimposed bacterial, viral, fungal, or parasitic infection occur in patients with underlying IBD with or without concurrent use of immunosuppressive drugs. Fortunately, the majority of patients with primary infectious bowel diseases are self-limited entities. Conventional stool culture, and ova and parasites are commonly performed to identify pathogens. Multipathogen molecular panels on diarrheal stool samples and, in some cases, rectal swabs have emerged as an important diagnostic modality for the identification of various bacterial, viral, and parasitic pathogens.

Chronic infection with certain pathogens (such as tuberculosis) can lead to chronic bowel inflammation, posing a challenge in the differential diagnosis with true IBD. While upper and lower gastrointestinal (GI) endoscopies offer limited diagnostic value in acute enteritis or acute colitis, they are valuable tools for the evaluation of chronic bowel symptoms (>weeks of duration). Some of infectious enteritis or colitis have characteristic features on endoscopy. In

Atlas of Endoscopy Imaging in Inflammatory Bowel Disease. DOI: https://doi.org/10.1016/B978-0-12-814811-2.00025-6
© 2020 Elsevier Inc. All rights reserved.

addition, endoscopy provides access to tissue sampling. In this chapter, endoscopic features of common bacteria, virus, and parasite-associated enteritis or colitis, which mimics IBD are described (Table 25.1). Superimposed infection in IBD is discussed in a separate chapter (Chapter 23: Superimposed infections in inflammatory bowel diseases).

Bacteria-associated bowel diseases

A variety of bacterial agents can cause enteritis and colitis. Some of them, particularly, intracellular bacteria can cause chronic enteritis or colitis, mimicking IBD.

Intestinal tuberculosis

The small and large bowel is a common site for extrapulmonary tuberculosis resulting from *Mycobacterium tuberculosis*. The causal agents can be *M. tuberculosis* or *Mycobacterium bovis*. The pathogenetic route of infection of the latter agent is through drinking contaminated dairy products. The main routes of infection of intestinal tuberculosis (ITB) are GI tract, blood flow, or adjacent tuberculous lesions. The gold standard for the diagnosis of ITB is identification of the bacteria in the stool or GI tissue specimens. However, the detection of the bacterial agent can be challenging. Skin and serological tests can only provide supportive evidence, and their use in the differential diagnosis of ITB and Crohn's disease (CD) has been limited, especially in endemic areas of tuberculosis.

ITB and CD share many aspects of clinical, endoscopic, radiographic, and histologic examinations. Both can be presented with low-grade fever, night sweats, abdominal pain, diarrhea, or abdominal mass. Both have a predilection of involvement of the terminal ileum and ileocecal valve (ICV). Parenteral manifestations and intestinal fistula, however, appear to be more common in CD than ITB.

Endoscopy can play an important role in the differential diagnosis of ITB and CD (Table 25.2). Generally, the endoscopic features of CD include aphthous ulcers, asymmetrical longitudinal ulcers, skip lesions, and the presence of strictures, fistulae, or perianal disease. The ICV often has ulcers, stricture, or deformity in patients with distal ileum ITB. In contrast, ulcers in ITB are often circumferentially distributed. While mucosal nodularity of the ileum is more common in ITB, cobblestoning of the ileal mucosa is more often seen in CD (Figs. 25.1 and 25.2) [2−4]. Histologic distinction between ITB and CD is discussed in a separate chapter (Chapter 34: Histology correlation with common endoscopic

TABLE 25.1 Common gastrointestinal pathogen-associated bowel diseases mimicking inflammatory bowel diseases.

Bacteria	*Mycobacterium tuberculosis, Clostridium difficile, Salmonella, Shigella, Yersinia, Enterohemorrhagic Escherichia coli, Campylobacter, Aeromonas, Plesiomonas*
Viruses	Human immunodeficiency virus, *Cytomegalovirus*, Epstein−Barr virus
Fungi	*Candida albicans, Histoplasma, Coccidioides, Blastomyces dermatitidis*
Parasites	*Giardia, Cryptosporidium lamblia, Entamoeba histolytica, Strongyloides stercoralis, Schistosoma*

TABLE 25.2 Comparison of endoscopic features of intestinal tuberculosis and Crohn's disease.

	Intestinal tuberculosis	Crohn's disease
Orientation of ulcers	Traverse	Longitudinal
Aphthous erosions	+/−	++
Cobblestoning mucosa	+	++
Nodular mucosa	++	+
Mucosal bridge	+/−	+
Stricture	+/−	+++
Fistula	+	++
Ileocecal valve feature	Patulous	Ulcerated, strictured, or deformed
Colon and rectum involvement	+/−	+
Perianal disease	+/−	++
Skip lesion	+/−	+++

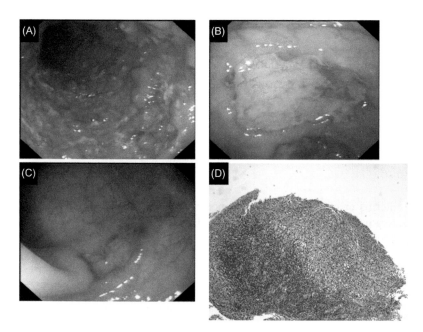

FIGURE 25.1 Tuberculosis in the colon: (A) Nodularity of colon mucosa; (B and C) discrete small and large superficial ulcers in the colon; (D) tissue biopsy showed granulomas.

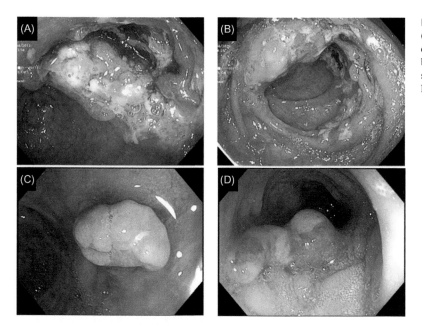

FIGURE 25.2 Patterns of intestinal tuberculosis: (A) Ulcerated, patent ICV; (B) multiple ulcers in a circumferential fashion in the cecum and ICV on the background of mucosal edema; (C and D) tuberculosis in the terminal ileum in remission with pseudopolyps. *ICV*, Ileocecal valve.

abnormalities). The overlapping features between ITB and CD can make the differential diagnosis difficult, which has mandated empiric anti-ITB therapy in some patients.

Intracellular enteric pathogens

Invasive, intracellular bacteria, such as *Salmonella enterica serotype typhi* and *Yersinia enterocolitica* can affect GI tract, with the former causing gastroenteritis and the latter leading to enterocolitis. These foodborne illnesses normally result in acute, self-limited GI diseases. However, some of the patients may develop chronic enteritis or colitis with phenotypes overlapping with that of CD. Their chronic disease process and transmural involvement can have inflammatory, stricturing, or penetrating phenotypes, with manifestations of low-grade fever, night sweat, nausea, bloating, abdominal pain, diarrhea, hematochezia, abdominal mass, or even ascites. On endoscopy, there can be nodularity, ulcers, and strictures (Fig. 25.3) [5,6]. Its distinction from IBD, particularly CD, is critical, as the management of the two disease entities is different.

Clostridioides (formerly Clostridium) difficile

Clostridioides difficile infection can involve the colon, small bowel [7], as well as ileal pouches [8]. Patients may or may not have concurrent underlying IBD. Superimposed *C. difficile* infection in patients with underlying IBD is discussed in a separate chapter (Chapter 23: Superimposed infections in inflammatory bowel diseases). *C. difficile* colitis in non-IBD patients has a wide spectrum of endoscopic features, ranging from diffuse bowel wall edema to classic pseudomembranous colitis. Chronic or recurrent *C. difficile* colitis can even develop inflammatory polyps (Fig. 25.4). Pseudomembranous are rarely found in patients with superimposed *C. difficile* infection on underlying IBD.

Virus-associated bowel diseases

Infection of some viruses may cause chronic enteritis or colitis in immune competent or immunocompromised hosts. Those virus-associated chronic bowel diseases are a part of differential diagnosis of IBD.

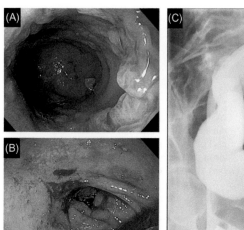
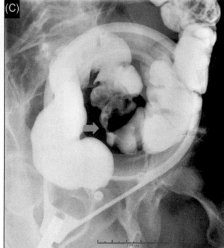

FIGURE 25.3 Sigmoid colon stricture with stool cultures positive for *Salmonella* Group C and *Blastocystis hominis*: (A) Distal proctitis with ulcerated and nodular mucosa; (B) stricture of the sigmoid colon; (C) stricture at the sigmoid colon on gastrografin enema (*green arrow*).

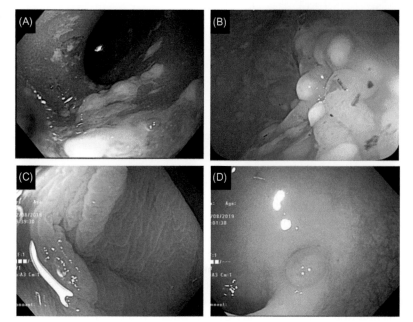

FIGURE 25.4 Patterns of *Clostridioides difficile* infection in the small and large bowel: (A) *C. difficile* enteritis in a patient with an ileostomy. There were ulcers covered with pseudomembranes; (B) classic *C. difficile* colitis showing diffuse erosions covered with pseudomembranes; (C) *C. difficile* colitis presenting only with diffuse bowel wall edema; (D) chronic *C. difficile* infection with inflammatory polyp. *Endoscopy image (B): Courtesy Dr. Mei Wang of Affiliated Hospital of Yangzhou University.*

Human immune deficiency virus

Primary or secondary immune deficiency can affect GI tract, causing gastritis, enteritis, and colitis, which share common endoscopic and histologic features with CD and ulcerative colitis (UC).

Acquired immune deficiency syndrome (AIDS) is a serious disease caused by the infection of human immunodeficiency virus (HIV). More than 90% of patients of AIDS have some forms of GI infection, due to severe immunodeficiency. The clinical manifestations of HIV-associated bowel disease include persistent fever, refractory diarrhea, progressive emaciation, opportunistic infections, and even malignancy.

T cells, as well as epithelial cells, in GI tract can be affected by HIV. The early GI manifestation of HIV-associated bowel disease is usually the presence of colitis. Endoscopic appearance of HIV-associated colitis can be similar to that in UC, including edema, erythema, and ulcers (Fig. 25.5). Concurrent infection of HIV and *Cytomegalovirus* (CMV) can cause more severe and more extensive GI disorders, with diffuse ulceration and even perforation.

Cytomegalovirus

CMV is the type 5 of human herpes virus, belonging to the subfamily of herpes virus. It is a linear double-stranded DNA virus. Human is the only host of CMV. The finding of the "owl eye cells" in the pathology is diagnostic. While primary CMV colitis in immune competent patients is rare, CMV colitis as a part of opportunistic infection in the setting of immune suppression is common. CMV infection in the immunocompetent host is generally asymptomatic or may manifest mononucleosis syndrome. In contrast, CMV infection in immunocompromised patients can result in significant morbidities and even mortality. CMV colitis is almost always the case of reactivation of latent CMV in the setting of immune suppression.

Cytomegalovirus colitis or CMV infection in patients with underlying IBD is common, especially in those with a long-standing treatment of glucocorticoids or other immunosuppressants. Superimposed CMV colitis in IBD is discussed in a separate chapter (Chapter 23: Superimposed infections in inflammatory bowel diseases). The endoscopic manifestations of IBD complicated with CMV infection include scattered multiple ulcers with various configurations, ranging from deep chiseling ulcers, irregular or geographic ulcers to punched-out ulcers. In severe cases a large area of mucosa or submucosa can be denuded. Those types of ulceration are different from those of UC or Crohn's colitis. Strictures in CMV colitis are not common [9]. The hematoxylin and eosin with "owl eye" nuclear inclusion bodies and immunohistochemistry of mucosal biopsy, as well as the CMV-DNA quantification of blood, are helpful for the diagnosis (Figs. 25.6—25.8).

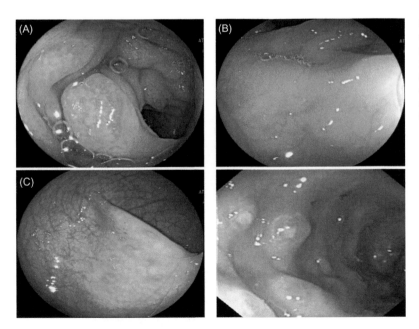

FIGURE 25.5 Patterns of HIV-associated colitis: (A) Edema, erythema, and polypoid lesions in the terminal ileum; (B) edema and nodularity of the mucosa in the descending colon; (C) patchy erythema and erosions in the sigmoid colon; (D) nodular mucosa with erosions and exudates surrounded by diffuse mucosal edema in the sigmoid colon. *HIV*, Human immunodeficiency virus.

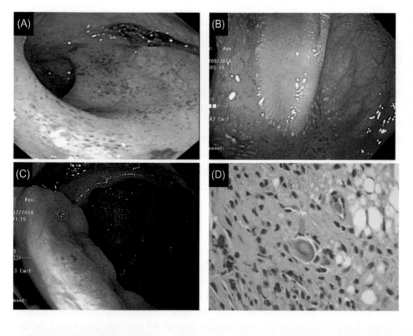

FIGURE 25.6 Patterns of CMV-associated colitis: (A) CMV colitis after liver transplantation with diffuse punctate erythema throughout colon; (B) CMV colitis with patchy erythema and friable mucosa; (C and D) CMV colitis manifested as large, discrete, clean-based ulcer with raised edge. Histology of biopsy of the edge of the ulcer showed characteristic giant cell with inclusion bodies (*green arrow*). *CMV*, Cytomegalovirus.

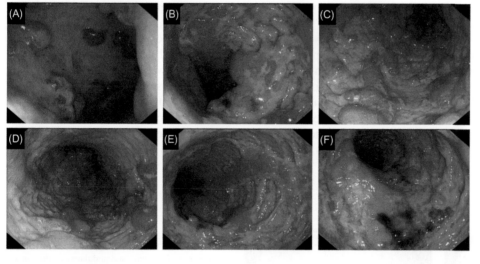

FIGURE 25.7 Ulcerative colitis complicated with cytomegalovirus infection: (A–F) Diffuse hyperemia, edema, ulcers, and denuded mucosa and submucosa from the descending colon to rectum. There were islets of hyperplastic nodules in between.

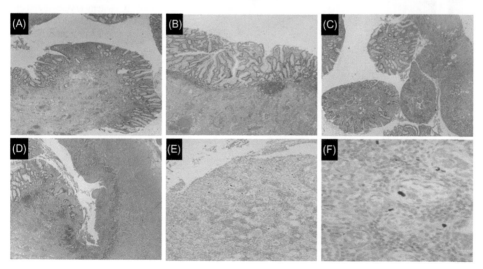

FIGURE 25.8 Ulcerative colitis complicated with CMV infection: (A and B) Diffuse atrophy and crypt distortion in the intestinal mucosa; (C) multiple inflammatory polyps; (D) ulcer involving superficial muscle layer; (E and F) CMV-stained mononuclear cells on low- and high-power fields. *CMV*, Cytomegalovirus.

Epstein—Barr virus

Epstein—Barr virus (EBV) is a human herpes virus with a complex genome. After adult is infected with EBV, the clinical manifestations and outcomes are diverse, including primary acute infection, chronic active infection, lymphoproliferative disease, and tumor. Chronic active EBV infection can involve the GI tract, which often results from immune suppression. IBD patients have an increased risk of EBV infection, which may aggravate IBD and increase the risk for lymphoma, especially with the use of thiopurines. Endoscopic features of EBV-associated enteritis or colitis are not specific, with multiple discrete ulcers in the small intestine or colon, accompanied by erythema, edema, and erosion of mucosa, which is difficult to distinguish from underlying active IBD. Histologic evaluation with in situ hybridization is diagnostic (Figs. 25.9 and 25.10).

Human papilla virus

Sexually transmitted diseases, such as condyloma acuminatum of the anus caused by epidermotropic human papilloma virus may mimic the presentation of distal ulcerative proctitis or perianal CD (Fig. 25.11).

Fungus-associated bowel diseases

Fungi are a component of the intestinal microecology, accounting for approximately 0.1%. Candida is one of the normal floras, which exists in human body and environment widely. Candida is the most common opportunistic pathogen, which is located in various organs of the human body, including oropharynx, nasopharynx, GI tract, urethra, and vagina. Common colonized *Candida* species in humans are *Candida albicans*, *Candida glabrata*, and *Candida parapsilosis*. There is a competitive relationship between fungi and bacteria in the gut, and the use of antibiotics or

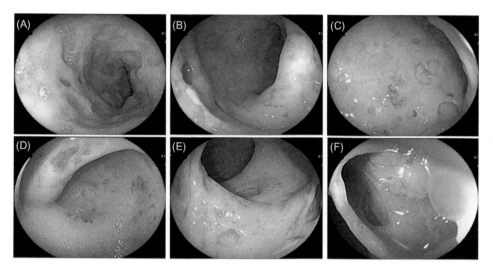

FIGURE 25.9 Chronic intestinal EBV infection. Scattered congestion and erythema of the intestinal mucosa with punched-out ulcers in the (D) the normal mucosa was observed in the descending colon (A and B), sigmoid colon (C and D), and rectum (E and F). The patient underwent colectomy for refractory disease. *EBV*, Epstein—Barr virus.

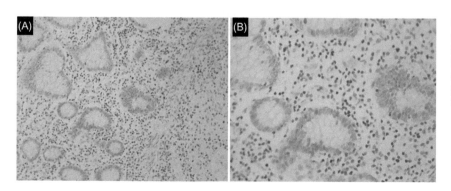

FIGURE 25.10 Histopathology of colectomy specimen of EBV infection. The tissue specimen is collected from the above patient (Fig. 25.9). EBV-infected lymphocytes highlighted with chromogen in situ hybridization in low-power (A) and high-power (B) fields. *EBV*, Epstein—Barr virus.

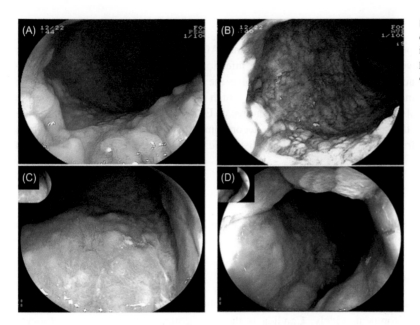

FIGURE 25.11 Condyloma acuminatum of the anus caused by epidermotropic HPV, the pattern mimicking distal ulcerative proctitis (A−D). *HPV,* Human papilloma virus. *Photos courtesy Dr. Yubei Gu of Shanghai Jiaotong University Ruijin Hospital.*

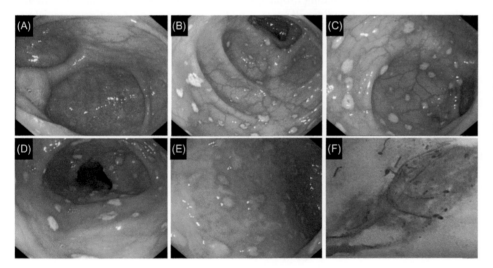

FIGURE 25.12 Fungal colitis in an immunocompromised host. Diffuse erosions with overlying white plaques in the cecum (A), ascending colon (B), transverse colon (C), descending colon (D), and rectum (E); (F) fungal hyphae and spores are found in the smear from the white plaques.

immunosuppressive agents can lead to fungal overgrowth. Fungal enteritis or colitis is more common in infants, pregnant women, the elderly and the infirm and immunodeficiency, especially in the patients of long-term use of antibiotics, glucocorticosteroids, and other immunosuppressive agents. Therefore fungal enteritis or colitis can occur in patients with underlying IBD for which immunosuppressive drugs are commonly used. Fungal enteritis or colitis is often accompanied by thrush in oral cavity, tongue, and throat. Colonoscopy often reveals patchy erythema, superficial ulcers, or white plaques. Candida and mycelia can be found in the brushed white patches (Fig. 25.12).

Histoplasmosis is an infection caused by *Histoplasma*. It can occur in immune competent or immunocompromised hosts. Histoplasmosis can involve multiple organs, with the GI tract being one of them. The GI presentation and endoscopic features (such as nodular mucosa and ulcers) of histoplasmosis may mimic that of IBD (Fig. 25.13).

Parasite-associated bowel diseases

Primary parasite-associated bowel diseases mainly occur in endemic areas. However, some of the infections can occur in immunocompromised patients in developed countries. There are overlapping endoscopic features between chronic parasite-associated bowel diseases and IBD.

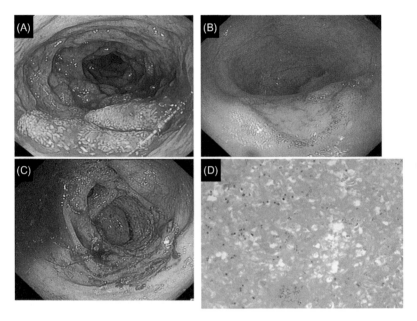

FIGURE 25.13 Systemic histoplasmosis involving the gastrointestinal tract. The female patient represented with fever, cervical lymphadenopathy, bloating, and chronic diarrhea: (A) Duodenum, ulcers, and friable mucosa; (B) discrete shallow, clean ulcers in the terminal ileum; (C) ulcers, nodularity, and spontaneous bleeding in the proximal colon; (D) fungal spores on mucosal biopsy from the terminal ileum lesion (silver stain). *Endoscopy images: Courtesy Drs. Qunyin Wang, Jin Ding, Chong Lu of Jinhua Hospital of Zhejiang University.*

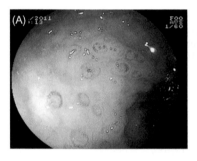
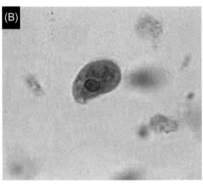

FIGURE 25.14 Intestinal amebiasis resulting from *Entamoeba histolytica* infection: (A) Oval erosion with overlying yellowish exudates, surrounded by red halos in the colon; (B) the smear of ulcerative exudate showed amebic cyst.

Entamoeba histolytica

Intestinal amebiasis is a disease caused by *Entamoeba histolytica* parasitic in the colon, mainly involving the proximal colon and cecum. The common endoscopic manifestations are bowel wall edema, isolated, deep ulcers with central indentations, ulcerated mass, and purulent or dark red secretions. The margins are hyperemic and the mucosa between the ulcers is usually normal [10]. *E. histolytica* trophozoites or cysts can be found in the fecal or ulcerative exudates (Fig. 25.14). There has been a concern on the risk of perforation from deep ulcers of intestinal amebiasis by colonoscopy [11].

Schistosoma

Schistosomiasis is a disease caused by infection of one of three major species *Schistosoma mansoni*, *Schistosoma japonicum*, and *Schistosoma haematobium*. Colonic schistosomiasis is caused by the deposition of large numbers of schistosome eggs on the wall of the colon. The patient often has a large liver, spleen, and a history of water exposure in the endemic area. Colonic schistosomiasis has a segmental distribution, mainly affecting sigmoid colon and rectum. Colonoscopy may show hyperemia, edema, flaky hemorrhage, superficial ulcers in the early stage; pale, atrophic, and rough mucosa in the later stage; and fibrotic scarring can be found in the intestinal wall. The presence of schistosome eggs in feces is diagnostic (Figs. 25.15−25.17).

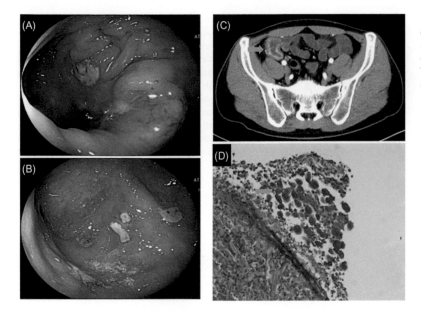

FIGURE 25.15 Amebiasis in the cecum; (A and B) edema, erythema, ulcers, and exudates in the cecum; (C) thickened cecum with mucosal hyperenhancement on computed tomography enterography; (D) H&E of biopsy from the cecum showing ameba trophozoites (*green arrow*). *Photo courtesy: Dr. Yue Li of Peking Union College Medicine Hospital.*

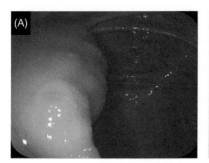

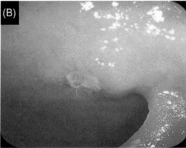

FIGURE 25.16 Colitis associated with schistosomiasis: (A and B) Discrete erosions and ulcers throughout the colon. *Endoscopy images: Courtesy Dr. Mei Wang of Affiliated Hospital of Yangzhou University.*

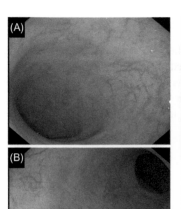

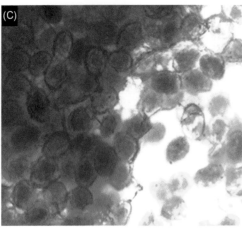

FIGURE 25.17 Colonic schistosomiasis. Patchy, flake, yellowish mucosa in the sigmoid colon (A) and rectum (B). The pattern is similar to mucosal scars in treated ulcerative colitis or Crohn's colitis. The smear of colon mucosa showed schistosome eggs (C).

Summary and recommendations

Acute infectious enteritis or colitis is usually self-limited, which is commonly foodborne in etiology or occur in endemic areas or in the immunocompromised host. However, some bacterial, fungal, viral, or parasitic bowel diseases can occur in immune competent patients and canrun a prolonged or chronic disease course. Chronic infectious enteritis or colitis resembles IBD in disease course, endoscopic, histologic, and radiographic features. These infectious agents can cause superimposed diseases on top of underlying IBD. Colonoscopy and tissue biopsy provides a valuable tool for the diagnosis and differential diagnosis, with documentation of disease extent, disease distribution, disease pattern, and disease severity.

Disclosure

The authors declared no financial conflict of interest.

References

[1]   Ng SC, Shi HY, Hamidi N, Underwood FE, Tang W, Benchimol EI, et al. Worldwide incidence and prevalence of inflammatory bowel disease in the 21st century: a systematic review of population-based studies. Lancet 2018;390:2769—78.

[2]   Miao YL, Ouyang Q, Chen DY. The role of endoscopic and pathologic examination in the differential diagnosis of Crohn's disease and intestinal tuberculosis. Chin J Dig Endosc 2002;19:9—12 [Chinese].

[3]   Limsrivilai J, Shreiner AB, Pongpaibul A, Laohapand C, Boonanuwat R, Pausawasdi N, et al. Meta-analytic Bayesian model for differentiating intestinal tuberculosis from Crohn's disease. Am J Gastroenterol 2017;112:415—27.

[4]   Jung Y, Hwangbo Y, Yoon SM, Koo HS, Shin HD, Shin JE, et al. Predictive factors for differentiating between Crohn's disease and intestinal tuberculosis in Koreans. Am J Gastroenterol 2016;111:1156—64.

[5]   Dagash M, Hayek T, Gallimidi Z, Yassin K, Brook JG. Transient radiological and colonoscopic features of inflammatory bowel disease in a patient with severe *Salmonella gastroenteritis*. Am J Gastroenterol 1997;92:349—51.

[6]   Matsumoto T, Iida M, Matsui T, Sakamoto K, Fuchigami T, Haraguchi Y, et al. Endoscopic findings in *Yersinia enterocolitica* enterocolitis. Gastrointest Endosc 1990;36:583—7.

[7]   Kochhar G, Edge P, Blomme C, Wu XR, Lopez R, Ashburn J, et al. *Clostridium difficile* enteropathy is associated with a higher risk for acute kidney injury in patients with an ileostomy—a case-control study. Inflamm Bowel Dis 2018;24:402—9.

[8]   Kistangari G, Lopez R, Shen B. Frequency and risk factors of *Clostridium difficile* infection in hospitalized patients with pouchitis: a population-based study. Inflamm Bowel Dis 2017;23:661—71.

[9]   Iida T, Ikeya K, Watanabe F, Abe J, Maruyama Y, Ohata A, et al. Looking for endoscopic features of cytomegalovirus colitis: a study of 187 patients with active ulcerative colitis, positive and negative for cytomegalovirus. Inflamm Bowel Dis 2013;19:1156—63.

[10] Lee KC, Lu CC, Hu WH, Lin SE, Chen HH. Colonoscopic diagnosis of amebiasis: a case series and systematic review. Int J Colorectal Dis 2015;30:31—41.

[11] de Leijer JH, Tan ACITL, Mulder B, Zomer SF. Unexpected amebic colitis presenting with rectal bleeding and perforation after biopsy. Gastrointest Endosc 2018;88:565—6.

Chapter 26

Inflammatory bowel disease—like conditions: other immune-mediated gastrointestinal disorders

Ying-Hong Wang[1], Yan Chen[2], Xiaoying Wang[2] and Bo Shen[3]

[1]Department of Gastroenterology, University of Texas MD Anderson Cancer Center, Houston, TX, United States, [2]Department of Gastroenterology, The Second Affiliated Hospital, School of Medicine, Zhejiang University, Hangzhou, P.R. China, [3]Center for Inflammatory Bowel Diseases, Columbia University Irving Medical Center-New York Presbyterian Hospital, New York, NY, United States

Chapter Outline

Abbreviations

AID autoinflammatory disease
AIE autoimmune enteropathy
ANA antinuclear antigen
CTLA-4 cytotoxic T-cell lymphocyte-4
CC collagenous colitis
CeD celiac disease
CD Crohn's disease
FMF familial Mediterranean fever
GI gastrointestinal
IBD inflammatory bowel disease
ICPi immune checkpoint inhibitor
IMD immune-mediated disorder
LC lymphocytic colitis
MC microscopic colitis
NAID NOD2-associated autoinflammatory diseases
NSAID nonsteroidal antiinflammatory drug
PD-1 programmed cell death 1 protein
PSC primary sclerosing cholangitis

Atlas of Endoscopy Imaging in Inflammatory Bowel Disease. DOI: https://doi.org/10.1016/B978-0-12-814811-2.00026-8
© 2020 Elsevier Inc. All rights reserved.

RA rheumatoid arthritisTNF tumor necrosis factor
TNF tumor necrosis factor
SLE systemic lupus erythematosus
UC ulcerative colitis

Introduction

Immune-mediated disorders (IMDs) comprise a wide array of gastrointestinal (GI) and extraintestinal phenotypes. Inflammatory bowel disease (IBD), that is, Crohn's disease (CD) and ulcerative colitis (UC) are the primary forms. Current theories hold that IMD results from dysregulated innate and adaptive immune response to environmental or microbiological factors in genetically susceptible hosts. Multiple layers of the interaction of these factors lead to various forms of IMD. For example, abnormal innate immunity plays a dominant role in the pathogenesis of autoinflammatory diseases (AIDs), while dysregulated adaptive immunity exerts a major role in the development of autoimmune enteropathy (AIE). Abnormalities of both innate and adaptive immunities impact the pathogenesis of IBD.

The etiopathogenetic pathways, clinical presentations, endoscopic features, and histologic characteristics overlap between classic IBD, autoimmune disorders, and AIDs. In addition, IMD can be triggered by factors, such as bowel altering surgery and medications. For example, bariatric surgery [1] or liver or kidney transplantation [2,3] may be associated with the development of de novo IBD. Immune checkpoint inhibitors (ICPis) are known to cause IBD-like IMD.

The association between GI IMD and extraintestinal IMD is mutual or multilayered. Patients with IBD often present with extraintestinal immune-mediated systemic disorders such as erythema nodosum, pyoderma gangrenosum, and primary sclerosing cholangitis (PSC). IBD may also have concurrent autoimmune disorders, such as rheumatoid arthritis (RA) and psoriasis. On the other hand, systemic autoimmune disorders or AID may affect the GI tract. For example, patients with systemic lupus erythematosus (SLE) may have lupus enteritis resulting from lupus vasculitis.

Within GI manifestations of IMD, there are overlaps of disease phenotypes too. For example, IBD and celiac disease (CeD) or microscopic colitis (MC) may coexist. The coexisting disease phenotypes may develop concurrently or sequentially. For example, lymphocytic colitis (LC) may progress to CD. The complexity of interactions between GI and extraintestinal IMD has led to a new classification of broad-sensed IBD, as outlined in Chapter 1, Introduction and classification of inflammatory bowel diseases.

Endoscopic features of these IMDs are discussed in this chapter, which include ICPi-associated colitis. Other oncology medicine-associated IBD-like conditions are discussed in a separate chapter (Chapter 29: Inflammatory bowel disease−like conditions: medication-induced enteropathy). Immune-mediated vasculitis-associated GI disorders are discussed in a separate chapter (Chapter 28: Inflammatory bowel disease−like conditions: ischemic bowel diseases and vasculitides).

Celiac disease

CeD is also called gluten-sensitive enteropathy and nontropical sprue. There is a gradient in the decreasing disease severity from the proximal to the distal small intestine. Therefore duodenal biopsy plays a critical role in the diagnosis and differential diagnosis. Common endoscopic features of CeD are atrophic mucosa with loss of folds or sparse folds, mosaic patterns, visible fissures, granularity, nodularity, scalloping, and prominent submucosal vascularity (Figs. 26.1−26.4). However, these endoscopic features suggestive of CeD yield a low diagnostic sensitivity and high specificity. These endoscopic features can also be seen in CD or UC involving the duodenum, or AIE. Histology is more reliable for the diagnosis of CeD than endoscopy and serology. At least four biopsies of postbulbar duodenum along with 1−2 biopsies of the bulb should be taken and separately labeled. Primary histologic features are mucosal inflammation of lymphocytes, crypt hyperplasia, and villous atrophy (Fig. 26.3D).

Collagenous sprue, a rare, little-understood disorder, is characterized by subepithelial collagen deposition. Patients with CeD carry a risk for the development of enteropathy-associated T-cell lymphoma, which shares some of the endoscopic features of duodenal CD, including edema, erythema, granularity, and ulcers of the duodenum (Fig. 26.5).

There is an association between CeD and IBD [4,5]. The risk of IBD in patients with CeD was elevated [6,7]. In contrast, the risk for CeD in IBD patients was comparable to controls [8]. Patients may have coexisting CeD and IBD [9,10]. Patients with coexisting IBD and CeD have been shown to have a higher frequency of PSC, extensive UC, and family history of CeD than those with IBD alone (Fig. 26.4) [10]. Evaluation of CeD should be performed in IBD patients with persistent iron deficiency anemia which poorly responds to iron supplement therapy. IBD-associated surgeries, such as ileal pouch−anal anastomosis, may trigger the development of de novo CeD (Fig. 26.3) [11].

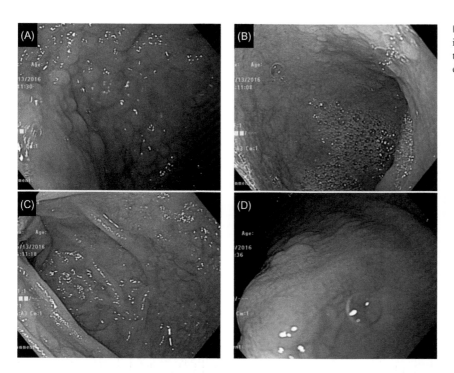

FIGURE 26.1 Abnormal duodenum mucosa in celiac disease. (A–D) Nodular mucosa of the duodenum. Notice that the lumen of the duodenum was dilated.

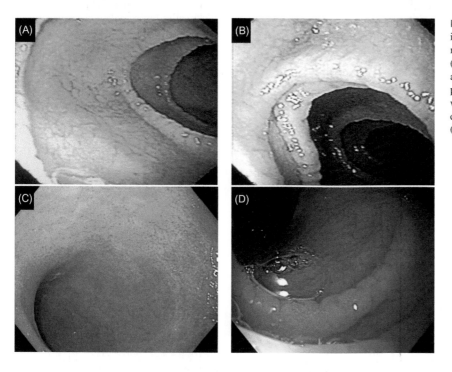

FIGURE 26.2 Abnormal duodenum mucosa in celiac disease. (A and B) Visible fissures of mucosa and scalloping of the duodenum folds; (C and D) de novo celiac disease developed after restorative proctocolectomy with ileal pouch—anal anastomosis in a separate patient with underlying ulcerative colitis. Flatten duodenum mucosa with loss (C) and scalloping (D) of the duodenum folds.

Microscopic colitis

MC consists of two primary phenotypes, LC and collagenous colitis (CC). Patients with MC often present with chronic nonbloody diarrhea. MC is named for the disease process which is detected under light microscope in the absence of obvious endoscopic abnormalities. However, endoscopic inflammation is common in MC. Association between IBD and MC has several folds: (1) patients with MC can evolve into IBD which is named IBD transformer [12]; (2) IBD in

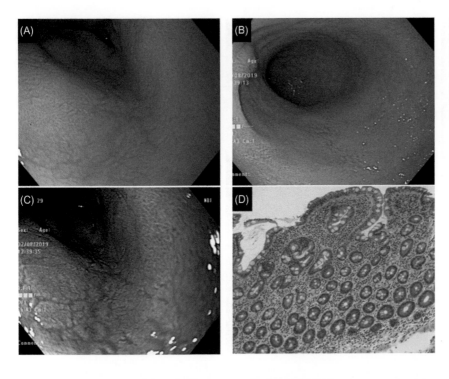

FIGURE 26.3 De novo celiac disease after restorative proctocolectomy and ileal pouch—anal anastomosis for ulcerative colitis. (A and B) Flat and fissured duodenum mucosa with loss of folds; (C) the mucosal fissures highlighted with narrow-band imaging; (D) villous atrophy and lymphocyte infiltration in the epithelia and lamina propria on histology.

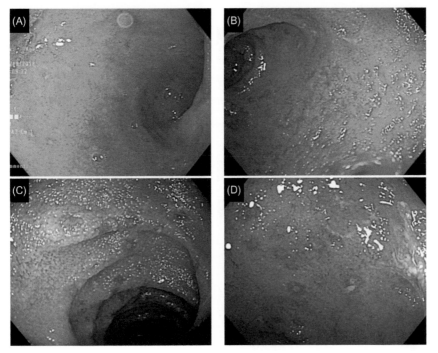

FIGURE 26.4 Coexisting Crohn's disease and celiac disease in a 43-year-old patient. (A and B) Granular and flat duodenum mucosa with loss of folds; (C and D) multiple aphthous erosions in the terminal ileum with chronic active enteritis on histology.

remission may present in MC pattern on histology; and (3) concurrent IBD and MC may also be encountered with concomitant CC and UC being the most common pattern [13].

Lymphocytic colitis

Colonoscopy often demonstrates normal-appearing mucosa. However, some patients may present with mild edema, erythema, loss of vascular pattern, mucosal fissures, friability, exudates, granularity, or nodularity (Figs. 26.6 and 26.7). The characteristic histologic feature is intraepithelial lymphocytosis (Fig. 26.8).

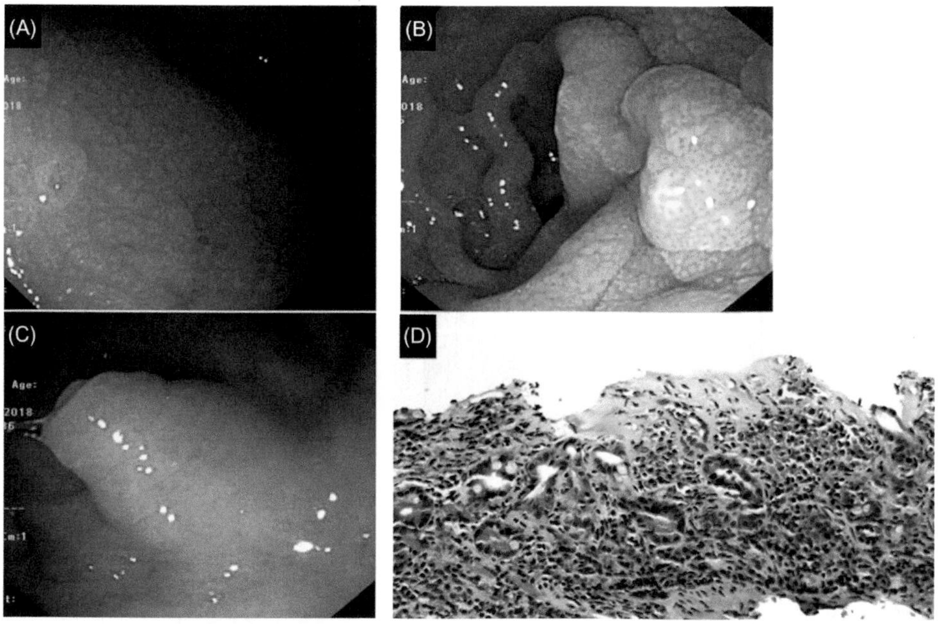

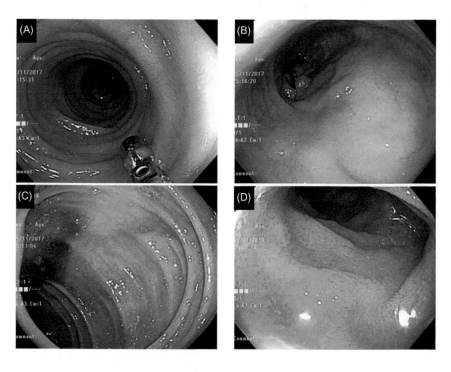

FIGURE 26.5 Collagenous sprue. (A) Granular mucosa of the duodenum; (B) edematous mucosa of the duodenum; (C) prominent duodenum papilla; (D) histologic features an absence of villi, surface epithelial injury with depletion, thickening of the subepithelial collagen layer, intraepithelial lymphocytosis, and chronic inflammation. *(D) Courtesy Dr. Xiuli Liu of University of Florida.*

FIGURE 26.6 Patterns of lymphocytic colitis. (A and B) Mucosal edema of the colon; (C) normal colon mucosa. (D) Normal terminal ileum in a separate patient with lymphocytic colitis.

Collagenous colitis

Patients with CC often present with some forms of endoscopic abnormalities, ranging from pseudomembranes, edema, erythema, hemorrhagic spots, loss of vascular pattern, mosaic pattern ("honeycomb"), scalloping, to aphthae or erosions, ulcers, mucosal laceration or tear, granularity or nodularity, or mucosal scars [14—16] (Figs. 26.9—26.11). Image-enhanced endoscopy, such as magnifying colonoscopy, narrow-band imaging, and chromoendoscopy, may enhance the visualization of the macroscopic inflammation [8]. The disease process may not be limited to the colon, as the immune-mediated disease process can affect the stomach or small intestine (Fig. 26.9A). Histologic features of CC include thickened subepithelial collagen band, trapped capillaries in the band, sloughing of the epithelia, and intraepithelial lymphocytosis (Fig. 26.12).

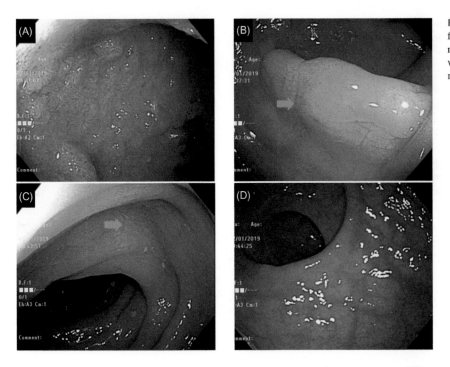

FIGURE 26.7 Diffuse lymphocytic colitis from the cecum to rectum. (A) Normal terminal ileum; (B and C) slightly raised mucosa with fissures (*green arrows*); (D) normal rectal mucosa.

 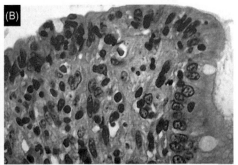

FIGURE 26.8 Histology of lymphocytic colitis. (A and B) Intraepithelial lymphocytosis along with lymphocytic infiltrates of the lamina propria. There is also mucin depletion of the intestinal epithelia.

Autoimmune enteropathy

Autoimmune enteropathy, a rare IMD, is characterized by severe diarrhea, hypokalemia, metabolic acidosis, and immune-mediated small intestinal villous atrophy. AIE is more common in children than adult. Patients with AIE do not typically respond to gluten free diet. They often have other evidence of autoimmunity, such as the presence of anti-enterocyte antibodies, antinuclear antibody (ANA), antigliadin antibody, antineutrophil cytoplasmic antibody, antiextractable nuclear antibodies (Ro, La, Sm, RNP, Jo-1, Scl-70, dsDNA), antiparietal cell antibody, antimitochondrial antibody, anti–liver–kidney microsomal antibody, antismooth muscle antibody, rheumatoid factor, antipancreatic islet cell antibodies, anticyclic citrullinated peptide antibody, and antiglutamic acid decarboxylase antibody. The proposed diagnosis criteria for (1) diarrhea >6 weeks, and being is refractory to treatment with diet modification or antimotility agents; (2) malabsorption; (3) histologic features of villous atrophy, apoptotic bodies, and crypt lymphocytic proliferation; and (4) exclusion of CeD or other disorders with similar presentations [17].

The presence of villous atrophy is a diagnostic requirement. In addition, a wide range of histologic features has been described. Based on the predominant histologic patterns of the duodenum, AIE has been classified as (1) active chronic duodenitis with villous blunting, infiltration of mononuclear cells (mainly plasma cells) and neutrophilic cryptitis with or without crypt abscesses. There may be increased apoptosis in the crypt epithelium; (2) CeD-like with villous blunting and marked increase in intraepithelial lymphocytes (>40 intraepithelial lymphocytes/100 enterocytes); (3) graft-versus-host-like, with increased apoptosis in crypt epithelium (>1 apoptotic figure per 10 crypts). There is or

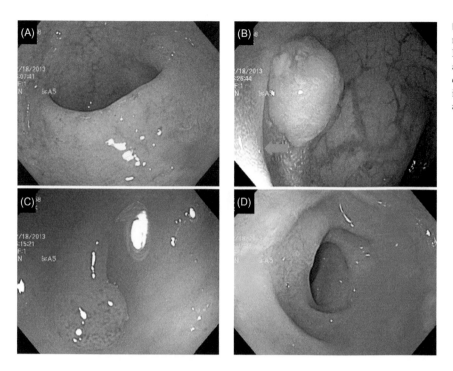

FIGURE 26.9 Disease process of collagenous colitis can affect the small bowel. (A) Flatten and fissured mucosa in the duodenum as well as the large bowel; (B−D) mucosal edema of the colon (*green arrow*) with an incidental finding of inflammatory polyps (B and C).

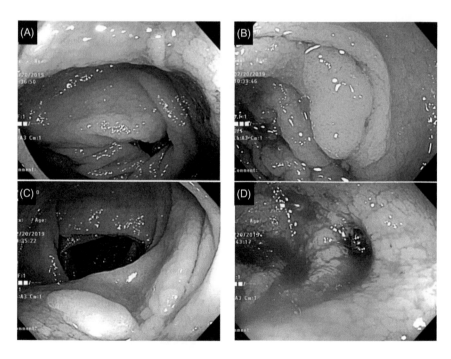

FIGURE 26.10 Collagenous colitis with pseudopolyps. (A−D) Granularity of colon mucosa with pseudopolyps (B and C). The mucosa was also friable (D).

there is not crypt dropout with minimal inflammation; and (4) mixed/no predominant pattern [18]. AIE can also involve other parts of the GI tract.

Endoscopic presentations of AIE are not specific, and some patients may have normal-appearing gastric, duodenal, or colonic mucosa. Villous atrophy in AIE may be detected with high-definition white-light endoscopy, image-enhanced endoscopy, video capsule endoscopy [19,20], and enteroscopy [4]. Other endoscopic features include duodenal scalloping, fissuring, mosaic pattern, edema, erythema, granularity, nodularity, cobblestoning, or aphthous ulcers (Figs. 26.13 and 26.14) [4,21−23]. Transmural disease, strictures, or fistulas are rare in AIE.

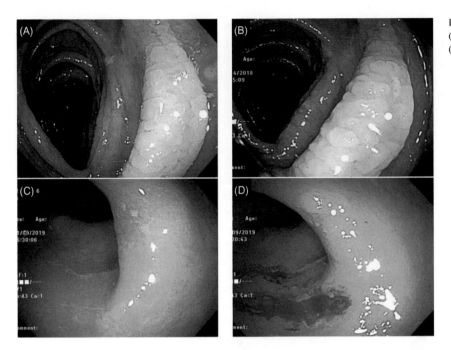

FIGURE 26.11 Patterns of collagenous colitis. (A and B) Scalloping of the colon mucosa; (C and D) edematous and friable colon mucosa.

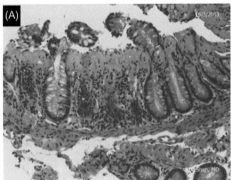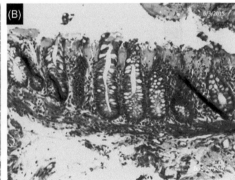

FIGURE 26.12 Histology of collagenous colitis. (A) Sloughing of the colon epithelia. There is a thick subepithelial collagen band with trapped capillaries on hematoxylin and eosin stain; (B) The think subepithelial collagen band is highlighted with trichrome stain.

Autoimmune disorders with gastrointestinal or extraintestinal involvement

It has been difficult to sort out the difference between various IBD-associated extraintestinal disorders (such as CD with concurrent SLE vs SLE with concurrent ileitis). These disease entities may represent phenotypes in a wide spectrum of immune-mediated GI and extraintestinal disorders (Chapter 1: Introduction and classification of inflammatory bowel diseases). Patients with SLE, systemic sclerosis, or RA often present with GI symptoms and endoscopic or histologic abnormalities. First of all, nonsteroidal antiinflammatory drug (NSAID) use in these patients should be excluded [24].

Concurrent AIE and IBD may occur. In addition, IBD may coexist with IMD of intestinal or extraintestinal organs. For example, UC can be concomitantly seen in patients with autoimmune hepatitis (Fig. 26.15); autoimmune gastropathy and duodenopathy may coexist with pouchitis (Figs. 26.16 and 26.17); and CD may be concomitant with multiple sclerosis (Fig. 26.18) or psoriasis (Fig. 26.19).

Lupus enteritis

Lupus-associated vasculitis involves small- and medium-sized vessels and can affect the GI tract, that is, lupus enteritis. The latter is defined as either vasculitis or inflammation of the small intestine, with supportive image and/or histologic

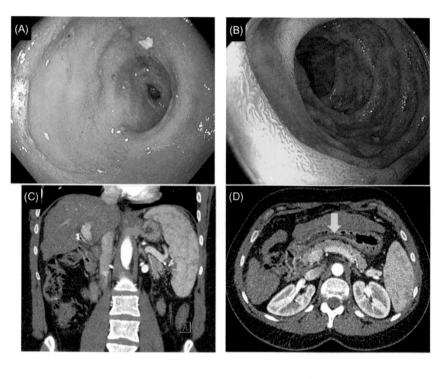

FIGURE 26.13 Autoimmune gastropathy and enteropathy in a patient with concurrent autoimmune hepatitis. (A) Diffuse edematous and erythematous gastric mucosa; (B) patchy erythema of the duodenal mucosa; (C) thickened gastric wall (*green arrow*); (D) m hyperenhancement of the duodenal mucosa (*yellow arrow*).

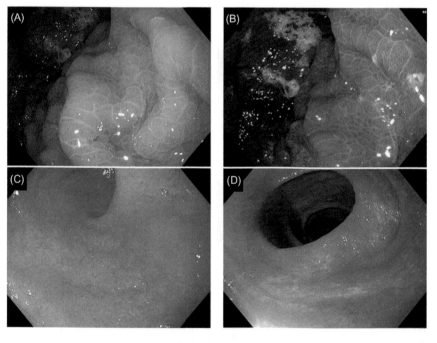

FIGURE 26.14 Autoimmune enteropathy and gastropathy in a female patient with concurrent collagenous colitis, positive celiac serology, and failure to respond to gluten free diet. (A and B) White-light endoscopy and narrow band imaging showed mosaic pattern of gastric mucosa; (C and D) villous atrophy of the duodenal mucosa.

findings [25]. Vasculitis may cause erosions, ulcers, bleeding, or strictures formation of the GI tract, intestinal pseudoobstruction, or perforation from ischemia and infarction (Figure 28.13) [26].

Systemic sclerosis

Systemic sclerosis, or scleroderma, is a multisystem connective tissue disorder characterized by fibrotic arteriosclerosis of peripheral and visceral vasculature. The speculated disease theory is a combination of autoimmune and autonomic (with the enteric nervous system) etiology. One of the primary targets for scleroderma in the GI tract is the muscularis propria. The esophagus is involved in approximately 50%—90% of cases, although any parts of the GI tract can be

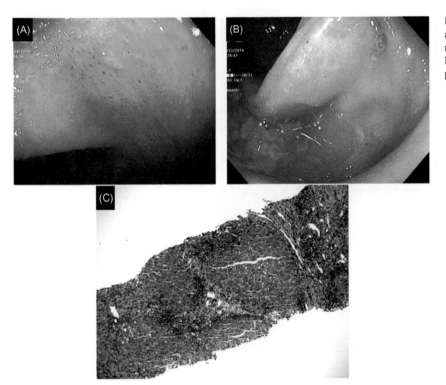

FIGURE 26.15 Concurrent ulcerative colitis and autoimmune hepatitis. (A and B) Diffuse mild colitis with erythema and loss of vascularity; (C) autoimmune hepatitis with lymphoplasmacytosis in the portal triad.

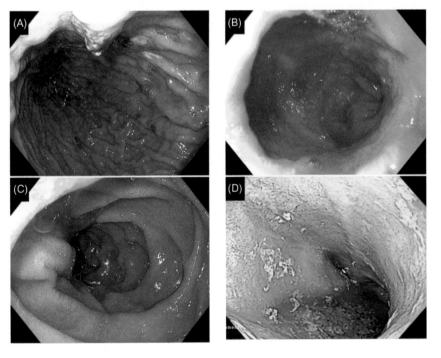

FIGURE 26.16 Autoimmune gastropathy and bulbar duodenopathy in a patient with restorative proctocolectomy and ileal pouch—anal anastomosis. (A) Autoimmune gastritis with erythema, friability, and bleeding; (B) diffuse erythema and nodularity of the duodenum bulb; (C) normal second portion of the duodenum; (D) loss of vascularity of the ileal pouch body.

affected (Fig. 26.20) [27—29]. Patients with scleroderma can have gastroesophageal reflux l even severe esophagitis [30,31]. The patients may also present with gastroparesis, small intestinal bacterial overgrowth, chronic constipation or diarrhea, or fecal incontinence. Skin ulcers occur more in the toes and fingers [32] but can also be seen in the perianal area, mimicking perianal CD (Fig. 26.21).

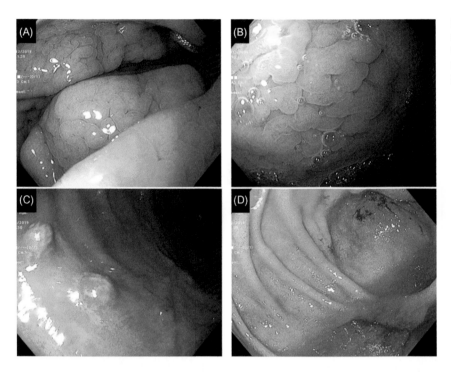

FIGURE 26.17 Autoimmune gastropathy and autoimmune pouchitis. (A and B) Granularity of gastric mucosa; (C and D) mild mucosal inflammation with inflammatory polyps (C) of the pouch body.

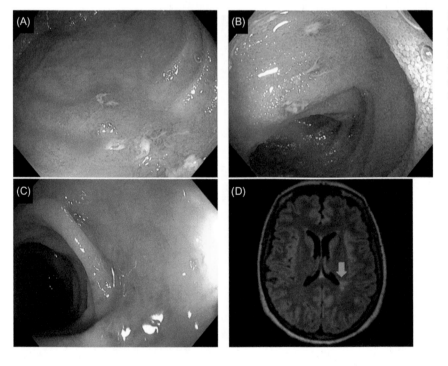

FIGURE 26.18 Crohn's disease with concurrent multiple sclerosis and uveitis (not shown) in a 23-year-old male patient. (A and B) Multiple small ulcers in the terminal ileum; (C) patchy erythema of mucosa at the sigmoid colon; (D) white plaques (*green arrow*) in magnetic resonance imaging.

Sjögren's syndrome

Sjögren's syndrome is a chronic inflammatory autoimmune disease involving the exocrine glands. Patients with Sjögren's syndrome are often found multiple autoimmune antibodies positive and accompanied with other autoimmune disorders. ANA is the most commonly found and anti-Ro/SS-A is the most specific antibody for the disease. One third of the patients have concomitant SLE or RA. Small bowel ulcers or strictures can be seen in patients with Sjögren's syndrome. It is difficult to differentiate Sjögren's syndrome from small bowel disease from CD with Sjögren's syndrome (Fig. 26.22).

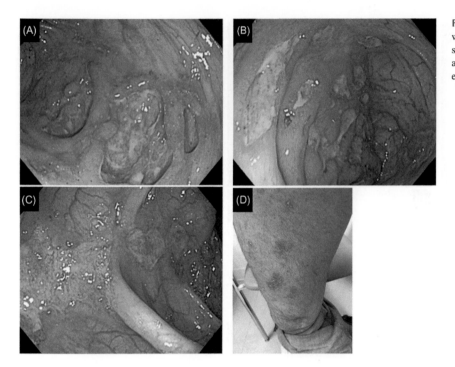

FIGURE 26.19 Crohn's disease of the colon with concurrent psoriasis: (A–C) multiple small and large irregular ulcers in the colon and (D) the skin lesions in the left lower extremity.

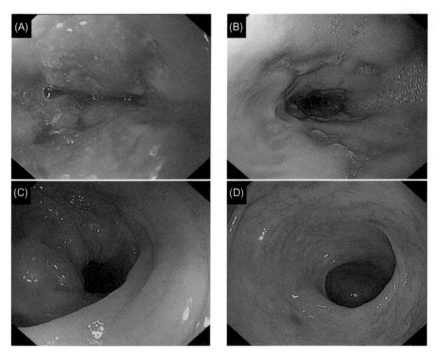

FIGURE 26.20 Patterns of mucosal inflammation in systemic sclerosis: (A and B) erosions and ulcers in the esophagus and (C and D) edema and congestion in the descending colon.

Rheumatoid arthritis

RA can affect the GI tract. Visceral arteritis, a form of rheumatoid vasculitis can affect the bowel, along with peripheral nerve, lungs, heart, spleen, and other organs. In severe cases, patients may present with bowel ulcers, bleeding, even infarction. On endoscopy, longitudinal ulcers may be seen (Fig. 26.23). In addition to NSAIDs, Tocilizumab, a monoclonal antibody targeting the IL-6 receptor for RA may also have intestinal side effects, including ulcers and perforation [33].

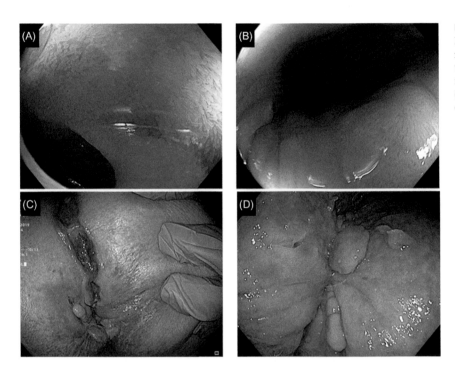

FIGURE 26.21 Small intestine and perianal lesions in a female patient with systemic sclerosis: (A) a tight, nonulcerated stricture at the terminal ileum; (B) normal anal canal; (C) mid-line long, deep ulcers in the perianal area; (D) external hemorrhoids due to chronic constipation.

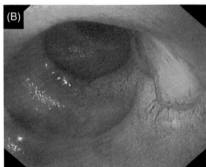

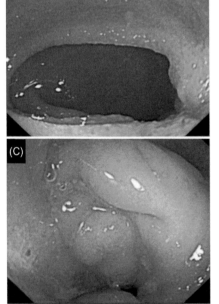

FIGURE 26.22 Patterns of ulcers in patients with concurrent CD and Sjögren's syndrome on colonoscopy and signal balloon enteroscopy: (A) a longitudinal ulcer in the ileum, (B) a clean-based ulcer in the ileum, and (C) ulcerated inflammatory stricture in the descending colon. *CD*, Crohn's disease.

Autoinflammatory diseases

Autoinflammatory diseasess consist of are a group of genetically heterogeneous inflammatory disorders, primarily resulting from abnormal *innate* immunity in genetically susceptible hosts [34]. In contrast, traditional autoimmune disorders are characterized by the presence of dysregulated adaptive immune response and autoantibodies. Clinically, AIDs present with hereditary periodic fever syndromes. Representative phenotypes of AID are familial Mediterranean fever (FMF), tumor necrosis factor (TNF) receptor-1 associated periodic syndrome, hyperimmunoglobulin D

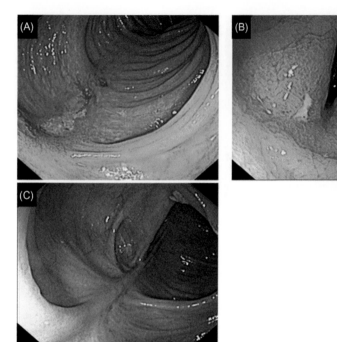

FIGURE 26.23 Enterocolitis with ulcers and strictures in rheumatoid arthritis: (A) longitudinal ulcers in the ascending colon, (B) mucosal congestion with in the transverse colon, and (C) the healed ulcer with the formation of mucosal scars in the ascending colon after medical therapy.

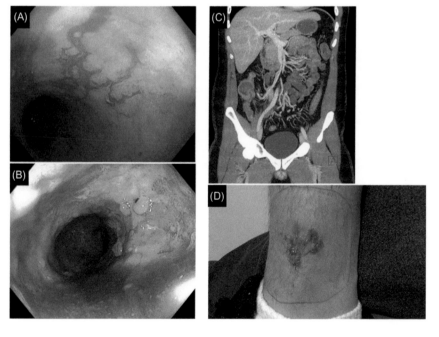

FIGURE 26.24 NOD2-associate diffuse colitis with edematous, erythematous, and nodular mucosa with histology showing auto-inflammatory diseases of the colon and skin. Parts (A) and (B) showing chronic active colitis with ulceration; (C) mildly thickened colon wall; (D) suppurative folliculitis and panniculitis fat necrosis. The patient had heterozygous c.2848A > C alteration in exon 9 of NOD2/CARD15 genes. Homozygous for polymorphisms c.802T > C and c.1377T > C. *NOD2*, Nucleotide-binding oligomerization domain 2.

syndrome, cryopyrin-associated periodic syndromes, periodic fever with aphthous stomatitis, pharyngitis, and adenitis syndrome. Other forms including Blau syndrome and Yao syndrome, similarly to CD, are linked to the nucleotide-binding oligomerization domain 2 (NOD2) gene encoding NOD2 [35−38]. Together they are named NOD2-associated AIDs (NAID) [39]. Associated NOD2 variants were primarily IVS8+158 or compound IVS8+158 and R702W for NAID.

Clinical manifestations of AID include fever, rash, serositis, arthritis, uveitis, meningitis, uveitis, lymphadenopathy, and splenomegaly. The majority of patients develop their first disease manifestations in childhood, which mimics the first peak of age onset in of IBD. In patients with FMF, functional GI disorders [40] and mucosal

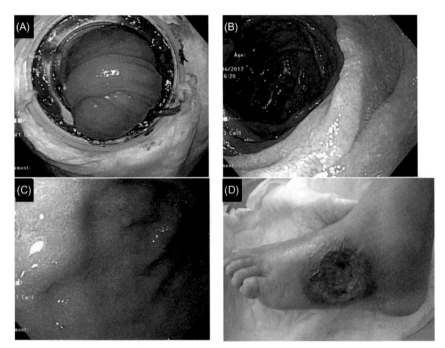

FIGURE 26.25 NOD2-associated autoinflammatory disease in a patient with failed ileal pouch—anal anastomosis and pericardial effusion (not shown): (A) ileostomy that was created due to pouch failure, (B and C) mild diffuse duodenitis with edema, and (C) large pyoderma gangrenosum of the left foot. The patient had mutations of c.3020insC and IVS8$^+$158C > T of NOD2/CARD15 genes. *NOD2*, Nucleotide-binding oligomerization domain 2.

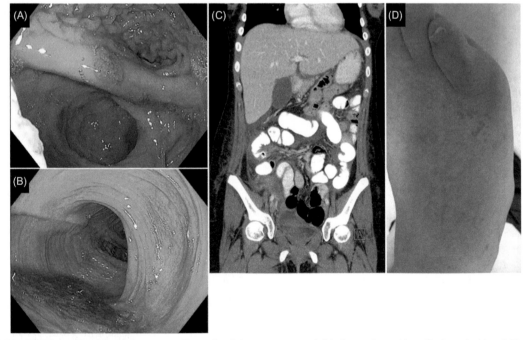

FIGURE 26.26 NOD2-associated autoinflammatory disease involving mesentery and skin in a patient with an ileal pouch: (A and B) normal pouch body mucosa, (C) mesenteric straining, and (D) skin lesions on foot with histology of skin biopsy showing spongiosis dermatitis. The patient had homozygous for c.534C > G, c.802T > C, c.1377T > C. *NOD2*, Nucleotide-binding oligomerization domain 2.

inflammation [41] in the GI tract are common. Patients may present with mucosal edema, erosions, and ulceration in the jejunum and ileum (Figs. 26.24—26.26) [42]. Microscopic inflammation of the GI tract may be more extensive, involving the esophagus, stomach, or colon, in addition to the small bowel [43]. In a large cohort of 143 patients clinically suspected of NAID, 67 (47%) carry NOD2 variants; the genotype frequency was significantly higher among our cohort than in the historical healthy controls. Of the 67 carriers of NOD2 variants, 54 were diagnosed as having NAID. A total of 39 (72%) patients complained of recurrent and intermittent abdominal pain, and diarrhea, of whom 27 underwent further workups with computed tomography enterography, esophagogastroduodenoscopy, and colonoscopy. Three patients had nonspecific colitis and one had granulomatous changes [28]. Overlapping FMF and CD have been reported [44].

Immune checkpoint inhibitor—associated colitis

Immune-mediated colitis also includes ICPi-associated diarrhea and inflammation of the small and large bowel. Commonly used checkpoint inhibitors for the treatment of melanoma are Ipilimumab [a monoclonal antibody against cytotoxic T-cell lymphocyte-4 (CTLA-4) on the T-cell surface], Pembrolizumab, and Nivolumab [monoclonal antibodies directed against the programmed cell death 1 protein (PD-1)]. Pembrolizumab has also been approved for nonsmall cell lung cancer and head and neck squamous cell cancer, while Nivolumab is approved for advanced nonsmall cell lung cancer, Hodgkin lymphoma, and renal cell carcinoma. These ICPis are often associated with adverse events of the GI tract.

Diarrhea and colitis are the most common GI presentations, which usually occur approximately 6 weeks after the initiation of the treatment [45]. The incidence of diarrhea is higher in patients receiving CTLA-4-blocking antibodies than those on PD-1 inhibitors. Endoscopic findings reveal mucosal edema with biopsies demonstrating neutrophilic, lymphocytic, or mixed neutrophilic-lymphocytic infiltration [46–48].

The severity of ICPi-associated colitis ranges from mild bowel illness to fulminant colitis even perforation. The most common endoscopic findings were erythema, friability, congestion, and ulcers [49,50]. The spectrum of endoscopic features ranges from normal-appearing colon to nonulcerative inflammation with edema, erythema, exudates, loss of vascularity, friability, or spontaneous bleeding to mucosal ulcers in various sizes, shapes, and depths [46]. The distribution of colitis can be left-sided, extensive, or involvement of the distal ileum. The patterns of inflammation can be diffuse or segmental (Figs. 26.27—26.31) [46]. The majority of patients with ICPi-associated colitis have the disease in the rectum and sigmoid colon, and the distal small bowel involvement is rare. Therefore sigmoidoscopy may be sufficient for evaluation [45].

Histologic features in IPCi-associated colitis are acute inflammation and chronic inflammation [45]. There are neutrophilic or eosinophilic infiltrates, cryptitis, crypt abscess, apoptosis, intraepithelial lymphocytosis, basal lymphocytic infiltrate, cryptic architecture distortion, and Paneth cell metaplasia [46]. Endoscopic and

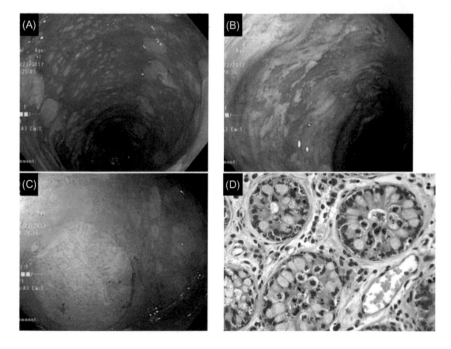

FIGURE 26.27 Immune checkpoint inhibitor—associated colitis with the patient being exposed to ipilimumab for malignant malenoma: (A and B) diffuse erythema and exudates of the colon mucosa, (C) small and large superficial round ulcers of the colon mucosa, (D) active colitis with marked apoptosis in the epithelia. *(D) Courtesy Dr. Ana E Bennett of Cleveland Clinic.*

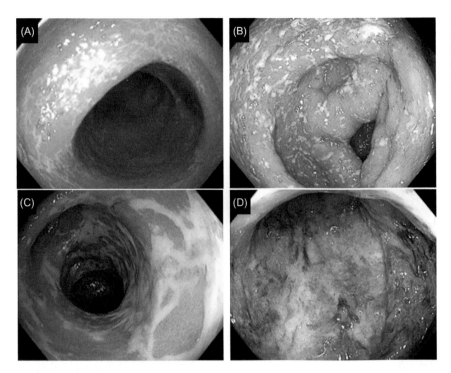

FIGURE 26.28 Immune checkpoint inhibitor—associated colitis with the patient being exposed to nivolumab: (A and B) diffuse erythema, edema, and exudates, (C) mucopurulent exudates coating the colon mucosa, and (D) large ulceration with bleeding.

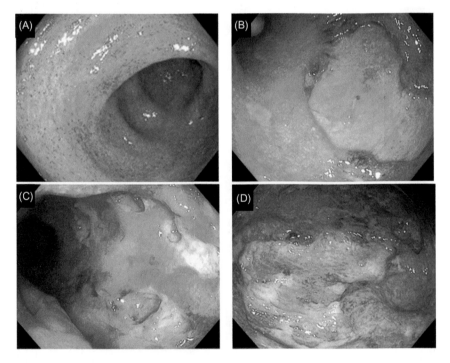

FIGURE 26.29 Immune checkpoint inhibitor—associated colitis with the patient being exposed to ipilimumab: (A) diffuse erythema of the colon mucosa and (B—D) various shaped and sized ulcers.

histologic features of ICPi-associated colitis are listed in Table 26.1. These features almost completely overlap with that in IBD.

Sarcoidosis

GI manifestations of sarcoidosis are discussed in Chapter 32, Inflammatory bowel disease—like conditions: miscellaneous (Fig. 32.12). Small bowel and large bowel ulcers and nonulcerative inflammation may be presented (Fig. 26.32).

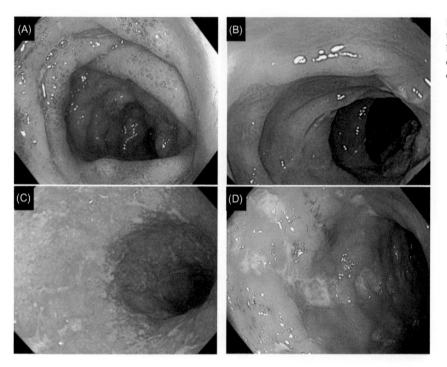

FIGURE 26.30 Immune checkpoint inhibitor–associated diffuse colitis with the patient being exposed to ipilimumab: (A and B) diffuse edema and erythema, (C) diffuse erythema and exudates, and (D) longitudinal ulcers.

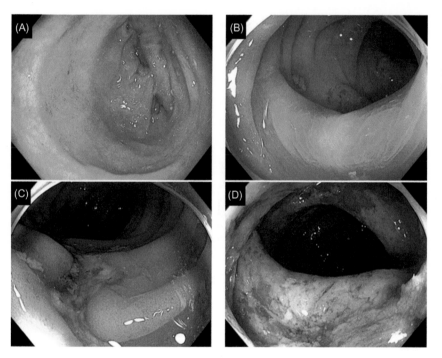

FIGURE 26.31 Immune checkpoint inhibitor–associated colitis with exposure to atezolizumab: (A–D) range of patterns of colitis with loss of vascularity (A), erythema (B), and longitudinal and circumferential ulcers with exudates (C, D).

Amyloidosis

Amyloidosis is a clinical syndrome in which the deposition of amyloid in various organs of the body [51]. Common forms of amyloidosis are AL (primary) and AA (secondary) types. Secondary amyloidosis is an epiphenomenon of underlying causes, including IMD, chronic inflammation, certain genetic diseases, or tumors. Secondary amyloidosis is associated with the deposition of serum amyloid protein, which is an acute-phase reactant during chronic inflammatory or infectious diseases, such as RA and CD [52]. The deposits in the GI tract are common.

TABLE 26.1 Common endoscopic and histologic features of immune checkpoint inhibitor—associated colitis.

Endoscopic features	Histologic features
Edema	Erosion and ulcers
Erythema	Cryptitis and crypt abscess
Friability	Crypt distortion and branching
Granularity and nodularity	Prominent crypt apoptosis
Exudates	Intraepithelial lymphocytosis
Loss of vascular pattern	Infiltration of mononuclear cells in the lamina propria
Mucosal scars	Basal lymphoplasmacytosis
Skip or continuous inflammation	Intranuclear inclusion bodies (concurrent cytomegalovirus infection)
Spontaneous bleeding	Rare transmural inflammation
Various forms of erosions and ulcers	Noncaseating granulomas

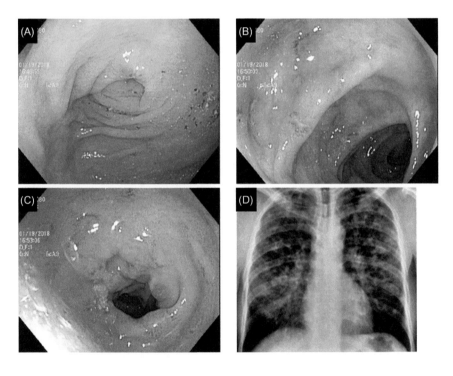

FIGURE 26.32 Colitis in a patient with sarcoidosis: (A) mild erythema around the appendiceal orifice, (B and C) diffuse erythema and small ulcers in the colon, and (D) pulmonary sarcoidosis with multiple nodules on chest X-ray.

Clinical presentations depend on the location of amyloid deposits. Endoscopic findings of amyloidosis in the GI tract are nonspecific, with features including erythema, erosions, ulcerations, granularity, friability, polypoid protrusions, and submucosal lesions (Fig. 26.33).

Summary and recommendations

The diagnosis of IMDs covers a long list of GI and extraintestinal disease phenotypes, ranging from CD and UC to CeD, MC, AIE, and AID. There are overlaps in the etiopathogenesis, clinical presentations, and endoscopic and histologic features. The assessment of both distinction and associations among IMD is important for the management and prognosis. The distinction and association can be moving targets, which require periodic disease evaluation and

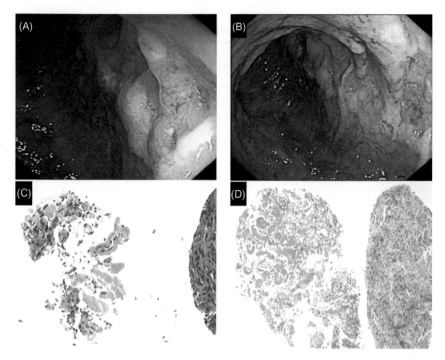

FIGURE 26.33 Patterns of ulcers in amyloidosis on colonoscopy: (A and B) Erythema, telangiectasia, erosions, ulcers, nodularity. with unclear boundary in the ascending colon; (C) Amorphous, eosinophilic, extracellular materials (H&E 100×); and (D) Amorphous extracellular material orange-red which is apple green (not photographed) under polarized microscope, confirming amyloid deposition (*Congo red* 100×). *Colonoscopy photos courtesy Drs. Yan Chen, MD and Xiao-Ying Wang of the 2nd Affiliated Hospital of Zhejiang University; Histology photos courtesy Dr. Xiuli Liu of University of Florida.*

monitoring. The shared and distinctive endoscopic features may help diagnosis and differential diagnosis. It is important to put endoscopic findings in systemic context.

References

[1] Braga Neto MB, Gregory M, Ramos GP, Loftus Jr EV, Ciorba MA, Bruining DH, et al. De-novo inflammatory bowel disease after bariatric surgery: a large case series. J Crohns Colitis 2018;12:452−7.

[2] Kochhar G, Singh T, Dust H, Lopez R, McCullough AJ, Liu X, et al. Impact of de novo and preexisting inflammatory bowel disease on the outcome of orthotopic liver transplantation. Inflamm Bowel Dis 2016;22:1670−8.

[3] Fernandes MA, Braun HJ, Evason K, Rhee S, Perito ER. De novo inflammatory bowel disease after pediatric kidney or liver transplant. Pediatr Transpl 2017;21(1) Epub 2016 Nov 11.

[4] Breen EG, Coghlan G, Connolly EC, Stevens FM, McCarthy CF. Increased association of ulcerative colitis and coeliac disease. Ir J Med Sci 1987;156:120.

[5] Pascual V, Dieli-Crimi R, López-Palacios N, Bodas A, Medrano LM, Núñez C. Inflammatory bowel disease and celiac disease: overlaps and differences. World J Gastroenterol 2014;20:4846−56.

[6] Leeds JS, Höroldt BS, Sidhu R, Hopper AD, Robinson K, Toulson B, et al. Is there an association between coeliac disease and inflammatory bowel diseases? A study of relative prevalence in comparison with population controls. Scand J Gastroenterol 2007;42:1214−420.

[7] Kocsis D, Tóth Z, Csontos ÁA, Miheller P, Pák P, Herszényi L, et al. Prevalence of inflammatory bowel disease among coeliac disease patients in a Hungarian coeliac centre. BMC Gastroenterol 2015;15:141.

[8] Kobayashi M, Hoshi T, Morita SI, Kanefuji T, Suda T, Hasegawa G, et al. Magnifying image-enhanced endoscopy for collagenous colitis. Endosc Int Open 2017;5:E1069−73.

[9] Casella G, Di Bella C, Salemme M, Villanacci V, Antonelli E, Baldini V, et al. Celiac disease, non-celiac gluten sensitivity and inflammatory bowel disease. Minerva Gastroenterol Dietol 2015;61:267−71.

[10] Tse CS, Deepak P, De La Fuente J, Bledsoe AC, Larson JJ, Murray JA, et al. Phenotype and clinical course of inflammatory bowel disease with co-existent celiac disease. J Crohns Colitis 2018;12:973−80.

[11] Lian L, Remzi FH, Kiran RP, Fazio VW, Shen B. Clinical implication of false-positive celiac serology in patients with ileal pouch. Dis Colon Rectum 2010;53:1446−51.

[12] Li J, Yan Y, Meng Z, Liu S, Beck PL, Ghosh S, et al. Microscopic colitis evolved into inflammatory bowel diseases is characterized by increased th1/tc1 cells in colonic mucosal lamina propria. Dig Dis Sci 2017;62:2755−67.

[13] Wickbom A, Bohr J, Nyhlin N, Eriksson A, Lapidus A, Münch A, et al. Swedish Organisation for the Study of Inflammatory Bowel Disease (SOIBD). Microscopic colitis in patients with ulcerative colitis or Crohn's disease: a retrospective observational study and review of the literature. Scand J Gastroenterol 2018;53:410−16.

[14] Koulaouzidis A, Saeed AA. Distinct colonoscopy findings of microscopic colitis: not so microscopic after all? World J Gastroenterol 2011;17:4157—65.

[15] Saito S, Tsumura T, Nishikawa H, Takeda H, Nakajima J, Kanesaka T, et al. Clinical characteristics of collagenous colitis with linear ulcerations. Dig Endosc 2014;26:69—76.

[16] Shiratori Y, Fukuda K. Collagenous colitis diagnosed by endoscopically induced mucosal tears. BMJ Case Rep 2019;12(5). Available from: https://doi.org/10.1136/bcr-2019-230570 pii: e230570.

[17] Murray JA, Rubio-Tapia A. Diarrhoea due to small bowel diseases. Best Pract Res Clin Gastroenterol 2012;26:581—600.

[18] Masia R, Peyton S, Lauwers GY, Brown I. Gastrointestinal biopsy findings of autoimmune enteropathy: a review of 25 cases. Am J Surg Pathol 2014;38:1319—29.

[19] Gram-Kampmann EM, Lillevang ST, Detlefsen S, Laursen SB. Wireless capsule endoscopy as a tool in diagnosing autoimmune enteropathy. BMJ Case Rep 2015;2015.

[20] Akram S, Murray JA, Pardi DS, Alexander GL, Schaffner JA, Russo PA, et al. Adult autoimmune enteropathy: Mayo Clinic Rochester experience. Clin Gastroenterol Hepatol 2007;5:1282—90 quiz 1245.

[21] Unsworth DJ, Walker-Smith JA. Autoimmunity in diarrhoeal disease. J Pediatr Gastroenterol Nutr 1985;4:375—80.

[22] Gentile NM, Murray JA, Pardi DS. Autoimmune enteropathy: a review and update of clinical management. Curr Gastroenterol Rep 2012;14:380—5.

[23] Villanacci V, Lougaris V, Ravelli A, Buscarini E, Salviato T, Lionetti P, et al. Clinical manifestations and gastrointestinal pathology in 40 patients with autoimmune enteropathy. Clin Immunol 2019;207:10—17.

[24] Steen KS, Nurmohamed MT, Visman I, Heijerman M, Boers M, Dijkmans BA, et al. Decreasing incidence of symptomatic gastrointestinal ulcers and ulcer complications in patients with rheumatoid arthritis. Ann Rheum Dis 2008;67:256—9.

[25] Isenberg DA, Rahman A, Allen E, Farewell V, Akil M, Bruce IN, et al. BILAG 2004. Development and initial validation of an updated version of the British Isles Lupus Assessment Group's disease activity index for patients with systemic lupus erythematosus. Rheumatology (Oxford) 2005;44:902—6.

[26] Ebert EC, Hagspiel KD. Gastrointestinal and hepatic manifestations of systemic lupus erythematosus. J Clin Gastroenterol 2011;45:436—41.

[27] Lock G, Holstege A, Lang B, Schölmerich J. Gastrointestinal manifestations of progressive systemic sclerosis. Am J Gastroenterol 1997;92:763—71.

[28] Rohrmann Jr CA, Ricci MT, Krishnamurthy S, et al. Radiologic and histologic differentiation of neuromuscular disorders of the gastrointestinal tract: visceral myopathies, visceral neuropathies, and progressive systemic sclerosis. AJR Am J Roentgenol 1984;143:933—41.

[29] Zaninotto G, Peserico A, Costantini M, Zaninotto G, Peserico A, Costantini M, et al. Oesophageal motility and lower oesophageal sphincter competence in progressive systemic sclerosis and localized scleroderma. Scand J Gastroenterol 1989;24:95—102.

[30] Rose S, Young MA, Reynolds JC. Gastrointestinal manifestations of scleroderma. Gastroenterol Clin North Am 1998;27:563—94.

[31] Murphy JR, McInally P, Peller P, Shay SS. Prolonged clearance is the primary abnormal reflux parameter in patients with progressive systemic sclerosis and esophagitis. Dig Dis Sci 1992;37:833—41.

[32] Giuggioli D, Manfredi A, Lumetti F, Colaci M, Ferri C. Scleroderma skin ulcers definition, classification and treatment strategies our experience and review of the literature. Autoimmun Rev 2018;17:155—64.

[33] Gout T, Ostör AJ, Nisar MK. Lower gastrointestinal perforation in rheumatoid arthritis patients treated with conventional DMARDs or tocilizumab: a systematic literature review. Clin Rheumatol 2011;30:1471—4.

[34] Aksentijevich I. Update on genetics and pathogenesis of autoinflammatory diseases: the last 2 years. Semin Immunopathol 2015;37:395—401.

[35] Cho JH. The Nod2 gene in Crohn's disease: implications for future research into the genetics and immunology of Crohn's disease. Inflamm Bowel Dis 2001;7:271—5.

[36] Yao Q. Nucleotide-binding oligomerization domain containing 2: structure, function, and diseases. Semin Arthritis Rheum 2013;43:125—30.

[37] Yao Q, Zhou L, Cusumano P, Bose N, Piliang M, Jayakar B, et al. A new category of autoinflammatory disease associated with NOD2 gene mutations. Arthritis Res Ther 2011;13:R148.

[38] Yao Q, Shen B. A systematic analysis of treatment and outcomes of NOD2-associated autoinflammatory disease. Am J Med 2017;130:365. e13—18.

[39] Yao Q, Shen M, McDonald C, Lacbawan F, Moran R, Shen B. NOD2-associated autoinflammatory disease: a large cohort study. Rheumatology (Oxford) 2015;54:1904—12.

[40] Börekci E, Celikbilek M, Soytürk M, Akar S, Börekci H, Günaydin I. Functional gastrointestinal disorders in patients with familial Mediterranean fever. Int J Rheum Dis 2017;20:2101—5.

[41] Gurkan OE, Dalgic B. Gastrointestinal mucosal involvement without amyloidosis in children with familial Mediterranean fever. J Pediatr Gastroenterol Nutr 2013;57:319—23.

[42] Demir A, Akyüz F, Göktürk S, Evirgen S, Akyüz U, Örmeci A, et al. Small bowel mucosal damage in familial Mediterranean fever: results of capsule endoscopy screening. Scand J Gastroenterol 2014;49:1414—18.

[43] Agin M, Tumgor G, Kont A, Karakoc GB, Altintas DU, Yilmaz M. Endoscopic findings in patients with familial Mediterranean fever and dyspeptic symptoms. Prz Gastroenterol 2018;13:234—41.

[44] Witten J, Siles R, Shen B, Yao Q. Triple disease combination: familial Mediterranean fever, Crohn's disease, and chronic idiopathic urticaria with angioedema. Inflamm Bowel Dis 2016;22:E12—13.

[45] Weber JS, Dummer R, de Pril V, Lebbé C, Hodi FS, MDX010-20 Investigators. Patterns of onset and resolution of immune-related adverse events of special interest with ipilimumab: detailed safety analysis from a phase 3 trial in patients with advanced melanoma. Cancer 2013;119:1675–82.

[46] Berman D, Parker SM, Siegel J, Chasalow SD, Weber J, Galbraith S, et al. Blockade of cytotoxic T-lymphocyte antigen-4 by ipilimumab results in dysregulation of gastrointestinal immunity in patients with advanced melanoma. Cancer Immun 2010;10:11 [E journal].

[47] Maker AV, Phan GQ, Attia P, Yang JC, Sherry RM, Topalian SL, et al. Tumor regression and autoimmunity in patients treated with cytotoxic T lymphocyte-associated antigen 4 blockade and interleukin 2: a phase I/II study. Ann Surg Oncol 2005;12:1005–16.

[48] Khan F, Funchain P, Bennett A, Hull TL, Shen B. How should we diagnose and manage checkpoint inhibitor-associated colitis? Cleve Clin J Med 2018;85:679–83.

[49] Wright AP, Piper MS, Bishu S, Stidham RW. Systematic review and case series: flexible sigmoidoscopy identifies most cases of checkpoint inhibitor-induced colitis. Aliment Pharmacol Ther 2019;49:1474–83.

[50] Wang Y, Abu-Sbeih H, Mao E, Ali N, Qiao W, Trinh VA, et al. Endoscopic and histologic features of immune checkpoint inhibitor-related colitis. Inflamm Bowel Dis 2018;24:1695–705.

[51] Gertz MA, Rajkumar SV. Primary systemic amyloidosis. Curr Treat Options Oncol 2002;3:261–71.

[52] Petre S, Shah IA, Gilani N. Review article: gastrointestinal amyloidosis: clinical features, diagnosis and therapy. Aliment Pharmacol Ther 2008;27:1006–16.

Chapter 27

Inflammatory bowel disease—like conditions after organ transplantation

Bo Shen

Center for Inflammatory Bowel Diseases, Columbia University Irving Medical Center-New York Presbyterian Hospital, New York, NY, United States

Chapter Outline

Abbreviations

CD Crohn's disease
CMV cytomegalovirus
DAMP disease-associated molecular pattern
EBV Epstein—Barr virus
GI gastrointestinal
GVHD graft-versus-host disease
HSCT hematopoietic stem cell transplantation
IBD inflammatory bowel disease
MMF mycophenolate mofetil
OLT orthotopic liver transplantation
OT organ transplantation
PSC primary sclerosing cholangitis
SOT solid organ transplantation
UC ulcerative colitis

Introduction

The reported prevalence of diarrhea is up to 72% after solid-organ transplantation (SOT) or hematopoietic stem cell transplantation (HSCT) [1]. Some of them have acute and/or chronic endoscopic and histologic of the gastrointestinal (GI) tract, with some having features of inflammatory bowel disease (IBD). Immunological factors exert a key role in the process of OT. Similar factors are important in the etiopathogenesis of IBD. The association between OT and IBD is complex. The patient undergoes OT may have existing IBD, especially in those with primary sclerosing cholangitis (PSC). Solid-organ transplant (SOT) or HSCT may develop de novo IBD or IBD-like conditions. The classic example is cord colitis syndrome [2]. The use of antirejection medicines can cause IBD or IBD-like conditions, with a typical example of mycophenolate mofetil (MMF)—associated colitis [3]. Antirejection medications with their immunosuppressive effect may lead to bacterial (such as *Clostridium difficile*) [4], viral [such as cytomegalovirus (CMV)] [4], fungal, or parasitic infections, neutropenia, and immune deficiency, mimicking IBD. Besides, graft-versus-host disease

Atlas of Endoscopy Imaging in Inflammatory Bowel Disease. DOI: https://doi.org/10.1016/B978-0-12-814811-2.00027-X
© 2020 Elsevier Inc. All rights reserved.

(GVHD) of the GI tract can develop after HSCT or SOT with endoscopic and histologic features resembling that in IBD.

Patients with de novo IBD and IBD-like conditions commonly present with diarrhea after SOT or HSCT. The frequency of de novo IBD appears to be more common in orthotopic liver transplantation (OLT) than that of other solid organ transplantations. De novo IBD or IBD-like conditions can develop despite the use of antirejection immunosuppressive medications [5]. It is speculated that damage-associated molecular patterns (DAMP) and pathogen-associated molecular patterns and their associated ongoing inflammation in the transplanted organ, as well as the recipients' small and large bowel, are possible etiologies.

The characterization of posttransplantation de novo IBD or IBD-like conditions is important for the diagnosis, management, and prognosis of the affected GI organs as well as the maintenance of healthy transplanted organs.

De novo inflammatory bowel disease or inflammatory bowel disease—like conditions after organ transplantation

De novo IBD occurs more often after liver and/or kidney transplantation than that in other solid organs [6]. The use of tacrolimus or MMF or CMV mismatch is identified risk factors for de novo IBD in post SOT patients. Patients with OLT for PSC have a higher risk for the development of ulcerative colitis (UC). In contrast to SOT, de novo IBD after HSCT appeared to be rare and there are only case reports with de novo UC [7,8] or de novo Crohn's disease (CD) after HSCT [9].

De novo IBD may be presented as UC- or CD-like, affecting mainly the stomach, ileum, and colon [6]. The endoscopic features include diffuse left-sided colitis, extensive colitis, terminal ileitis [1], mucosal ulceration with skip lesions, nodularity, friability, inflammatory strictures, and rectal sparing (Figs. 27.1–27.4) [6]. Histologic features include acute and chronic changes, such as eosinophil infiltration, gastric intestinal metaplasia, crypt distortion, Paneth cell metaplasia, and noncaseating granulomas [6]. The severity of endoscopic and histologic inflammation may not correlate [6,10,11].

GI symptoms are common after HSCT [12]. In post-HSCT, diarrhea often results from acute or chronic GVHD (see next) [13,14]. Infectious enteritis or colitis, especially *C. difficile* and CMV infections, are common [15]. Other rare etiologies include cord colitis syndrome developed after cord bloodHSCT [2]. Susceptibility gene polymorphisms for CD in the recipient nonhematopoietic cells may be a risk factor for the development of de novo CD [9]. Post-HSCT de novo CD may have endoscopic features such as purulent discharge, contact bleeding, loss of vascular pattern, and ulcers. Histology shows ulceration, mucin depletion, crypt distortion, crypt abscess, and in some cases noncaseating

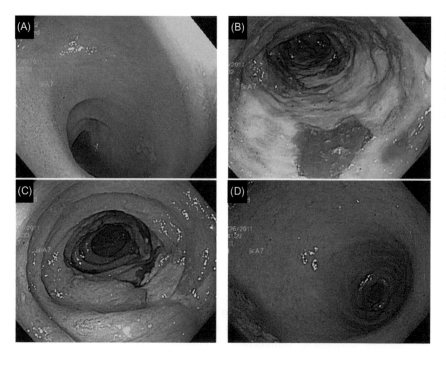

FIGURE 27.1 De novo inflammatory bowel disease after liver transplantation for crytogenic liver cirrhosis in a 70-year-old male patient: (A) normal terminal ileum; (B and C) large longitudinal ulcers in the ascending and descending colon, with histology showing ulcerated mucosa and ischemic-type injury; and (D) disease-spared rectum.

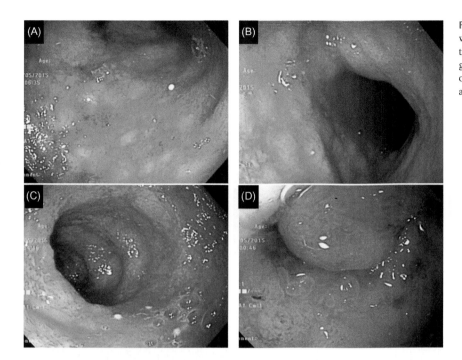

FIGURE 27.2 De novo ulcerative colitis with stricture 6 years after orthotopic liver transplantation for primary sclerosing cholangitis: (A–C) Diffuse erythema and nodularity of the colon and (D) an inflammatory stricture at the transverse colon.

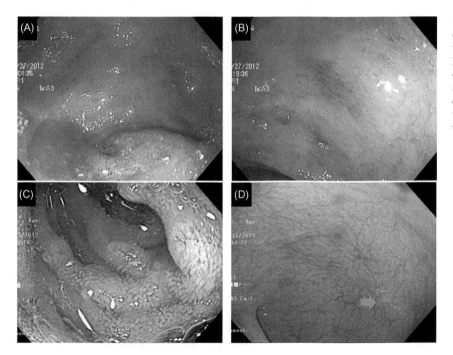

FIGURE 27.3 De novo ulcerative colitis developed 5 years orthotopic liver transplantation for primary sclerosing cholangitis: (A and B) diffuse mild colitis with erythema throughout the colon with histology showing chronic active colitis; (C and D) follow-up colonoscopy showed patchy erythema and erosions in the terminal ileum; and (D) mucosal scars in the colon (*green arrow*).

epithelioid granulomas. Crypt epithelial apoptosis, crypt destruction, crypt distortion, and intraepithelial lymphocytosis, and viral inclusion bodies are rare in de novo IBD [9,10]. Interestingly, various forms of stem cells have been administered to treat refractory CD, at least partially due to myeloablative or immunoablative conditioning [16–18].

Refractory CD is the second most common indication for small bowel transplantation in adults, with an improved survival outcome due to recent surgical innovation, novel immunosuppression, and better postoperative care [19]. However, the main challenge has been the disease recurrence after the transplantation with some patients developing noncaseating granulomas of the transplanted small bowel immediately after the surgery. The recurrent CD in the transplanted small bowel may have the same endoscopic and histologic appearance before and after the transplantation

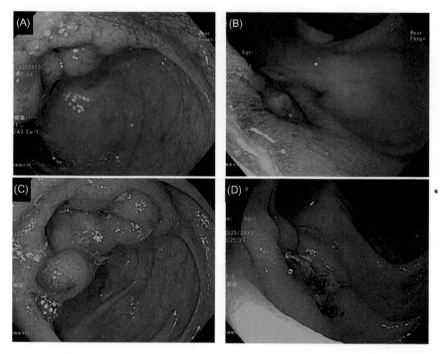

FIGURE 27.4 Crohn's disease—like condition in the colon after renal transplantation for end-stage kidney disease: (A) nodularity of the ileo-cecal valve, (B) medium-sized bleeding ulcer in the proximal colon, (C and D) follow-up colonoscopy showed persistent lesions in the same location. Histology of colon biopsy showed mucosal ulceration and architectural distortion. Cytomegalovirus immunohistochemistry was negative.

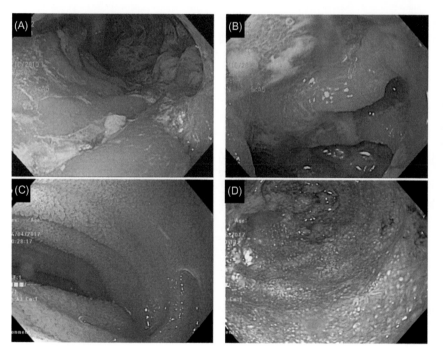

FIGURE 27.5 Recurrent Crohn's disease after a small bowel transplantation in a 28-year-old female patient: (A) diffuse colitis with edema and ulcers, (B) diffuse enteritis with edema and ulcers, (C) normal transplanted small bowel, and (D) recurrent Crohn's disease of the transplanted small bowel. Biopsy of inflamed large and small bowel before and after transplantation all showed small noncaseating granulomas suggesting recurrent Crohn's disease.

(Fig. 27.5). Cautions should be taken during endoscopy and tissue biopsy, as the mucosal and submucosal tissues are friable and easy to bleed. Therefore, a gentle, superficial biopsy is performed (Fig. 27.6).

Small bowel transplantation for non-IBD patients can also lead to inflammation in the transplanted small bowel (Fig. 27.7). But the endoscopic pattern of bowel inflammation is different from that in those with underlying preoperative diagnosis of CD.

Neutropenic enterocolitis or neutropenic typhlitis can occur in HSCT, SOT, leukemia with chemotherapy, and disease-associated immunosuppression status. Contributing factors for neutropenic colitis include neutropenia, cytotoxic drug—associated mucosal injury. The cecum is uniformly affects, while the disease may affect the terminal ileum or

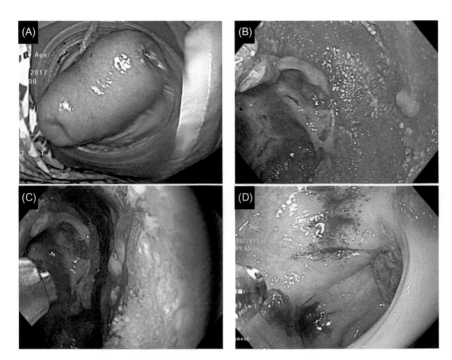

FIGURE 27.6 Recurrent Crohn's disease after small bowel transplantation: (A) loop ileostomy and (B–D) ulcerated and friable mucosa with a predilection to bleeding even after a superficial biopsy. The bleeding was controlled by spraying 50% glucose (C and D). The histology of mucosal biopsy showed noncaseating granulomas.

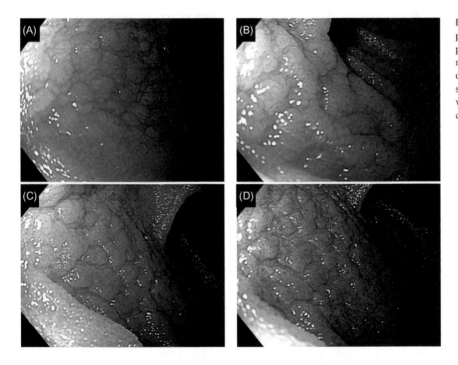

FIGURE 27.7 Nodular mucosa of the transplanted small bowel. The transplantation was performed for small bowel necrosis from mesenteric vein thrombosis: (A–C) nodularity of the distal small bowel on ileoscopy via stoma and (D) the nodularity was highlighted with narrow-band imaging. Histology showed chronic active enteritis.

other parts of the large bowel. It is speculated that the disease's predilection to the cecum may be explained by the thin thickness of the bowel wall and limited blood supply. Thickened bowel wall on abdominal imaging is common in neutropenic colitis. The diagnosis of neutropenic enterocolitis is based on a thickened bowel wall in neutropenic patients with fever and abdominal pain. Colonoscopy is not indicated with a concern of procedure-associated perforation. However, if colonoscopy is incidentally performed, the endoscopic features of neutropenic colitis resemble that in acute infectious colitis and IBD (Fig. 27.8).

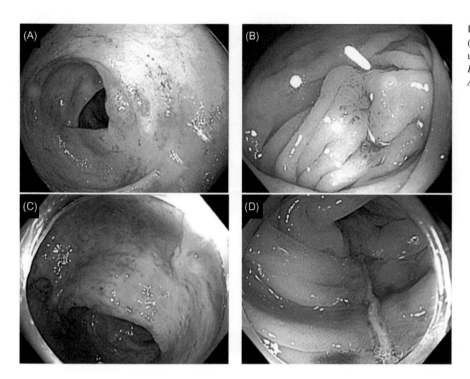

FIGURE 27.8 Neutropenic colitis: (A–D) patchy erythema, edema, and linear ulcers of the colon. *Courtesy of Dr. Ying-Hong Wang of University of Texas MD Anderson Cancer Center.*

Cytomegalovirus- and Epstein–Barr virus–associated inflammatory bowel disease–like conditions after organ transplantation

CMV infection can affect any part of the GI tract with a predilection to the large bowel. CMV mismatch has been shown to a risk factor for the development of both de novo and recurrent IBD in OT [20]. CMV-associated enterocolitis after OT may present with sole CMV infection or superimposed CMV infection on de novo or preexisting IBD. To complicate the matter even further, CMV infection may exist with other viral infections [such as Epstein–Barr virus (EBV) infection] and/or GVHD, making differential diagnosis difficult. In OT patients, endoscopic features overlap among sole CMV enterocolitis, concurrent CMV and IBD, concurrent CMV and GVHD, sole GVHD, and concurrent CVHD and IBD. Endoscopic features in CMV colitis include patchy erythema, exudates, erosions, ulcers with various shapes, sizes, and depths, inflammatory pseudopolyps (Fig. 27.9). A combined assessment of clinical, endoscopic, microbiological, and histologic features is required.

EBV-associated enterocolitis after OT should be carefully evaluated for the development of lymphoproliferative disorders. Endoscopic features of EBV-associated lymphoproliferative disease orders are described in Chapter 12, Inflammatory bowel disease–associated neoplasia.

Graft-versus-host disease of the gastrointestinal tract

GVHD can have acute and chronic forms or mixed forms, involving multiple organs, which is a common presentation of allogeneic HSCT. The cutoff for dividing acute and chronic GVHD is the time of onset of 100 days after HSCT. Acute and chronic GVHD frequently involves both upper and lower GI tract. Common GI clinical presentations of GVHD include persistent nausea, emesis, abdominal cramps, and diarrhea with or without hematochezia.

The diagnosis is confirmed by pathological evaluation of tissue through upper endoscopy or ileocolonoscopy. Endoscopy in most patients with acute GVHD in the GI tract is normal [21]. Acute GVHD manifests as spotted erythema, aphthous lesions, and denudation of the mucosa on endoscopy (Figs. 27.10–27.14) [22,23]. The Freiburg Criteria for the endoscopic grading of acute intestinal GVHD has been proposed, consisting of Grade 1—normal; Grade 2—spotted erythema, initial aphthous lesions; and Grade 3—aphthous lesions (CD-like) or focal erosions [17].

Rectal biopsy often suffices the diagnosis of acute GVHD [13–15]. Histologic hallmark of GVHD is the presence of gland apoptosis, which is not explained by other inflammatory or infectious etiologies. Histologic features of acute GVHD are as follows: Grade 1—isolated apoptotic epithelial cells without crypt loss; Grade 2—loss of isolated crypts

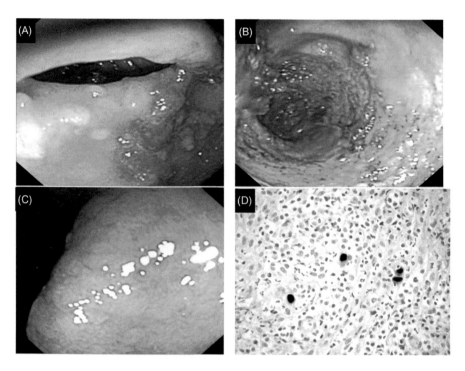

FIGURE 27.9 Cytomegalovirus colitis after orthotopic liver transplantation: (A) small nodules on the ileocecal valve; (B and C) diffuse colitis with edema, granularity, and friability; (D) immunohistochemistry showed nuclear stains of the histiocytes for cytomegalovirus.

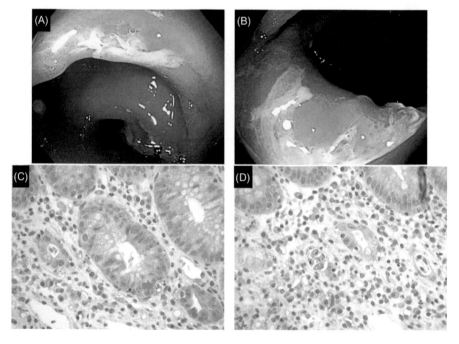

FIGURE 27.10 Acute graft-versus-host disease of the colon after orthotopic liver transplantation: (A and B) semicircumferential ulcers on the folds covered with mucopurulent exudates and (C and D) apoptic bodies in the epithelia of the colon (*green arrow*). *Histology* images: *courtesy of Dr. Scott Robertson of Cleveland Clinic.*

without the loss of contiguous crypts Grade 3—loss of two or more contiguous crypts; and Grade 4—extensive crypt loss with mucosal denudation [24].

Chronic GVHD of the GI tract presents with dry oral mucosa and ulcerations and sclerosis; exudates, erosions, ulceration of mucosa of the GI tract; and a rising serum bilirubin concentration [25]. It can involve the stomach and small and large intestines. The National Institutes of Health consensus on the diagnosis of chronic GVHD is the presence of esophageal web or strictures in the upper to mid-third of the esophagus, as a part of the criteria [26]. Endoscopic features of chronic GVHD in the GI tract vary, ranging from loss of vascular pattern, mild erythema to severe erythema, edema, exudates, erosions, ulceration, and strictures (Figs. 27.15 and 27.16) [16]. Histology of chronic

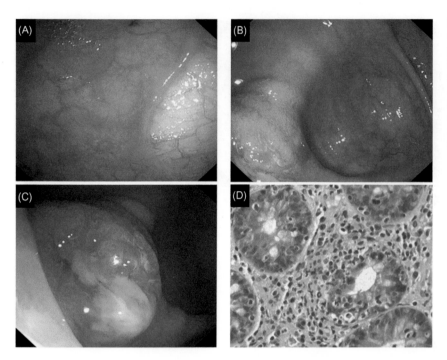

FIGURE 27.11 Concurrent acute graft-versus-host disease and cytomegalovirus infection of the colon in hematopoietic stem cell transplantation: (A–C) erythema and ulcerated nodules in the colon and (D) representative histology of graft-versus-host disease with extensive crypt apoptosis. *Endoscopy images courtesy Dr. Guodong Chen, Beijing University Ren Min Hospital; Histology image: courtesy of Dr. Ana E. Bennett of Cleveland Clinic.*

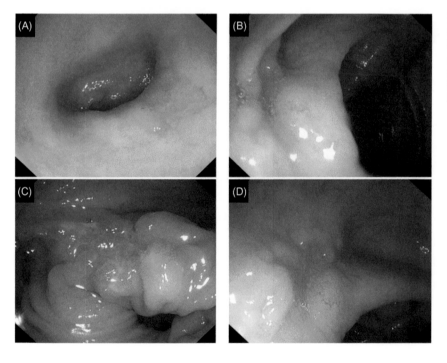

FIGURE 27.12 Concurrent acute graft-versus-host disease and Epstein–Barr virus infection of the colon: (A–D) diffuse edema and patchy ulceration with adjacent nodular mucosa throughout the colon. *Courtesy of Dr. Guodong Chen, Beijing University Ren Min Hospital.*

GVHD, similar to that in acute GVHD, may be categorized into Grade 1—increased crypt apoptosis, Grade 2—apoptosis with crypt abscesses, Grade 3—individual crypt, or Grade 4—total denudation of the mucosa [27].

Mycophenolate mofetil-associated inflammatory bowel disease–like conditions

While MMF has been explored for the treatment of refractory IBD [28], MMF is known to cause de novo IBD or IBD-like conditions. Among SOT patients on MMF undergoing colonoscopy, 9% were found to have MMF colitis [3].

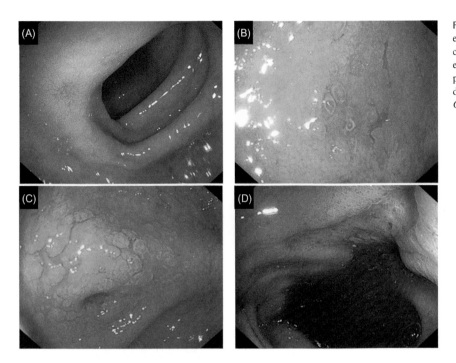

FIGURE 27.13 Acute graft-versus-host disease of the colon after hematopoietic stem cell transplantation: (A) patchy erythema and erosions, (B and C) friable mucosa with mosaic patterns, and (D) deep irregular-shaped and demarcated ulcer. *Courtesy* of *Dr. Guodong Chen, Beijing University Ren Min Hospital.*

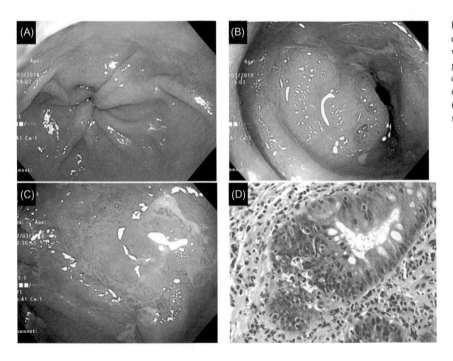

FIGURE 27.14 Acute graft-versus-host disease of the gastrointestinal tract: (A) gastritis with linear erythema of the antrum; (B) mild granularity of the mucosa of the ascending colon; (C) edema, nodularity, erosions, and exudates of the descending colon mucosa; and (D) representative histology of colon biopsy showing extensive crypt apoptosis.

In fact, MMF used OLT was found to increase a 3.3-fold risk for the development of de novo IBD as compared with OLT patients without use [5]. The risk for MMF after renal transplant is even higher than OLT [24]. Common endoscopic features of MMF colitis include normal appearance, erythema, erosions, and ulcers (Fig. 27.17) [24]. Segmental inflammation and rectal sparing are common [24]. The histologic hallmark of MMF-associated colitis is the presence of increased epithelial apoptosis [29]. However, histologic findings are classified in acute colitis-like, IBD-like, ischemia-like, and GVHD-like features [24,30,31].

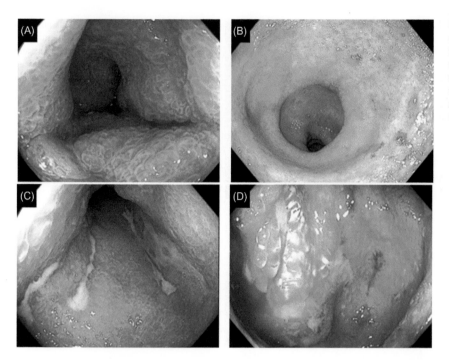

FIGURE 27.15 Chronic graft-versus-host disease of the colon: (A) diffuse edema, erythema, and granularity of the ileum mucosa; (B) loss of vascular pattern of the colon; and (C and D) edema, erythema, and longitudinal ulcers of the colon. *Photos courtesy Dr. Ying-Hong Wang of University of Texas Anderson Cancer Center.*

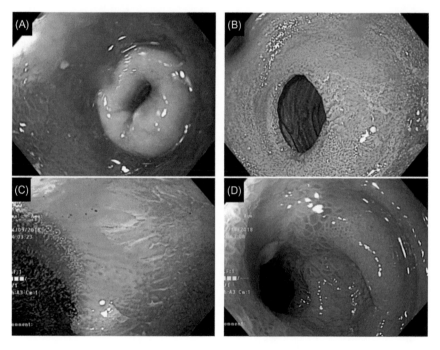

FIGURE 27.16 Concurrent chronic graft-versus-host disease and cytomegalovirus infection of the small bowel after orthotopic liver transplantation for hepatitis C-associated liver cirrhosis. The patient underwent diverting ileostomy due to persistent colitis. (A and B) Circumferential strictures at the neo-distal small bowel on ileoscopy via stoma; and (C and D) patchy erythema, friability, and ulceration of the small bowel with an inflammatory stricture.

Summary and recommendations

Multiple aspects of the association between SOT or HSCT and IBD exit. Patients with transplantation may exist IBD or develop de novo IBD after the surgery. In addition, GVHD, which commonly occurs after HSCT, less often in SOT, can mimic clinical, endoscopic, radiographic, and histologic presentation. Also, transplanted patients are at the risk for CMV and EBV infection. These infections can coexist with GVHD (Figs. 27.11 and 27.12) and IBD. Finally, antirejection medications, especially MMF, can cause IBD-like conditions. Endoscopy with histologic evaluation plays a key role in the diagnosis and differential diagnosis of transplantation-associated GI disorders. Clinical, microbiological, and radiographic evaluation is also important.

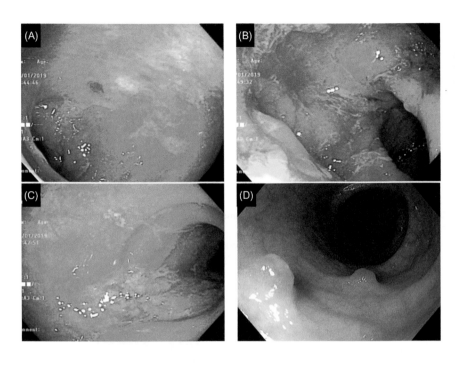

FIGURE 27.17 Mycophenolate mofetil—associated active colitis in a patient with autoimmune hepatitis who was evaluated for liver transplantation: (A—C) diffuse colitis with edema, erythema, large superficial ulcers, and mucopurulent exudates and (D) near resolution of colitis with residual inflammatory polyps after discontinuation of the agent except few inflammatory polyps. Histology showed atrophic crypts, ulceration, and granulation tissue (not shown).

Conflict of interest

The author declared no financial conflict of interest.

References

[1] Worns MA, Lohse AW, Neurath MF, Croxford A, Otto G, Kreft A, et al. Five cases of de novo inflammatory bowel disease after orthotopic liver transplantation. Am J Gastroenterol 2006;101:1931—7.

[2] Herrera AF, Soriano G, Bellizzi AM, Hornick JL, Ho VT, Ballen KK, et al. Cord colitis syndrome in cord-blood stem-cell transplantation. N Engl J Med 2011;365:815—24.

[3] Calmet FH, Yarur AJ, Pukazhendhi G, Ahmad J, Bhamidimarri KR. Endoscopic and histological features of mycophenolate mofetil colitis in patients after solid organ transplantation. Ann Gastroenterol 2015;28:366—73.

[4] Wong NA, Bathgate AJ, Bellamy CO. Colorectal disease in liver allograft recipients — a clinicopathological study with follow-up. Eur J Gastroenterol Hepatol 2002;14:231—6.

[5] Mouchli MA, Singh S, Boardman L, Bruining DH, Lightner AL, Rosen CB, et al. Natural history of established and de novo inflammatory bowel disease after liver transplantation for primary sclerosing cholangitis. Inflamm Bowel Dis 2018;24:1074—81.

[6] Fernandes MA, Braun HJ, Evason K, Rhee S, Perito ER. De novo inflammatory bowel disease after pediatric kidney or liver transplant. Pediatr Transplant 2017;21 Epub 2016 Nov 11.

[7] Spiers AS. Ulcerative colitis after bone-marrow transplantation for acute leukemia. N Engl J Med 1984;311:1259.

[8] Koike K, Kohda K, Kuga T, Nakazawa O, Ando M, Takayanagi N, et al. Ulcerative colitis after autologous peripheral blood stem cell transplantation for non-Hodgkin's lymphoma. Bone Marrow Transplant 2001;28:619—21.

[9] Sonwalkar SA, James RM, Ahmad T, Zhang L, Verbeke CS, Barnard DL, et al. Fulminant Crohn's colitis after allogeneic stem cell transplantation. Gut 2003;52:1518—21.

[10] Nepal S, Navaneethan U, Bennett AE, Shen B. De novo inflammatory bowel disease and its mimics after organ transplantation. Inflamm Bowel Dis 2013;19:1518—27.

[11] Kochhar G, Singh T, Dust H, Lopez R, McCullough AJ, Liu X, et al. Impact of *de novo* and preexisting inflammatory bowel disease on the outcome of orthotopic liver transplantation. Inflamm Bowel Dis 2016;22:1670—8.

[12] Iqbal N, Salzman D, Lazenby AJ, et al. Diagnosis of gastrointestinal graft-versus-host disease. Am J Gastroenterol 2000;95:3034—8.

[13] Cox GJ, Matsui SM, Lo RS, et al. Etiology and outcome of diarrhea after marrow transplantation: a prospective study. Gastroenterology 1994;107:1398—407.

[14] Aslanian H, Chander B, Robert M, Cooper D, Proctor D, Seropian S, et al. Prospective evaluation of acute graft-versus-host disease. Dig Dis Sci 2012;57:720—5.

[15] Ross WA, Ghosh S, Dekovich AA, Liu S, Ayers GD, Cleary KR, et al. Endoscopic biopsy diagnosis of acute gastrointestinal graft-versus-host disease: rectosigmoid biopsies are more sensitive than upper gastrointestinal biopsies. Am J Gastroenterol 2008;103:982—9.

[16] Oyama Y, Craig RM, Traynor AE, Quigley K, Statkute L, Halverson A, et al. Autologous hematopoietic stem cell transplantation in patients with refractory Crohn's disease. Gastroenterology 2005;128:552−63.

[17] Hommes DW, Duijvestein M, Zelinkova Z, Stokkers PC, Ley MH, Stoker J, et al. Long-term follow-up of autologous hematopoietic stem cell transplantation for severe refractory Crohn's disease. J Crohns Colitis 2011;5:543−9.

[18] Panés J, García-Olmo D, Van Assche G, Colombel JF, Reinisch W, Baumgart DC, et al. Expanded allogeneic adipose-derived mesenchymal stem cells (Cx601) for complex perianal fistulas in Crohn's disease: a phase 3 randomised, double-blind controlled trial. Lancet 2016;388:1281−90.

[19] Nyabanga C, Kochhar G, Costa G, Soliman B, Shen B, Abu-Elmagd K. Management of Crohn's disease in the new era of gut rehabilitation and intestinal transplantation. Inflamm Bowel Dis 2016;22:1763−76.

[20] Haagsma EB, Van Den Berg AP, Kleibeuker JH, et al. Inflammatory bowel disease after liver transplantation: the effect of different immuno-suppressive regimens. Aliment Pharmacol Ther 2003;18:33−44.

[21] Sultan M, Ramprasad J, Jensen MK, Margolis D, Werlin S. Endoscopic diagnosis of pediatric acute gastrointestinal graft-versus-host disease. J Pediatr Gastroenterol Nutr 2012;55:417−20.

[22] Cruz-Correa M, Poonawala A, Abraham SC, Wu TT, Zahurak M, Vogelsang G, et al. Endoscopic findings predict the histologic diagnosis in gastrointestinal graft-versus-host disease. Endoscopy 2002;34:808−13.

[23] Kreisel W, Dahlberg M, Bertz H, Harder J, Potthoff K, Deibert P, et al. Endoscopic diagnosis of acute intestinal GVHD following allogeneic hematopoietic SCT: a retrospective analysis in 175 patients. Bone Marrow Transplant 2012;47:430−8.

[24] Washington K, Jagasia M. Pathology of graft-versus-host disease in the gastrointestinal tract. Hum Pathol 2009;40:909−17.

[25] Jacobsohn DA, Kurland BF, Pidala J, Inamoto Y, Chai X, Palmer JM, et al. Correlation between NIH composite skin score, patient-reported skin score, and outcome: results from the Chronic GVHD Consortium. Blood 2012;120:2545−52.

[26] Filipovitch AH, Weisdorf D, Pavletic S, Socie G, Wingard JR, Lee SJ, et al. National Institutes of Health consensus development project on criteria for clinical trials in chronic graft-versus-host disease: 1. Diagnosis and staging working group report. Biol Blood Marrow Transplant 2005;11:945−56.

[27] Snover DC, Weisdorf SA, Vercellotti GM, Rank B, Hutton S, McGlave P. A histopathologic study of gastric and small intestinal graft-versus-host disease following allogeneic bone marrow transplantation. Hum Pathol 1985;16:387−92.

[28] Smith MR, Cooper SC. Mycophenolate mofetil therapy in the management of inflammatory bowel disease—a retrospective case series and review. J Crohns Colitis 2014;8:890−7.

[29] Liapis G, Boletis J, Skalioti C, Bamias G, Tsimaratou K, Patsouris E, et al. Histological spectrum of mycophenolate mofetil-related colitis: association with apoptosis. Histopathology 2013;63:649−58.

[30] de Andrade LG, Rodrigues MA, Romeiro FG, Garcia PD, Contti MM, de Carvalho MF. Clinicopathologic features and outcome of mycophenolate-induced colitis in renal transplant recipients. Clin Transplant 2014;28:1244−8.

[31] Izower MA, Rahman M, Molmenti EP, Bhaskaran MC, Amin VG, Khan S, et al. Correlation of abnormal histology with endoscopic findings among mycophenolate mofetil treated patients. World J Gastrointest Endosc 2017;9:405−10.

Chapter 28

Inflammatory bowel disease—like conditions: ischemic bowel diseases and vasculitides

Xiaoying Wang[1], Yan Chen[1] and Bo Shen[2]

[1]*Department of Gastroenterology, The Second Affiliated Hospital, School of Medicine, Zhejiang University, Hangzhou, P.R. China,* [2]*Center for Inflammatory Bowel Diseases, Columbia University-New York Presbyterian Hospital, NY, United States*

Chapter Outline

Abbreviations

ANCA	antineutrophil cytoplasmic antibody
BD	Behçet disease
CD	Crohn's disease CT computed tomography
EGPA	eosinophilic granulomatosis with polyangiitis
IBD	inflammatory bowel disease
GI	gastrointestinal Ig immunoglobulin
PAN	polyarteritis nodosa
SLE	systemic lupus erythematosus
UC	ulcerative colitis

Introduction

Diagnosis and differential diagnosis of inflammatory bowel disease (IBD) are important for the management and prognosis. Among mimics of IBD a group of diseases with underlying ischemia, vasculitis, or vasculopathy share similar clinical presentations and abdominal imaging with that of IBD. Treatment strategies of ischemic bowel disease are different from that of IBD. Even the management plan of immune-mediated vasculitis is not necessarily the same to that for Crohn's disease (CD) or ulcerative colitis (UC). Underlying disease processes in ischemic bowel disease and systemic vasculitides, such as tissue ischemia, tissue hypoxia, hypercoagulability, and vascular inflammation, may also contribute to the etiopathogenesis and disease exacerbation or progression of IBD [1,2]. On the other hand, granulomatous inflammation may be present in vasculitis-associated gastrointestinal (GI) disorders, such as Behçet disease (BD) [3]. Patients with IBD carry a high risk for concurrent vasculitis [4]. It is speculated that these

Atlas of Endoscopy Imaging in Inflammatory Bowel Disease. DOI: https://doi.org/10.1016/B978-0-12-814811-2.00028-1
© 2020 Elsevier Inc. All rights reserved.

immune-mediated GI diseases may represent phenotypes of a wide spectrum of IBD (Chapter 1: Introduction and classification of inflammatory bowel diseases).

Endoscopic evaluation with tissue biopsy plays a key role in the diagnosis and differential diagnosis. A combined assessment of clinical, endoscopic, histologic, serological, and radiographic features is needed. Endoscopic evaluation should include documentation of disease distribution and severity and features of inflammation and ulcerations. It should be pointed out that endoscopy should be performed with caution, due to the increased risk for perforation, in a patient with suspected GI involvement from an acute flare of ischemic colitis or vasculitis (Chapter 2: Setup and principle of endoscopy in inflammatory bowel disease).

Ischemic bowel diseases

Intestinal ischemia can result from acute arterial embolic or thrombotic occlusion, venous thrombosis, or hypoperfusion of the mesenteric vasculature. Colonic ischemia is the most common form.

Colonic ischemia

The etiology of colonic ischemia includes nonocclusive colonic ischemia, embolic and thrombotic arterial occlusion, and Mesenteric vein thrombosis. It can present with acute or chronic forms. Nonocclusive colonic ischemia predominantly affects watershed areas, such as the splenic flexure and rectosigmoid junction. Segmental distribution is one of the hallmarks of colonic ischemia. Colonic ischemia can present with acute or chronic forms. Endoscopic features of acute colonic ischemia are edematous, erythema, friable mucosa, spontaneous bleeding, erosions, ulcers, and in severe cases, necrosis and perforation (Figs. 28.1–28.4). Patients with colonic ischemia may have mucosal scars, mucosal bridges, pseudopolyps, and ulcerated or nonulcerated strictures on endoscopy (Figs. 28.5 and 28.6). These endoscopic features overlap with some of CD and UC.

Phlebosclerotic colitis

Phlebosclerotic colitis is a rare form of ischemic colitis, which develops as a result of mesenteric venous calcification and calcification in the invaded colonic submucosal layer [5]. The etiology and pathogenesis are not clear, but the use of herbal medications was observed in some patients. Common presentations are abdominal pain, intestinal obstruction, diarrhea, and GI bleeding, but sometimes no symptoms [6]. Varying degrees of calcifications of the colon and mesenteric venous can be found on computed tomography (CT). The lesions mainly located in the transverse and ascending colon [1]. Colonoscopy shows characteristic dark purple mucosa and sometimes multiple ulcers without clear boundaries mainly in the ascending colon to the transverse colon (Fig. 28.7). The histopathological changes include thickening of the colon wall, fibrotic degeneration of the submucosa, and thickening, fibrosis, and calcification of the vein wall [7].

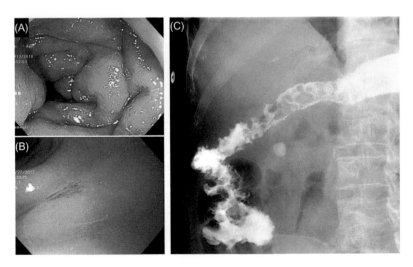

FIGURE 28.1 Acute ischemic colitis with edema and erythema. (A) Diffuse severe edema of the transverse colon, (B) mucosal edema and patchy erythema, and (C) thumb printing of the ascending and transverse colon and hepatic flexure on contrasted enemas.

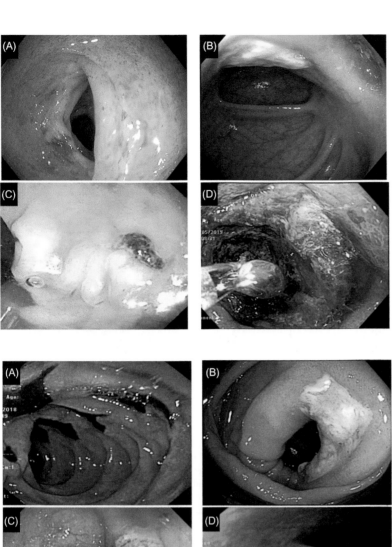

FIGURE 28.2 Patterns of acute colonic ischemia. (A) Patchy erythema of mucosa at the rectosigmoid junction; (B) discrete deep ulcers at the splenic flexure; (C) small ulcer with stigmata of bleeding at the splenic flexure; and (D) diffuse mucosal edema, erythema, and bleeding at the hepatic flexure.

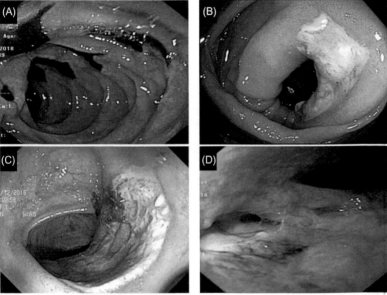

FIGURE 28.3 Ulcer patterns in duodenal and colonic ischemia. (A) Series of ischemic ulcers covered with black plaques in the folds of the second part of the duodenum; (B and C) discrete, large, clean-based ulcers in the right colon; and (D) anastomotic ulcers with a leak at the colo-colonic anastomosis after left partial colectomy for ischemic colitis.

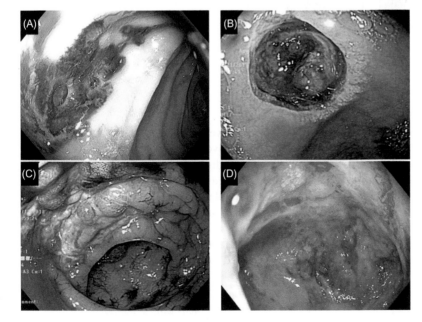

FIGURE 28.4 Marked ulcers with necrosis in acute bowel ischemia, the pattern is rare in Crohn's disease or ulcerative colitis. (A) Large ischemic ulcer with superficial necrosis at the transverse colon; (B) deep ischemic ulcer with necrosis at the duodenum bulb resulting from angiogram embolization of prior duodenum ulcer; (C) circumferential deep ulcers with nodularity of the surround sigmoid colon mucosa; and (D) deep, large ischemic ulcer with necrosis in the cecum, with a high risk of bowel perforation.

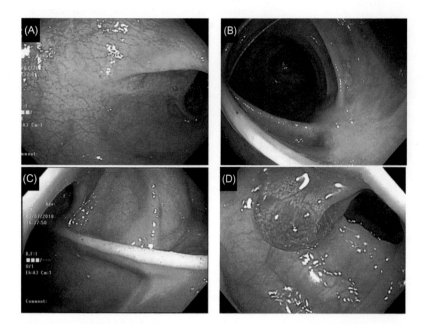

FIGURE 28.5 Mucosal patterns of chronic colonic ischemia. (A and B) Linear, longitudinal mucosal scars in the hepatic flexure; (C) long mucosal bridge at the splenic flexure; and (D) inflammatory polyp at the rectal sigmoid junction. These patterns are also commonly seen in ulcerative colitis or Crohn's colitis with mucosal healing.

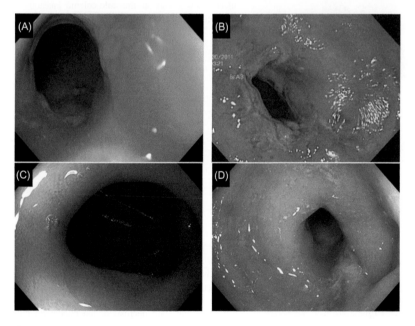

FIGURE 28.6 Colonic strictures in the ischemic colon. (A and B) Ulcerated strictures at the splenic flexure; (C and D) nonulcerated strictures in the recto-sigmoid junction. The splenic flexure and rectosigmoid colon are common locations for nonocclusive colonic ischemia.

Vasculitides

Vasculitides are characterized by the presence of inflammatory leukocytes in vessel walls with reactive damage to mural structures. They often involve the GI tract. Common vasculitides with GI involvement include large-vessel vasculitis [e.g., Takayasu arteritis, giant cell (temporal) arteritis], medium-vessel vasculitis [e.g. polyarteritis nodosa (PAN)], small-vessel vasculitis [antineutrophil cytoplasmic antibody (ANCA)-associated vasculitis, immunoglobulin (Ig) A vasculitis (Henoch−Schönlein purpura), and cryoglobulinemic vasculitis], and variable-vessel vasculitis. ANCA-associated vasculitis consists of microscopic polyangiitis, granulomatosis with polyangiitis, and eosinophilic granulomatosis with polyangiitis (EGPA). One of the classic examples of variable-vessel vasculitis is BD.

Common endoscopic presentations of vasculitis are erosions, petechiae, ulcers, mucosal or submucosal hemorrhage, edema, and nodularity. Vasculitis-induced mesenteric ischemia may be accessed by CT angiogram and less commonly catheter-based angiogram.

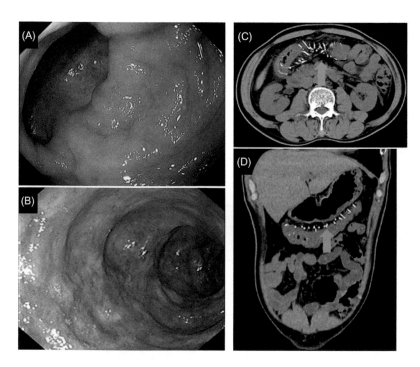

FIGURE 28.7 Patterns of mucosa in idiopathic mesenteric venous phlebosclerotic colitis on colonoscopy. (A and B) Characteristic dark purple mucosa of the transverse colon; (C and D) calcifications of mesenteric veins on CT enterography (*green arrows*). *CT,* Computed tomography.

Behçet disease

BD or Behçet syndrome is chronic, relapsing, multisystem vasculitis of unknown etiology. BD is featured with recurrent oral aphthae, genital aphthae, ocular disease, and GI involvement. Other extraintestinal manifestations include vascular disease, skin lesions, neurologic disease, and arthritis. The underlying disease process in BD is vasculitis, affecting small, medium, and large blood vessels on both the arterial and venous sides of the circulation. The classic histopathologic lesion of BD is necrotizing leukocytoclastic obliterative perivasculitis and venous thrombosis with lymphocytic infiltration of capillaries, veins, and arteries of all sizes.

The changes involving the GI tract are mainly ulcers, which are common in the esophagus, and the involvement of the small intestine is similar to CD. There are great overlaps in oral, perianal, and GI lesions between IBD (particularly CD) and BD [3]. However, some distinguishing features of BD exit, including large round or oval ulcers, the involvement of a single-segment of the bowel (particularly the ileocecal area), and a deformed ileocecal valve [8]. Colonic ulcers are more common in the transverse colon, ascending colon, ileocecal area, with a risk of bleeding and perforation [9]. Based on colonoscopy appearance, ulcers have been described as three types: (1) volcano ulcers, (2) geographic ulcers, and (3) the aphthous type. Volcano ulcers are predominant ones, which are characterized by a deep and discreet margin [10]. In contrast to CD, strictures and fistulas are rare in BD (Table 28.1; Figs. 28.8—28.10) [11,12].

IgA vasculitis (Henoch—Schönlein purpura)

IgA vasculitis, an immune-mediated disease associated with IgA deposition, is the most common form of systemic vasculitis in children. In addition to the skin lesion, the patient may also present with arthralgia, arthritis, and renal disease. Patients have leukocytoclastic vasculitis or proliferative glomerulonephritis, with predominant IgA deposition. Abdominal pain is a common clinical presentation, which sometimes results from its GI involvement. The common locations of GI involvement are the small bowel and stomach. Endoscopic features include edema, submucosal hemorrhage, and ulcers. Intestinal intussusception, obstruction, or perforation can occur. Purpuric lesions may be seen, which are commonly located at the descending duodenum, stomach, colon, jejunum, and terminal ileum (Figs. 28.11 and 28.12). CD should be a differential diagnosis. Coexisting IgA vasculitis and UC has been reported [13,14].

Systemic lupus erythematosus

Systemic lupus erythematosus (SLE) is a chronic or relapsing inflammatory disease that can affect the skin, joints, kidneys, lungs, nervous system, serous membranes, and other organs. Vasculitis associated with SLE can small- and

TABLE 28.1 Endoscopic and histologic features between Behçet disease and Crohn's disease.

		Behçet disease	Crohn's disease
Endoscopic features	**Number of lesions**	**Smaller**	**Greater**
	Ulcer shape	Discrete, well-demarcated round or oval, with a sharp edge; circumferential distribution of ulcers	Multiple, linear, prominent ulcers with raised edges along the mesentery site
	Deep ulcer with a risk of free perforation	Free perforation from deep ulcers	Contained perforation from fistula
	Cobble-stoning	Rare	Common
	Solitary ulcer in the ileocecal area	Common	Uncommon
	Single-segment bowel involvement	Common	Uncommon
	Ileocecal valve	Often deformed	Often strictured
	Stricture	Rare	Common
	Fistula	Rare	More common
Histology		Vasculitis	Granulomas

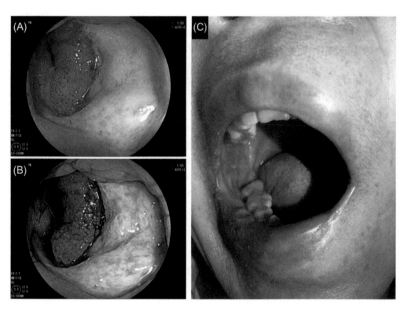

FIGURE 28.8 Behçet disease of the large bowel. A 40-year-old patient with recurrent abdominal pain. (A) Deep, well-demarcated, clean-based ulcer at the cecum; (B) the ulcer viewed with chromoendoscopy; and (C) ulcers on lip and oral cavity. *Photo courtesy: Dr. Xinbo Ai, Jinan University Zhuhai Hospital.*

medium-sized vessels, which affect the GI tract. SLE-related GI disorders, including lupus enteritis [15,16], esophagitis, intestinal pseudo-obstruction, protein-losing enteropathy [17], mesenteric vasculitis, or ischemia (Fig. 28.13). Common GI presentations include abdominal pain, blood in the stool, and diarrhea. Acute abdominal pain is often caused by bowel ischemia secondary to lupus mesenteric vasculitis [18]. Lupus enteritis and protein-losing enteropathy may present with diffuse small bowel edema and erosions. Ischemic changes depend on the layers of bowel wall involved,

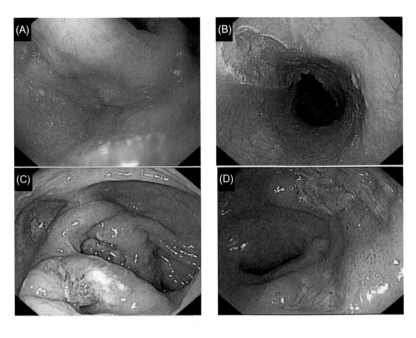

FIGURE 28.9   Patterns of ulcers in Behçet's disease. (A) Mild mucosal swelling with surface erosion in the left piriform fossa; (B) discrete, linear, clean-based ulcer with a clear boundary in the lower and middle parts of the esophagus; (C) discrete deep ulcers and ulcerated stricture at the ileocecum; and (D) multiple irregular ulcers in the transverse colon.

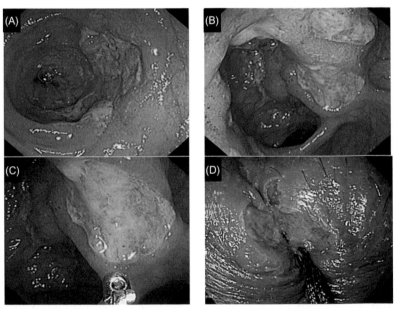

FIGURE 28.10   Behçet disease. (A–C) Discrete large ulcers in the ileocecal lesions and (D) large deep perianal ulcers.

ranging mucosal ulceration and hemorrhage, submucosal edema, and intestinal pseudo-obstruction due to a muscular injury, and ascites and perforation due to serosal damage (Fig. 28.13) [18].

Polyarteritis nodosa

PAN is a rare systemic necrotizing vasculitis, affecting medium-sized muscular arteries and occasionally involving small muscular arteries. Proximately 50% of patients with PAN presented with GI symptoms, including postprandial abdominal pain, discomfort, nausea, vomiting, and bleeding [19]. Main underlying disease process for GI presentation results from mesenteric arteritis and vasculitis. Angiography showing evidence of aneurysms is diagnostic in most patients, and confirmation may need histopathological evaluation [20]. Endoscopic features include erosions, ulcers,

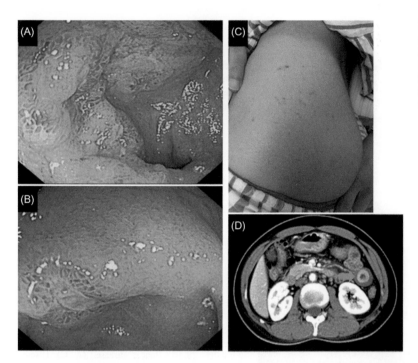

FIGURE 28.11 IgA vasculitis (Henoch–Schönlein syndrome) involving the distal ileum. (A and B) Mucosal edema and ulceration at the terminal ileum; (C) the purpura lesions at the right hip; and (D) diffuse distal small bowel edema with mucosal hyperenhancement on CT enterography. *CT*, Computed tomography.

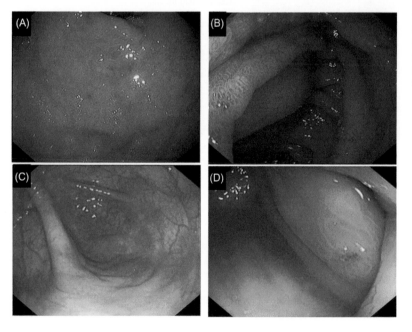

FIGURE 28.12 Patterns of mucosa in IgA vasculitis (Henoch–Schönlein purpura). (A) Mucosa hyperemia in the fundus; (B) diffuse mucosal hyperemia in the descending portion of the duodenum; (C) mucosal congestion at the cecal base; and (D) mucosal congestion with surface erosion in the descending colon.

and mucosal or submucosal hemorrhage in the small and large bowel. Bowel necrosis and perforation may occur in severe cases (Fig. 28.14). Of note, patients with PAN have a higher risk for colonoscopy perforation [21].

Eosinophilic granulomatosis with polyangiitis

EGPA is also named Churg–Strauss syndrome or allergic granulomatosis and angiitis. EGPA is a rare autoimmune disease involving multiple systems. It is mainly characterized by eosinophilia, infiltration, and small and medium vascular necrotizing granulomatous inflammation in peripheral tissues and tissues. The disease is associated with anti-ANCA-mediated injury to systemic blood vessels. EGPA commonly affects the sinus, lungs, nervous system,

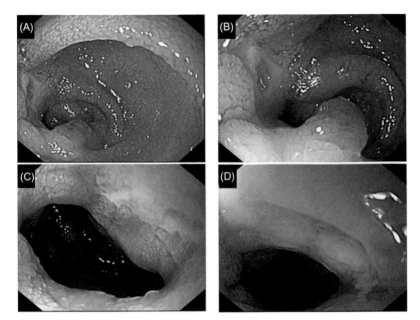

FIGURE 28.13   Patterns of small bowel disease in a patient with systemic lupus erythematosus and Sjögren's syndrome on balloon-assisted enteroscopy. (A and B) Ulcerated strictures at the jejunum; (C and D) semicircumferential ulcers with exudates at the jejunum.

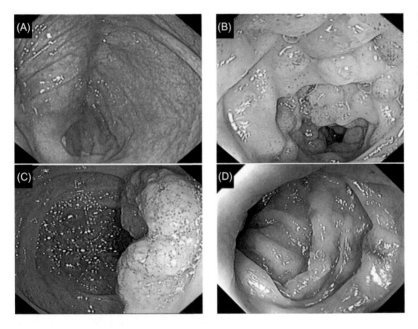

FIGURE 28.14   Patterns of mucosa in eosinophilic granulomatosis with polyangiitis on colonoscopy. (A) Mucosal erythema congestion in the cecum; (B) nodularity with petechiae of the terminal ileum; (C) nodularity and edema of the ileocecal valve; and (D) mucosal edema with erythema and scattered erosions in the descending colon.

kidneys, and GI tract. GI presentations include abdominal pain, diarrhea, bleeding, intestinal obstruction or perforation, and peritonitis [22].

EGPA develops in the prodromal, eosinophilic, and vasculitic phases. Underlying vasculitis process affects small- and medium-sized arteries. Although EGPA commonly affects the lungs and skin, it also causes GI presentations. The initial GI presentations result from eosinophilic gastroenteritis with edema, erythema, erosions, and solitary or multiple ulcers. In the vasculitic phase, patients can present with bleeding or even perforation from severe enteritis and colitis (Fig. 28.15) [23,24].

Summary and recommendations

Ischemic bowel diseases and systemic vasculitides involving the GI tract often share similar clinical, endoscopic, and imaging presentations or underlying pathogenetic pathways with that in CD and UC. Careful endoscopic and histologic

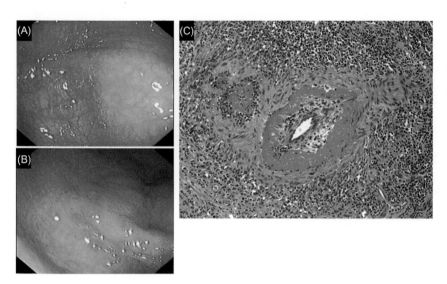

FIGURE 28.15 Patterns of mucosal inflammation in polyarteritis nodosa. (A and B) Congestion and erosions in the sigmoid colon on colonoscopyand (C) bowel resection specimen in a separate patient showing vasculitis with a large number of lymphocytic infiltration in the vascular wall and perivascular space on histology.

evaluation plays a key role in the differential diagnosis. It is important to document disease distribution and severity, along with features of inflammation and ulcers. Endoscopy may be avoided in the acute phase of ischemic bowel diseases or vasculitis-associated GI diseases, due to the risk for perforation.

Acknowledgment

The authors would like to thank Dandan Zhong, Yan Li, Wen Hu, and Dingting Xu for providing some of images.

Disclosure

The authors declared no financial conflict of interest.

References

[1] Kuy S, Dua A, Chappidi R, Seabrook G, Brown KR, Lewis B, et al. The increasing incidence of thromboembolic events among hospitalized patients with inflammatory bowel disease. Vascular 2015;23:260−4.

[2] Guntas G, Sahin A, Duran S, Kahraman R, Duran I, Sonmez C, et al. Evaluation of ischemia-modified albumin in patients with inflammatory bowel disease. Clin Lab 2017;63:341−7.

[3] Valenti S, Gallizzi R, De Vivo D, Romano C. Intestinal Behçet and Crohn's disease: two sides of the same coin. Pediatr Rheumatol Online J 2017;15:33 (electronic journal).

[4] Sy A, Khalidi N, Dehghan N, Barra L, Carette S, Cuthbertson D, et al. Vasculitis in patients with inflammatory bowel diseases: a study of 32 patients and systematic review of the literature. Semin Arthritis Rheum 2016;45:475−82.

[5] Kusanagi M, Matsui O, Kawashima H, Gabata T, Ida M, Abo H, et al. Phlebosclerotic colitis: imaging-pathologic correlation. AJR Am J Roentgenol 2005;185:441−7.

[6] Chen W, Zhu H, Chen H, Shan G, Xu G, Chen L, et al. Phlebosclerotic colitis: our clinical experience of 25 patients in China. Medicine (Baltimore) 2018;97:e12824.

[7] Iwashita A, Yao T, Schlemper RJ, Yao T, Iida M, Matsumoto T, et al. Mesenteric phlebosclerosis: a new disease entity causing ischemic colitis. Dis Colon Rectum 2003;46:209−20.

[8] Sakane T, Takeno M, Suzuki N, Inaba G. Behçet's disease. N Engl J Med 1999;341:1284−91.

[9] Fujita H, Kiriyama M, Kawamura T, Ii T, Takegawa S, Dohba S, et al. Massive hemorrhage in a patient with intestinal Behçet's disease: report of a case. Surg Today 2002;32:378−82.

[10] Kim JS, Lim SH, Choi IJ. Prediction of the clinical course of Behçet's coli-tis according to macroscopic classification by colonoscopy. Endoscopy 2000;32:635−40.

[11] Ye JF, Guan JL. Differentiation between intestinal Behçet's disease and Crohn's disease based on endoscopy. Turk J Med Sci 2019;49:42−9.

[12] Li J, Li P, Bai J, Lyu H, Li Y, Yang H, et al. Discriminating potential of extraintestinal systemic manifestations and colonoscopic features in Chinese patients with intestinal Behçet's disease and Crohn's disease. Chin Med J (Engl) 2015;128:233−8.

[13] LaConti JJ, Donet JA, Cho-Vega JH, Sussman DA, Ascherman D, Deshpande AR. Henoch-Schönlein purpura with adalimumab therapy for ulcerative colitis: a case report and review of the literature. Case Rep Rheumatol 2016;2016:2812980.

[14] Lu B, Niu LL, Xu XG, Yao SL, Tan XY. Ulcerative colitis in an adult patient mimicking Henoch-Schönlein purpura: a case report. Medicine (Baltimore) 2018;97:e12036.

[15] Smith LW, Petri M. Lupus enteritis: an uncommon manifestation of systemic lupus erythematosus. J Clin Rheumatol 2013;19:84−6.

[16] Lee HA, Shm HG, Seo YH, Choi SJ, Lee BJ, Lee YH, et al. Panenteritis as an initial presentation of systemic lupus erythematosus. Korean J Gastroenterol 2016;67:107−11.

[17] Murali A, Narasimhan D, Krishnaveni J, Rajendiran G. Protein losing enteropathy in systemic lupus erythematosus. J Assoc Physicians India 2013;61:747−9.

[18] Ju JH, Min JK, Jung CK, Oh SN, Kwok SK, Kang KY, et al. Lupus mesenteric vasculitis can cause acute abdominal pain in patients with SLE. Nat Rev Rheumatol 2009;5:273−81.

[19] Guillevin L, Lhote F, Gallais V, Lhote F, Jarrousse B, Casassus P. Gastrointestinal tract involvement in polyarteritis nodosa and Churg-Strauss syndrome. Ann Med Interne (Paris) 1995;146:260−7.

[20] Schmidt WA. Use of imaging studies in the diagnosis of vasculitis. Curr Rheumatol Rep 2004;6:203−11.

[21] Levine SM, Hellmann DB, Stone JH. Gastrointestinal involvement in polyarteritis nodosa (1986-2000): presentation and outcomes in 24 patients. Am J Med 2002;112:386−91.

[22] Mir O, Nazal EM, Cohen P, Krivitzky A, Christoforov B, Jian R, et al. Esophageal involvement as an initial manifestation of Churg-Strauss syndrome. Presse Med 2007;36(1 Pt1):57−60.

[23] Kawasaki K, Eizuka M, Murata O, Ishida K, Nakamura S, Sugai T, et al. Eosinophilic granulomatosis with polyangiitis involving the small intestine: radiographic and endoscopic findings. Endoscopy 2015;47(Suppl. 1):E492−4.

[24] Pagnoux C, Mahr A, Cohen P, Guillevin L. Presentation and outcome of gastrointestinal involvement in systemic necrotizing vasculitides: analysis of 62 patients with polyarteritis nodosa, microscopic polyangiitis, Wegener granulomatosis, Churg-Strauss syndrome, or rheumatoid arthritis-associated vasculitis. Medicine (Baltimore) 2005;84:115−28.

Chapter 29

Inflammatory bowel disease—like conditions: medication-induced enteropathy

Sara El Ouali[1] and Bo Shen[2]

[1]Department of Gastroenterology, Hepatology and Nutrition, the Cleveland Clinic, Cleveland, OH, United States, [2]Center for Inflammatory Bowel Diseases, Columbia University Irving Medical Center-New York Presbyterian Hospital, New York, NY, United States

Chapter Outline

Abbreviations

5-FU 5-fluorouracil
ACEIs angiotensin-converting enzyme inhibitors
CD Crohn's disease
CMV cytomegalovirus
COX cyclooxygenase
CTLA-4 anticytotoxic T-lymphocyte-association protein-4
GI gastrointestinal
IBD inflammatory bowel disease
ICIs immune checkpoint inhibitors
MMF mycophenolate mofetil
NSAIDs nonsteroidal antiinflammatory drugs
OAE olmesartan-associated enteropathy
PD-1 programmed cell death protein-1
PERT pancreatic enzyme replacement therapy
SPS sodium polystyrene sulfonate

Introduction

Many conditions may share the clinical and endoscopic features of inflammatory bowel disease (IBD), and it is, therefore, essential to make differential differentiate and initiate the appropriate management plan. An important condition to recognize is drug-induced enteropathy as it is ubiquitous. A multitude of medications have been associated with the development of gastrointestinal (GI) lesions with ulcerative or nonulcerative inflammation and strictures, nonsteroidal antiinflammatory drugs (NSAIDs) being the most common ones. Indeed, more than 700 medications are found to be

Atlas of Endoscopy Imaging in Inflammatory Bowel Disease. DOI: https://doi.org/10.1016/B978-0-12-814811-2.00029-3
© 2020 Elsevier Inc. All rights reserved.

associated with enterocolitis [1]. Clinical manifestations of drug-induced enteropathy or colopathy resemble those of IBD and may include nausea, vomiting, abdominal pain, diarrhea, or rectal bleeding.

Drug-induced GI injury may occur through topical or systemic mechanisms. Physiologically or pathologically narrowed lumens of the GI tract, particularly the esophagus pylorus, and terminal ileum, are at risk. Commonly reported agents causing esophageal injury include NSAIDs, tetracycline, emepronium, potassium chloride, and quinidine sulfate [2]. In addition to causing mucosal damage through a variety of pathways, drugs may also lead to vascular and ischemic complications, impact microbiota, or disrupt mechanisms regulating electrolyte and fluid balance [1].

Drug-induced GI complications not only include mucosal injury, inflammation, strictures, or perforation but can also lead to other disease entities such as microscopic colitis (further discussed in Chapter 26: Inflammatory bowel disease—like conditions: other immune-mediated gastrointestinal disorders) or exacerbation of underlying IBD. A classic example is the use of NSAIDs. It is also important to note that medications may induce other conditions mimicking IBD through complications from their underlying mechanism of action. This is, for example, the case with antibiotics allowing *Clostridium difficile* infection, or immunosuppressants leading to infectious enterocolitis from cytomegalovirus (CMV) [3].

Drug-induced enteropathy remains a diagnosis of exclusion, and care has to be taken to rule out any other contributing conditions. Conditions that mimic IBD are extensively described in this book, such as immune-mediated conditions, immune deficiency—associated GI disorders, malignancies, infectious etiologies (such as tuberculosis), or radiation-induced GI injury. They may share common clinical, endoscopic, histologic, and imaging features. This chapter is to describe common drug-induced GI complications for their clinical, endoscopic, and histologic features as well as a brief guide to management.

Nonsteroidal antiinflammatory drugs

NSAIDs are widely used in the management of pain and inflammatory conditions. These molecules work by inhibiting cyclooxygenase (COX) activity with subsequent downstream effects. Through inhibiting COX activity, NSAIDs inhibit the production of prostaglandins, which play an important role in mucosal healing and repair [4].

NSAIDs have been associated with a variety of GI lesions. Although peptic ulcer disease is the most widely described complication initially, the advent of small bowel imaging and endoscopy i.e.balloon-assisted enteroscopy and video capsule endoscopy has demonstrated that small bowel lesions are frequently found in patients on NSAID therapy [4].

NSAIDs inhibit the two isoforms of the COX enzyme, namely, COX-1 and COX-2 [1]. Selective COX-2 inhibitors have been developed in the hope of reduced GI damage. Although this category of medications have similar therapeutic effects to nonselective NSAIDs, they initially did appear to lead to less GI injury [5]. However, recent data have shown COX-2 inhibitors to be associated with GI lesions similarly to nonselective NSAIDs [6].

NSAID-induced GI lesions are common. In a prospective cohort, small bowel mucosal breaks were found in 55% of patients receiving naproxen along with a proton-pump inhibitor for 2 weeks [5]. NSAIDs may cause a variety of GI complications ranging from erosions, ulcers, strictures, friability, or perforation, which may manifest by different clinical presentations.

Risk factors for NSAID-induced gastroduodenal injury include history of peptic ulcer disease, advanced age, combination of NSAIDs and corticosteroids or anticoagulants, comorbidities, and *Helicobacter pylori* infection [7]. Risk factors for NSAID-related small bowel injury are different and include the use of "oxicam" type of NSAIDs (such as meloxicam), diclofenac, the presence of comorbidities, and the combination of NSAIDs and aspirin [7]. In a retrospective study assessing NSAID-related small bowel lesions with video capsule endoscopy, most lesions were found in the jejunum (52.8%), followed by the ileum (27.9%) [8]. In a retrospective analysis of NSAID-associated GI injury, most patients had ulcers in the ileocecal location. The ulcers were semicircumferential or circumferential with sharp demarcation [9].

In addition to causing inflammation in the GI tract, the use of NSAIDs contributes to worsening or exacerbation of underlying Crohn's disease (CD), ulcerative colitis [10], or pouchitis [11]. Commonly over-the-counter or prescribed NSAIDs are used to treat IBD-associated arthralgia or arthropathy. The use of NSAIDs can also cause microscopic colitis or diverticular complications such as perforation and fistula [6,12]. Furthermore, chronic use of NSAIDs can lead to diaphragm disease in the lumen of GI tract with formation of short intestinal strictures. These strictures occur most commonly in the small bowel and right colon. Patients with diaphragm disease frequently present with GI bleeding or obstructive symptoms, anywhere from 2 months to more than 20 years into NSAID therapy [13].

The clinical presentation of NSAID-induced GI lesions varies according to disease type and location. NSAID-induced lesions may be asymptomatic or present as iron deficiency anemia or hypoalbuminemia [14]. Obscure or overt

GI bleeding is common presentation [12]. Other clinical manifestations include abdominal pain, diarrhea, nausea, and vomiting. Obstructive symptoms may also occur [13]. Symptoms may arise anytime while on NSAIDs but tend to occur in a median time of 3 months after the start of therapy [12].

Endoscopic features vary, depending on the location of GI jury, duration and dosage of NSAIDs, concurrent use of ulcer-prone medications, and comorbidities. Differential diagnosis should be made between NSAID-induced GI tract injury, IBD, and NSAID-associated IBD flare-up. Although commonly seen in the stomach and duodenum, the endoscopic lesions are also found in the esophagus, small bowel, and colon [15]. Petechiae, erythema, erosions, and ulcerations are common. Erosions and ulcers can be small or large, and round, linear or irregular (Figs. 29.1—29.8A and B) [5,6,16]. Active ulcer bleeding may be seen, as well as perforation [5]. Other endoscopic features include mucosal scars (Fig. 29.8C), strictures (Figs. 29.8D—29.10A and B), diverticuli (Fig. 29.10C and D), and diaphragms (Fig. 29.11). The latter is characterized by the presence of thin membranous strictures in the small bowel and colon [13,17].

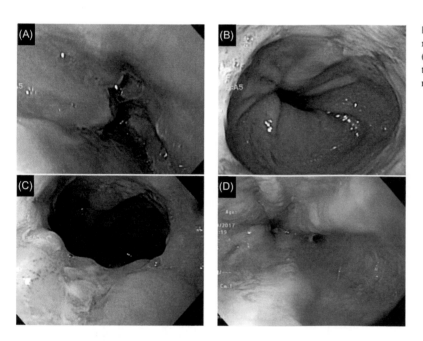

FIGURE 29.1 Patterns of nonsteroidal antiinflammatory drug-induced esophagitis: (A) esophagitis; (B) linear ulcerations at the gastroesophageal junction, with a small hiatal hernia; (C) and (D) longitudinal ulcerations at the gastroesophageal junction.

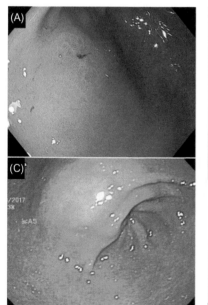
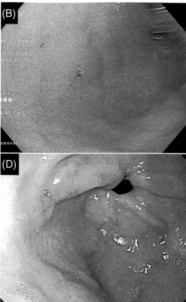

FIGURE 29.2 Patterns of nonsteroidal antiinflammatory drug-induced gastritis: (A) and (B) erosions with hematin spots in the antrum; (C) erosions, erythema, and edematous folds in the gastric antrum and prepyloric area; and (D) linear erosions in the stomach body with stigmata of recent bleeding.

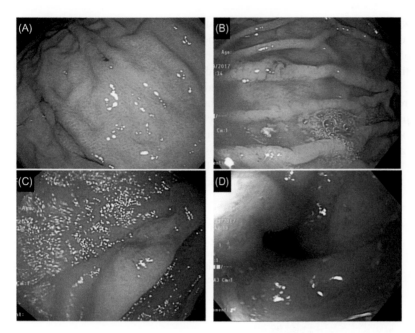

FIGURE 29.3 Injury to the stomach and duodenum from nonsteroidal antiinflammatory drugs: (A) and (B) erosive gastritis with linear erythema and erosions; (C) discrete ulcer in the duodenum; and (D) circumferential inflammatory stricture with normal overlying mucosa at the third portion of the duodenum.

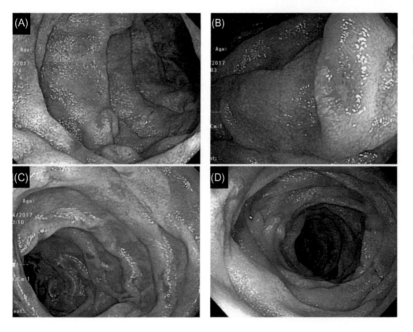

FIGURE 29.4 Nonsteroidal antiinflammatory drug-induced duodenitis: (A)–(D) multiple erosions and aphthous ulcers throughout the duodenum with normal intervening mucosa.

NSAIDs have also been associated with nonulcerative inflammation, ulcers, friability, spontaneous bleeding, and strictures in the afferent limb, pouch body, or cuff in patients with restorative proctocolectomy and ileal pouch-anal anastomosis (Figs. 29.12–29.17). The afferent limb ulcers and strictures may thereby mimic CD [18,19]. Histologic features of NSAID-induced enteritis include reactive epithelial changes, subepithelial fibrosis, crypt apoptosis, cytolysis, tissue eosinophilia, and overall increased number of inflammatory cells [7,20].

The first step in the management of NSAID-induced enteropathy is the discontinuation of the drug. In a retrospective analysis, improvement in the endoscopic ulcers upon discontinuation of NSAIDs for 3 to 10 weeks was documented [9]. Further management modalities vary and depend on the underlying type of lesion. Endoscopic hemostasis may be required in the setting of active GI bleeding. Endoscopic intervention such as balloon dilation, endoscopic stricturotomy, or surgery might be required in the case of refractory, symptomatic stricturing disease (Fig. 29.10A and B) [13].

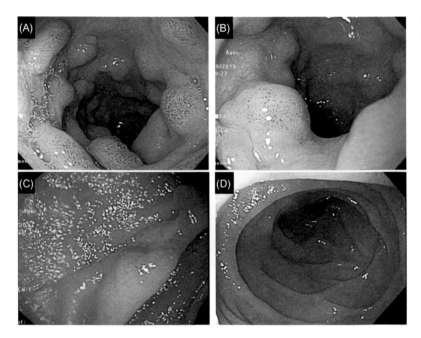

FIGURE 29.5 Nonsteroidal antiinflammatory drug-induced duodenitis: (A)–(C) erythematous, nodular, and congested duodenum mucosa; (D) normal distal duodenum.

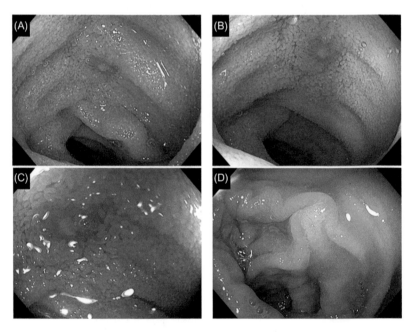

FIGURE 29.6 Nonsteroidal antiinflammatory drug-induced erosions in the ileum and colon in a patient with otherwise quiescent Crohn's disease in the same patient: (A)–(C) aphthous ulcers with normal intervening mucosa; (D) single erosion in the ascending colon.

Potassium agents

Oral potassium agents, particularly slow-release potassium chloride tablets, can cause ulcerations in the esophagus, stomach, or small intestine [21]. Oral potassium agents may also cause mucosal injury to GI tract in patients with CD (Fig. 29.18).

Angiotensin-converting enzyme inhibitors and receptor blockers

Angiotensin-converting enzyme inhibitors (ACEIs) are medications used in the treatment of hypertension and heart failure. ACEi inhibits the breakdown of bradykinin, resulting in vasodilation [22]. ACEIs have been associated with the

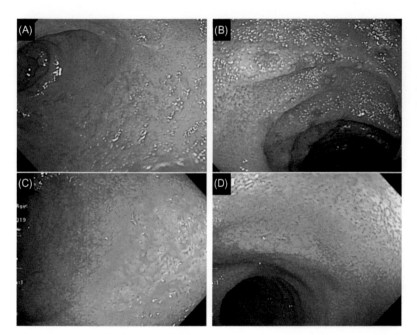

FIGURE 29.7 Nonsteroidal antiinflammatory drug-induced injury in the terminal ileum in a patient with small bowel Crohn's disease: (A) and (B) erythema, granularity, and erosions in the terminal ileum; (C) and (D) resolution of inflammation of the cessation of nonsteroidal antiinflammatory drug use without adjustment of Crohn's disease medications.

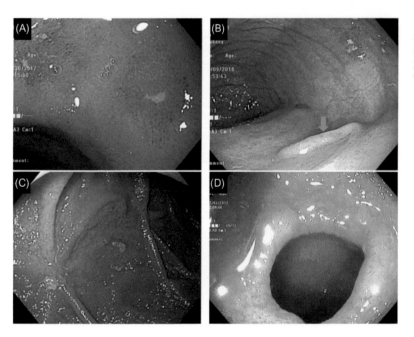

FIGURE 29.8 Patterns of nonsteroidal antiinflammatory drug injury to the small bowel: (A) discrete aphthous erosions; (B) erosions, ulcers, and elongated polypoid lesion (*green arrow*); (C) mucosal scars; and (D) circumferential stricture.

development of angioedema involving the face and upper respiratory tract [23]. Small bowel angioedema has also been reported in association with the use of ACEI mimicking CD [24].

Clinical presentation may vary. The onset of symptoms ranges from a few days to several years after starting ACEI [22]. Abdominal pain is the most common symptom and can be chronic, intermittent, and severe. Other manifestations include diarrhea, nausea, and vomiting [24]. Surgical abdomen with peritoneal signs on examination has also been reported [25].

The endoscopic investigation may be necessary to rule out other causes such as IBD or malignancy. Endoscopic appearance of ACEI-induced small bowel angioedema has rarely been described in published case series and may include normal findings or mild atrophy (Fig. 29.19) [22,26]. On cross-sectional imaging, ACEI-induced angioedema may show features such as bowel wall thickening, luminal narrowing, submucosal edema, mesenteric edema, and

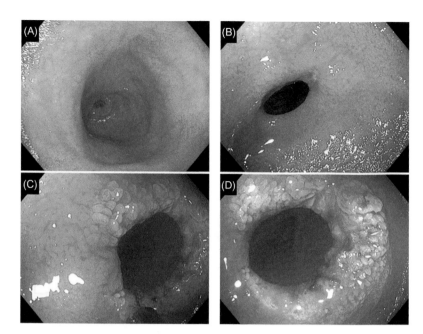

FIGURE 29.9 Ileal stricture from the use of non-steroidal antiinflammatory drugs in a patient with Crohn's disease: (A) and (B) circumferential stricture with a superficial erosion in the distal ileum. The ileum mucosa was otherwise normal. Crohn's disease was in remission; (C) and (D) the stricture was treated with endoscopic stricturotomy with an insulated-tip knife and cauterization effect.

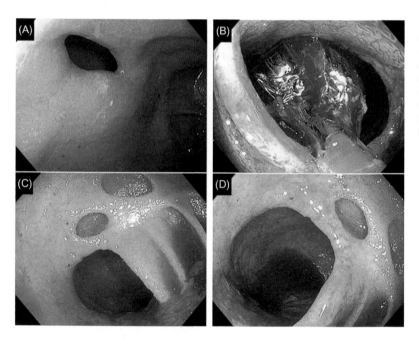

FIGURE 29.10 Nonsteroidal antiinflammatory drug-induced ileocecal valve stricture and diverticuli of the ileum in a patient with small bowel Crohn's disease before the initiation of medical therapy: (A) A short ileocecal valve nonulcerated stricture (*green arrow*); (B) endoscopic balloon dilation of the stricture; (C) and (D) prestenotic dilation of the lumen of the terminal ileum and diverticuli found after endoscopically dilating and traversing stricture.

ascites (Fig. 29.19E) [24]. In one case series, bowel wall thickness ranged from 4 to 9 mm over 20 to 60 cm segments [27]. Resolution of symptoms and imaging abnormalities have been reported upon discontinuation of the ACEI [27] and may occur as early as 1—4 days after cessation of the drug [28].

Olmesartan is an angiotensin II receptor blocker used in the treatment of hypertension. The agent is associated with a rare but severe type of "sprue-like" enteropathy [29]. Patients with olmesartan-associated enteropathy (OAE) most commonly present with chronic diarrhea and weight loss [30]. Celiac serology is typically negative, but several patients have been found to have antienterocyte antibodies as well as antinuclear antibodies [30,31]. On imaging, diffuse small bowel edema and bowel wall thickening have been described [30]. Endoscopic findings in OAE include nodularity of the mucosa, as well as ulceration and villous atrophy, involving the duodenum, jejunum, or ileum [30,31]. Histopathological features of OAE include intraepithelial lymphocytosis, villous atrophy, crypt hypertrophy, and mucosal inflammation [29,31]. Lymphocytic or collagenous gastritis and colitis have also been described [29].

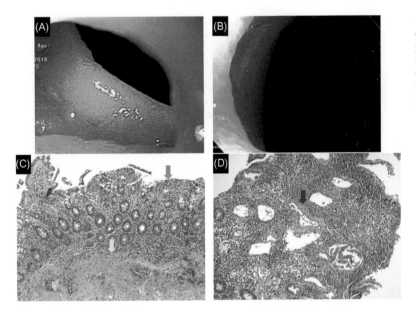

FIGURE 29.11 Nonsteroidal antiinflammatory drug-induced duodenal diaphragm: (A) and (B) diaphragm at the duodenum cap which was not traversable to a gastroscope; (C) and (D) histology of endoscopic biopsy showed erosions (*green arrow*), fibrosis of the muscularis mucosae (*yellow arrow*), and hyalinized vessels (*blue arrow*).

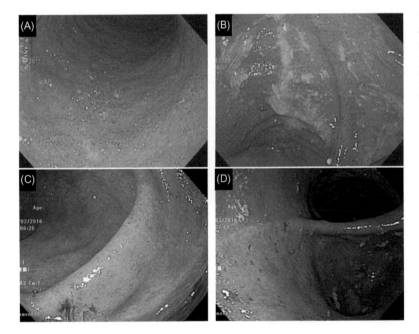

FIGURE 29.12 Nonsteroidal antiinflammatory drug-induced enteritis and pouchitis in a patient with restorative proctocolectomy for ulcerative colitis: (A) mild diffuse inflammation with small erosions and loss of vascularity of the afferent limb; (B)–(D) diffuse pouchitis with exudates, erythema, edema, and spontaneous bleeding of the pouch body.

Sodium polystyrene sulfonate

Sodium polystyrene sulfonate (SPS) is used in the management of mild hyperkalemia. It is a cation-exchange resin and promotes potassium loss in the stool [32]. Case reports and case series have found SPS to be associated with different types and locations of GI tract injury.

The underlying mechanism of injury is unclear. GI injury from SPS was initially thought to be related to the sorbitol component. However, more recent data have continued to report concerning findings of GI injury despite the removal of sorbitol from the preparation [33]. A systematic review found that 71% of patients with GI injury due to SPS had a history of chronic kidney disease [33].

Clinical presentation ranges from abdominal pain, diarrhea, nausea, and vomiting to GI bleeding and acute abdomen. Symptom onset appears to occur approximately 2 days after SPS administration [33]. Although most patients with

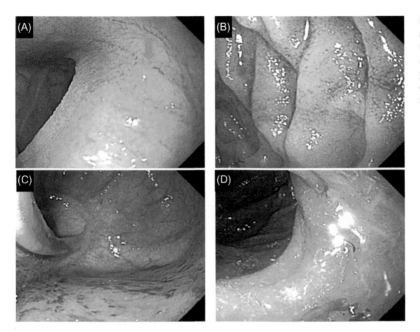

FIGURE 29.13 Enteritis, pouchitis, and cuffitis associated with the use of nonsteroidal antiinflammatory drugs in an ulcerative colitis patient with restorative proctocolectomy: (A) and (B) enteritis with prominent linear mucosal vasculature (*green arrow*), edema, erythema, and mucosal breaks; (C) hemorrhagic pouchitis; (D) edema of the rectal cuff.

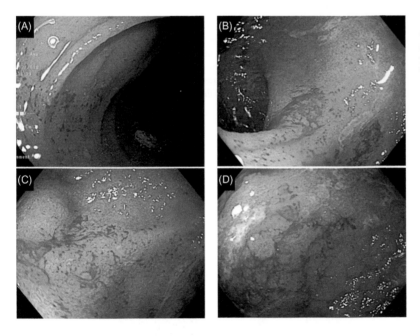

FIGURE 29.14 Nonsteroidal antiinflammatory drug-induced hemorrhagic enteritis and pouchitis: (A) spontaneous bleeding in the afferent limb; (B)—(D) friable mucosa and spontaneous bleeding of the pouch body.

SPS-induced enteropathy present with lower GI tract injury, upper GI involvement has also been reported affecting the esophagus, stomach, and duodenum.

Endoscopic features range from erosions and ulcers the upper GI tract (Fig. 29.20) to esophageal plaques (similar to *Candida* esophagitis) and gastric bezoars, as seen in 11 patients using SPS [34]. Gastric serpiginous ulcers have also been reported, similar to CD [35]. Cases of emphysematous gastritis, gastric necrosis, and gastric perforation have been reported [36,37].

The colon is the most frequently affected location. Ileocolonic injury, necrosis, and subsequent death have all been reported in the setting of SPS administration [33,38]. Endoscopic findings in the lower GI tract include edema, friability, erosions, ulcerations, and necrosis [39—41].

The classic histopathological finding is the presence of crystals in 90% of patients [33]. These are basophilic on hematoxylin and eosin stain. SPS crystals have a characteristic crystalline mosaic pattern, which helps to distinguish

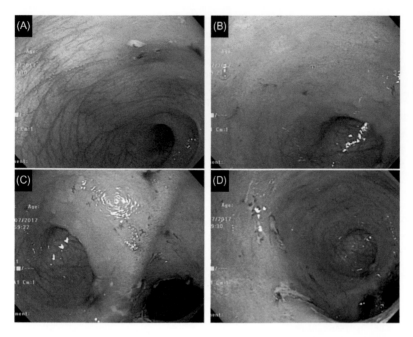

FIGURE 29.15 Nonsteroidal inflammatory drug-associated ulcers in the ileal pouch-anal anastomosis: (A) discrete erosions and ulcers with stigmata of bleeding in the afferent limb; (B)–(D) discrete erosions and ulcers with stigmata of bleeding in the pouch body.

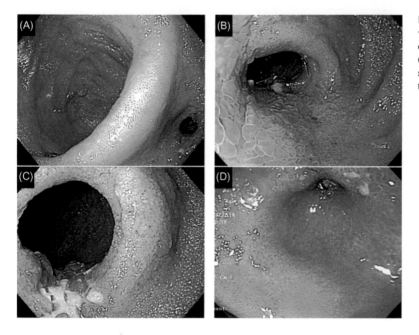

FIGURE 29.16 Ileal pouch strictures associated with the regular use of nonsteroidal antiinflammatory drugs in patients with restorative proctocolectomy: (A) normal pouch body; (B) and (C) two ulcerated strictures at the afferent limb; and (D) ulcerated stricture at the inlet.

them from cholestyramine crystals (Fig. 29.20) [34]. Other histopathologic findings include inflammatory changes, erosion, ulceration, and necrosis [33,34].

Mycophenolate mofetil

Mycophenolic acid is an immunosuppressive agent that is commonly used in several autoimmune diseases and in a posttransplantation setting to reduce the risk of rejection.

The mechanism of action is believed to be a decrease in the replication of B and T lymphocytes by inhibiting purine synthesis [42]. Mycophenolic acid has two forms of medications: mycophenolate mofetil (MMF) and enteric-coated mycophenolate sodium.

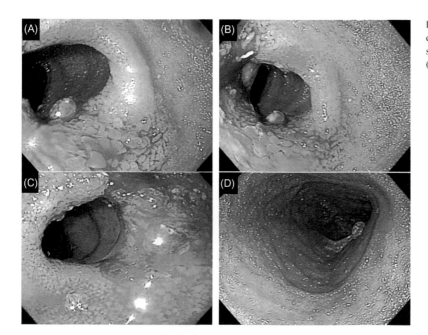

FIGURE 29.17 Nonsteroidal antiinflammatory drug-induced pouch inlet stricture: (A)—(C) ulcerated stricture at the pouch inlet with stigmata of bleeding; (D) normal afferent limb.

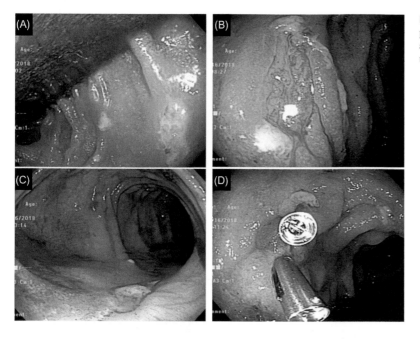

FIGURE 29.18 Patient with Crohn's colitis using potassium citrate: (A) multiple colonic ulcers; (B)—(C) actively oozing ulcers; and (D) hemostasis achieved using endoclips.

GI side effects from MMF therapy are common affecting up to 85% of patients [43]. Diarrhea is the most frequent GI symptom, followed by abdominal pain, nausea, and vomiting [44].

Endoscopic features of MMF toxicity vary and may involve the upper and lower GI tract [45,46]. Endoscopic findings range from erythema and edema to erosions and ulcerations [47,48] (Fig. 29.21). In a study of video capsule endoscopy, erythema, erosions, and ulcers were visualized in the stomach and small bowel [49]. Duodenal scalloping and villous atrophy have also been described [43,46]. Some have documented the resolution of villous atrophy and normalization of the mucosa after discontinuation of MMF [46,50].

Various histopathological features have been described in MMF-induced GI injury. Similar to IBD, crypt abscesses can be present. However, crypt abscesses are typically apoptotic in MMF-associated injury, whereas they are more frequently neutrophilic in IBD [51]. In a case series of patients receiving MMF for renal transplantation, CD-like histologic features were found including focal crypt distortion and patchy, focal inflammation [47]. In a retrospective study

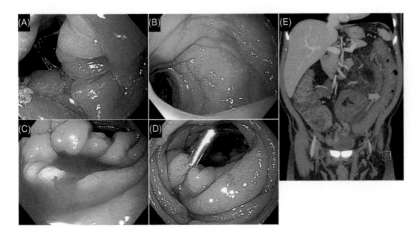

FIGURE 29.19   ACEI-induced angioedema: (A)–(B) congested, edematous folds with loss of vascular pattern in the sigmoid and descending colon; (C) bleeding after biopsy; (D) hemostasis achieved with placement of endoclip; (E) edema of the left colon on computed tomography (*green arrow*). *ACEI*, Angiotensin-converting enzyme inhibitor.

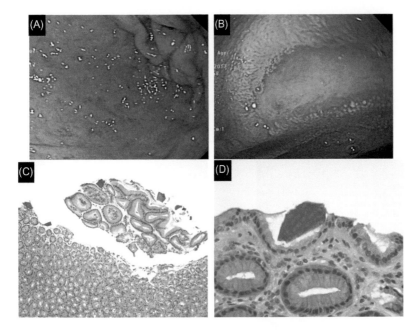

FIGURE 29.20   Gastroduodenitisfrom the use of sodium polystyrene sulfonate: (A) erosions in the gastric antrum; (B) ulcer in the prepylorus area; and (C) and (D) basophilic (purple) Kayexalate crystals in luminal/extracellular space. The other important feature is that the crystals have a so-called mosaic or fish scale appearance on higher magnification (D).

of colon biopsies in patients receiving MMF, histologic features are apoptosis, architectural distortion, and other forms of chronic inflammation. [48].

Pancreatic enzyme replacement therapy

Pancreatic enzymes are widely used in the management of pancreatic insufficiency. The supplements are occasionally used in patients with concurrent IBD and chronic autoimmune pancreatitis. Several cases of "fibrosing colonopathy" have been reported in patients with cystic fibrosis on pancreatic enzyme replacement therapy (PERT) [52,53]. However, this phenomenon has also been described in patients without underlying cystic fibrosis [54]. Fibrosing colonopathy refers to the occurrence of colonic stricturing disease in patients on PERT [54]. Most cases have been reported in children and are thought to be dose dependent, with an increased risk of fibrosis with higher PERT dosing [55].

Presenting symptoms may include abdominal pain, nausea, vomiting, and abdominal distention, as well as constipation [52].

Endoscopic features such as luminal narrowing, strictures, friability, erythema, and ulceration have been described (Fig. 29.22) [56].

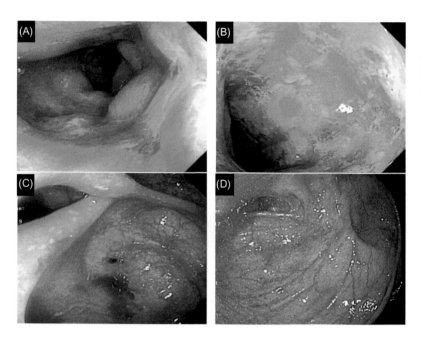

FIGURE 29.21 Mycophenolate mofetil-induced colitis: (A)–(B) diffuse inflammation with erythema, edema, exudate, friability, and erosions; (C) inflamed cecum with erosions and friability; and (D) normal cecum in the same patient prior to mycophenolate mofetil-induced therapy.

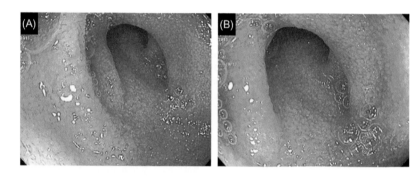

FIGURE 29.22 Distal ileal stricture in a patient using pancreatic enzymes (A) and (B). The short, moderate distal ileal nonulcerated stricture.

Histologic features may include submucosal fibrosis, cryptitis, eosinophilia, chronic inflammatory changes, and epithelial regeneration, some of which may resemble CD [57]. Imaging findings may include increased bowel wall thickness, luminal narrowing, loss of normal haustration, and shortening of the colon [57]. Ileal involvement has also been described [54].

Immune checkpoint inhibitor—associated colitis

Immune checkpoint inhibitors (ICIs) are a group of immunotherapeutic agents used in the management of metastatic melanoma, metastatic nonsmall-cell lung cancers, and other malignancies [58]. ICIs include anticytotoxic T-lymphocyte-association protein-4 (CTLA-4) antibodies, such as ipilimumab and tremelimumab, as well as antiprogrammed cell death protein-1 (PD-1)antibodies, such as pembrolizumab and nivolumab. These molecules have led to remarkable improvement in oncological outcomes [58]. However, ICIs can cause a variety of side effects and immune-mediated complications. Up to 62% of patients receiving anti-CTLA-4 therapy are thought to have immune-related adverse events which appear to be correlated with response to therapy [59]. Multiple organ systems may be involved, with the GI tract and the skin being most commonly affected[58].

GI complications are common. Clinical manifestations are variable and may include diarrhea, abdominal pain, rectal bleeding, nausea, and vomiting [60]. Symptoms, however, appear to correlate poorly with endoscopic findings [61].

Endoscopic features may range from normal mucosa, erythema, edema and decreased vascular pattern to friability, erosions, and ulcerations in the upper or lower GI tract (Figs. 29.23–29.25) [60,62]. In a retrospective study of 53

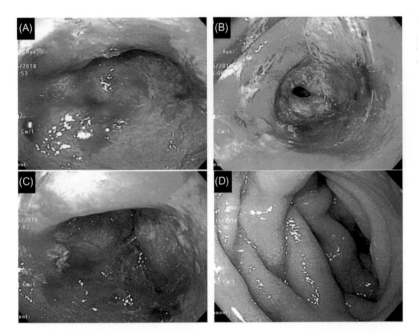

FIGURE 29.23 Checkpoint inhibitor—induced gastroduodenitis: (A)—(C) hemorrhagic gastritis with congestion, ulcers, and active diffuse bleeding; (D) edematous duodenal folds.

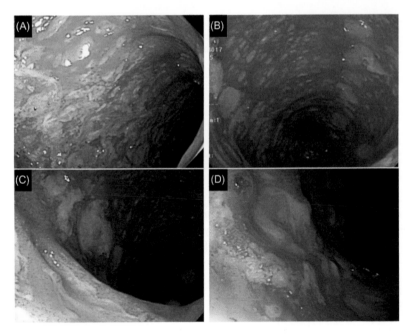

FIGURE 29.24 Checkpoint inhibitor—induced colitis in a patient using PD-1 inhibitor therapy: (A) and (B) diffuse inflammation with erythema, absent vascular pattern, superficial and deep ulcerations; (C) and (D) deep ulcerations. *PD-1,* Programmed cell death protein-1.

patients receiving ICIs and undergoing colonoscopy or sigmoidoscopy for GI symptoms, 40% of patients had ulcerations, 42% had nonulcerative inflammation, and 19% had no inflammation. In terms of location of disease, 14% had ileocolonic inflammation, whereas 43% had left-sided colitis, and 40% had extensive colitis. Only 2% of the patients had isolated ileal inflammation. The presence of colon ulcers was found to be associated with a more severe, steroid-refractory disease course [63,64]. The timing of colonoscopy is important for the diagnosis and early intervention. In a retrospective analysis, early colonoscopy (within 7 days of symptom onset) was associated with a shorter duration of symptoms, shorter steroid course, and a decreased length of stay in hospital [61].

From the histopathological standpoint, acute inflammatory changes are the most commonly described ICI toxicity with heterogeneous findings of cryptitis, crypt abscesses, and intraepithelial lymphocytic infiltration [61]. Inflammatory changes are usually diffuse. Studies have also reported the presence of chronic inflammatory changes, basal plasmacytosis and crypt architectural distortion in up to 40% of patients [60,63].

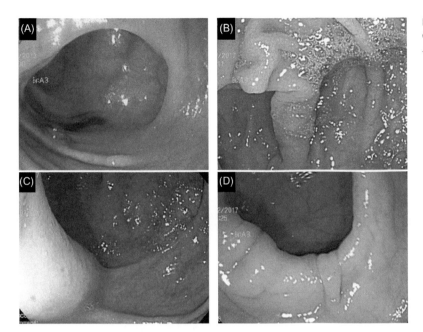

FIGURE 29.25 Normal colon in the same patient (Fig. 29.24), prior to PD-1 inhibitor therapy (A)−(D). *PD-1,* Programmed cell death protein-1.

Management starts with proper diagnosis and exclusion of other disease conditions, such as infectious complications from *C. difficile,* CMV or others. Early detection is essential as ICI-related complications are associated with significant morbidities [58]. Patients should be managed according to the type and severity of the presentation. Treatment may include the cessation of the offending drug, and the initiation of corticosteroids or biological therapy with infliximab or vedolizumab. Guidelines to further help with the management of ICI-related complications have been published [58].

Further information on ICI-induced colitis is found in Chapter 26, Inflammatory bowel disease—like conditions: other immune-mediated gastrointestinal disorders, along with endoscopic images.

Enterocolitis due to miscellaneous agents

Several chemotherapy agents have been associated with enterocolitis, including 5-fluorouracil (5-FU), irinotecan, and capecitabine [12,65]. The use of 5-FU can cause inflammation in the upper or lower GI tract [12]. Small bowel toxicity is presented with severe abdominal pain and diarrhea. Endoscopic features include erosions and ulcerations [66]. Management strategies are supportive care along with the cessation of the offending drug.

Gold compounds, cyclosporine, herbal medicines (Fig. 29.26), and a multitude of other medications, have been associated with the development of enterocolitis [67]. Sometimes, oral iron supplements in IBD patients may cause a melena-like picture in the colon (Fig. 29.27).

Distal colonic injury has also been described in the setting of enemas or drugs applied topically, such as ergotamine suppositories, radiocontrast agents, and herbal substances [12,67]. Clinical presentations range from abdominal pain to diarrhea and rectal bleeding. Endoscopic features include erythema, friability, erosions, and ulcers. Histologic findings may vary and usually reveal acute inflammatory changes [68].

Summary and recommendations

Medications may induce a variety of GI complications, sometimes mimicking the clinical, endoscopic, or histologic features of IBD. A detailed history with special attention to current or prior medical therapies, is essential to make the proper diagnosis. Drug-induced GI lesions may involve the upper or lower GI tract and cause a range of different endoscopic patterns. Histologic evaluation may be helpful for the diagnosis and differential diagnosis. Recognizing drug-induced enteropathy and identifying the culprit medication is the first step toward management. Cessation of the offending drug is essential. Further management according to the type and severity of the presentation may be required.

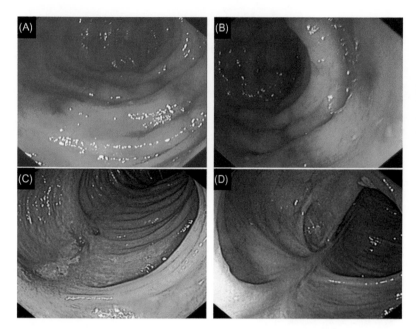

FIGURE 29.26 Patterns of injury to colon mucosa by herbal medicines induced (including indirubin and indigo) in two patients: (A) and (B) patchy erythema and edema throughout the colon in patient 1; (C) and (D) longitudinal ulcer in the colon (C) which was healed with the formation of mucosal scar (D) by discontinuation of the herbal medicine. *Photo courtesy Dr. Yan Chen and Dr. Xiao-Ying Wang of the Second Affiliated Hospital of Zhejiang University.*

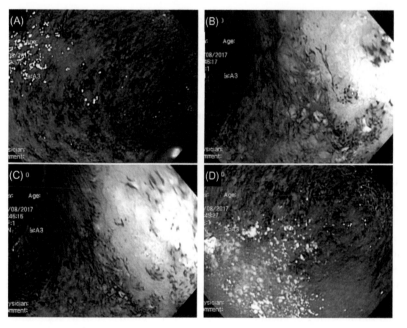

FIGURE 29.27 Black colon in a patient with ulcerative colitis who was taking oral iron supplements mimicking bleeding: (A)−(D) diffuse colitis covered with black deposits mimicking melena.

Disclosure

The authors declared no financial conflict of interest.

References

[1] Grattagliano I, Ubaldi E, Portincasa P. Drug-induced enterocolitis: prevention and management in primary care. J Dig Dis 2018;19(3):127−35. Available from: https://doi.org/10.1111/1751-2980.12585.

[2] Eng J, Sabanathan S. Drug-induced esophagitis. Am J Gastroenterol 1991;86(9):1127−33 <http://www.ncbi.nlm.nih.gov/pubmed/1882789> [accessed 28.10.19].

[3] Price AB. Pathology of drug-associated gastrointestinal disease. Br J Clin Pharmacol 2003;56(5):477−82. Available from: https://doi.org/10.1046/j.1365-2125.2003.01980.x.

[4] Wallace JL. Mechanisms, prevention and clinical implications of nonsteroidal anti-inflammatory drug-enteropathy. World J Gastroenterol 2013;19(12):1861−76. Available from: https://doi.org/10.3748/wjg.v19.i12.1861.

[5] Goldstein JL, Eisen GM, Lewis B, Gralnek IM, Zlotnick S, Fort JG. Video capsule endoscopy to prospectively assess small bowel injury with celecoxib, naproxen plus omeprazole, and placebo. Clin Gastroenterol Hepatol 2005;3(2):133−41. Available from: https://doi.org/10.1016/S1542-3565(04)00619-6.

[6] Maiden L, Thjodleifsson B, Seigal A, et al. Long-term effects of nonsteroidal anti-inflammatory drugs and cyclooxygenase-2 selective agents on the small bowel: a cross-sectional capsule enteroscopy study. Clin Gastroenterol Hepatol 2007;5(9):1040−5. Available from: https://doi.org/10.1016/j.cgh.2007.04.031.

[7] Ishihara M, Ohmiya N, Nakamura M, et al. Risk factors of symptomatic NSAID-induced small intestinal injury and diaphragm disease. Aliment Pharmacol Ther 2014;40(5):538−47. Available from: https://doi.org/10.1111/apt.12858.

[8] Shim KN, Song EM, Jeen YT, et al. Long-term outcomes of NSAID-induced small intestinal injury assessed by capsule endoscopy in Korea: a nationwide multicenter retrospective study. Gut Liver 2015;9(6):727−33. Available from: https://doi.org/10.5009/gnl14134.

[9] Kurahara K, Matsumoto T, Iida M, Honda K, Yao T, Fujishima M. Clinical and endoscopic features of nonsteroidal anti-inflammatory drug-induced colonic ulcerations. Am J Gastroenterol. 2001;96(2):473−80. Available from: https://doi.org/10.1016/S0002-9270(00)02323-6.

[10] Kaufmann HJ, Taubin HL. Nonsteroidal anti-inflammatory drugs activate quiescent inflammatory bowel disease. Ann Intern Med 1987;107(4):513−16. Available from: https://doi.org/10.7326/0003-4819-107-4-513.

[11] Shen B, Fazio VW, Remzi FH, et al. Effect of withdrawal of nonsteroidal anti-inflammatory drug use on ileal pouch disorders. Dig Dis Sci 2007;52(12):3321−8. Available from: https://doi.org/10.1007/s10620-006-9710-3.

[12] Thorsen AJ. Noninfectious colitides: collagenous colitis, lymphocytic colitis, diversion colitis, and chemically induced colitis. Clin Colon Rectal Surg 2007;20(1):47−57. Available from: https://doi.org/10.1055/s-2007-970200.

[13] Wang YZ, Sun G, Cai FC, Yang YS. Clinical features, diagnosis, and treatment strategies of gastrointestinal diaphragm disease associated with nonsteroidal anti-inflammatory drugs. Gastroenterol Res Pract 2016;2016. Available from: https://doi.org/10.1155/2016/3679741 (Dd).

[14] Bjarnason I, Prouse P, Smith T, et al. Blood and protein loss via small-intestinal inflammation induced by non-steroidal anti-inflammatory drugs. Lancet 1987;330(8561):711−14. Available from: https://doi.org/10.1016/S0140-6736(87)91075-0.

[15] Leighton JA, Pasha SF. Inflammatory disorders of the small bowel. Gastrointest Endosc Clin N Am 2017;27(1):63−77. Available from: https://doi.org/10.1016/j.giec.2016.08.004.

[16] Graham DY, Opekun AR, Willingham FF, Qureshi WA. Visible small-intestinal mucosal injury in chronic NSAID users. Clin Gastroenterol Hepatol 2005;3(1):55−9. Available from: https://doi.org/10.1016/S1542-3565(04)00603-2.

[17] Grattan BJ, Bennett T, Starks MR. Diaphragm disease: NSAID-induced small bowel stricture. Case Rep Gastroenterol 2018;12(2):327−30. Available from: https://doi.org/10.1159/000489301.

[18] Shen B, Fazio VW, Remzi FH, et al. Risk factors for diseases of ileal pouch-anal anastomosis after restorative proctocolectomy for ulcerative colitis. Clin Gastroenterol Hepatol 2006;4(1):81−9. Available from: https://doi.org/10.1016/j.cgh.2005.10.004.

[19] Wolf JM, Achkar J-P, Lashner BA, et al. Afferent limb ulcers predict Crohn's disease in patients with ileal pouch-anal anastomosis. Gastroenterology 2004;126:1686−91. Available from: https://doi.org/10.1053/j.gastro.2004.02.019.

[20] Püspök A, Kiener HP, Oberhuber G. Clinical, endoscopic, and histologic spectrum of nonsteroidal anti- inflammatory drug-induced lesions in the colon. Dis Colon Rectum 2000;43(5):685−91. Available from: https://doi.org/10.1007/BF02235589.

[21] Leijonmarck C, Raf L. Ulceration of the small intestine due to slow-release potassium chloride tablets. Acta Chir Scand 1985;151(3):273−8.

[22] Vallabh H, Hahn B, Bryan C, Hogg J, Kupec JT. Small bowel angioedema from angiotensin-converting enzyme: changes on computed tomography. Radiol Case Rep 2018;13(1):55−7. Available from: https://doi.org/10.1016/j.radcr.2017.09.014.

[23] Abdelmalek MF, Douglas DD. Lisinopril-induced isolated visceral angioedema. Review of ACE-inhibitor-induced small bowel angioedema. Dig Dis Sci 1997;42(4):847−50. Available from: https://doi.org/10.1023/A:1018884702345.

[24] Scheirey CD, Scholz FJ, Shortsleeve MJ, Katz DS. Angiotensin-converting enzyme inhibitor-induced small-bowel angioedema: clinical and imaging findings in 20 patients. Am J Roentgenol 2011;197(2):393−8. Available from: https://doi.org/10.2214/AJR.10.4451.

[25] Bloom AS, Schranz C. Angiotensin-converting enzyme inhibitor-induced angioedema of the small bowel − a surgical abdomen mimic. J Emerg Med 2015;48(6):e127−9. Available from: https://doi.org/10.1016/j.jemermed.2015.01.016.

[26] Schmidt TD, McGrath KM. Angiotensin-converting enzyme inhibitor angioedema of the intestine: a case report and review of the literature. Am J Med Sci 2002;324(2):106−8. Available from: https://doi.org/10.1097/00000441-200208000-00011.

[27] Vallurupalli K, Coakley KJ. MDCT features of angiotensin-converting enzyme inhibitor-induced visceral angioedema. Am J Roentgenol 2011;196(4):405−11. Available from: https://doi.org/10.2214/AJR.10.4856.

[28] Wilin KL, Czupryn MJ, Mui R, Renno A, Murphy JA. ACE inhibitor-induced angioedema of the small bowel: a case report and review of the literature. J Pharm Pract 2018;31(1):99−103. Available from: https://doi.org/10.1177/0897190017690641.

[29] Rubio-Tapia A, Herman ML, Ludvigsson JF, et al. Severe spruelike enteropathy associated with olmesartan. Mayo Clin Proc 2012;87(8):732−8. Available from: https://doi.org/10.1016/j.mayocp.2012.06.003.

[30] Choi EYK, McKenna BJ. Olmesartan-associated enteropathy a review of clinical and histologic findings. Arch Pathol Lab Med 2015;139(10):1242−7. Available from: https://doi.org/10.5858/arpa.2015-0204-RA.

[31] Marthey L, Cadiot G, Seksik P, et al. Olmesartan-associated enteropathy: results of a national survey. Aliment Pharmacol Ther 2014;40(9):1103−9. Available from: https://doi.org/10.1111/apt.12937.

[32] Tapia C, Schneider T, Manz M. From hyperkalemia to ischemic colitis: a resinous way. Clin Gastroenterol Hepatol 2009;7(8). Available from: https://doi.org/10.1016/j.cgh.2009.02.030.

[33] Harel Z, Harel S, Shah PS, Wald R, Perl J, Bell CM. Gastrointestinal adverse events with sodium polystyrene sulfonate (Kayexalate) use: a systematic review. Am J Med 2013;126(3):264.e9–264.e24. Available from: https://doi.org/10.1016/j.amjmed.2012.08.016.

[34] Abraham SC, Bhagavan BS, Lee LA, Rashid A, Wu TT. Upper gastrointestinal tract injury in patients receiving Kayexalate (sodium polystyrene sulfonate) in sorbitol: clinical, endoscopic, and histopathologic findings. Am J Surg Pathol 2001;25(5):637–44. Available from: https://doi.org/10.1097/00000478-200105000-00011.

[35] Roy-Chaudhury P, Meisels IS, Freedman S, Steinman TI, Steer M. Combined gastric and ileocecal toxicity (serpiginous ulcers) after oral Kayexalate in sorbital therapy. Am J Kidney Dis 1997;30(1):120–2. Available from: https://doi.org/10.1016/S0272-6386(97)90574-6.

[36] Hajjar R, Sebajang H, Schwenter F, Mercier F. Sodium polystyrene sulfonate crystals in the gastric wall of a patient with upper gastrointestinal bleeding and gastric perforation: an incidental finding or a pathogenic factor? J Surg Case Rep 2018;6:1–3. Available from: https://doi.org/10.1093/jscr/rjy138.

[37] Usta Y, Ramirez C, Dennert B. Image of the month: emphysematous gastritis and necrosis as a result of orally ingested sodium polystyrene sulfonate (Kayexalate) in sorbitol. Am J Gastroenterol 2016;111(3):309. Available from: https://doi.org/10.1038/ajg.2015.250.

[38] Trottier V, Drolet S, Morcos MW. Ileocolic perforation secondary to sodium polystyrene sulfonate in sorbitol use: a case report. Can J Gastroenterol 2009;23:689–90.

[39] Edhi AI, Cappell MS, Sharma N, Amin M, Patel A. One oral dose of sodium polystyrene sulfonate associated with ischemic colitis and crystal deposition in colonic mucosa. ACG Case Rep J 2018;5:e74. Available from: https://doi.org/10.14309/crj.2018.74.

[40] Bomback AS, Woosley JT, Kshirsagar AV. Colonic necrosis due to sodium polystyrene sulfate (Kayexalate). Am J Emerg Med 2009;27 (6):753.e1–2. Available from: https://doi.org/10.1016/j.ajem.2008.10.002.

[41] Thomas A, James BR, Landsberg D. Colonic necrosis due to oral Kayexalate in a critically-ill patient. Am J Med Sci 2009;337(4):305–6. Available from: https://doi.org/10.1097/MAJ.0b013e31818dd715.

[42] Behrend M. Adverse gastrointestinal effects of mycophenolate mofetil: aetiology, incidence and management. Drug Saf 2001;24:645–63.

[43] Jehangir A, Shaikh B, Hunt J, Spiegel A. Severe enteropathy from mycophenolate mofetil REPORTS J acgcasereports.gi.orgACG Case Rep J 2016;3(2):101–3. Available from: https://doi.org/10.14309/crj.2016.13.

[44] Arns W. Noninfectious gastrointestinal (GI) complications of mycophenolic acid therapy: a consequence of local GI toxicity? Transplant Proc 2007;39(1):88–93. Available from: https://doi.org/10.1016/j.transproceed.2006.10.189.

[45] Nguyen T, Park JY, Scudiere JR, Montgomery E. Mycophenolic acid (cellcept and myofortic) induced injury of the upper GI tract. Am J Surg Pathol 2009;33(9):1355–63. Available from: https://doi.org/10.1097/PAS.0b013e3181a755bd.

[46] Kamar N, Faure P, Dupuis E, et al. Villous atrophy induced by mycophenolate mofetil in renal-transplant patients. Transpl Int 2004;17 (8):463–7. Available from: https://doi.org/10.1111/j.1432-2277.2004.tb00471.x.

[47] Maes BD, Dalle I, Geboes K, et al. Erosive enterocolitis in mycophenolate mofetil-treated renal-transplant recipients with persistent afebrile diarrhea. Transplantation 2003;75(5):665–72. Available from: https://doi.org/10.1097/01.TP.0000053753.43268.F0.

[48] Selbst MK, Ahrens WA, Robert ME, Friedman A, Proctor DD, Jain D. Spectrum of histologic changes in colonic biopsies in patients treated with mycophenolate mofetil. Mod Pathol 2009;. Available from: https://doi.org/10.1038/modpathol.2009.44.

[49] Bunnapradist S, Sampaio MS, Wilkinson AH, et al. Changes in the small bowel of symptomatic kidney transplant recipients converted from mycophenolate mofetil to enteric-coated mycophenolate sodium. Am J Nephrol 2014;40(2):184–90. Available from: https://doi.org/10.1159/000365360.

[50] Filiopoulos V, Sakellariou S, Papaxoinis K, et al. Celiac-like enteropathy associated with mycophenolate sodium in renal transplant recipients. Transplant Direct 2018;4(8):e375. Available from: https://doi.org/10.1097/TXD.0000000000000812.

[51] Talmon G, Manasek T, Miller R, Muirhead D, Lazenby A. The apoptotic crypt abscess an underappreciated histologic finding in gastrointestinal pathology. Am J Clin Pathol 2017;148(6):538–44. Available from: https://doi.org/10.1093/AJCP/AQX100.

[52] Smyth RL. Fibrosing colonopathy in cystic fibrosis. Arch Dis Child 1996;74:464–8. Available from: https://doi.org/10.1136/adc.74.5.464.

[53] Smyth RL, van Velzen D, Smyth AR, Lloyd DA, Heaf DP. Strictures of ascending colon in cystic fibrosis and high-strength pancreatic enzymes. Lancet (London, England) 1994;343(8889):85–6. Available from: https://doi.org/10.1016/s0140-6736(94)90817-6.

[54] Bansi DS, Price A, Russell C, Sarner M. Fibrosing colonopathy in an adult owing to over use of pancreatic enzyme supplements. Gut 2000;46:283–5.

[55] Fitzsimmons SC, Burkhart GA, Borowitz D, et al. High-dose pancreatic-enzyme supplements and fibrosing colonopathy in children with cystic fibrosis. N Engl J Med 1997;336(18):1283–9. Available from: https://doi.org/10.1056/NEJM199705013361803.

[56] Terheggen G, Dieninghoff D, Rietschel E, Drebber U, Kruis W, Leifeld L. Successful non-invasive treatment of stricturing fibrosing colonopathy in an adult patient. Eur J Med Res 2011;16(9):411–14. Available from: https://doi.org/10.1186/2047-783x-16-9-411.

[57] Lloyd-Still JD, Beno DW, Kimura RM. Cystic fibrosis colonopathy. Curr Gastroenterol Rep 1999;1(3):231–7 <http://www.ncbi.nlm.nih.gov/pubmed/10980955> [accessed 11.10.19].

[58] Brahmer JR, Lacchetti C, Schneider BJ, et al. Management of immune-related adverse events in patients treated with immune checkpoint inhibitor therapy: American Society of Clinical Oncology Clinical Practice Guideline. J Clin Oncol 2018;36(17):1714–68. Available from: https://doi.org/10.1200/JCO.2017.77.6385.

[59] Downey SG, Klapper JA, Smith FO, et al. Prognostic factors related to clinical response in patients with metastatic melanoma treated by CTL-associated antigen-4 blockade. Clin Cancer Res 2007;13(22):6681–8. Available from: https://doi.org/10.1158/1078-0432.CCR-07-0187.

[60] Soularue E, Lepage P, Colombel JF, et al. Enterocolitis due to immune checkpoint inhibitors: a systematic review. Gut 2018;67(11):2056−67. Available from: https://doi.org/10.1136/gutjnl-2018-316948.

[61] Abu-Sbeih H, Ali FS, Luo W, Qiao W, Raju GS, Wang Y. Importance of endoscopic and histological evaluation in the management of immune checkpoint inhibitor-induced colitis 11 Medical and Health Sciences 1103 Clinical Sciences. J Immunother Cancer 2018;6(1):1−11. Available from: https://doi.org/10.1186/s40425-018-0411-1.

[62] Yip RHL, Lee LH, Schaeffer DF, Horst BA, Yang HM. Lymphocytic gastritis induced by pembrolizumab in a patient with metastatic melanoma. Melanoma Res 2018;28(6):645−7. Available from: https://doi.org/10.1097/CMR.0000000000000502.

[63] Wang Y, Abu-Sbeih H, Mao E, et al. Endoscopic and histologic features of immune checkpoint inhibitor-related colitis. Inflamm Bowel Dis 2018;24(9):1695−705. Available from: https://doi.org/10.1093/ibd/izy104.

[64] Jain A, Lipson EJ, Sharfman WH, Brant SR, Lazarev MG. Colonic ulcerations may predict steroid-refractory course in patients with ipilimumab-mediated enterocolitis. World J Gastroenterol 2017;23(11):2023−8. Available from: https://doi.org/10.3748/wjg.v23.i11.2023.

[65] Barcenas CH, Ibrahim NK. Chemotherapy-induced colitis, <www.intechopen.com> *Open Access Peer-reviewed Chapter*. 2012, 115−128 [accessed 24.10.19].

[66] Fata F, Ron IG, Kemeny N, O'Reilly E, Klimstra D, Kelsen DP. 5-Fluorouracil-induced small bowel toxicity in patients with colorectal carcinoma. Cancer 1999;86(7):1129−34. Available from: https://doi.org/10.1002/(SICI)1097-0142(19991001)86:7 < 1129::AID-CNCR5 > 3.0.CO;2-4.

[67] Sheibani S, Gerson LB. Chemical colitis. J Clin Gastroenterol 2008;42(2):115−21. Available from: https://doi.org/10.1097/MCG.0b013e318151470e.

[68] Cappell MS. Colonic toxicity of administered drugs and chemicals. Am J Gastroenterol 2004;99(6):1175−90. Available from: https://doi.org/10.1111/j.1572-0241.2004.30192.x.

Chapter 30

Inflammatory bowel disease—like conditions: immune deficiencies

Bo Shen

Center for Inflammatory Bowel Diseases, Columbia University Irving Medical Center-New York Presbyterian Hospital, New York, NY, United States

Chapter Outline

Abbreviations

CD Crohn's disease
CVID common variable immune deficiency
GI gastrointestinal
IBD inflammatory bowel disease
Ig immunoglobulin
IVIG intravenous immunoglobulin
SIgAD selective IgA deficiency
UC ulcerative colitis

Introduction

Primary or secondary immune defects can occur in both children and adults. Primary immune defects are mainly seen in pediatric patients. In contrast, immune defects in adults often result from secondary causes. Secondary immune disorders can be caused by various medical conditions or treatments for these conditions, such as diabetes mellitus, cirrhosis, hemoglobinopathy, malnutrition, nephrotic syndrome, protein-losing states (such as enteropathies, severe exudative skin disease, and peritoneal dialysis), splenectomy, malignancy, and radiation therapy. One of the common etiologies of secondary immune deficiency is the use of immunosuppressive therapy, including cytotoxic chemotherapy for malignancy, immunomodulator therapy for autoimmune diseases, antirejection therapy for solid organ or bone marrow transplantation, bone marrow ablation before transplantation, and management of graft-versus-host disease. Corticosteroids, immunomodulators, and immunosuppressive biological therapy are widely used in the management of Crohn's disease (CD) and ulcerative colitis (UC). Patients with inflammatory bowel disease (IBD) may develop primary or secondary treatment failure to these agents.

Components of immune deficiency include defects in immunoglobulins (Ig), complements, granulocytes, and cell-mediated immunity. There is reciprocal relationship between immune deficiency and infection. While viral, bacterial, and fungal infections, such as human immunodeficiency virus (HIV) can result in immune defects, secondary immune deficiency may cause infections from these pathogens. The predominant presentations of immunodeficiency are stereotypic patterns of recurrent infections. The pattern of recurrent infections may suggest the deficient component of

Atlas of Endoscopy Imaging in Inflammatory Bowel Disease. DOI: https://doi.org/10.1016/B978-0-12-814811-2.00030-X
© 2020 Elsevier Inc. All rights reserved.

immune system is involved. Gastrointestinal (GI) tract infections from viral, bacterial, fungal, and parasitic agents are common presentations in patients with primary or secondary immune deficiency. In addition, immune deficiency may present with acute and chronic inflammation in the GI tract with or without identifiable pathogens.

Evaluation of immune deficiency

A detailed evaluation of family history is important for the detection of primary immunodeficiencies. Testing for HIV is a necessary next step. Selective primary immunodeficiencies, including mutations in IL10/IL10 receptor, IL-21, NADPH oxidase complex, XIAP, LRBA, and CTLA-4, are found to be associated with early-onset or pediatric IBD [1,2]. Primary immune deficiency disorders as a cause for IBD or IBD-like conditions appears to be restricted to those with pediatric-onset and severe disease [3]. Monogenetic, early-onset IBD conditions are discussed in Chapter 31, Inflammatory bowel disease—like conditions: monogenic gastrointestinal (GI) disorders.

Granulocytes and lymphocytes can be measured by complete blood counts with white blood cell counts. Chronic granulomatous disease can be assessed flow cytometric testing of dihydrorhodamine. Screening complement testing with C3, C4, C5 through C9, and CH50 may be performed.

The evaluation of humoral immunity consists of serum levels of IgG, IgA, IgM, and IgE as screening tests for antibody defects. Flow cytometry is used to quantify T- and B-cell subsets. NK-cell functional assessment can also be performed.

A GI evaluation may be of diagnostic for some immune deficient disorders, such as duodenal biopsy for the diagnosis of common variable immune deficiency (CVID).

Gastrointestinal infections

Patients with primary or secondary humoral or cellular immune deficiencies or deficiencies in the innate immunity are prone to the development of various forms of GI infections. IBD-like GI infections are discussed in Chapter 23, Superimposed infections in inflammatory bowel diseases, and Chapter 25, Inflammatory bowel disease—like conditions: infectious.

Clostridioides difficile colitis and enteritis

Relapsing and/or recurrent *Clostridioides* (or *Clostridium*) *difficile* colitis can occur in both immune-competent and immune-compromised patients. *Clostridioides difficile* can even occur the small bowel or ileal pouch, mainly in patients with underlying IBD [4,5]. The classic pseudomembranes appear to be more common in patients with *C. difficile* colitis than *C. difficile* enteritis; and more common in those with sole *C. difficile* colitis than *C. difficile* superimposed on underlying IBD (Fig. 30.1B and C).

Enterocolitis from pathogens

Relapsing, recurrent, and/or progressive infectious enterocolitis from *Giardia lamblia*, enteroviruses, cytomegalovirus (Fig. 30.2), and *Campylobacter* are common in primary or secondary humoral and T-cell immunodeficiencies. Enterocolitis may be caused by other uncommon pathogens, such as *Isospora*, *Microsporidia*, and *Cyclospora* (Fig. 30.3).

Immune deficiency-associated gastrointestinal inflammation

Acute or chronic inflammation in the GI tract can be a part of multisystem presentations of primary or secondary humoral or T-cell immune deficiencies or deficiencies in innate immunity.

Cellular immune deficiency

Deficiency in cellular immune can result in inflammatory conditions in the GI tract, with HIV infection is a classic example (Fig. 30.1A). HIV-associated GI disorders are discussed in Chapter 23, Superimposed infections in inflammatory bowel diseases, and Chapter 25.

Common variable immunodeficiency

Common variable immunodeficiency, the most prevalent form of severe antibody deficiency, is a primary immunodeficiency, which can affect both pediatric and adult populations. CVID is characterized by impaired B-cell differentiation

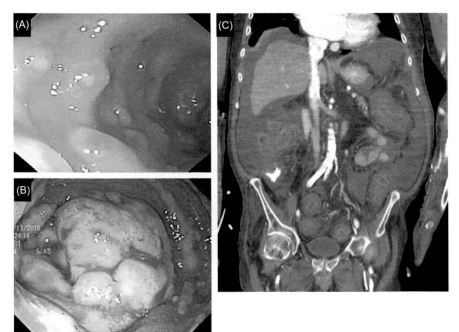

FIGURE 30.1 Infectious colitis due to secondary immune deficiency. (A) Colitis with edema and erosions, and mucopurulent exudates in a patient with human immunodeficiency virus; (B and C) *Clostridium difficile* infection with pseudomembranes on proctoscopy and diffuse inflamed colon on computed tomography in a patient with liver cirrhosis, end-stage renal disease, and hemodialysis.

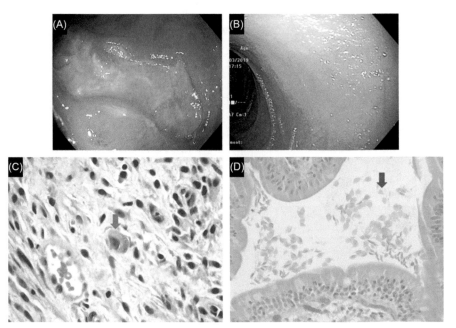

FIGURE 30.2 Infectious colitis and enteritis in secondary immune deficiency. (A and C) Deep discrete colon ulcers with cytomegalovirus-infect monocytes (*green arrow*) in a patient with ulcerative colitis on long-term antitumor necrosis factor therapy; (B and D) *Giardia lamblia* infection dilated lumen and flatten folds of the duodenum and enteroparasites (*blue arrow*) in a patient with cystic fibrosis. *Histology images: Courtesy of Drs. Ilyssa Gordon and Ana E. Bennett of Cleveland Clinic.*

with defective Ig production. CVID represents a spectrum of hypogammaglobulinemia syndromes caused by many genetic defects. The diagnostic criteria are markedly reduced serum concentrations of IgG and low levels of IgA and/or IgM, poor or absent response to immunizations, and an absence of other defined immunodeficiency state [6]. Patients with CVID may present with recurrent infections, autoimmunity, various forms of inflammatory disorders, and malignancy. It was estimated that 15% of patients with CVID had inflammatory diseases of the GI tract [7]. Patients with CVID are at risk for the development of acute diarrhea from infections with norovirus, *Campylobacter jejuni*, or *Salmonella*, or chronic diarrhea from cytomegalovirus, norovirus, *G. lamblia* (Fig. 30.4), cryptosporidium, or strongyloides (Fig. 30.5), with speculated theories in deficiencies in both cellular immunity and antibody deficiencies.

In addition to infectious etiology, chronic GI symptoms in patients with CVID may result from the disease per se. Patients with CVID may have inflammation of the GI tract without concurrent microbial infections. The inflammation

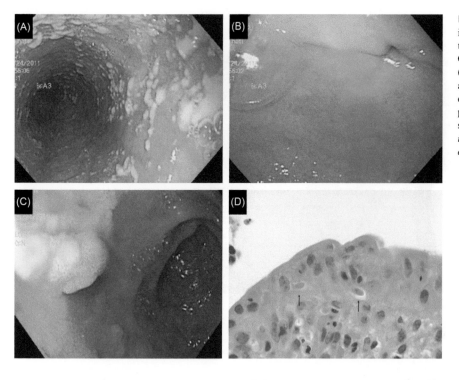

FIGURE 30.3 *Isospora belli* and candida infection in a patient with eosinophilic gastroenteritis receiving corticosteroids. (A) Candida esophagitis with white plaques; (B) mild antral gastritis with atrophic appearance; (C) nodular mucosa of the duodenum; (D) numerous subnuclear cytoplasmic inclusions (*thin black arrows*), suggestive of protozoal infection. *Histology image: Courtesy of Dr. Erinn Downs-Kelly of Cleveland Clinic.*

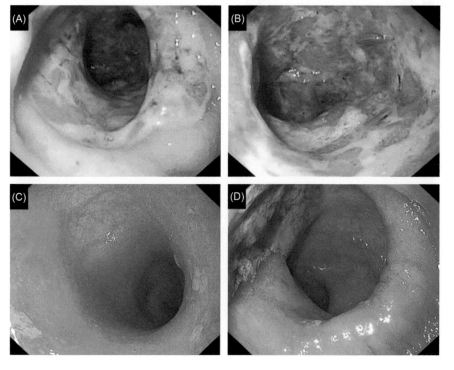

FIGURE 30.4 Common variable immune deficiency with concurrent infections of *Giardia lamblia* and cytomegalovirus. (A and B) Severe colitis with diffuse ulcers and mucopurulent exudates before antipathogen therapy; (C and D) resolving colitis after the therapy. *Photo courtesy: Dr. Pei Tang of Peking Union Medical College Hospital.*

in this setting ranges from mild (Fig. 30.6) to severe enteritis or colitis (Fig. 30.7). Interestingly, CVID is often associated with various autoimmune disorders in the GI tract [8,9]. Enteropathy in CVID was found to be associated with autoimmunity and with low serum IgM, but not with low serum IgA [10]. From the perspective of clinical presentation, endoscopy, and histology, these autoimmune disorders resemble IBD (CD-like or UC-like), celiac disease, microscopic colitis, and protein-losing enteropathy. Patients with CVID may have symptoms of malabsorption, small intestinal bacterial overgrowth, or GI lymphoma. Noncaseating granulomas may be found in the intestine as well as other

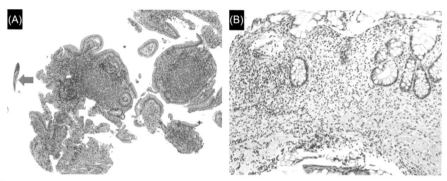

FIGURE 30.5 Duodenum and colon biopsies in a patient with common variable immune deficiency. (A) Duodenal follicular lymphoid hyperplasia, and chronic strongyloides infestation (*green arrow*). There are no plasma cells. (B) Colonic mucosa in the same patient showing chronic colitis with crypt distortion and loss, mixed inflammatory cellular infiltration, including many eosinophils and lack of plasma cells. *Histology image: Courtesy Dr. Shu-Yuan Xiao of University of Chicago Medical Center.*

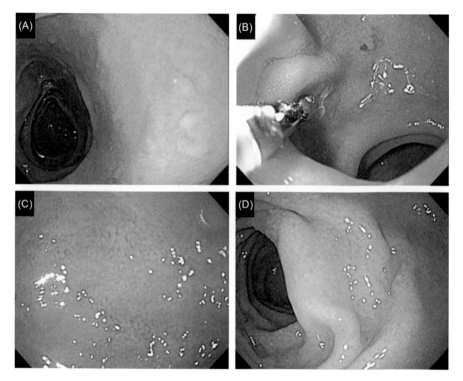

FIGURE 30.6 Duodenum and colon in common variable immune deficiency without identifiable infectious pathogens. (A) Nodular duodenum mucosal with flatten folds; (B) mild erythema of duodenum mucosa; (C and D) mild erythema of the colon mucosa.

extraintestinal tissues in patients with CVID [11−13]. Differential diagnosis should be made between CVID and CD in this setting. Unlike CD, patients with CVID and granulomatous disease commonly present with lymphadenopathy, splenomegaly, and pulmonary symptoms (Fig. 30.8).

IgA deficiency

Selective IgA deficiency (SIgAD), the most common immunological defect in humans,which is defined as an isolated deficiency of serum IgA in the setting of normal serum levels of IgG and IgM in an individual >4 years of age, with the exclusion of other etiologies of hypogammaglobulinemia. SIgAD in some of the patients may evolve into CVID. Patients may be asymptomatic or present with recurrent infections or autoimmune disorders. IgA consists of two forms: monomeric IgA in the serum and dimeric secretory IgA in secretions. The exact etiopathogenesis of SIgAD is not clear, with speculated defects in B cells or defective interactions between B and T cells. Common presentations include recurrent sinopulmonary infections, allergic disorders, anaphylactic transfusion reactions, *G. lamblia* infection and other

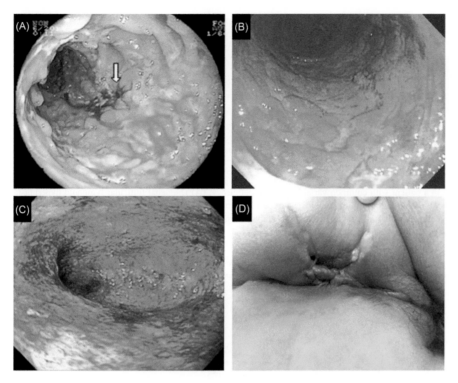

FIGURE 30.7 Diffuse colitis and perianal disease in common variable immunodeficiency in IgG, IgM, and IgA. (A–C) Diffuse colitis with ulcers (*arrow*) and nodularity at three different occasions. Histologic evaluation of colon biopsy showed chronic active colitis; (D) perianal fistulae and abscesses. *Photo courtesy: Dr. Bolin Yang of Jiangsu Province Hospital of Chinese Medicine and Affiliated Hospital of Nanjing University of Chinese Medicine.*

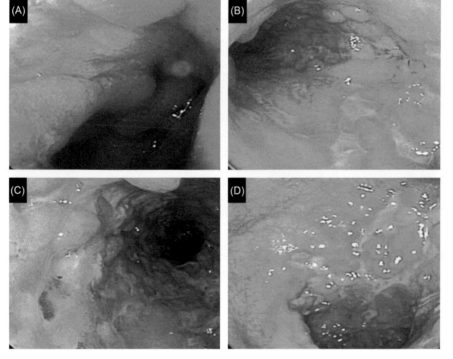

FIGURE 30.8 Secondary IgG deficiency in a patient refractory ileocolitis from extensive use of various biological agents, including infliximab, adalimumab, vedolizumab, and ustekinumab. (A and B) Severe colitis in the descending and sigmoid colon with ulcers and nodularity; (C and D) Persistent, even worsen colitis after the use of the two biological agents. The patient was found to have secondary deficient IgG, who responded to intravenous immunoglobulin infusion.

intestinal disorders, and autoimmune disorders. Sinopulmonary infections appear to be more common than GI infections, as partially compensated secreted IgM secretion is more prominent in the GI tract than the respiratory tract [14]. While the symptomatology of the GI tract in the patients is largely associated with infections of pathogens, gut inflammation resulting from deficient IgA per se is uncommon. Some patients may however have nodular mucosa of the small intestine and colon (Fig. 30.9). Patients with severe SIgAD can experience infusion reactions to blood products containing small amounts of IgA, typically in plasma, which may be related to the presence of anti-IgA antibody. Therefore all blood products, including the use of intravenous Ig (IVIG), should be used with caution in these patients to anaphylaxis.

SIgAD should be differentiated from secondary IgA deficiency, in which drugs are common causes. Drug-induced secondary IgA deficiency can result from the use of anticonvulsants, sulfasalazine, and cyclosporine.

Immune deficiency resulting from inflammatory bowel disease or inflammatory bowel disease therapy

Immune dysregulation has a key role in the pathogenesis of IBD. Immunosuppressive therapy is the mainstay for the treatment of IBD. Current clinical practice guidelines with precision medicine have been focused on the therapeutic drug monitoring and enhancement of the therapy with an increased dose of biologics or a combination of biologics and immunomodulators. On the other hand, immunosuppressive medication may trigger the development, exacerbation, and refractory disease course of IBD, through mechanisms ranging from immune deficiency to immune dysregulation. For example, the use of antitumor necrosis factor agents can trigger the development of IBD [15]. Immunosuppressive medications (such as mycophenolate mofetil) are used to treat IBD [16], while they can cause de novo IBD or IBD-like conditions [17].

IBD may coexist with antibody deficiencies, which were reported in five adult patients, four with CVID and one with X-linked agammaglobulinemia [18]. Our group and others have reported secondary Ig deficiency in patients with

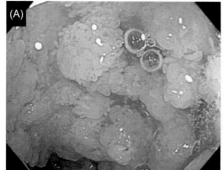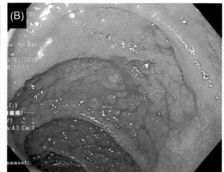

FIGURE 30.9 Small and large bowel in an adult patient with selective IgA deficiency who had chronic diarrhea and abdominal pain. (A) Nodular mucosa of the terminal ileum; (B) nodular mucosa of the colon.

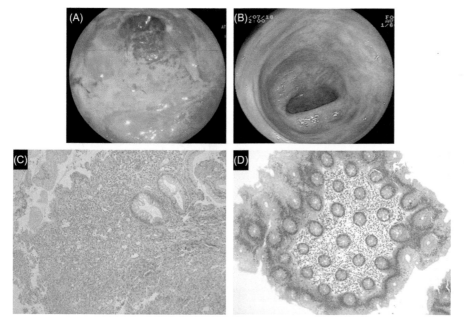

FIGURE 30.10 Intravenous immunoglobulin therapy in a patient with refractory ulcerative colitis (to corticosteroids, immunomodulators, and infliximab) and secondary immunoglobulin deficiency. (A and C) Endoscopic and histologic features before the therapy with deep large ulcers; (B and D) mucosal healing with scars on endoscopy (B) and near resolution of inflammatory infiltration on histology (D). *Photo courtesy: Dr. Yubei Gu of Shanghai Jiao-tong University Ruijin Hospital.*

refractory IBD (Fig. 30.9) [19,20]. Most patients had therapy with immunosuppressive agents, which, along with malnutrition and protein-losing enteropathy, may explain the low level of Ig or Ig subclasses. Low IgG/G1 levels were found to be associated with IBD-related surgery in CD [2]. IVIG infusion has been used for the treatment-refractory IBD with a promising outcome (Figs. 30.10−30.12). Concomitant *Clostridium difficile* infection was a risk factor for the treatment failure of IVIG for refractory IBD [21].

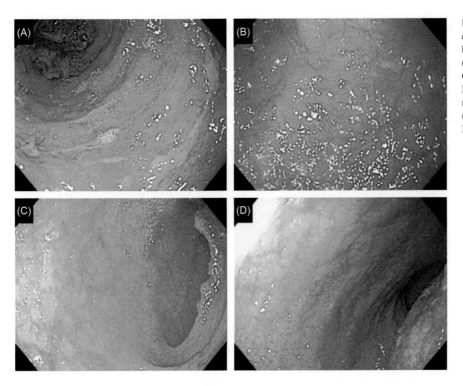

FIGURE 30.11 Refractory Crohn's disease of the ileal pouch with IgG deficiency before and after intravenous Ig infusion. (A and B) Chronic antibiotic-refractory Crohn's disease with ulcers in the afferent limb (A) and pouch body (B); (C and D) normal afferent limb (C) and pouch body (D) after one dose of intravenous Ig. *Ig*, Immunoglobulin.

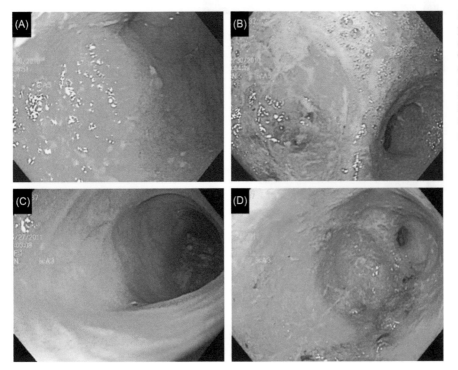

FIGURE 30.12 Antibiotic-refractory pouchitis with IgG deficiency before and after intravenous Ig infusion. (A and B) Chronic antibiotic-refractory pouchitis with diffuse ulcers and exudates in the distal afferent limb (A) and pouch body (B); (C and D) normal afferent limb (C) and improved inflammation in the pouch body (D) after two doses of intravenous Ig therapy. *Ig*, Immunoglobulin.

Summary and recommendations

Immune deficiency status is common in a clinical setting, which can involve the innate or adaptive immunity. Immune deficiency can be primary or secondary, leading to multisystem presentations. The GI tract is a common organ system of the involvement. The GI manifestations can result from microbial infections, concurrent autoimmune conditions, or immune deficiency per se. Immune deficiency, especially antibody deficiencies, is common in patients with IBD, resulting from the underlying disease process of IBD, IBD medications, malabsorption, or malnutrition. Differential diagnosis between GI manifestations of immune deficiency, immune deficiency—associated infections, IBD, concurrent IBD, and immune deficiency relies on a combined assessment of clinical presentation, endoscopy, histology, microbiology, serology, and immunogenetic testing.

References

[1]   Bégin P, Patey N, Mueller P, Rasquin A, Sirard A, Klein C, et al. Inflammatory bowel disease and T cell lymphopenia in G6PC3 deficiency. J Clin Immunol 2013;33:520—5.

[2]   Salzer E, Kansu A, Sic H, Májek P, Ikincioğullari A, Dogu FE, et al. Early-onset inflammatory bowel disease and common variable immunodeficiency-like disease caused by IL-21 deficiency. J Allergy Clin Immunol 2014;133:1651—9.

[3]   Kelsen JR, Sullivan KE. Inflammatory bowel disease in primary immunodeficiencies. Curr Allergy Asthma Rep 2017;17:57 [E journal].

[4]   Kochhar G, Edge P, Blomme C, Wu XR, Lopez R, Ashburn J, et al. *Clostridium difficile* enteropathy is associated with a higher risk for acute kidney injury in patients with an ileostomy-a case-control study. Inflamm Bowel Dis 2018;24:402—9.

[5]   Seril DN, Ashburn JH, Lian L, Shen B. Risk factors and management of refractory or recurrent Clostridium difficile infection in ileal pouch patients. Inflamm Bowel Dis 2014;20:2226—33.

[6]   Conley ME, Notarangelo LD, Etzioni A. Diagnostic criteria for primary immunodeficiencies. Representing PAGID (Pan-American Group for Immunodeficiency) and ESID (European Society for Immunodeficiencies). Clin Immunol. 1999;93:190—7.

[7]   Resnick ES, Moshier EL, Godbold JH, Cunningham-Rundles C. Morbidity and mortality in common variable immune deficiency over 4 decades. Blood 2012;119:1650—7.

[8]   Wang J, Cunningham-Rundles C. Treatment and outcome of autoimmune hematologic disease in common variable immunodeficiency (CVID). J Autoimmun 2005;25:57—62.

[9]   Boileau J, Mouillot G, Gérard L, Carmagnat M, Rabian C, Oksenhendler E, et al. Autoimmunity in common variable immunodeficiency: correlation with lymphocyte phenotype in the French DEFI study. J Autoimmun 2011;36:25—32.

[10] Gathmann B, Mahlaoui N, Ceredih GL, Oksenhendler E, Warnatz K, Schulze I, et al. Clinical picture and treatment of 2212 patients with common variable immunodeficiency. J Allergy Clin Immunol 2014;134:116—26.

[11] Mechanic LJ, Dikman S, Cunningham-Rundles C. Granulomatous disease in common variable immunodeficiency. Ann Intern Med 1997;127:613—17.

[12] Ardeniz O, Cunningham-Rundles C. Granulomatous disease in common variable immunodeficiency. Clin Immunol 2009;133:198—207.

[13] Boursiquot JN, Gérard L, Malphettes M, Fieschi C, Galicier L, Boutboul D, et al. Granulomatous disease in CVID: retrospective analysis of clinical characteristics and treatment efficacy in a cohort of 59 patients. J Clin Immunol 2013;33:84—95.

[14] Macpherson AJ, McCoy KD, Johansen FE, Brandtzaeg P. The immune geography of IgA induction and function. Mucosal Immunol 2008;1:11—22.

[15] Bieber A, Fawaz A, Novofastovski I, Mader R. Antitumor necrosis factor-α therapy associated with inflammatory bowel disease: three cases and a systematic literature review. J Rheumatol 2017;44:1088—95.

[16] Smith MR, Cooper SC. Mycophenolate mofetil therapy in the management of inflammatory bowel disease—a retrospective case series and review. J Crohns Colitis 2014;8:890—7.

[17] Papadimitriou JC, Cangro CB, Lustberg A, Khaled A, Nogueira J, Wiland A, et al. Histologic features of mycophenolate mofetil-related colitis: a graft-versus-host disease-like pattern. Int J Surg Pathol 2003;11:295—302.

[18] Conlong P, Rees W, Shaffer JL, Nicholson D, Jewell D, Heaney M, et al. Primary antibody deficiency and Crohn's disease. Postgrad Med J 1999;75:161—4.

[19] Horton N, Wu X, Philpott J, Garber A, Achkar JP, Brzezinski A, et al. Impact of low immunoglobulin G levels on disease outcomes in patients with inflammatory bowel diseases. Dig Dis Sci 2016;61:3270—7.

[20] Merkley SA, Beaulieu DB, Horst S, Duley C, Annis K, Nohl A, et al. Use of intravenous immunoglobulin for patients with inflammatory bowel disease with contraindications or who are unresponsive to conventional treatments. Inflamm Bowel Dis 2015;21:1854—9.

[21] Horton N, Kochhar G, Patel K, Lopez R, Shen B. Efficacy and factors associated with treatment response of intravenous immunoglobulin in inpatients with refractory inflammatory bowel diseases. Inflamm Bowel Dis 2017;23:1080—7.

Chapter 31

Inflammatory bowel disease—like conditions: monogenic gastrointestinal disorders

Ying Huang, Yuhuan Wang, Zifei Tang and Ziqing Ye

Department of Gastroenterology, Children's Hospital of Fudan University, Shanghai, P.R. China

Chapter Outline

Abbreviations

CD	Crohn's disease
GI	gastrointestinal
IBD	inflammatory bowel disease
IL10RA	interleukin 10 receptor-α
IPEX	immune dysregulation, polyendocrinopathy, enteropathy, X-linked
LRBA	LPS-responsive beige-like anchor protein
VEOIBD	very early—onset inflammatory bowel disease

Introduction

Very early—onset inflammatory bowel disease (IBD) (VEOIBD) refers to patients who have onset of disease before 6 years of age [1]. The majority of patients in this group have monogenic defects. To date, there are approximately 50 genetic mutations that were found to be associated with monogenic IBD-like diseases, including mutations of *IL10*, *IL10RA*, *IL10RB*, *XIAP*, *LRBA*, and *TTC7A*. On the other hand, pediatric patients with primary immunodeficiencies can have clinical presentations mimicking that in IBD.

Very early—onset inflammatory bowel disease

Various gene mutations are associated with monogenic VEOIBD (Table 31.1). Monogenic VEOIBD is most frequently caused by mutations in *IL10*, *IL10RA*, and *IL10RB*. These patients present with refractory diarrhea in early infancy, with severe perianal disease, that is, abscesses and fistulae [2,3]. Lower gastrointestinal (GI) endoscopy often reveals a predominant involvement of the large bowel with severe deep ulceration being the most common finding (Figs. 31.1—31.6). Hematopoietic stem-cell transplantation, with either bone marrow transplantation or umbilical cord blood stem-cell transplantation, usually provides a cure for patients with *IL10*, *IL10RA*, and *IL10RB* mutations. Follow-up endoscopy after the transplantation often shows mucosal healing.

Atlas of Endoscopy Imaging in Inflammatory Bowel Disease. DOI: https://doi.org/10.1016/B978-0-12-814811-2.00031-1
© 2020 Elsevier Inc. All rights reserved.

TABLE 31.1 Endoscopic and histologic manifestations of common monogenic very early–onset inflammatory bowel disease.

Genetic defects	Endoscopic findings	Pathological findings
IL10, IL10RA	Crohn's disease–like penetrating, fistulae, and deep ulcerations	Crypt distortion, microabscess
FOXP3 (IPEX)	Small intestine involvement more common with ulcerations	Villous atrophy, crypt hyperplasia
LRBA	Crohn's disease–like severe ulcerations	Nonspecific
TTC7A	Intestinal atresia	Epithelial apoptosis

IPEX, Immune dysregulation, polyendocrinopathy, enteropathy, X-linked.

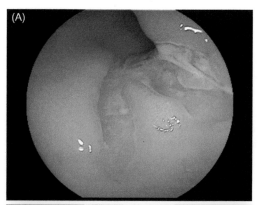

FIGURE 31.1 Very early–onset IBD due to *IL10RA* mutations. A 7-month-old girl had recurrent fever, hematochezia, anal fistula, and rectal vaginal fistula. There was irregular ulceration with elevated edematous mucosa between ulcers in the transverse colon and descending colon. The patient was found to have *IL10RA* c.269T > C/ c.537G > A mutation on whole-exome sequencing (A)(B). The patient received umbilical cord blood transplantation at the age of 13 months after birth. *IBD*, Inflammatory bowel disease.

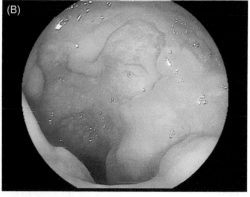

FIGURE 31.2 Very early–onset IBD due to *IL10RA* mutations. This patient had recurrent diarrhea and hematochezia at 4 days of age and recurrent fever since 20 days postnatal. There were irregular, severe, and deep ulcers to the level of the muscular layer with exudates and surrounding edematous mucosa in the ascending colon. The patient was found to have homozygous *IL10RA* c.301C > T mutation on targeted gene panel. *IBD*, Inflammatory bowel disease.

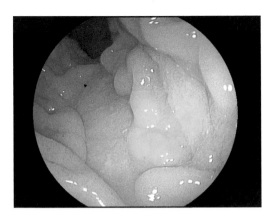

FIGURE 31.3 Very early–onset IBD due to *IL10RA* mutations. This patient had recurrent fever, diarrhea, and hematochezia at 7 days after birth and perianal abscess and fistula at the age of 9 months. There were deep, linear, and longitudinal ulcerations in the descending colon (*pictured*). The patient was found to have *IL10RA* c.301C > T/ c.349C > T mutation on whole-exome sequencing. *IBD*, Inflammatory bowel disease.

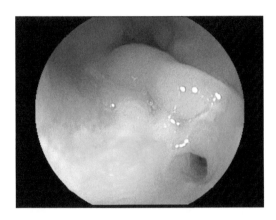

FIGURE 31.4 Very early–onset IBD due to *IL10RA* mutations. This patient had recurrent diarrhea 4 days after birth. He suffered from bowel perforation at the age of 6 months who was found to have a fistula opening at the distal rectum. Water infusion test confirmed the presence of recto-urethral fistula. The patient had *IL10RA* c.634C > T/c.537G > A mutation on whole-exome sequencing. *IBD*, Inflammatory bowel disease.

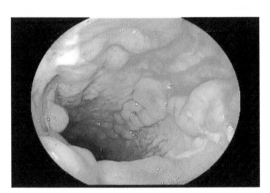

FIGURE 31.5 Very early–onset IBD due to *IL10RA* mutations. This patient presented with recurrent fever, oral ulcers, anal skin tag, and fissures 8 days after birth. There were serpiginous ulcers with pus, cobblestoning, and pseudopolyps in the descending colon. The patient was found to have *IL10RA* c.301C > T/c.537G > A mutation on targeted panel sequencing. *IBD*, Inflammatory bowel disease.

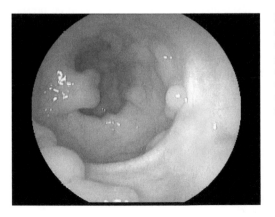

FIGURE 31.6 Very early–onset IBD due to *LL10RA* mutations. This patient presented with diarrhea and fever since 8 days' postnatal. She was also found to have perianal abscess and fistula at the age of 1.5 months. Colonoscopy showed inflammatory polyps. Colostomy was performed for rectal stricture and ileus. She had *IL10RA* c.301C > T/c.299T mutation on whole-exome sequencing, for which she received umbilical cord blood transplantation 6 months after birth. Hyperplastic and granulomatous lesions in the ascending colon were observed 5 months after umbilical cord blood transplantation. *IBD*, Inflammatory bowel disease.

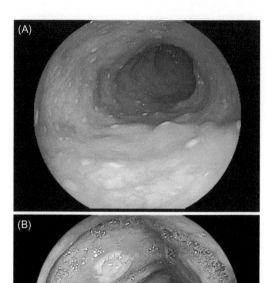

FIGURE 31.7 Very early—onset IBD due to *TNFAIP3* mutations. This child had recurrent fever and diarrhea 2 months after birth and subsequently developed arthralgia at the age of 1.5 years. Scattered ulcers (A) and erosions (B) in the ascending colon were demonstrated. The patient was found to have *TNFAIP3* c.1108het_delC mutation on whole-exome sequencing. *IBD*, Inflammatory bowel disease.

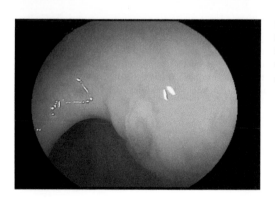

FIGURE 31.8 Very early—onset IBD due to *LRBA* mutations. This child had recurrent bloody bowel movements since 10 months after birth. Ulcers with surrounding mucosal edema in the sigmoid colon were shown. The patient had homozygous *LRBA* c.1570G > A mutation on targeted panel sequencing. *IBD*, Inflammatory bowel disease; *LRBA*, lipopolysaccharide (LPS)-responsive beige-like anchor protein.

Patients with immune dysregulation, polyendocrinopathy, enteropathy, X-linked (IPEX), due to *FOXP3* mutations, develop enteropathy in early infancy. One of the endoscopic and histologic hallmarks for IPEX is the presence of villous atrophy, although nonspecific lesions, including ulcers, crypt hyperplasia, and crypt abscesses can also be present [4]. *TNFAIP3* mutations may result in colitis in susceptible hosts with extraintestinal manifestations (Fig. 31.7). *LRBA* deficiency may present with predominant IBD-like phenotypes. Common endoscopic findings include severe circular inflammation in the rectum, mucosal edema and erythema, and numerous ulcerations in the colon (Fig. 31.8). Histopathological findings are not specific, with villous atrophy being reported in some patients [5].

Primary immunodeficiency

Patients with primary immunodeficiencies can have diarrhea, rectal bleeding, and abdominal cramps with histopathological changes in the GI tract mimicking that of IBD (Table 31.2). In fact, approximately 40% of patients with chronic granulomatous disease develop Crohn's disease (CD)-like GI tract inflammation. These patients have been found to have mutations in *NCF* or *CYBB* (Figs. 31.9—31.12). Common endoscopic findings include mucosal edema and

TABLE 31.2 Endoscopic and histologic manifestations of primary deficiency mimicking inflammatory bowel disease.

Disorder	Genetic defects	Endoscopic findings
Chronic granulomatous disease	NCF2, CYBB, CYBA	CD-like, edema and erythema, and erosions
X-linked ectodermal immunodeficiency	IKBKG	CD-like
Leukocyte adhesion deficiency 1	ITGB2	CD-like and severe ulcerations
Agammaglobulinemia	BTK	CD-like
Wiskott—Aldrich syndrome	WAS	UC-like

CD, Crohn's disease; UC, ulcerative colitis.

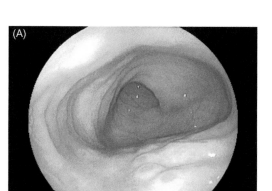

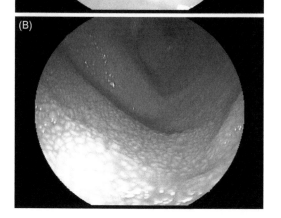

FIGURE 31.9 Very early—onset chronic granulomatous disease—associated colitis with *CYBB* mutation. This patient had recurrent fever at the age of 19 days; pulmonary fungal infection at the age of 4 months; and recurrent diarrhea at the age of 10 months, after birth. (A) Marked nodularity in the sigmoid colon with edema at the surrounding mucosa and (B) diffuse white spots from lymphotelangiectasia in the terminal ileum.

erythema in mild cases and extensive erosions and ulcers in severe cases [6]. Skip lesions are also common. Histopathology often demonstrates the presence of multiple granulomas.

Patients with X-linked ectodermal immunodeficiency, due to *IKBKG* mutations, develop CD-like manifestations, such as mucoal edema, exudates, and ulcers of the GI tract [7]. Wiskott—Aldrich syndrome is primary immunodeficiency, characterized by microthrombocytopenia, eczema, and recurrent infections due to cellular and humoral immune defects. Many patients present with ulcerative colitis—like symptomatology in early infancy [8]. The phenotypes of the GI disorders are not specific. Genetic analysis and immunological investigations are required for accurate diagnosis.

Infectious colitis in infants

Infectious colitis can occur in infant patients with primary or secondary immune deficiency disorders, with pathogenic agents such as *Clostridium difficile* and cytomegalovirus (Fig. 31.13). These infectious etiologies can be superimposed on underlying IBD-like conditions.

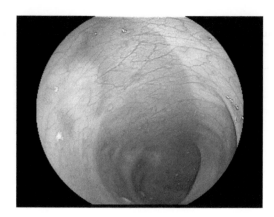

FIGURE 31.10 Very early—onset chronic granulomatous disease—associated colitis with *CYBB* mutation. This child had recurrent fever and diarrhea 3 days after birth and developed perianal abscess at the age 4 months. He received umbilical cord blood transplantation 17 months after birth for chronic granulomatous disease. There were marked diffuse edema and erythema in the sigmoid colon and descending colon.

(A)
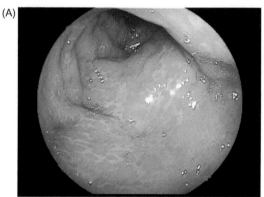

(B)
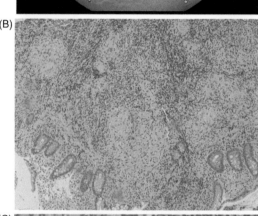

(C)
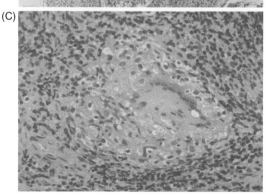

FIGURE 31.11 Very early—onset chronic granulomatous disease—associated colitis with *CYBB* mutation. This young boy was diagnosed as having chronic granulomatous disease with recurrent fever and pulmonary infection since 6 months after birth and blood in stool with mucus and pus since age of 6.5 years. (A) Colonoscopy showed the lesions, involving the rectum and entire colon, with marked erythema and superficial erosions in the sigmoid colon; (B) low-power histology showed multiple granulomas with mononuclear cell infiltration in the colon; and (C) high-power histology showed a large granuloma.

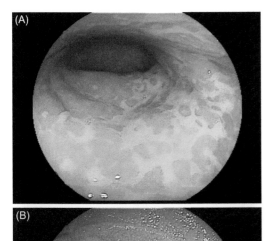

FIGURE 31.12 Very early—onset chronic granulomatous disease—associated colitis. This patient had neck abscess at 6 months after birth; recurrent respiratory infection accompanied by pulmonary fungal infection around 1 year old; and symptoms of mucus in the stool at the age of 7 years. He was diagnosed as having chronic granulomatous disease (*CYBB* c.1413G T. hemi) (P.E467X mutation). (A) Diffuse mucosal erosions and edema involved the rectum and entire colon, the rectum and sigmoid colon and (B) diffuse white spots from lymphotelangiectasia in the terminal ileum in the same patient.

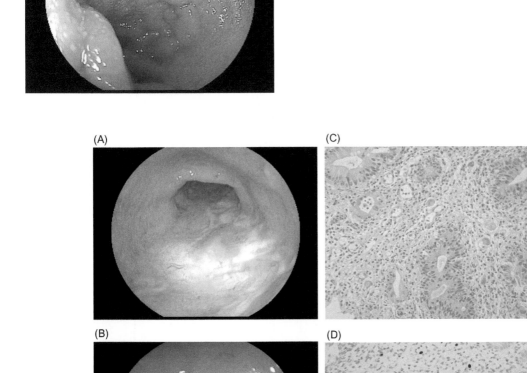

FIGURE 31.13 CMV colitis. This male patient presented with recurrent diarrhea since 10 days. He was tested positive for CMV IgM, IgG, and CMV DNA. (A) Irregular ulcerations and (B) skip lesions with exudates in the ascending colon; (C) histopathology showed active inflammation and multiple inclusion bodies, distorted and atrophic crypts; and (D) positive for CMV in immunohistochemistry. *CMV*, Cytomegalovirus.

Summary and recommendations

Monogenic IBD-like diseases are associated with early-onset, severe presentations, and unfavorable outcomes. Endoscopic and histologic findings in monogenic IBD-like diseases are often nonspecific, and some patients might have primary immunodeficiencies. Therefore other diagnostic tests, including immunological investigation and genetic testing (e.g., whole-exome sequencing), are required. Accurate diagnosis will enable physicians to carry out timely and precise treatment, such as hematopoietic stem-cell transplantation in patients with certain genetic defects, for example, *IL10*, *IL10RA*, *IL10RB*, *XIAP*, *NCF2*, and *CYBB*, and treatment with CTLA4 fusion protein in those with *LRBA* mutations.

Disclosure

The authors declared no financial conflict of interest.

References

[1] Uhlig HH, Schwerd T, Koletzko S, Shah N, Kammermeier J, Elkadri A, et al. The diagnostic approach to monogenic very early onset inflammatory bowel disease. Gastroenterology 2014;147:990–1007.

[2] Huang Z, Peng K, Li X, Zhao R, You J, Cheng X, et al. Mutations in interleukin-10 receptor and clinical phenotypes in patients with very early onset inflammatory bowel disease: a Chinese VEO-IBD collaboration group survey. Inflamm Bowel Dis 2017;23:578–90.

[3] Ye Z, Zhou Y, Huang Y, Wang Y, Lu J, Tang Z, et al. Phenotype and management of infantile-onset inflammatory bowel disease: experience from a tertiary care center in China. Inflamm Bowel Dis. 2017;23:2154–64.

[4] Barzaghi F, Amaya Hernandez LC, Neven B, Ricci S, Kucuk ZY, Bleesing JJ, et al. Long-term follow-up of IPEX syndrome patients after different therapeutic strategies: an international multicenter retrospective study. J Allergy Clin Immunol 2018;141:1036–49.

[5] Alkhairy OK, Abolhassani H, Rezaei N, Fang M, Andersen KK, Chavoshzadeh Z, et al. Spectrum of phenotypes associated with mutations in LRBA. J Clin Immunol 2016;36:33–45.

[6] Khangura S, Kamal N, Ho N, Quezado M, Zhao X, Marciano B, et al. Gastrointestinal features of chronic granulomatous disease found during endoscopy. Clin Gastroenterol Hepatol 2016;14:395–402.

[7] Cheng LE, Kanwar B, Tcheurekdjian H, Grenert JP, Muskat M, Heyman MB, et al. Persistent systemic inflammation and atypical enterocolitis in patients with NEMO syndrome. Clin Immunol 2009;132:124–31.

[8] Catucci M, Castiello MC, Pala F, Bosticardo M, Villa A. Autoimmunity in Wiskott-Aldrich syndrome: an unsolved enigma. Front Immunol 2012;3:209.

Chapter 32

Inflammatory bowel disease—like conditions: miscellaneous

Yan Chen[1], Xiaoying Wang[1], Yu-Bei Gu[2] and Bo Shen[3]

[1]Department of Gastroenterology, The Second Affiliated Hospital, School of Medicine, Zhejiang University, Hangzhou, P.R. China, [2]Department of Gastroenterology, Ruijin Hospital of Shanghai Jiaotong University, Shanghai, P.R. China, [3]Center for Inflammatory Bowel Diseases, Columbia University Irving Medical Center-New York Presbyterian Hospital, New York, NY, United States

Chapter Outline

Abbreviations

CD	Crohn's disease
CSC	cat scratch colon
EGID	eosinophilic gastrointestinal disorders
CMUSE	cryptogenic multifocal ulcerous stenosing enteritis
IBD	inflammatory bowel disease
UC	ulcerative colitis

Introduction

A long list of conditions can present inflammation or ulcers or strictures in the gastrointestinal (GI) tract. These conditions include eosinophilic GI disorders (EGID), cryptogenic multifocal ulcerous stenosing enteritis (CMUSE), sclerosing mesenteritis, diverticular colitis, and rectal prolapse. They mimic inflammatory or fibrostenotic small bowel or large bowel Crohn's disease (CD) or ulcerative colitis (UC). Careful endoscopic examination and documentation, along with histologic assessment, is the key for the diagnosis and differential diagnosis.

Eosinophilic gastrointestinal disorders

EGID represent a disease spectrum eosinophilic esophagitis, eosinophilic gastritis, eosinophilic gastroenteritis, enteritis, and eosinophilic colitis. The etiology of EGID is unknown, with allergic factors be contributing factors. EGID are characterized by eosinophilic infiltration, edema, and thickening of the GI tract, leading to various symptoms. There are mucosal, muscular layer, and subserosal phenotypes, which necessitate the differential diagnosis from CD and UC.

Atlas of Endoscopy Imaging in Inflammatory Bowel Disease. DOI: https://doi.org/10.1016/B978-0-12-814811-2.00032-3
© 2020 Elsevier Inc. All rights reserved.

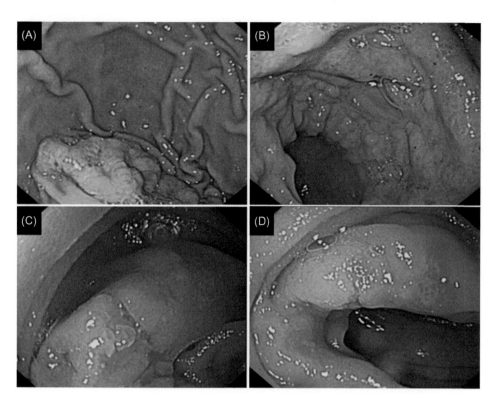

FIGURE 32.1 Eosinophilic gastroenteritis. (A) Patchy erythema at the gastric body and nodularity of the gastric antrum and (B–D) similar patterns with mucosal edema and nodularity of the colon.

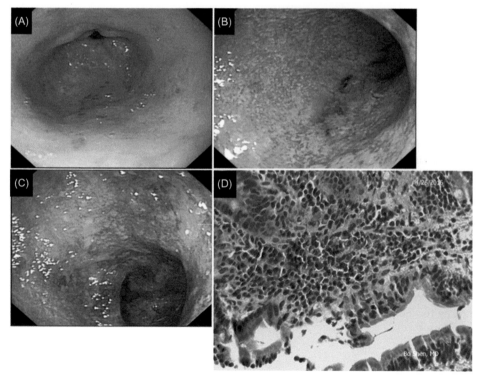

FIGURE 32.2 Eosinophilic gastroenteritis involving colon. (A) Gastritis with linear erythema and loss of vascularity of mucosa; (B and C) hyperemia, edema, and erosions of colon mucosa; and (D) histopathology showed inflammation of the lamina propria of the colon with numerous plasma cells and eosinophils. *Endoscopy image courtesy Ying-Lei Miao, MD, PhD, of Kunming Medical University.*

Endoscopic findings of mucosal disease are nonspecific, including nodular or polypoid mucosa, thickened (gastric) folds, loss of vascularity, erythema, mucosal erosions, and ulceration (Figs. 32.1–32.3). Mucosal and transmural biopsies may be helpful. Some patients may require laparoscopic full-thickness biopsy. Of note, peripheral and tissue eosinophilia can also be present in patients with inflammatory bowel disease (IBD) [1,2].

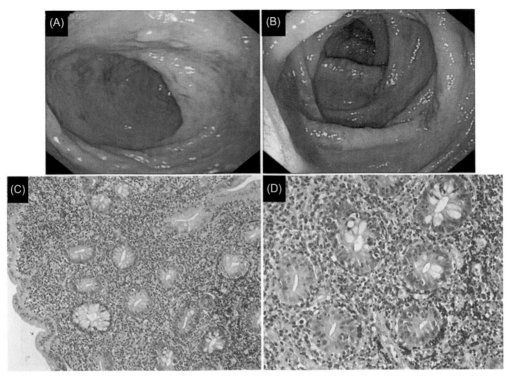

FIGURE 32.3   Eosinophilic colitis. (A and B) Diffuse mild edema of the colon with patchy erythema and erosions and(C and D) diffuse infiltration of eosinophils in the colon epithelia and lamina propria on H&E histology. *Courtesy Xinbo Ai, MD, PhD, of Jinan University.*

Cryptogenic multifocal ulcerous stenosing enteritis

CMUSE with unknown etiology is characterized by chronic and intermittent bouts of bowel obstruction resulting from multiple short ulcerated strictures in the jejunum and/or ileum. CMUSE mainly affects middle-aged patients. CMUSE lesions are limited to the mucosa and submucosa, without granuloma formation. Overlapping is observed between CMUSE and small bowel stricturing CD (L_1B_2 in the Montreal classification) [3] in clinical presentation and imaging [4]. Endoscopy may provide information for diagnosis and differential diagnosis [5]. The role of small intestine wireless capsule endoscopy in the diagnosis of CMUSE is yet to be defined, due to the concern of retained capsule (Figs. 32.4−32.6).

Small bowel ulcers resulting from adhesion and strangulation

Adhesions and strangulation of the small bowel can result in small bowel ulcers in addition to obstruction [6]. In small bowel CD, persistent small bowel inflammation can lead to the formation of intrinsic strictures and subsequent obstruction. In contrast, adhesions or hernias can result in distortion of the lumen of the small bowel, extrinsic obstruction, and mucosal ulcers. Small bowel ulcers are speculated to be caused by mesenteric compression (Figs. 32.6 and 32.7).

Sclerosing mesenteritis

The disease entity is a nonneoplastic inflammatory and fibrotic disease affecting the mesentery and subsequently impacts the integrity of the lumen of the GI tract and mesenteric vessels via a mass effect. Occasionally, ileocolonoscopic evaluation may show loss of vascularity, nodularity, and extrinsic mass in the lumen (Fig. 32.8).

Cat scratch colon

The hallmark of cat scratch colon (CSC) is bright red longitudinally linear tears in the colon mucosa on colonoscopy. In most cases, CSC is an incidental finding, which has a limited clinical implication (Fig. 32.9). Differential diagnoses include Crohn's colitis, UC, and ischemic colitis.

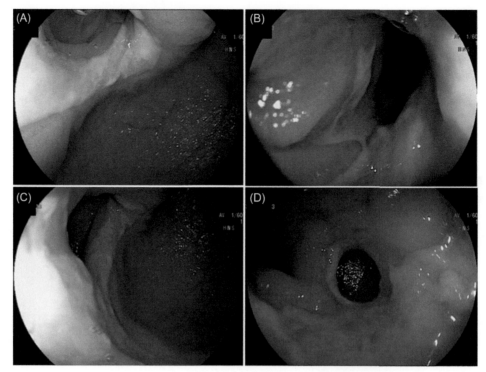

FIGURE 32.4   Cryptogenic multifocal ulcerous stenosing enteritis. (A—D) Discrete ulcerated strictures in the jejunum and ileum with normal mucosa of the small bowel segments in between. *Courtesy Yang Hong, MD, PhD, of Peking Union Medical College Hospital.*

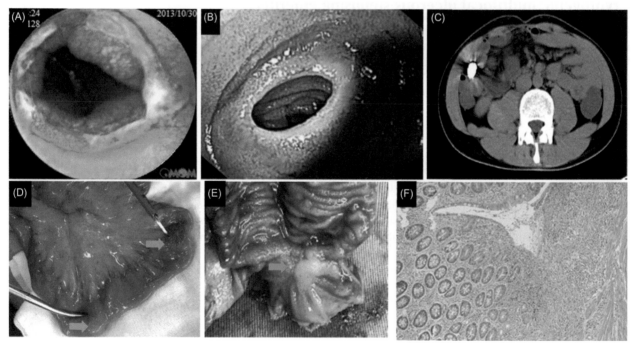

FIGURE 32.5   Another patient with CMUSE in the jejunum. (A) Ulcerated stricture at the distal jejunum on capsule endoscopy; (B) a thin and circumferential stricture in the proximal jejunum on balloon-assisted enteroscopy; (C) transiently trapped capsule on CT due to the small bowel strictures; (D and E) the small bowel ring-shaped strictures on surgical resection specimen (*green arrows*); and (F) histopathology shows ulceration and infiltration of chronic inflammatory cells in the lamina propria and muscularis mucosae, in the absence of transmural inflammation. *CMUSE,* Cryptogenic multifocal ulcerous stenosing enteritis.

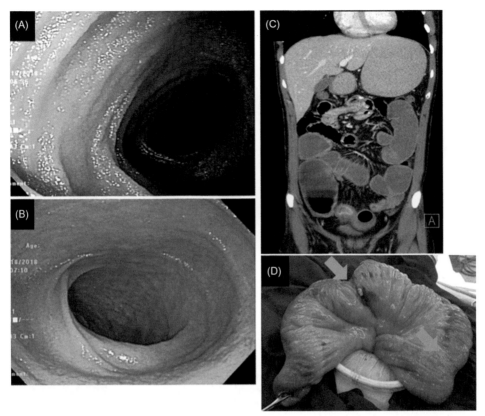

FIGURE 32.6 CMUSE missed by conventional ileocolonoscopy. The patient was mid-diagnosed as having Crohn's disease of the small bowel. (A and B) Normal mucosa of the distal ileum; (C) multiple small bowel strictures with dilated small bowel loops in between; and (D) intraoperative photography showed strictured and dilated small bowel (*green arrows*). *CMUSE*, Cryptogenic multifocal ulcerous stenosing enteritis.

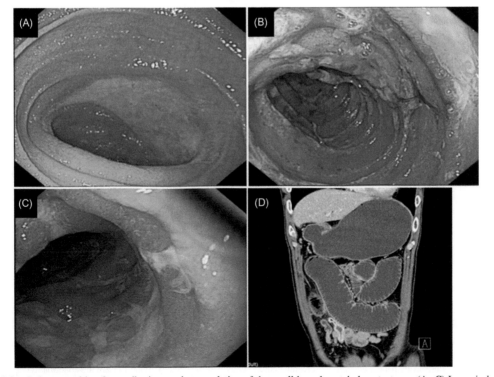

FIGURE 32.7 Jejunal ulcers resulting from adhesions and strangulation of the small bowel postcholecystectomy. (A–C) Large ischemic ulcers with nodular mucosa 100 cm after the pylorus on balloon-assisted enteroscopy and (D) dilated loops of the jejunum on CT.

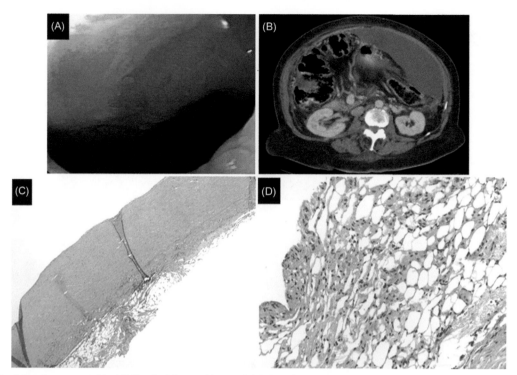

FIGURE 32.8   Sclerosing mesenteritis. (A) Terminal ileum with superficial ulceration and mild stenotic ileocecal valve; (B) fatty and vessel straining and haziness of the mesentery on CT enterography. Ascites resulted concurrent liver cirrhosis; and (C and D) peritoneal biopsy showed dense fibrosis with chronic inflammation and fat necrosis. *Histology photo courtesy Anna Bennett, MD, of Cleveland Clinic.*

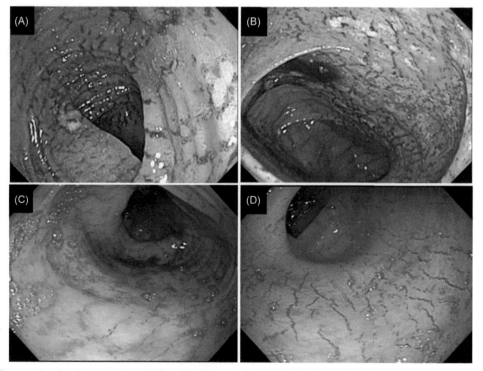

FIGURE 32.9   Cat scratch colon in two patients. Differential diagnoses should include microscopic colitis, Crohn's colitis, and ulcerative colitis. (A and B) The patient presented with chronic diarrhea and abdominal pain. Colonoscopy showed longitudinally linear erythema and tears in the ascending colon; (C and D) the asymptomatic patient undergoing screening colonoscopy was incidentally found to have the abnormality in the ascending colon.

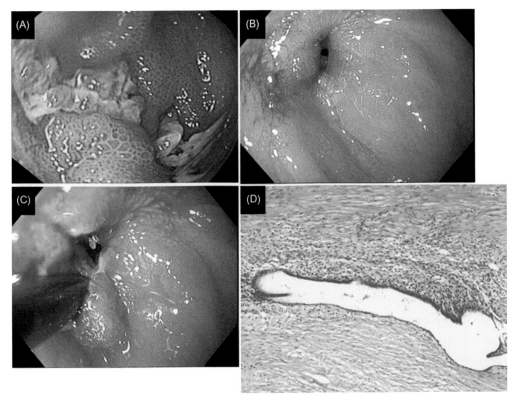

FIGURE 32.10   Endometriosis with ileocolonic ulcers treated with ileocolonic resection. The 40-year-old patient presented with chronic intermittent abdominal pain, diarrhea, and bloating. (A) Mucosal edema, ulcers, and exudates in the terminal ileum; (B and C) ileocolonic anastomosis stricture treated with endoscopic stricturotomy; and (D) surgical pathology of the resected specimen showed endometrium islet.

Endometriosis

Endometriosis refers to the presence of endometrioid tissue outside the uterus. The invasion and destruction of normal surrounding tissues from ectopic endometrium produce corresponding clinical symptoms. The most commonly involved organs are ovary and pelvic peritoneum, followed by the GI tract. In GI the most often involved ones are the rectum and sigmoid colon, followed by the ileum. Endometriosis in the ileum often leads to repeated small bowel obstruction. Involved bowel may have mucosal edema, nodularity, submucosal mass, ulcers, and strictures. In authors' experience, anastomotic strictures are common after bowel resection for endometriosis (Fig. 32.10) [7].

Meckel's diverticulum

Ileitis can occur in Meckel's diverticulum, which may share similar clinical and endoscopic presentations of CD (Fig. 32.11) [8]. Interestingly, Meckel's diverticulitis can occur in patients of CD [9].

Sarcoidosis

Sarcoidosis is a multisystem granulomatous disorder of unknown etiology. The presence of noncaseating granulomas in the GI tract necessitates differential diagnosis from CD. GI tract involvement is not common in sarcoidosis. However, symptoms result from mucosal, muscular, or myenteric involvement of the GI tract. The endoscopic appearance of upper GI sarcoidosis is nonspecific, including discrete, gray, plaque-like lesions, mucosal hyperemia, nodularity, thickened mucosa, and benign- or malignant-appearing ulcers (Fig. 32.12). In colonic sarcoidosis, endoscopic presentations include multiple nodules, polyps, stenosis, obstructive lesions, aphthous erosions, and small punctuated bleeding sites [10]. Of note, sarcoidosis is occasionally reported in association with CD or UC. Sarcoidosis has been reported in IBD patients after antitumor necrosis factor or antiintegrin therapy [11,12].

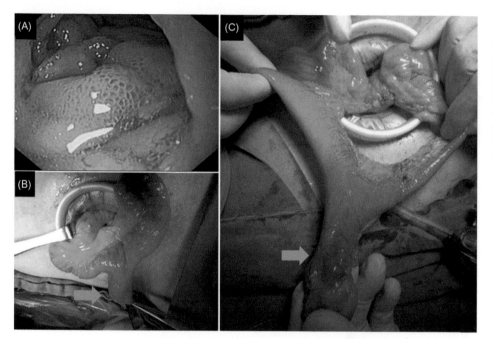

FIGURE 32.11 Inflamed Meckel's diverticulum compressed by the mesentery. A 22-year-old male presented with abdominal pain, nausea, and vomiting. (A) Inverted and inflamed Meckel's diverticulum at the distal ileum, 60 cm from the ileocecal valve, mimicking Crohn's disease and (B and C) surgical resection of the diverticulum (*green arrows*).

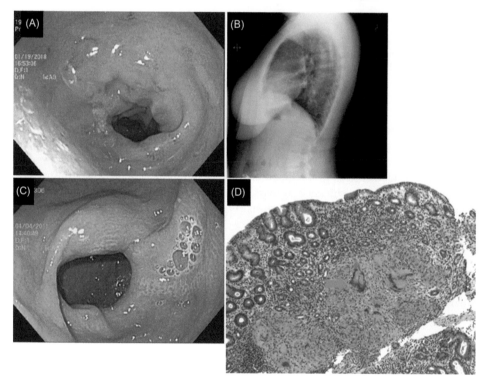

FIGURE 32.12 Sarcoidosis in the gastrointestinal tract. (A and B) Nodular and ulcerated mucosa of the colon (A) and hilar lymphadenopathy and interstitial lung disease on chest X-ray (B) and (C and D) gastric sarcoidosis with edematous and prominent pylorus (C) in another patient. Histology of gastric biopsy showed multiple large noncaseating granulomas (D) (*green arrow*).

Hypoalbuminemia

Severe nutrition and hypoalbuminemia can cause diffuse small and large bowel wall thickening, presenting as "enteritis" or "colitis" on cross-sectional imaging. Ileocolonoscopy may show classic bowel wall edema in the absence of mucosal inflammation (Fig. 32.13).

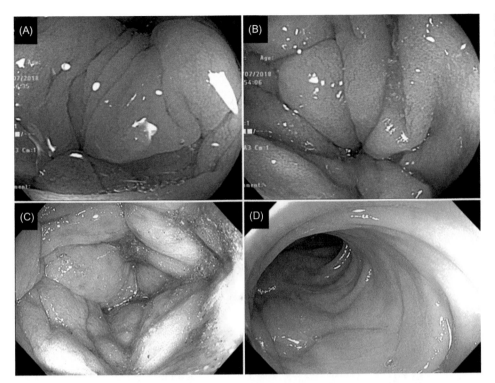

FIGURE 32.13 Colon edema from severe hypoalbuminemia in two patients. (A–D) Severe protein malnutrition can present "colitis" or "enteritis" with diffuse bowel wall thickening on abdominal imaging. Colonoscopy demonstrates diffuse bowel wall edema.

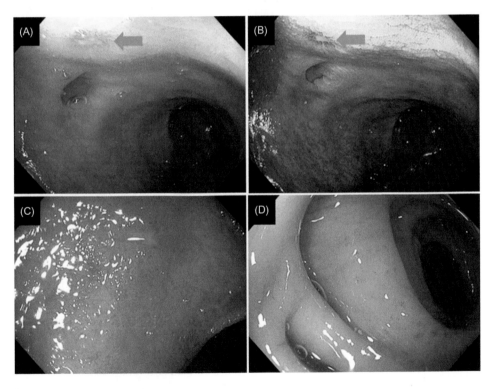

FIGURE 32.14 Diverticular colitis in the sigmoid colon, with differential diagnoses of Crohn's disease and ischemic colitis. (A and B) Peri-diverticulum orifice erosion on white-light and narrowband imaging (*green arrows*) and (C and D) ulcers and erythema distally and close to the diverticulum in another patient.

Diverticular colitis

Diverticular colitis is characterized by the presence of inflammation around or adjacent to the orifice of diverticulum, in the absence of inflammation of the diverticulum per se. The disease entity is most commonly located at the sigmoid colon. Concurrent mucosal prolapse may exist in the sigmoid colon. Diverticular colitis has been classified into four

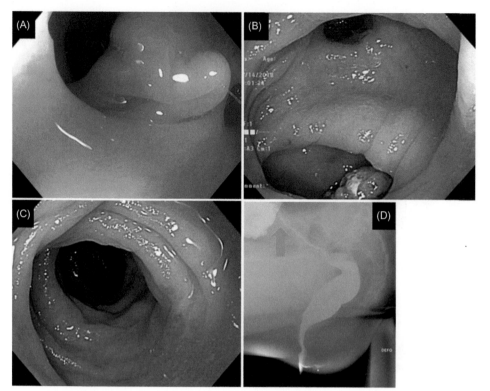

FIGURE 32.15 Diverticular colitis in the sigmoid colon in two patients. Differential diagnoses include Crohn's disease, ischemic colitis, and nonsteroidal antiinflammatory drug-induced injury. (A) Redundant sigmoid colon; (B and C) patchy erythema around the orifice of diverticuli; and (D) pseudostricture at the sigmoid colon (*green arrow*) with proximal bowel dilation on barium defecography.

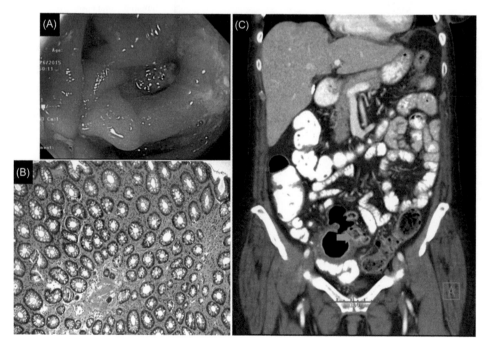

FIGURE 32.16 Diverticular colitis in the sigmoid colon. (A) Mucosal inflammation at the mouth of diverticulum but not within the diverticulum; (B) histology of the biopsy showed mucosal prolapse with intraepithelial fibrosis; and (C) mucosal hyperenhancement at the sigmoid colon, mimicking Crohn's colitis (*green arrow*).

types: Type A (crescentic fold)—reddish round lesions at the top of the colonic folds; Type B (mild-to-moderate UC-like)—loss of the submucosal vascular pattern, edema of the mucosa, hyperemia, and diffuse erosions; Type C (Crohn's colitis—like)—isolated aphthous ulcers; and Type D (severe UC-like)—loss of submucosa vascularity, intense hyperemia, diffuse ulcerations, and luminal narrowing (Figs. 32.14—32.17) [13].

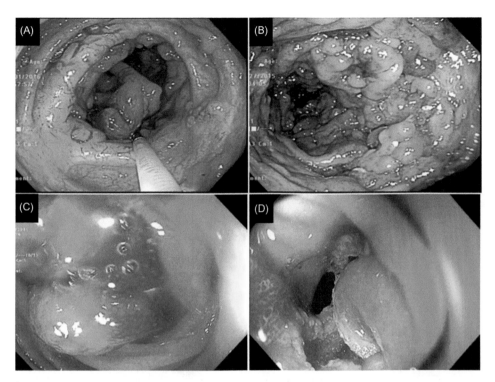

FIGURE 32.17 Chronic diverticular colitis with pseudopolyp formation in two patients, who presented with hematochezia and dyschezia. (A and B) Segmental colitis with multiple pseudopolyps and floppy mucosa at the sigmoid colon and (C and D) the patient 2 had a protruding polyp on an inflamed fold at the sigmoid colon with polypectomy.

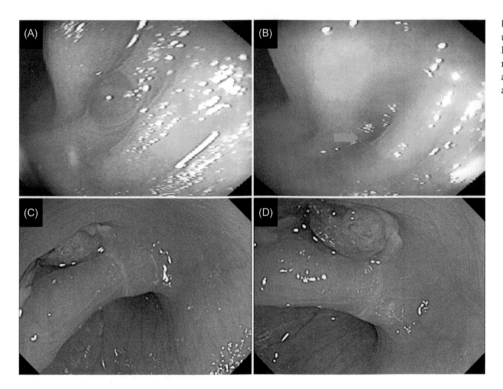

FIGURE 32.18 Inverted diverticulum and diverticulitis. (A and B) Inverted diverticulum at the sigmoid colon (*green arrow*) and (C and D) inflamed diverticulum with an inflammatory polyp.

Diverticular colitis or segmental colitis associated with diverticulosis should be differentiated from inverted diverticulum and diverticulitis. Diverticulitis, by definition, is inflammation of the diverticulum with normal surrounding mucosa (Fig. 32.18). Colonoscopy should be postponed during acute phase of diverticulitis.

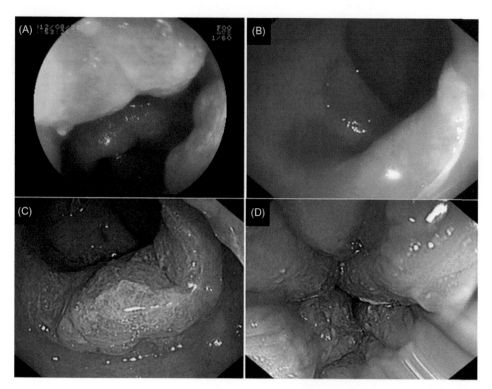

FIGURE 32.19 Progression of rectal prolapse in an 18-year-old male patient. The pattern of inflammation mimics that in ulcerative proctitis. (A) Distal rectum circumferential prolapse with edema and exudates; (B) colonoscopy 1 year later showed ulcers along the distal rectum folds; and (C and D) subsequent colonoscopy showed ulcerated and prolapsed distal rectal folds.

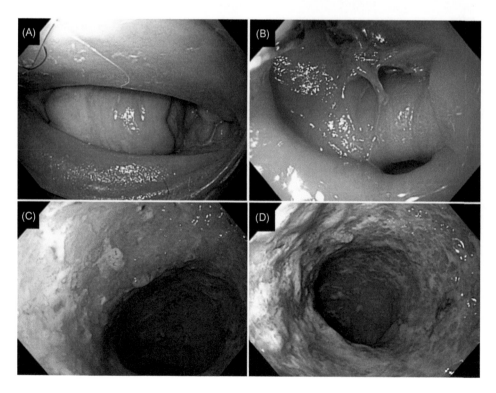

FIGURE 32.20 Rectal and uterine prolapse presenting with proctitis. (A) Uterine prolapse; (B) concurrent sigmoid diverticulosis; and (C and D) diffuse proctitis with edema, granularity, ulcers, and exudates resulting from rectal prolapse.

Rectal prolapse and solitary rectal ulcer syndrome

Rectal prolapse is more common in young adults. The prolapse can be mucosal or transmural, anterior, posterior, or circumferential. The disease process is mainly located at the distal rectum, which can resemble ulcerative proctitis and anorectal CD. The main clinical presentations include constipation, stool thinning, hematochezia, anorectal pain,

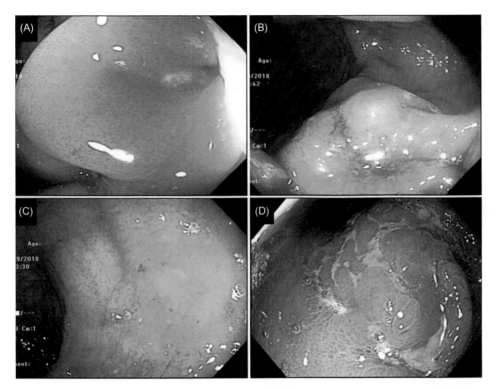

FIGURE 32.21 Anterior rectum and rectal cuff prolapse mimicking ulcerative proctitis, Crohn's proctitis, and rectal cuffitis (in ileal pouch—anal anastomosis). Mucosal inflammation is more prominent at the anterior wall of the rectum. (A–C) Intermittent anterior distal rectal wall prolapse (4–5 o'clock position) with erythema and granularity of the mucosa and (D) anterior cuff prolapse with ulcerated and nodular mucosa in a separate patient with ileal pouch—anal anastomosis.

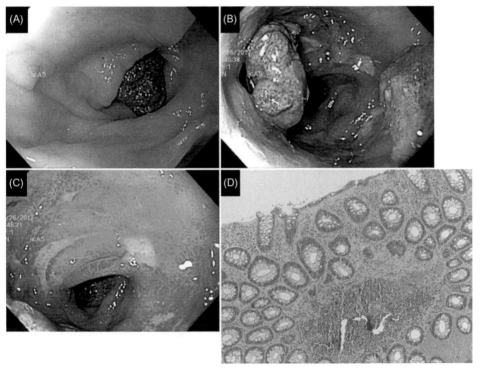

FIGURE 32.22 Circumferential rectal prolapse. Differential diagnoses include ulcerative proctitis and Crohn's disease. (A–C) Diverticuli with fecalith at the proximal rectum and circumferential ulcers and edema at the distal rectum and (D) histology of biopsy of the distal rectum showed classic prolapse feature, that is, intraepithelial fibrosis.

defecation, and rectal ulcer. Endoscopic features include edema, erosions, ulcers, and nodular or polypoid lesions (Figs. 32.19–32.22). Histologic features, such as elongation of epithelia and intraepithelial fibrosis, are valuable for the differential diagnosis between rectal prolapse and IBD. Anorectal manometry may show paradoxical contractions and failure for balloon expulsion.

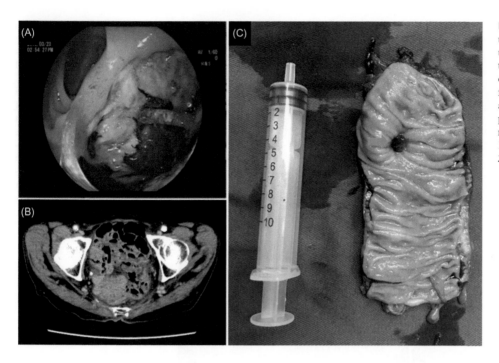

FIGURE 32.23 Solitary rectal ulcer syndrome. (A) A large deep ulcer at the anterior wall of the distal rectum; (B) the patient, later on, developed rectal perforation from the ulcer as shown on CT; and (C) partial proctectomy was performed and the ulcer was revealed (*green arrow*). *Courtesy Yi-Hong Fan, MD, PhD, of Zheijiang TCM Hospital.*

Solitary rectal ulcer syndrome typically presents with a discrete deep ulcer at the anterior wall of the rectum (Fig. 32.23). It should be differentiated from anorectal CD.

Summary and recommendations

IBD represents a disease spectrum phenotypically. On the other hands, there is a long list of IBD mimickers. Endoscopic and histologic evaluation is a key role in the diagnosis and differential diagnosis. However, some endoscopic features are subtle and easy to be missed.

References

[1] Canavese G, Villanacci V, Antonelli E, Cadei M, Sapino A, Rocca R, et al. Eosinophilia — associated basal plasmacytosis: an early and sensitive histologic feature of inflammatory bowel disease. APMIS 2017;125:179—83.

[2] Filippone RT, Sahakian L, Apostolopoulos V, Nurgali K. Eosinophils in inflammatory bowel disease. Inflamm Bowel Dis 2019;25:1140—51.

[3] Silverberg MS, Satsangi J, Ahmad T, Arnott ID, Bernstein CN, Brant SR, et al. Toward an integrated clinical, molecular and serological classification of inflammatory bowel disease: report of a Working Party of the 2005 Montreal World Congress of Gastroenterology. Can J Gastroenterol 2005;19(Suppl. A):5A—36A.

[4] Hwang J, Kim JS, Kim AY, Lim JS, Kim SH, Kim MJ, et al. Cryptogenic multifocal ulcerous stenosing enteritis: radiologic features and clinical behavior. World J Gastroenterol 2017;23:4615—23.

[5] Sun YN, Zuo XL, Wang X, Ma T, Li Z, Li YQ. Cryptogenic multifocal ulcerous stenosing enteritis diagnosed with single-balloon enteroscopy. Endoscopy 2019;51:E94—5.

[6] Käser SA, Willi N, Maurer CA. Mandatory resection of strangulation marks in small bowel obstruction? World J Surg 2014;38:11—15.

[7] Tomiguchi J, Miyamoto H, Ozono K, Gushima R, Shono T, Naoe H, et al. Preoperative diagnosis of intestinal endometriosis by magnifying colonoscopy and target biopsy. Case Rep Gastroenterol 2017;11:494—9.

[8] Hamilton CM, Arnason T. Ileitis associated with Meckel's diverticulum. Histopathology 2015;67:783—91.

[9] Kassim T, Abdussalam A, Jenkins E. Meckel's diverticulum in Crohn's disease revisited: a case of Meckel's diverticulitis in a patient with stricturing Crohn's disease. Cureus 2018;10:e2865.

[10] Ebert EC, Kierson M, Hagspiel KD. Gastrointestinal and hepatic manifestations of sarcoidosis. Am J Gastroenterol. 2008;103:3184—92.

[11] Takahashi H, Kaneta K, Honma M, Ishida-Yamamoto A, Ashida T, Kohgo Y, et al. Sarcoidosis during infliximab therapy for Crohn's disease. J Dermatol 2010;37:471—4.

[12] Parisinos CA, Lees CW, Wallace WA, Satsangi J. Sarcoidosis complicating treatment with natalizumab for Crohn's disease. Thorax 2011;66:1109.

[13] Tursi A, Elisei W, Brandimarte G, Giorgetti GM, Lecca PG, Di Cesare L, et al. The endoscopic spectrum of segmental colitis associated with diverticulosis. Colorectal Dis 2010;12:464—70.

Chapter 33

Inflammatory bowel disease—like conditions: gastrointestinal lymphoma and other neoplasms

Xinbo Ai[1], Yan Chen[2], Yu-Bei Gu[3], Xiaoying Wang[2] and Bo Shen[4]

[1]Department of Gastroenterology, Zhuhai People's Hospital, Jinan University, Zhuhai, P.R. China, [2]Department of Gastroenterology, The Second Affiliated Hospital, School of Medicine, Zhejiang University, Hangzhou, P.R. China, [3]Department of Gastroenterology, Ruijin Hospital of Shanghai Jiao Tong University, Shanghai, P.R. China, [4]Center for Inflammatory Bowel Diseases, Columbia University Irving Medical Center-New York Presbyterian Hospital, New York, NY, United States

Chapter Outline

Abbreviations

CCS	Cronkhite—Canada syndrome
CD	Crohn's disease
DLBL	diffuse large B-cell lymphoma
EATL	enteropathy-associated T-cell lymphoma
FAP	familial adenomatous polyposis
H & E	hematoxylin-Eosin
HNPCC	hereditary nonpolyposis colorectal cancer
IBD	inflammatory bowel disease
LPD	lymphoproliferative disorders
MALT	mucosa-associated lymphoid tissue
MCL	mantle cell lymphoma
NK	natural killer
NET	neuroendocrine tumor
NHL	non-Hodgkin's lymphoma
PGIL	primary gastrointestinal lymphoma
PJS	Peutz—Jeghers syndrome
UC	ulcerative colitis

Introduction

The incidence of both gastrointestinal (GI) malignancies and inflammatory bowel disease (IBD) is rising worldwide. Endoscopic features of GI malignancies, particularly various phenotypes of lymphoma, can mimic those in Crohn's disease (CD) and ulcerative colitis (UC). GI lymphoma is among the top list of differential diagnosis of IBD; and their

Atlas of Endoscopy Imaging in Inflammatory Bowel Disease. DOI: https://doi.org/10.1016/B978-0-12-814811-2.00033-5
© 2020 Elsevier Inc. All rights reserved.

management is different. Great overlaps in endoscopic features exist between GI lymphoma and IBD, such as erythema, ulcers nodularity, polypoid lesions, and strictures, which make the differential diagnosis difficult. Other GI neoplasms can also present with nodularity, ulcers, and polypoid lesions, similar to that seen in IBD.

Patients with IBD undergoing medical therapy, especially the use of purine analogs, carry a the risk for development of GI or extra-intestinal lymphomas. Fortunately, lymphomas in this setting are largely extranodal and GI lymphomas are rare. Furthermore, radiation therapy for GI cancer may result in radiation enteritis, colitis, or proctitis, which also share some of the endoscopic features with those in IBD. Radiation-associated GI injury is discussed in a separate chapter (Chapter 22: Inflammatory bowel disease–like conditions: radiation injury of the gut).

Gastrointestinal lymphoma

Lymphoma can affect any organs with lymph tissue. Lymphoma can originate from nodal or extranodal source. GI involvement by lymphoma can be primary or secondary. Primary gastrointestinal lymphoma (PGIL), a separate disease entity, is initially presented with a predominance of GI lesion with only lymph nodes affected in the immediate vicinity, normal, total, and differential white blood cell count, the absence of lymphomatous involvement of the liver and spleen, peripheral lymphadenopathy, or enlarged mediastinal lymph nodes. The majority of PGILs are non-Hodgkin's lymphoma (NHL). The stomach is the most common location affected by PGILs followed by the small and large intestines. In the United States, gastric lymphoma is the most common extranodal site of lymphoma, with a majority being either extranodal marginal zone B-cell lymphoma of mucosa-associated lymphoid tissue (MALT) (Figs. 33.1 and 33.2) or diffuse large B-cell lymphoma (DLBL). Among primary intestinal lymphoma, the ileum and cecum are the most involved sites [1,2]. It is estimated that 1% of patients with PGIL have diffuse colonic involvement, mimicking UC [1,2]. Endoscopy and histology play a critical role in the diagnosis and differential diagnosis of PGIL. IBD, including CD and UC, are characterized by the presence of infiltration of lymphocytes, plasma cells, and monocytes. Benign lymphocyte infiltration in IBD should be differentiated from malignant lymphoproliferative process in lymphoma. On the other hand, risk factors for lymphoma include infection of *Helicobacter pylori* and immunosuppression, autoimmune diseases, and celiac disease. IBD patients on the long-term use of purine analogs are considered as carrying a risk for NHL.

Common forms of PGIL are DLBL (Figs. 33.3–33.12), marginal zone lymphoma (Fig. 33.13), mantle cell lymphoma (MCL) (Fig. 33.14), follicular lymphoma (Fig. 33.15), Burkitt lymphoma, peripheral T-cell lymphoma (Figs. 33.16–33.18), enteropathy-associated T-cell lymphoma (EATL; also called intestinal T-cell lymphoma) (Figs. 33.19–33.21), and indolent T-cell lymphoproliferative disorders of the GI tract (Fig. 33.22). EATL is classified into two subtypes: Type I-EATL mainly associated with celiac disease and Type II, also named monomorphic epitheliotropic intestinal T-cell lymphoma (Figs. 33.23 and 33.24) [3]. Other forms are natural killer (NK)/T-cell lymphoma (Figs. 33.25–33.28).

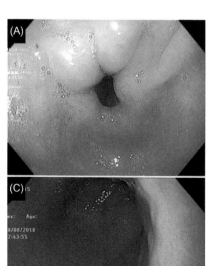
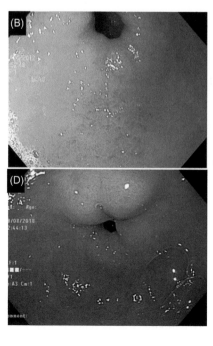

FIGURE 33.1 *Helicobacter pylori*–associated gastric-related gastric MALT lymphoma in a 50-year-old asymptomatic patient. (A and B) Mild granularity of the antrum, with biopsy-confirmed lymphoma and (C and D) anti–*H. pylori* therapy resulted in complete remission of lymphoma. *MALT*, Mucosa-associated lymphoid tissue.

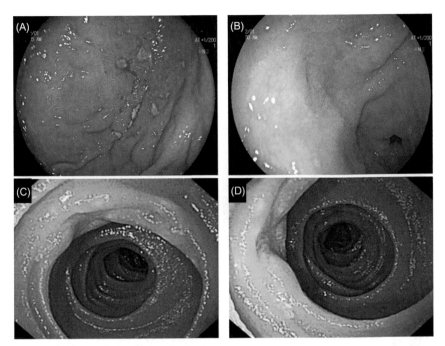

FIGURE 33.2 Primary MALT lymphoma of the stomach and jejunum. (A and B) Nodularity and ulcers in the stomach and (C and D) small, discrete ulcers in the proximal jejunum. *MALT*, Mucosa-associated lymphoid tissue. *Photo courtesy (A and B) Dr. Xin-Ying Wang, of Zhujiang Hospital of Southern Medical University; (C and D): Dr. Xiao-Ying Wang of the 2nd Hospital of Zhejiang University.*

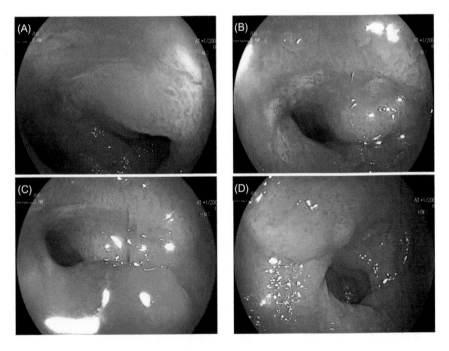

FIGURE 33.3 DLBL of the terminal ileum. (A–D) Edema, erythema, stricture, and polypoid mass in the terminal ileum. *DLBL*, Diffuse large B-cell lymphoma. *Photos courtesy Dr. Qing Guo of the 3rd Xiangya Hospital of Central Southern University.*

Gastroscopic features of gastric lymphoma include mucosal erythema (Fig. 33.1), mass or polypoid lesion with or without ulceration, gastric ulcers, nodularity (Fig. 33.2), strictures, and thickened, cerebroid gastric folds. Endoscopic features of PGIL include aphthous ulcers, longitudinal ulcers, irregular ulcers, nodularity, pseudopolyp-like lesions, luminal mass lesion, and strictures. In between types of PGIL, some may have predominant endoscopic features. For example, MCL often presents with nodularity or polyposis (2 mm to more than 2 cm in size) of the gut with or without normal intervening mucosa. This pattern is named lymphomatous polyposis (Fig. 33.14A−C). Follicular lymphoma commonly presents with small white nodules in the duodenum or jejunum (Fig. 33.15A and B) [4]. Patients with EATL or peripheral T-cell lymphoma often demonstrate large circumferential ulcers without overt tumor masses, or edema,

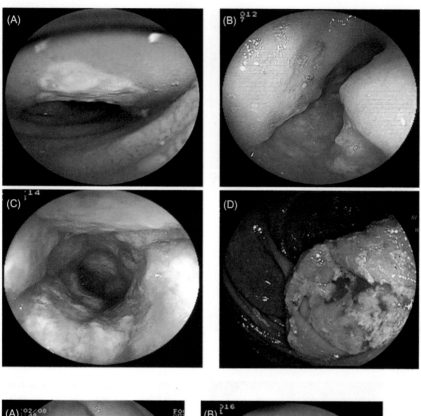

FIGURE 33.4 DLBL in the small and large bowel. (A–C) Small bowel lymphoma and (B and D) cecal lymphoma. *DLBL*, Diffuse large B-cell lymphoma. *Photo courtesy Dr. Yubei Yu and Dr. Jie Zhong of Ruijin Hospital of Shanghai Jiao Tong University.*

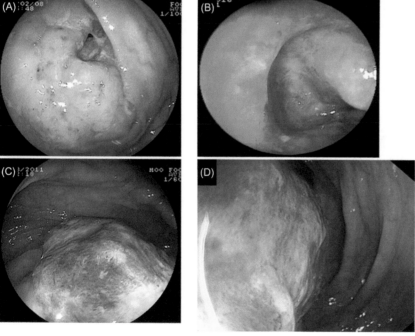

FIGURE 33.5 DLBL of the colon. (A–D) Various types of mass-like lesions with luminal narrowing. *DLBL*, Diffuse large B-cell lymphoma. *Photo courtesy Dr. Yubei Yu and Dr. Jie Zhong of Ruijin Hospital of Shanghai Jiao Tong University.*

granularity, diffuse superficial or deep ulcers, or mass-like lesions (Figs. 33.18, 33.20, 33.21, 33.23, and 33.24). Those with primary intestinal follicular lymphoma often demonstrate multiple small (1–5 mm) polypoid lesions in the second portion of the duodenum, the pattern of which is similar to that in some duodenum CDs (Fig. 4.6). Patients with large bowel lymphoma may present with diffuse mucosal nodularity, colitis-like changes with induration and ulceration, or a mass with or without ulceration (Fig. 33.16). NK/T-cell lymphoma often presents with various forms of ulcers ranging from ulcerative to ulceroinfiltrative appearance.

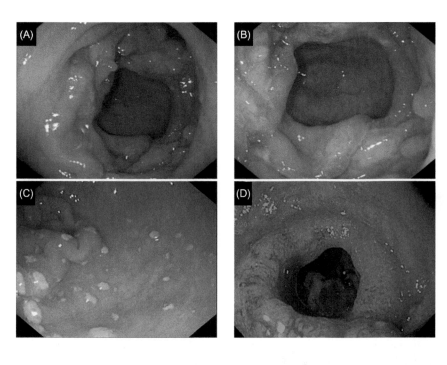

FIGURE 33.6 DLBL in the small and large bowel. (A and B) The ileocecal valve with irregular ulcers, nodularity, and deformity; (C) rectum involvement in a separate patient with diffuse small ulcers covered by muco-purulent exudates; and (D) circumferential lesion in distal ileum in a Crohn's disease patient with ileostomy. *DLBL*, Diffuse large B-cell lymphoma. *Photo courtesy (A and B) Dr. Danfeng Lan and Dr. Yinglei Miao of the 1ˢᵗ Hospital of Kunming Medical University; (C) Dr. Yu-Bei Gu and Dr. Jie Zhong of Ruijin Hospital of Shanghai Jiao Tong Hospital.*

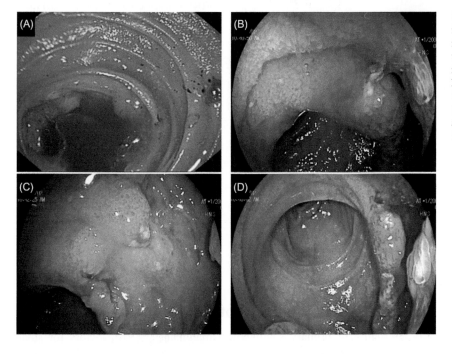

FIGURE 33.7 DLBL in the jejunum, distal ileum, and colon. (A) Circumferential nodularity in the jejunum and (B—D) large superficial ulcers with adjacent nodular mucosa in the ileum and colon. *DLBL*, Diffuse large B-cell lymphoma. *Photo courtesy (A): Dr. Xiao-Ying Wang of the 2nd Hospital of Zhejiang University and (B—D): Dr. Xing Guo, MD of the 3ʳᵈ Xiangya Hospital of Central South University.*

The earlier-mentioned endoscopic features are not specific and can be seen in patients with CD. While patients with CD more frequently have multiple GI site involvement and longitudinal ulcers, those with PGIL carry a higher chance to have irregular ulcers, and mass-like lesions were found to be more frequent in PGIL (Table 33.1). A distinguishing model consisting of endoscopic features along with clinical and imaging presentations has been proposed [5].

Esophagogastroduodenoscopy, push enteroscopy, ileocolonoscopy, and device-assisted enteroscopy may be used to assess the lesions. An endoscopic ultrasound helps determine the depth of invasion and the presence of extraluminal nodes. Multiple biopsies should be obtained from the abnormal-appearing lesions. Routine hematoxylin—eosin (H&E) stains and immunohistochemistry are performed.

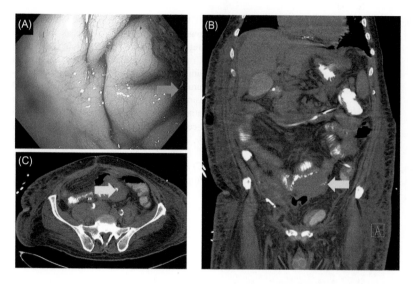

FIGURE 33.8 DLBL with a sigmoid colon stricture. (A) Extrinsic compression from the tumor, leading to a nonulcerated stricture and preventing passage of the scope (*green arrow*); notice that the colon mucosa was normal. (B and C) Large soft-tissue mass encasing the sigmoid colon, causing the stricture (*yellow arrows*). *DLBL*, Diffuse large B-cell lymphoma.

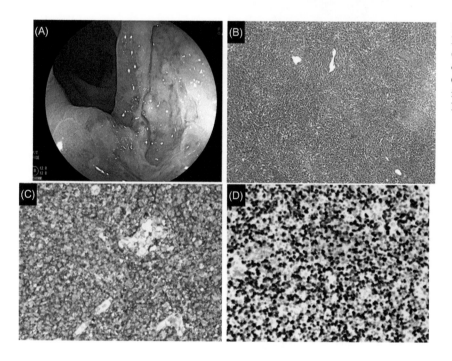

FIGURE 33.9 DLBL in the colon. (A) Discrete, large, clean-based ulcer in the colon; (B) homogeneous infiltration of lymphocytes on mucosal biopsy with H&E stain; (C) CD45RO immunohistochemistry; and (D) Ki-67 immunohistochemistry. *DLBL*, Diffuse large B-cell lymphoma; *H&E*, hematoxylin–eosin.

Polyposis, adenocarcinoma, and neuroendocrine tumor of the gastrointestinal tract

Adenocarcinomas comprise approximately 25%–40% of primary small bowel malignancies. Risk factors for small bowel adenocarcinoma include the presence of colorectal cancer [especially for hereditary nonpolyposis colorectal cancer (HNPCC), familial adenomatous polyposis (FAP), Cronkhite–Canada syndrome (CCS)], MUTYH-associated polyposis, and CD.

Dysplastic polyps in HNPCC, FAP (Figs. 33.29–33.32), CCS (Fig. 33.33), and MUTYH-associated polyposis can be found in the stomach, small intestine, or large intestine. Their endoscopic features include nodularity, sessile, semi-pedunculated, or pedunculated polyps with or without mucopurulent exudates on top. Some polyps may have erosions. The mucosa between the polyps can be edematous or erythematous. Nodularity and polypoid lesions of the stomach, small intestine, and large intestine can also be found in patients with IBD. Biopsy and polypectomy help make a differential diagnosis.

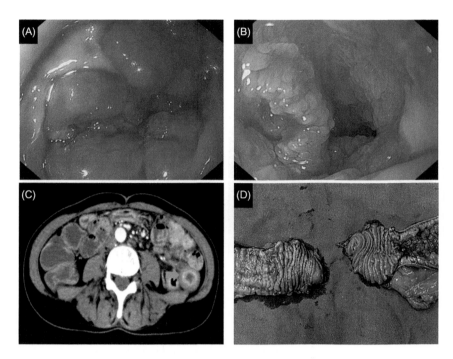

FIGURE 33.10 DLBL of the jejunum and ileum. (A and B) Ulcers, nodularity, edema, and inflammatory strictures at the jejunum; (C) thickened small bowel wall with mucosal hyperenhancement (*green arrow*); and (D) surgical specimens of the jejunum (left) and ileum (right) tumors. *DLBL*, Diffuse large B-cell lymphoma. *Photos courtesy Dr. Yi-Hong Fan of Zhejiang TCM University Hospital.*

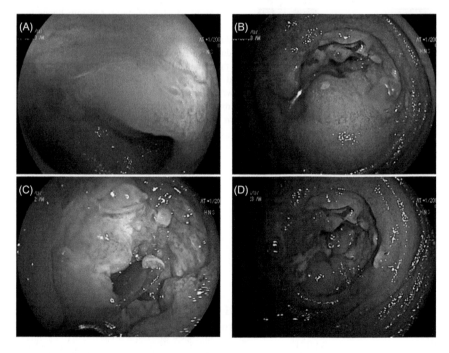

FIGURE 33.11 DLBL of the small bowel. (A) A tight stricture with surrounding superficial ulcers and (B—D) circumferentially ulcerated mass lesion blocking the lumen. *DLBL*, Diffuse large B-cell lymphoma. *Photos courtesy Dr. Qing Guo of the 3rd Xiangya Hospital of Central Southern University.*

Similarly, endoscopic presentation of hamartomatous polyposis syndromes can also mimic that in IBD. Hamartomatous polyposis syndromes include Peutz—Jeghers syndrome (PJS), Cowden syndrome, Bannayan—Riley—Ruvalcaba syndrome, and juvenile polyposis syndrome. These syndromes also have extraintestinal manifestations such as pigmented mucocutaneous macules in PJS. The polyps in these syndromes can be seen from the stomach to the rectum. Polyps vary in size and shapes with larger polyps may be multilobulated. The polyps may have a white exudate cap. Hamartomatous polyp syndromes occasionally develop adenomatous changes.

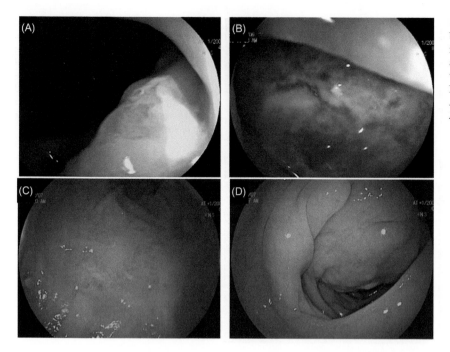

FIGURE 33.12 DLBL of the small and large bowel. (A and B) Lesions in the terminal ileum with ulcers and mucopurulent exudates and (C and D) lesions in the ascending colon with patchy erythema and mild mucosal nodularity. *DLBL,* Diffuse large B-cell lymphoma. *Photos courtesy Dr. Qing Guo of the 3rd Xiangya Hospital of Central South University.*

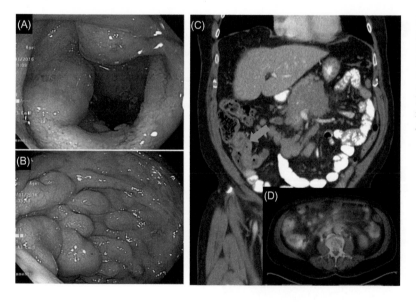

FIGURE 33.13 Gastrointestinal marginal zone lymphoma of small and large bowels. (A) Submucosal nodules in the terminal ileum; (B) long submucosal nodule in the ileocecal valve; and (C) thickened wall of the terminal ileum and proximal ascending colon; and (D) PET/CT hypermetabolic focus at the right lower quadrant at ileocecal region and mesenteric and retroperitoneal lymph nodes. *PET/CT, positron emission tomography/computed tomography*

A Brunner's gland adenoma is a rare small bowel neoplasm. It is caused by hyperplasia of the exocrine glands within the proximal duodenal mucosa. The pattern should be differentiated from duodenum CD or, occasionally, UC with duodenum involvement (Fig. 33.34).

GI adenocarcinomas can also present with ulcers, mucosal nodularity, strictures, as well as mass-like lesions (Figs. 33.35 and 33.36). Occasionally, peritoneal carcinomatosis may cause intestinal ulcers (Fig. 33.37). Primary GI adenocarcinoma and malignant stricture at the anastomosis should be distinguished from IBD-associated benign or malignant strictures, IBD cancer-associated anastomotic benign and malignant stricture. Therefore in IBD- or non-IBD patients, tissue biopsy should be taken during the index endoscopy and maybe subsequent endoscopy [6].

Carcinoid tumors are well-differentiated (grade 1 or 2) GI neuroendocrine tumors (NETs). Carcinoid syndrome describes those tumors' ability to secrete serotonin and other bioactive products. The GI tract is the most common location for NETs, with the majority of them at the small bowel, particularly the distal ileum within 60 cm of the ileocecal valve and appendix. It is believed that small bowel NETs originate from intraepithelial endocrine cells and appendiceal

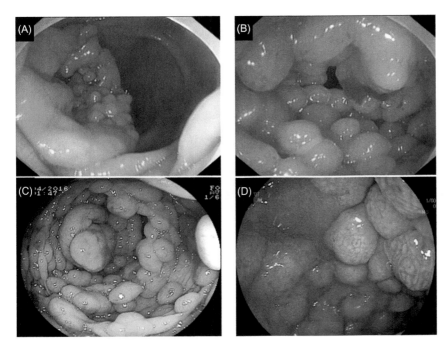

FIGURE 33.14 Mantle cell lymphoma and B-lymphoblastic lymphoma. (A and B) Mantle cell lymphoma with common presentation of multiple polypoid lesions (i.e., lymphoid of the colon in a 67-year-old patient); (C) Mantle cell lymphoma in a separate patient; and (D) lymphoblastic lymphoma in the duodenum. *Photos courtesy (A and B) Wen Tang, MD of the 2nd Hospital of Suzhou University; (C and D) Dr. Yu-Bei Gu and Dr. Jie Zhong of Ruijin Hospital of Shanghai Jiao Tong University.*

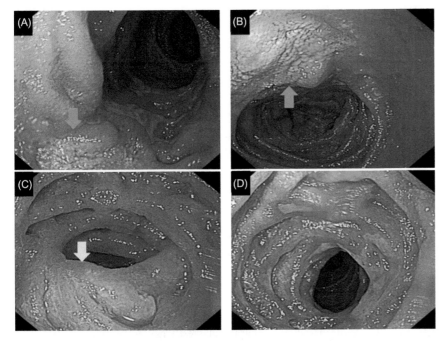

FIGURE 33.15 Follicular lymphoma of the jejunum. (A and B) Characteristic white patches of the mucosa (*green arrows*); (C) edematous mucosa and mild narrowing of the bowel lumen (*yellow arrow*); and (D) diffuse edema and circumferential superficial ulcers. *Photos courtesy Dr. Yan Chen and Dr. Xiao-Ying Wang of the 2nd Hospital of Zhejiang University.*

NETs arise from subepithelial endocrine cells. On endoscopy, small bowel and appendiceal NET may present as intraluminal mass, nodules, and intussusception (Fig. 33.38). NET can also affect the foregut and hind guts. Conventional endoscopy and endoscopy ultrasound play an important role in the diagnosis, differential diagnosis, and tumor staging.

Radiation can exert direct injury to the GI mucosa. Also, radiation can cause strictures at the GI anastomoses (Fig. 33.39). Differential diagnosis of radiation-associated stricture and IBD or surgical ischemia-associated stricture in patients with radiation exposure can be difficult. The details are discussed in Chapter 22, Inflammatory bowel disease—like conditions: radiation injury of the gut.

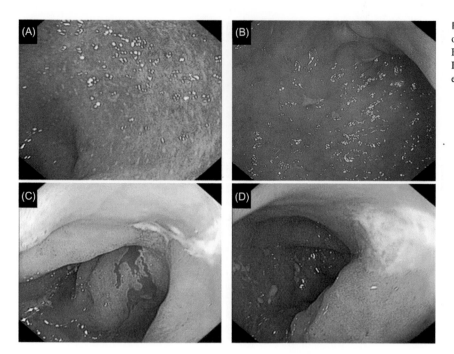

FIGURE 33.16 Peripheral T-cell lymphoma of the colon in a 37-year-old male patient with Epstein–Barr virus in tissue biopsy. (A–D) Diffuse mucosal inflammation with exudates, edema, exudates, and small irregular ulcers.

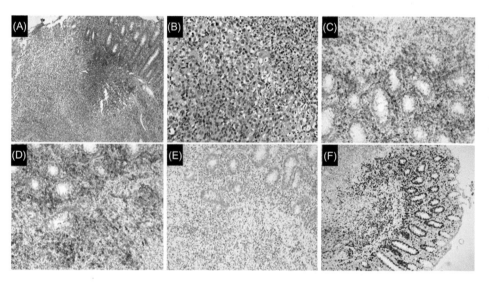

FIGURE 33.17 Peripheral T-cell lymphoma with H&E and immunohistochemistry in a 37-year-old patient. (A and B) Infiltration of medium-sized lymphocytes in the mucosa and submucosa with vascular invasion; (C) CD2$^+$; (D) CD56$^+$; (E) EBER$^+$; and (F) Ki-67$^+$ in 80% of cells. *H&E*, Hematoxylin–eosin.

Inflammatory bowel disease and gastrointestinal cancer

The association between IBD and lymphoproliferative disorders (LPD) has several folds. Small bowel and colorectum adenocarcinoma in CD and colorectal cancer in UC are discussed in a separate chapter (Chapter 12). Current literature does not support an increased risk for lymphoma in patients with IBD in reference to the general population. However, the risk for lymphoma in patients with CD is complicated by the concurrent use of immunosuppressive mediations, especially purine analogs. The majority of NHL in patients with IBD involves lymph nodes and lymphoid tissues with hepatosplenic T-cell lymphoma as a unique example. It appears that PGIL in patients with IBD is rare, with scant literature [7,8]. Otherwise, distinguishing CD from lymphoma and CD from concurrent CD and lymphoma can be challenging. The presence of LPD has been considered as a contraindication for the use of purine analogs and anti–tumor necrosis factor agents.

Inflammatory polyps in the colon and, much less commonly, in the small bowel are associated with CD and UC. These polyps include inflammatory pseudopolyps from mucosal ulceration and regeneration in response to

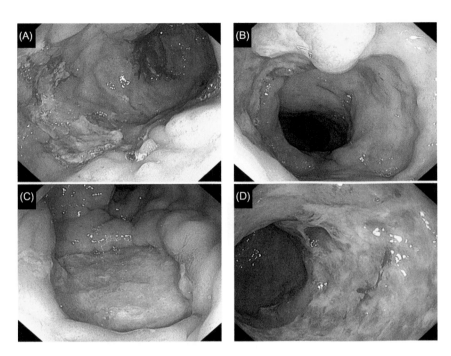

FIGURE 33.18 Peripheral T-cell lymphoma of the colon. Large, deep ulceroinfiltrative lesions with nodularity of the adjacent mucosa. (A) The ascending colon; (B) transverse colon; (C) descending colon; and (D) rectum. *Photos courtesy Dr. Min Chen of Zhongnan Hospital of Wuhan University.*

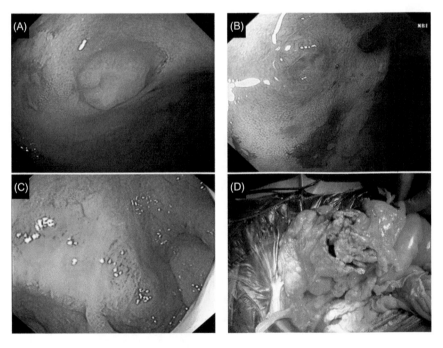

FIGURE 33.19 EATL of the colon, misdiagnosed as ulcerative colitis. (A and B) Discrete ulcer with adjacent edematous mucosa on white-light and narrow-band imaging; (C) erythema, edema, and nodularity of the colon mucosa; and (D) colonic perforation with transmural mass, resulting in right hemicolectomy. *EATL*, Enteropathy-associated T-cell lymphoma. *Photo Courtesy Dr. Xiao-Ying Wang and Dr. Yan Chen of the 2nd Hospital of Zhejiang University.*

inflammation and prolapse-type inflammatory polyps from traction, twisting, and distortion of the mucosa. On histology, inflammatory polyps consist of stromal and epithelial components and inflammatory cells. The prolapse-form inflammatory polyps have histologic features of localized ischemia and intramucosal fibrosis. The endoscopic features of IBD-associated inflammatory polyps and their clinical implications are discussed in Chapter 9 and Chapter 32, Inflammatory bowel disease—like conditions: miscellaneous. The distinction should be made between IBD-associated inflammatory polyps, cap polyposis syndrome, and intestinal polyposis.

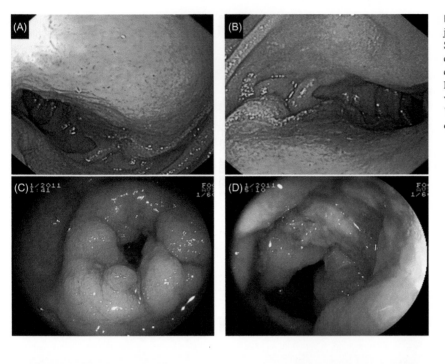

FIGURE 33.20 T-cell lymphoma of the jejunum and transverse colon. (A and B) Semicircumferential mass lesions with superficial ulceration in the jejunum and (C and D) circumferential and semicircumferential mass lesion with superficial ulceration in the transverse colon. *Photos courtesy Dr. Xiao-Ying Wang and Dr. Yan Chen of the 2nd Hospital of Zhejiang University.*

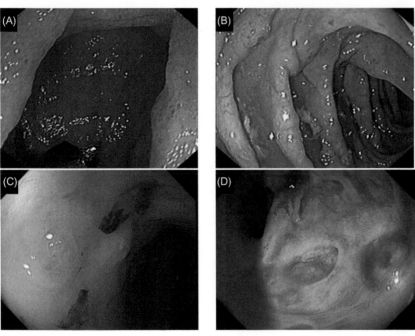

FIGURE 33.21 EATL with superimposed Epstein–Barr virus infection. This 27-year-old male patient was misdiagnosed as having ulcerative colitis with multiple colonoscopies and developed colonic perforation. Diagnosis was based on ileocolonic resection. (A) Erythema of the terminal ileum; (B) patchy erythema of the transverse colon; and (C and D) multiple ulcers in the colon with spontaneous bleeding. *EATL*, Enteropathy-associated T-cell lymphoma. *Photos courtesy Dr. Xiao-Ying Wang and Dr. Yan Chen of the 2nd Hospital of Zhejiang University.*

Summary and recommendations

GI neoplasms, particularly PGIL, can have similar endoscopic features to those seen in CD and UC. These disease entities also share clinical and imaging features. PGIL, with predominance of GI involvement and without obvious peripheral lymphocytosis, lymphadenopathy, and hepatosplenomegaly, makes differential diagnosis mainly based on endoscopic and histologic evaluations. The infiltration of chronic inflammatory cells in the gut tissues is present in all GI lymphomas and IBDs; histologic differential diagnosis, even with immunohistochemistry, can be challenging. Despite great overlaps, a careful endoscopic examination may still be able to demonstrate characteristics of GI lymphoma and IBD (Table 33.1). Other GI neoplasias may also share some endoscopic features with that seen in IBD.

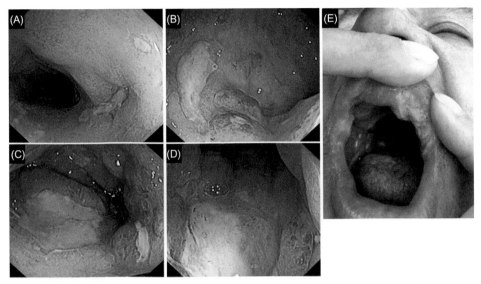

FIGURE 33.22   Indolent T-cell lymphoproliferative disorder of the left colon. (A—D) Multiple small and medium-sized, irregular, superficial ulcers with edematous, nodular, and erythematous surrounding mucosa and (E) large deep malignant ulcer of the upper lip. *Photos courtesy Dr. Yan Chen and Dr. Xiao-Ying Wang of the 2nd Hospital of Zhejiang University.*

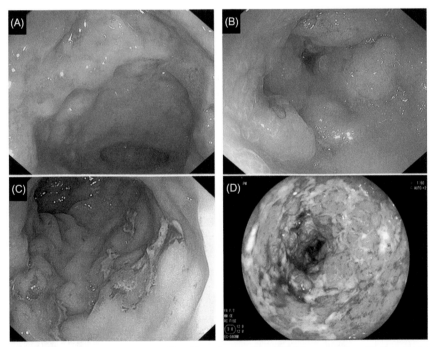

FIGURE 33.23   Various subtypes of T-cell lymphoma of small and large bowel. (A and B) Monomorphic epitheliotropic intestinal T-cell lymphoma, that is, type II EATL of the terminal ileum with ulcers, edema, nodularity, and strictures; (C) peripheral T-cell lymphoma of the colon with ulcers and friable mucosa; and (D) T-cell lymphoma of the colon with diffuse ulcers and nodularity. *EATL,* Enteropathy-associated T-cell lymphoma. *Photo courtesy (A—C) Dr. Yi-Hong Fan of Zhejiang TCM University Hospital; (C and D) Dr. Yubei Yu and Dr. Jie Zhong of Ruijin Hospital of Shanghai Jiao Tong University.*

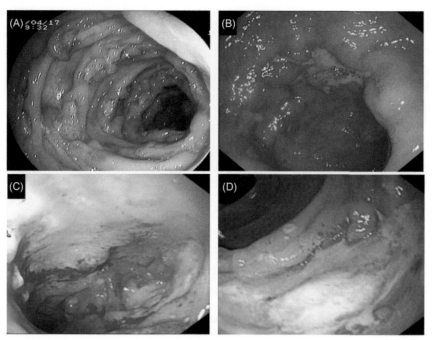

FIGURE 33.24 T-cell lymphoma and type II EATL of the colon. (A and B) T-cell lymphoma of the colon and (C and D) type II EATL (monomorphic, epitheliotropic, intestinal T-cell lymphoma) in the colon. *EATL,* Enteropathy-associated T-cell lymphoma. *Photo courtesy (A and B) Dr. Yubei Yu and Dr. Jie Zhong of Ruijin Hospital of Shanghai Jiao Tong University; (C and D) Dr. Min Chen of Zhongnan Hospital of Wuhan University.*

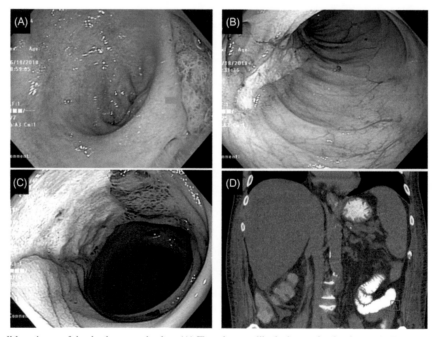

FIGURE 33.25 NK-cell lymphoma of the duodenum and colon. (A) Flat adenoma-like lesion at the duodenum bulb (*green arrow*); (B and C) longitudinal clean-based ulcers in the ascending colon; and (D) CT showing splenomegaly.

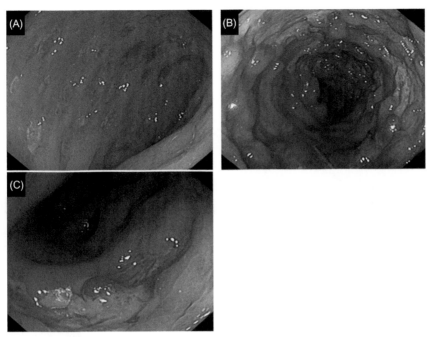

FIGURE 33.26 NK/T-cell lymphoma of the colon. (A—C) Diffuse nodular mucosa with ulcers. *Photos courtesy Dr. Xiao-Ying Wang of the 2nd Hospital of Zhejiang University.*

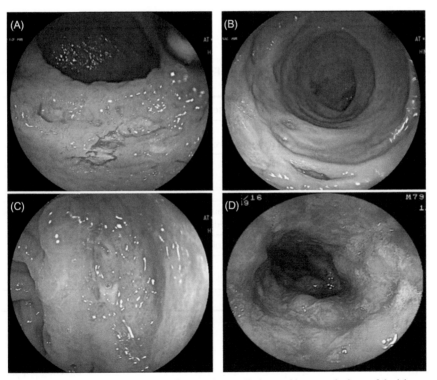

FIGURE 33.27 NK-cell lymphoma of the small intestine. (A—C) Irregular, small ulcers with mucosal edema of the jejunum and (D) diffuse superficial ulcers with mucopurulent exudates and nodular mucosa in the ileum. *Photo courtesy (A—C) Dr. Xing-Yin Wang of Zhujiang Hospital of Southern Medical University and (D) Drs. Yu-Bei Gu and Jie Zhong of Ruijin Hospital of Shanghai Jiao Tong University.*

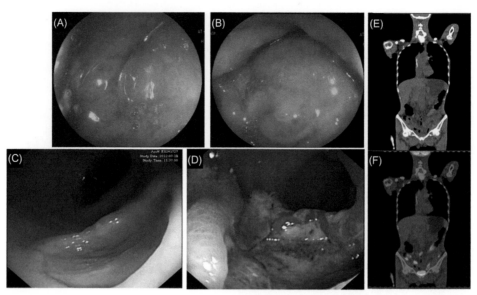

FIGURE 33.28   Extranodal NK/T-cell lymphoma in the small and large bowel. (A and B) Erosions in the terminal ileum in patient 1; (C and D) large ulcer craters in the sigmoid colon and transverse colon in patient 2. (E and F) CT/PET shows thickened wall of distal small bowel and left colon with increased FDG signals along with splenomegaly. *FDG, fluorodeoxyglucose Photos courtesy Dr. Juan Du of the 1st Hospital of Zhejiang University.*

TABLE 33.1 Comparison of endoscopic features between gastrointestinal lymphoma and inflammatory bowel disease.

	Gastrointestinal lymphoma	Crohn's disease	Ulcerative colitis
Disease extent	Stomach to rectum; rectal involvement is rare	Mouth to anus; segmental distribution with skip lesions; rectal sparing	Colon and rectum; rectal involvement is a norm; diffuse distribution
Disease depth	Can be transmural; intraluminal lesions are common	Transmural	Mucosa to superficial submucosa
Ulcers	Characteristic irregular ulcers with raised borders	Characteristic longitudinal ulcers along the mesentery edge	Diffuse small ulcers
Nodularity	Common	Characteristic cobblestoning appearance	Uncommon
Polypoid lesions	Common, multiple	Uncommon	Common
Mass	Common	Rare	Rare

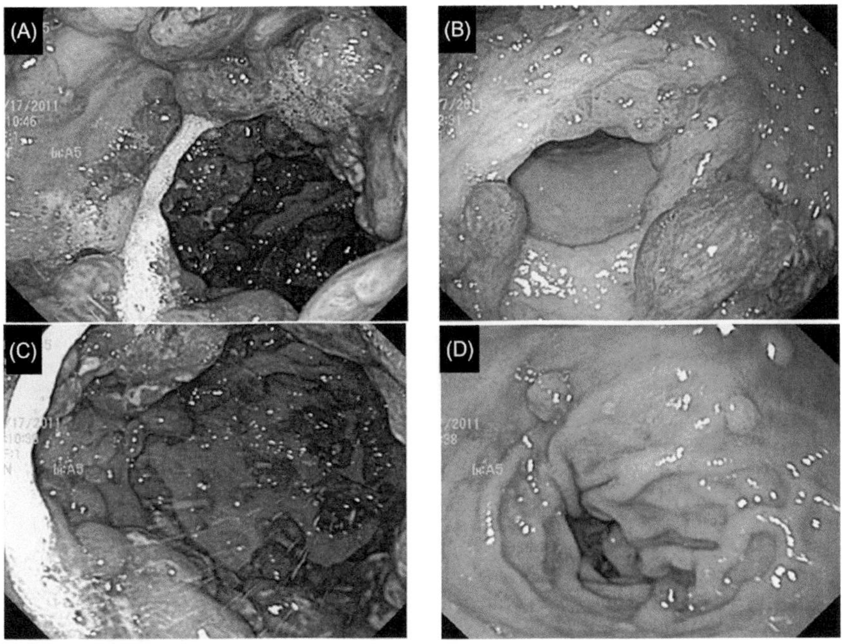

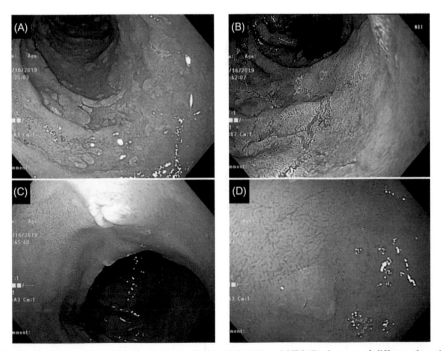

FIGURE 33.29 FAP in the stomach and proximal small bowel. (A—C) Diffuse inflammatory fundic-gland polyps with rich vasculature in the entire stomach and (D) small sessile polyps in the 3rd part of the duodenum. *FAP*, Familial adenomatous polyposis.

FIGURE 33.30 FAP in the duodenum. (A and B) Comparison of white-light (A) and NBI (B) features of diffuse polyposis flat lesions covering mucosa of the 2nd and 3rd parts of the duodenum; (C) small flat lesions around the papilla at the 2nd part of the duodenum; and (D) a flat polypoid lesion at the duodenum bulb, which should be distinguished from Brunner's gland hyperplasia. *FAP*, Familial adenomatous polyposis; *NBI*, narrow-band imaging.

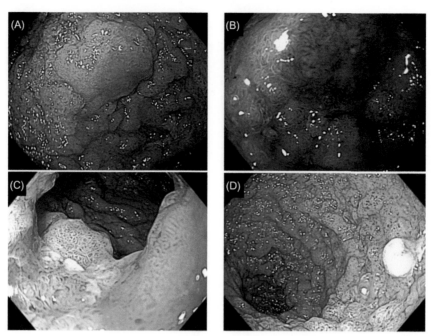

FIGURE 33.31   FAP in the colon. (A and B) Comparison of white-light (A) and NBI (B) features of diffuse polyposis with small and large nodules in the ascending colon. The pattern is highlighted with NBI; and (C and D) diffuse polyposis with mucosal nodularity carpeting the right colon. *FAP*, Familial adenomatous polyposis; *NBI*, narrow-band imaging.

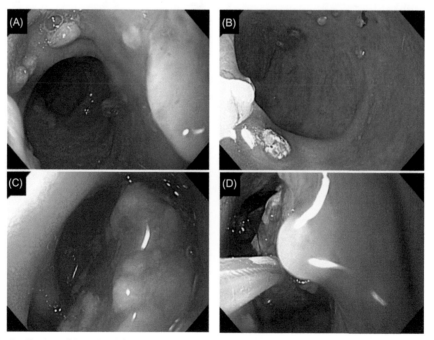

FIGURE 33.32   FAP in the distal small bowel and ileal pouch. (A and B) Small sessile polyps (A) treated with argon plasma coagulation; (C and D) large polyps with mucopurulent exudate on the top, treated with snare polypectomy (A and D). This should be differentiated from chronic pouchitis with inflammatory polyps in patients with underlying ulcerative patient and restorative proctocolectomy. *FAP*, Familial adenomatous polyposis.

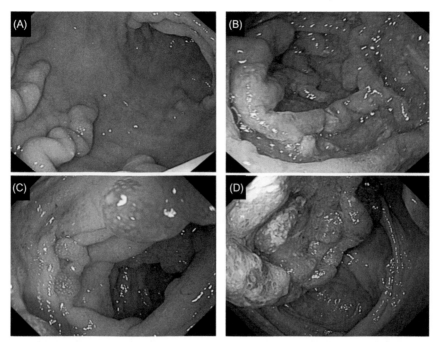

FIGURE 33.33   Patterns of mucosa in Cronkhite—Canada syndrome on gastroscope and colonoscope. (A) Prominent folds and mucosal nodularity of the stomach; (B and C) diffuse polyposis and prominent folds in the colon; and (D) a malignant appearing mass at the cecum. *Courtesy Dr. Yan Chen and Dr. Xiao-Ying Wang of the 2nd Hospital of Zhejiang University.*

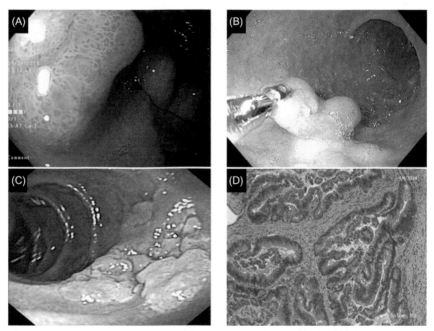

FIGURE 33.34   Differential diagnosis of duodenum polypoid lesions. (A) Duodenitis with inflammatory polyps at the 2nd part of the duodenum in a patient with ulcerative colitis; (B) a polypoid lesion at the 2nd part of the duodenum with histologic diagnosis of gastric ectopia in a patient with Crohn's disease and ileostomy; and (C and D) duodenum flat adenoma on endoscopy and histology.

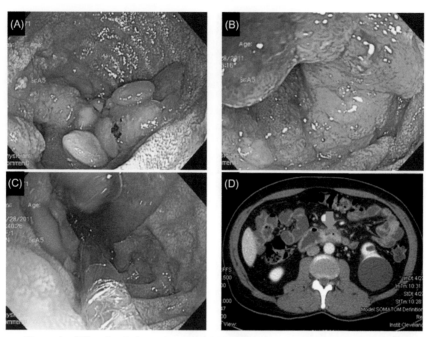

FIGURE 33.35 Jejunum with poorly differentiated signet ring cell adenocarcinoma in a 64-year-old male patient presented with nausea and vomiting. (A–C) Inflammatory stricture with edematous and nodular mucosa in the proximal jejunum; and (D) the stricture on CT enterography (*green arrow*).

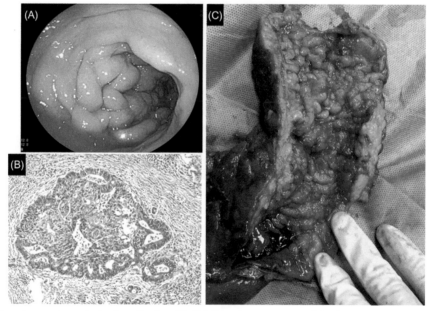

FIGURE 33.36 Invasive adenocarcinoma in the sigmoid colon. (A) Cobblestoning appearance of the sigmoid colon, mimicking that seen in Crohn's disease; (B) invasive adenocarcinoma in surgical pathology; (C) partial colectomy specimen showed the multiple nodules.

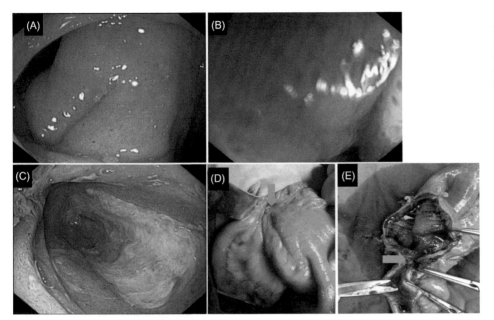

FIGURE 33.37 Small and large bowel disease resulting from peritoneal carcinomatosis from gastric cancer. (A and B) Colonic mucosal congestion and erythema with luminal bulging from compression of extraluminal malignant process and adhesions. (C) Benign linear ulcer in the jejunum; (D and E) intraoperative examination showed an adhesion band at the jejunum (*green arrows*). *Photo courtesy Dr. Yan Chen & D. Xiao-Ying Wang of the 2nd Hospital of Zhejiang University.*

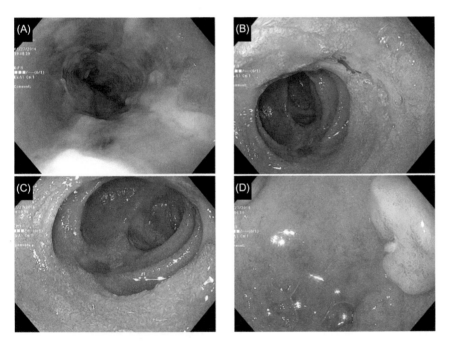

FIGURE 33.38 Well-differentiated neuroendocrine tumor (WHO—Grade 1). (A) Erosive esophagitis from partial gastric-outlet obstruction; (B—C) duodenal mucosa with ulceration; and (D) nodules in the bulb with histology showing well-differentiated neuroendocrine tumor.

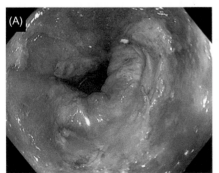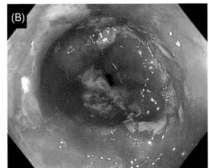

FIGURE 33.39 Radiation stricture colorectal anastomosis for rectal cancer. (A and B) Mixed inflammatory and fibrotic strictures with edema and friability. Contributing factors include radiation and surgical ischemia.

References

[1] Koch P, del Valle F, Berdel WE, Willich NA, Reers B, Hiddemann W, et al. Primary gastrointestinal non-Hodgkin's lymphoma: I. Anatomic and histologic distribution, clinical features, and survival data of 371 patients registered in the German Multicenter Study GIT NHL 01/92. J Clin Oncol 2001;19:3861−73.

[2] Papaxoinis G, Papageorgiou S, Rontogianni D, Kaloutsi V, Fountzilas G, Pavlidis N, et al. Primary gastrointestinal non-Hodgkin's lymphoma: a clinicopathologic study of 128 cases in Greece. A Hellenic Cooperative Oncology Group study (HeCOG). Leuk Lymphoma 2006;47:2140−6.

[3] Teras LR, DeSantis CE, Cerhan JR, Morton LM, Jemal A, Flowers CR. 2016 US lymphoid malignancy statistics by World Health Organization subtypes. CA Cancer J Clin 2016;66:443−59.

[4] Iwamuro M, Kondo E, Takata K, Yoshino T, Okada H. Diagnosis of follicular lymphoma of the gastrointestinal tract: a better initial diagnostic workup. World J Gastroenterol 2016;22:1674−83.

[5] Zhang TY, Lin Y, Fan R, Hu SR, Cheng MM, Zhang MC, et al. Potential model for differential diagnosis between Crohn's disease and primary intestinal lymphoma. World J Gastroenterol 2016;22:9411−18.

[6] Shen B, Kochhar G, Hull TL. Bridging medical and surgical treatment of inflammatory bowel disease: the role of interventional IBD. Am J Gastroenterol 2019;114:539−40.

[7] Castrellon A, Feldman PA, Suarez M, Spector S, Chua L, Byrnes J. Crohn's disease complicated by primary gastrointestinal Hodgkin's lymphoma presenting with small bowel perforation. J Gastrointest Liver Dis 2009;18:359−61.

[8] Hall Jr CH, Shamma M. Primary intestinal lymphoma complicating Crohn's disease. J Clin Gastroenterol 2003;36:332−6.

Chapter 34

Histology correlation with common endoscopic abnormalities

Jesse Kresak[1], Bo Shen[2] and Xiuli Liu[1]

[1]Department of Pathology, Immunology, and Laboratory Medicine, University of Florida School of Medicine, Gainesville, FL, United States, [2]Center for Inflammatory Bowel Diseases, Columbia University Irving Medical Center-New York Presbyterian Hospital, New York, NY, United States

Abbreviations

AFB	acid-fast bacillus
AIC	acute infectious colitis
AIE	autoimmune enteropathy
CD	Crohn's disease
DALM	dysplasia-associated lesion or mass
EMR	endoscopic mucosal resection
GI	gastrointestinal
GVHD	graft-versus-host disease
IBD	inflammatory bowel disease
ITB	intestinal tuberculosis
NSAIDs	Nonsteroidal antiinflammatory drugs
SRUS	solitary rectal ulcer syndrome
TNF	tumor necrosis factor
UC	ulcerative colitis

Introduction

Reaching an accurate diagnosis of inflammatory bowel disease (IBD) requires the synthesis of clinical, radiographic, endoscopic, and histological information. Endoscopy allows for the direct visualization of mucosa, assessment of extent of disease, and mucosal sampling for histologic examination. The biopsy samples obtained from endoscopy enable the

Atlas of Endoscopy Imaging in Inflammatory Bowel Disease. DOI: https://doi.org/10.1016/B978-0-12-814811-2.00034-7
© 2020 Elsevier Inc. All rights reserved.

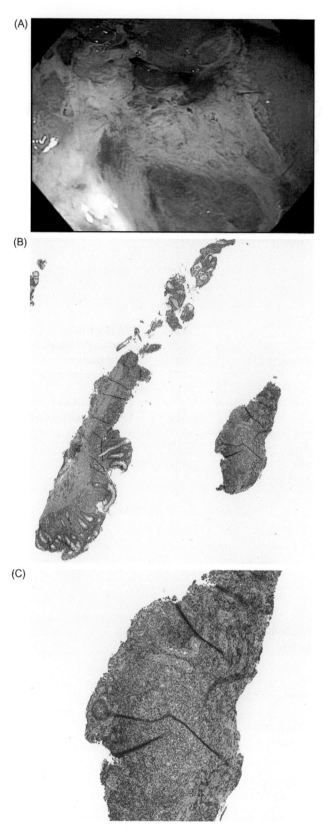

FIGURE 34.1 (A) UC can present with large deep ulcers. (B and C) Typical histology findings in UC include chronicity that is characterized by crypt distortion, basal lymphoplasmacytosis (B), and metaplasia and activity which is assessed by the neutrophil-associated epithelium injury from cryptitis, crypt abscess, and erosion to ulceration (C). *UC*, Ulcerative colitis.

pathologist to provide a histopathologic assessment of the chronicity and grade of colitis, neoplastic complications of IBD, superimposed pathologies, and an alternative diagnosis for the clinically and endoscopically presumed "IBD." This chapter aims to provide histologic examples of several common endoscopic abnormalities in patients undergoing ileocolonoscopy or colonoscopy.

Acute phase of inflammatory bowel disease

IBD is characterized with relapsing disease course, with intervening flare up and remission. Endoscopic and histologic examinations of the affected intestine help assess the severity of disease and its response to treatment [1].

Acute phase of ulcerative colitis

Endoscopic features of ulcerative colitis (UC) are as follows: Active inflammation in UC typically shows diffuse erythematous, friable, and granular mucosa with loss of the normal vascular pattern. The lesions begin at the anorectal junction and spread proximally in a homogenous fashion. The mucosa may show erosions, ulcerations, and spontaneous bleeding (Fig. 34.1A). Biopsy with histology is used to demonstrate the distribution and focality of disease throughout the colon. A minimum protocol should include biopsies from the terminal ileum, right colon, transverse colon, descending colon, sigmoid colon, and rectum.

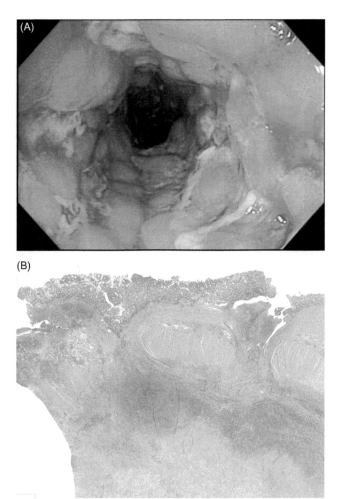

FIGURE 34.2   (A) CD presents with longitudinal ulcers and cobblestoning lesions.(B) Typical histologic findings are best demonstrated in resected segment affected by CD. The histology includes chronic active enterocolitis, fissuring ulceration, mural fibrosis and abscess, and serosal fibrous adhesions. *CD*, Crohn's disease.

Histologic features of UC are as follows: Corresponding to its endoscopic features, histology of UC is characterized by diffuse and homogenous chronic active colitis with near universal involvement of rectal mucosa. The inflammatory infiltrate of UC will involve predominantly the mucosa. Activity refers to the presence of neutrophilic inflammation within crypts and surface epithelium. The degree of activity is subjectively measured as mild, moderate, or severe depending on the density of neutrophilic infiltrate, presence of crypt abscesses, cryptitis, erosion, and ulceration (Fig. 34.1B and C). Chronicity is determined by architectural distortion such as crypt branching, shortening, and/or loss (Fig. 34.1B), basal lymphoplasmacytosis (Fig. 34.1B), and metaplasia, including the pyloric gland metaplasia in the ileum and colon and Paneth cell metaplasia in the left colon and rectum.

Acute phase of Crohn's disease

Endoscopic features of Crohn's disease (CD) are as follows: CD colitis is usually segmental and asymmetrical and spares the rectum. The typical endoscopic findings of CD include longitudinal ulcerations (predominantly along the mesentery side), cobblestoning lesions, and segmental distribution of disease (Fig. 34.2A). The patients with CD may present with aphthous ulcers in early stage.

Histologic features of CD are as follows: CD is also characterized by chronic active enteritis, colitis, and enterocolitis. CD involvement of the colon is typically noncontiguous, corresponding to skip lesions appreciated on endoscopy, with normal mucosa appearing between foci and segments of affected mucosa. In contrast to UC the inflammatory infiltrate of CD colitis is classically transmural, breaching the *muscularis mucosae* and extending into the submucosal space. Granulomas may also be seen, yet they are only found in a minority of biopsies and are not a requirement for diagnosis. The histology of CD is best appreciated on resected specimens (Fig. 34.2B). In this resected colon segment, there is chronic active colitis, mucosal ulceration, fissuring ulceration, mural

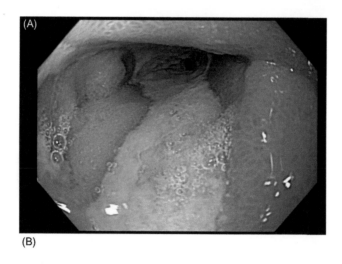

(A)

(B)

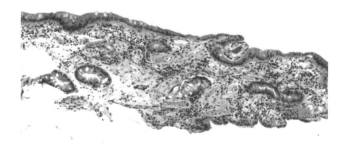

FIGURE 34.3 (A) CD presents with longitudinal ulcers and stenosis of small bowel lumen. (B) Biopsy from the stenotic mucosa shows mucosal fibrosis and dilated vessels. *CD*, Crohn's disease.

abscesses, mural fibrosis, and mural abscesses. The activity and chronicity of CD colitis on biopsies are assessed in the same manner as for UC.

Chronic phase or long-term complication of inflammatory bowel disease

Endoscopic features of mucosal scars, stenosis, and fistula in CD are as follows: prolonged and chronic CD may lead to mucosal scars and stenosis (Fig. 34.3A). In some cases, fistula may develop in a variety of sites. Endoscopic examination may identify fistula opening.

Histologic features of mucosal scars, stenosis, and fistula in CD are as follows: Chronic inflammation in CD can lead to mucosal scars and stenosis, with or without fistula. The histology of mucosal scar in CD may be subtle and non-specific and may only manifest as fibrosis (Fig. 34.3B) on biopsy and hypertrophy of *muscularis mucosae*. The histology of stenotic areas in CD is nonspecific. Most often the lamina propria contains fibrosis, and there is hypertrophy of the muscular layers and fibrosis of the intestinal wall (Fig. 34.2B). Fissures and fistulas are lined by histiocytic inflammation, occasionally with a giant-cell reaction, and are often accompanied by granulation tissue and fibrinopurulent debris (pus, clinically).

Endoscopic features of mucosal scars in ulcerative colitis

Prolonged mucosal inflammation or favorable response to effective medical therapy in UC may lead to extensive mucosal scars (Fig. 34.4A). This is particularly true in patients who are successfully treated with anti−tumor

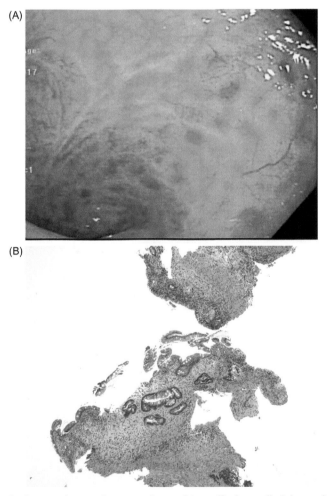

(A)

(B)

FIGURE 34.4 (A) Endoscopic examination reveals extensive mucosal scars due to effective medical therapy in prolonged UC. (B) Histologically, these biopsies demonstrate fibrosis of lamina propria, hypertrophy of *muscularis mucosae*, and marked architectural distortion and atrophy. *UC*, Ulcerative colitis.

necrosis factor-α (TNFα) or antiintegrin biological agents. While superficial fibrosis can cause mucosal scars, fibrosis in the deeper layers may lead to colonic stenosis in UC.

Histologic features of mucosal scars in UC are as follows: Prolonged course of UC may lead to mucosal scar. On histology, scars often show fibrosis of the lamina propria, crypt architectural distortion, and, sometimes, crypt dropout. There is often decreased inflammation, particularly a lack of active inflammation. The epithelium may show reactive changes such as prominent cytoplasmic eosinophilia and nucleoli. There may be hypertrophy, splitting, and fibrosis of *muscularis mucosae* (Fig. 34.4B). In UC colectomy specimens, submucosal fibrosis is often noted in areas affected by inflammation as well [2].

Short gut, obstruction, and lymphoma

With the transmural inflammation, mural fibrosis, and serosal fibrous adhesions, some patients with CD may develop obstruction. Patients with short gut syndrome may develop compensatory dilation of lymphatic ducts. Small bowel obstruction and short gut syndrome along with rare lymphoma in CD may lead to lymphangiectasia of small bowel. Lymphangiectasia manifests with diffuse tiny white spots of small bowel mucosa (Fig. 34.5A). Histologically, lymphangiectasia is characterized by dilated lymph lacteals in the mucosa (Fig. 34.5B). In addition, lymphoma can complicate IBD in patients receiving immunosuppressants, particularly, anti-TNFα agents. Mucosa taken from the area affected by lymphoma may show evidence of lymphoma or lymphangiectasia.

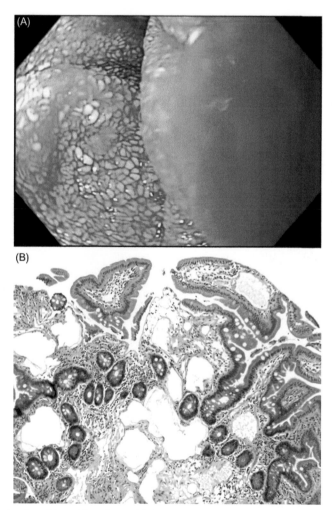

FIGURE 34.5   (A) Lymphangiectasia of small bowel in patients with ileostomy for CD presents with white granular mucosa. (B) Biopsy from these areas demonstrates dilated lacteals in the mucosa, features of lymphangiectasia. *CD*, Crohn's disease.

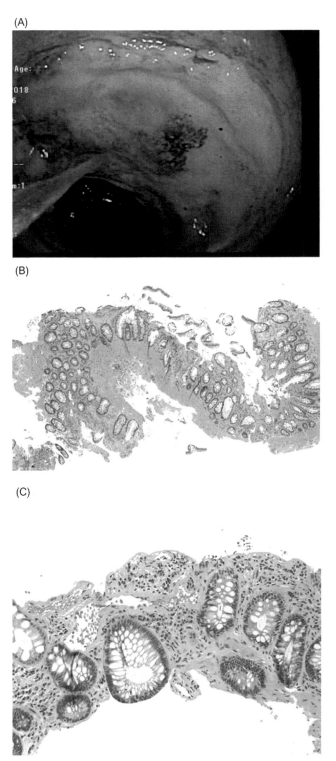

FIGURE 34.6 (A) Radiation of colitis with a large superficial ulcer with adjacent edematous mucosa. (B and C) Biopsy from radiation colitis typically shows sloughing of surface epithelium, lamina propria fibrosis, and ectatic vessels in the lamina propria.

Differential diagnosis of mucosal scar

Chronic ischemia: Chronic ischemia of the small bowel and colon represents one of the most challenging differential diagnoses of IBD. Histologically, chronic ischemic of the small bowel and colon may show architectural distortion, multifocal ulcers, and metaplasia in the mucosa, thus mimicking IBD [3]. Features that favor ischemic colitis over UC, including acute ischemic injury pattern such as withered epithelium and hyalinization of the lamina propria, ulceration in a "watershed" region, relative paucity of intraepithelial neutrophils, and chronic inflammation. In addition, clinical information, including old age and a myriad of other etiologies predisposing to ischemia, would help ascertain the diagnosis.

Radiation injury: Radiation colitis can have varying degrees of severity and distribution pattern. In the acute setting, endoscopic findings include edema, erythema, friability and erosions, and spontaneous bleeding. However in the chronic setting, radiation colitis may show loss of vascularity and small or large ulcers (Fig. 34.6A), strictures, and fistulas, thus mimicking IBD. Histologically, radiation-induced epithelial changes include hypereosinophilic cytoplasm and nuclear atypia, not to be misinterpreted as dysplasia. The lamina propria has an ischemic-type pattern with hyalinization and fibrosis (Fig. 34.6B) and crypt architectural distortion. There may be ectatic vessels in the lamina propria (Fig. 34.6C) or transversing the *muscularis mucosae*. The mural vessels will often be thickened with intimal hyperplasia and hyalinization of vascular walls.

Anastomosis: Various types of surgery are performed in patients with refractory CD or UC. The surgery involves suturing, stapling, and anastomosis. Anastomotic ulcers and strictures are common. On endoscopy, it has been difficult to differentiate disease-associated ulcers or strictures from surgical ischemia—associated anastomotic ulcers or strictures. Mucosal biopsy from a known intestinal anastomosis may show a variety of changes, including nonspecific, acute, and chronic inflammation, cryptitis, crypt abscess, erosion/ulceration, architectural distortion, metaplasia, and

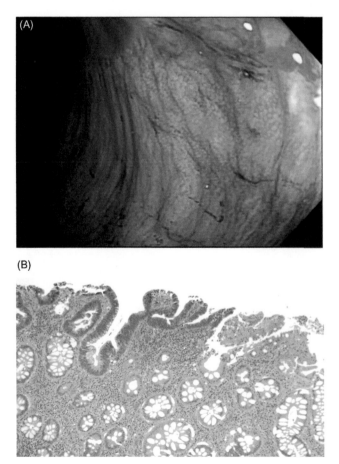

FIGURE 34.7 (A) Flat dysplasia in UC, which is appreciable on chromoendoscopy. (B) Biopsy from this region demonstrates villous epithelium lined by enlarged and hyperchromatic pencil-shaped nuclei, features of low-grade dysplasia. *UC*, Ulcerative colitis.

fibrosis, thus mimicking IBD. Knowing the clinical and surgical history would help ascertain that these changes are associated with anastomosis.

Neoplastic complications in inflammatory bowel disease patients

The risk of neoplasia arising in IBD increases with the duration of disease, the extent of disease, as well as the severity of inflammation. Histologic examination is critical for surveillance of dysplasia because not all IBD-associated dysplasia can be seen endoscopically, so-called flat dysplasia.

Flat dysplasia

Flat neoplasia, in contrast to polypoid or raised lesions, can only be identified based on its abnormal pit pattern that is visualized better on chromoendoscopy or narrow-band imaging than conventional white-light endoscopy. (Fig. 34.7A). Histologically, flat dysplasia can be classified as either low grade or high grade. Low-grade dysplasia is characterized by hyperchromatic and elongated nuclei while maintaining relative nuclear polarity (Fig. 34.7B). High-grade dysplasia shows more pleomorphic and rounded nuclei with loss of polarity and architectural complexity such as crowded crypts or cribriform patterns [4].

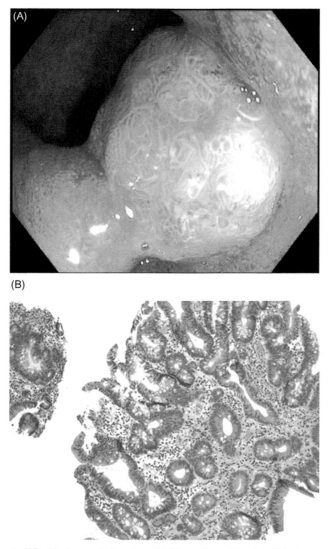

FIGURE 34.8 (A) A polypoid lesion in UC with abnormal pit pattern. (B) Biopsy from this polypoid lesion reveals low- and focally high-grade dysplasia with overall features resembling adenoma. *UC*, Ulcerative colitis.

Neoplasia visible on endoscopy

When a lesion or mass is seen endoscopically in patients with IBD and biopsies confirming the presence of dysplasia, it may then be referred to histologically as dysplasia-associated lesion/mass (DALM) or visible dysplasia [5,6]. The term "DALM" has become obsolete in recent years and currently, a term "polypoid neoplasia" is favored. It is further divided into two categories, endoscopically resectable or endoscopically unresectable. Endoscopically resectable polypoid dysplasia may resemble an adenoma in a sessile or pedunculated configuration (Fig. 34.8A). For endoscopically visible and resectable polypoid lesion, the endoscopist tends to remove it by polypectomy or endoscopic mucosal resection (EMR). Histologically, this type of lesion may resemble adenoma (Fig. 34.8B). The distinction between a polypoid dysplasia and a sporadic adenoma is difficult histologically if no background mucosa is examined and biopsied. When the dysplasia is in the context of colonic mucosa affected by ongoing chronic active colitis, it is most likely dysplasia arising due to IBD. In contrast, if the background mucosa is near normal, it is likely a sporadic adenoma. If no background mucosa is available for histologic examination, a diagnosis of "polypoid dysplasia" is usually rendered by pathologists and correlation with endoscopic impression is necessary. In either case the dysplasia will be classified as low or high grade using the same histologic criteria for flat dysplasia. For an endoscopically visible but unresectable lesion, the endoscopist, most likely, biopsies the lesion and the surrounding mucosa and submits them in separate container. These biopsies are evaluated by the pathologist using the same histologic criteria for flat dysplasia.

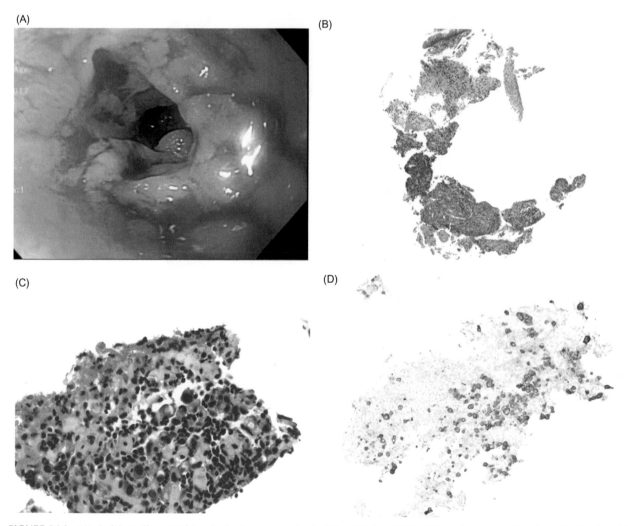

FIGURE 34.9 (A) A tight malignant stricture in the transverse colon in UC, which is visible but deemed unresectable endoscopically. (B and C) Biopsy from this endoscopically visible but unresectable lesion reveals infiltrating neoplastic cells with enlarged and hyperchromatic nuclei and intracellular mucin features diagnostic of adenocarcinoma. (D) These neoplastic cells are positive for cytokeratin AE1/AE3 by immunohistochemistry. *UC*, Ulcerative colitis.

If dysplasia is identified and confirmed by another pathologist with expertise in gastrointestinal (GI) pathology in the endoscopically unresectable lesion alone, the patient may be referred to an IBD center and subjected to chromoendoscopy or colonoscopy of high resolution to reassess its resectability and if it is deemed resectable, an EMR may be attempted. If the biopsy from the unresectable lesion, especially flat, depressed or strictured lesions (Fig. 34.9A) show invasive adenocarcinoma (Fig. 34.9B−D), colectomy is indicated [7].

Serrated lesions

Some of the small polypoid dysplasia lesions (Fig. 34.10A) in patients with chronic colitis may show serrated features. Those lesions are often endoscopically visible and resectable. Histologically, they can resemble typical sessile serrated polyp or traditional serrated adenoma (Fig. 34.10B).

Inflammatory pseudopolyps with or without white cap are often present in patients with long-term IBD. The polyps are often peduncular in shape with rich vasculature. The adjacent mucosa may or may not have active inflammation on endoscopic examination (Fig. 34.11A). Histologically, they are composed of granulation tissue, chronic active inflammation, and reactive epithelial changes (Fig. 34.11B).

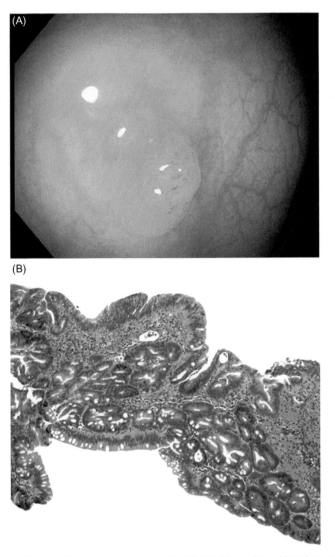

FIGURE 34.10   (A) A sessile polypoid lesion with abnormal pit pattern in UC. (B) Biopsy from this polyp reveals serrated epithelium with low-grade dysplasia, resembling traditional serrated adenoma. *UC*, Ulcerative colitis.

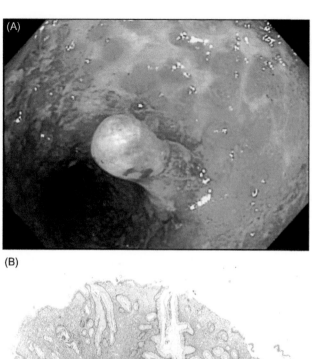

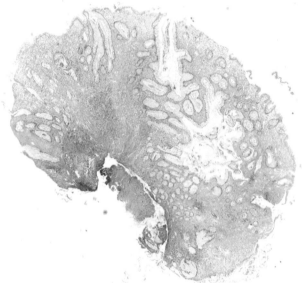

FIGURE 34.11   (A) A pedunculated inflammatory polyp in UC with adjacent mucosal inflammation. (B) This polyp shows inflammation, fibromuscular proliferation, reactive/hyperplastic changes, and focal erosion, which are features of inflammatory polyp. *UC*, Ulcerative colitis.

Endoscopic and histologic mimics of inflammatory bowel disease

Many etiologies can have similar endoscopic and histologic findings as IBD. When the clinical history, endoscopic findings, radiology (if applicable), and histology are evaluated in conjunction, a definitive diagnosis can often be reached. However, occasionally a differential diagnosis must be rendered until additional laboratory tests (such as bacterial and viral studies) or clinical queries (such as medication review) are preformed to rule out confounding scenarios.

Entities mimicking ulcerative colitis

Acute self-limiting infectious colitis (AIC): Endoscopically, AIC may appear as diffuse mild-to-severe inflammation with edema, erythema, exudates, friability, erosions, ulcers, and spontaneous bleeding (Fig. 34.12A). The disease course and prognosis of AIC are different from IBD. AIC is histologically characterized by neutrophilic inflammation of the epithelium and lamina propria in a background of preserved architecture (Fig. 34.12B). The changes may be focal and patchy or diffuse. Unlike IBD, AIC does not show architectural abnormalities, basal lymphoplasmacytosis, or metaplasia.

   C. difficile–associated colitis: The diagnosis of *C. difficile* colitis is often suspected clinically based on the endoscopic appearance and in the setting of a patient with recent antibiotic use. Endoscopically, *C. difficile* colitis shows

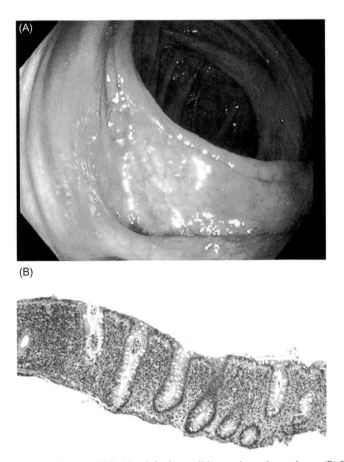

FIGURE 34.12 (A) Endoscopic examination of acute self-limiting infectious colitis reveals patchy erythema. (B) Biopsy reveals neutrophilic inflammation in the lamina propria and cryptitis and few crypt abscesses. There is no evidence of architectural distortion, basal lymphoplasmacytosis, or metaplasia.

yellow-to-white exudates, or pseudomembranes, which may bleed when manipulated. Diagnostic challenges arise because *C. difficile* colitis can be superimposed in patients with IBD. *C. difficile* colitis superimposed on IBD may or may not produce the classic pseudomembranes. The histology of *C. difficile* colitis is a more severe form of infectious colitis where the crypts are diffusely involved by extensive epithelial necrosis and often dilated by neutrophils and mucin. Pseudomembranes are often seen, composed of neutrophils, mucin, and fibrin and extending through the mucosa into the lumen in a mushroom or volcano shape [4]. Fig. 34.13 illustrates a case of *C. difficile* colitis superimposed on UC with pseudomembranes. Endoscopically, there are typical pseudomembranes (Fig. 34.13A). However, endoscopic features of *C. difficile* colitis range from completely normal colon mucosa to diffuse edema and erythema to severe inflammation with pseudomembranes. Histologically, there are dilated crypts with mucin, neutrophils, and pseudomembranes composed of mucin, fibrin, neutrophils, and cellular debris (Fig. 34.13B and C). The background or adjacent mucosa demonstrates crypt distortion, evidence of UC (Fig. 34.13B).

Ischemic colitis: Ischemic colitis can show erosions and ulcers of the mucosal surface or strictures mimicking an IBD (Fig. 34.14A). Ischemic colitis is often patchy with sharp demarcations between affected and nonaffected mucosa. The crypts are "withered" or small and hyperchromatic with hyalinization of the intervening lamina propria and denudation or reactive changes of the surface epithelium (Fig. 34.14B and C). Occasionally, fibrin thrombi can be identified within mucosal or submucosal vessels. In the cases of severe hypoperfusion or shock, the entire colon may be involved mimicking an ulcerative pancolitis.

Diverticular disease—associated colitis: On endoscopy the most common locations of diverticular disease—associated colitis are the sigmoid colon and rectosigmoid junction. Inflammation is present around, but not in, the diverticuli. There might be concurrent mucosal prolapse. Diverticular disease—associated colitis is histologically indiscernible from IBD, UC, or CD. The diagnosis is made based on the endoscopic correlation so that only the segment of

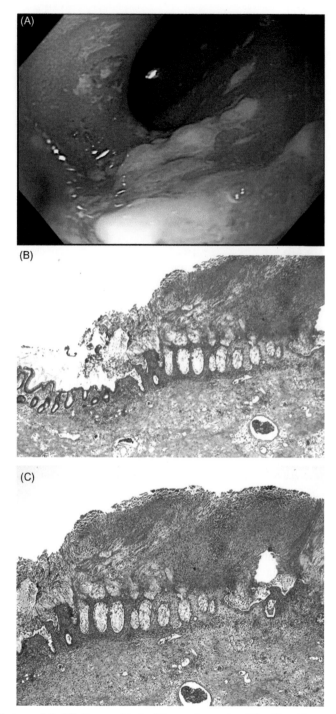

FIGURE 34.13 (A) Endoscopic features of *C. difficile*–associated colitis superimposed on UC with patchy pseudomembranes on ulcers. (B and C) Biopsy from *C. difficile*–associated colitis superimposed on UC with background marked crypt distortion (B) and pseudomembranes composed of mucin, fibrin, neutrophils, and cellular debris (B and C). *UC*, Ulcerative colitis.

colon with diverticulosis is involved by the colitis. Diverticular disease–associated colitis should nearly never involve the rectum.

Immunodeficiency-associated colitis: Immunodeficiency can affect the GI tract, particularly the small bowel, which manifests as classic autoimmune enteropathy (AIE), celiac disease–like, a mixed AIE/celiac disease–like, and acute graft-versus-host disease (GVHD)–like pattern. The endoscopic features of immunodeficiency-associated colitis range from edema, erythema, and erosions to ulcers. Luminal strictures are rare. Immunodeficiency can also affect the colon

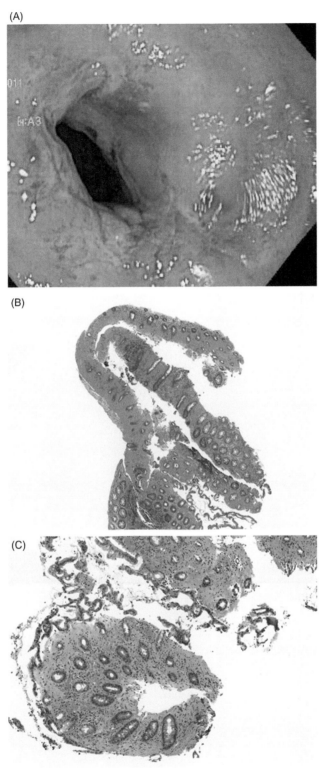

FIGURE 34.14 (A) Endoscopic examination reveals ulcerated stricture in ischemic colitis. (B) Biopsy from ischemic colitis is characterized by atrophic glands and hyalinization of the lamina propria. (C) In chronic ischemia, there may be fibrosis of the lamina propria and submucosa, which is corresponding to endoscopically evident stricture.

leading to a variable histology, including mixed active and chronic inflammation, chronic inflammation alone, intrae-pithelial lymphocytosis, and increased apoptosis resembling acute GVHD [8,9].

Solitary rectal ulcer syndrome (SRUS): SRUS is also known as mucosal prolapse syndrome. SRUS may not always present as a single ulcer as the name suggests. SRUS may have ulcerations (single or multiple), erosions, or polypoid or prolapsed areas, with the most common location at the anterior wall of the distal rectum. Given its rectal involvement, it may be mistaken for UC. Histologically, there is fibromuscular proliferation within the lamina propria. The fibrosis and smooth muscle may splay the crypts causing architectural distortion. Active inflammation in the form of cryptitis, erosions, and ulcerations may be present [4]. In contrast to UC, there is less lamina propria chronic inflammation and more fibromuscular hyperplasia. Attention to the location of the ulcer on the anterior wall of the distal rectum and pres-ence of fibromuscular obliteration of the lamina propria can help avoid an incorrect diagnosis of UC.

Nonsteroidal antiinflammatory drugs (NSAIDs) induced enteritis/colitis: This may show erosions, ulcerations, or strictures throughout the GI tract (Fig. 34.15A). In the colon, NSAID-induced colitis often presents as patchy active colitis with cryptitis and in some cases, with extensive ulceration (Fig. 34.15B).

Immune-regulatory agents (either suppressing or enhancing) associated enteritis/colitis: Gut immune homeostasis is maintained through a complex network in order to fight pathogens and to avoid uncontrolled inflammation. The use of immune-regulatory agents, either suppressing or enhancing, is associated with exacerbation or development of de novo IBD, which clinically, endoscopically, and histologically resembles either UC or CD [10]. The most commonly used categories are antiinflammatory group (such as anti-TNFα or antiintegrin biological agents), immune-enhancing

(A)

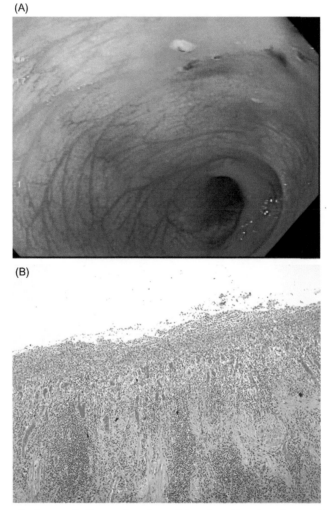

(B)

FIGURE 34.15 (A) Endoscopic examination reveals discrete ulcers in small bowel resulting from the use of nonsteroidal antiinflammatory drugs. (B) Biopsy from the ulcer reveals inflamed granulation tissue and neural hypertrophy, mimicking CD. *CD*, Crohn's disease.

antitumoral agents {such as immune checkpoint inhibitors [cytotoxic T-lymphocyte—associated antigen 4 or programmed death 1], or other antilymphoma agents [rituximab (an anti-CD20 antibody) and idelalisib (an inhibitor of the delta isoform of phosphatidylinositol 3-kinase). One case of idelalisib-associated colitis is illustrated in Fig. 34.16A and B. Histologically, there is cryptitis and deep intraepithelial lymphocytosis (Fig. 34.16B). If the colon is severely affected, the endoscopic and histologic features would be similar to UC, including crypt distortion and metaplasia in some cases [11].

Diversion colitis: Diversion colitis appears as erythematous, friable, granular, or nodular mucosa with exudates and spontaneous bleeding in the diverted colon or rectum endoscopically. At the microscopic level, diversion colitis presents as a chronic active colitis with crypt loss, architectural distortion, and varying degrees of activity. The key finding to support diversion colitis, aside from a history of fecal diversion, is the presence of large lymphoid follicles [4].

Entities mimicking Crohn's disease

Amebiasis: Though primarily seen in tropical or subtropical areas or in immunocompromised patients, amebic colitis is an important diagnostic consideration. It has a predilection for the cecum, yet all parts of the colon have been reported.

(A)

(B)

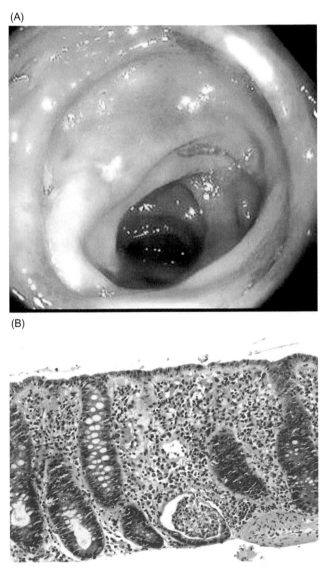

FIGURE 34.16 (A) Endoscopic examination of colitis induced by idelalisib, a phosphoinositide 3-kinase inhibitor (for the treatment of chronic lymphocytic leukemia and non-Hodgkin lymphoma), which is characterized by edema and ulcers. (B) Biopsy from the affected area reveals intraepithelial lymphocytosis in the surface epithelium and crypts, apoptosis of crypt epithelium, and neutrophilic cryptitis.

Endoscopically, it displays ulcers that may be small and scattered or coalesce to form large ulcers. In some cases the ulcers may be deep and flask-shaped, reminiscent of CD. Histologically, amebiasis has a brisk neutrophilic infiltrate, and the amebae can be seen in the lumen. The amebae resemble histiocytes with a distinct cell border, abundant foamy cytoplasm, and an eccentrically placed nuclei. In some cases, red blood cells may be seen in amebae.

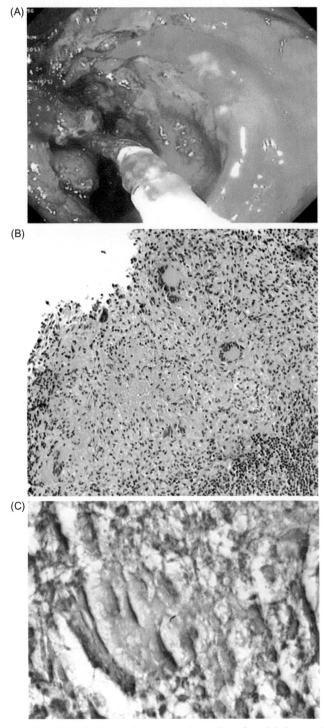

FIGURE 34.17   (A) Endoscopic examination of intestinal tuberculosis reveals large circumferential ulcers in the terminal ileum. (B) Biopsy shows large caseating granuloma. (C) AFB stain reveals few mycobacteria. *AFB*, Acid-fast bacillus.

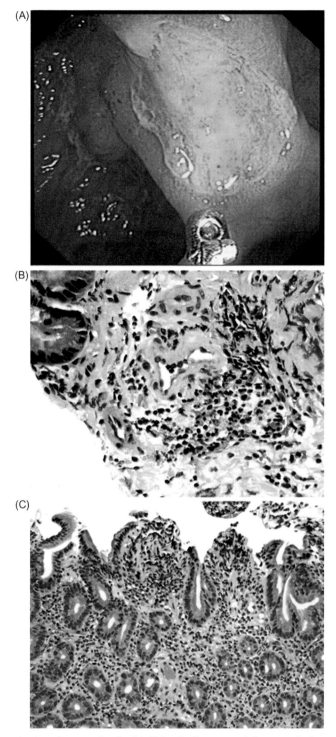

FIGURE 34.18   (A) Endoscopic examination of segment involved in Behçet's disease reveals large punched-out ulcer surrounded by normal mucosa. (B and C) Biopsy from the segment affected by Behçet's disease reveals the presence of perivascular inflammation with reactive endothelial cells in the submucosal venules and the mild inflammation in the mucosa.

Intestinal tuberculosis (ITB): Mycobacterial infections, involving the small bowel and colon, can be indistinguishable from CD on endoscopic (Fig. 34.17A) or radiographic studies. Common endoscopic features of ITB include circumferential ulcers of the small bowel and colon and deformed but patent ileocecal valve. While granulomas can be seen in both disease processes, the granulomas of ITB will often be increased in size and density, often caseating, may coalesce to form large granulomas (Fig. 34.17B), and may be cuffed by lymphocytes [12]. Chronic changes, such as architectural distortion and hyalinization of the lamina propria, can also be seen. An acid-fast bacillus stain can be helpful to highlight the organisms (Fig. 34.17C).

Behçet's disease: Behçet's disease is a rare vasculitic disorder that may affect in the small bowel and colon. Endoscopically, Behçet's disease of the intestine may manifest as punched-out large ulcers surrounded by normal circumferential mucosa (Fig. 34.18A). Mural vessels are affected by a lymphocytic or neutrophilic vasculitis with associated fibrinoid necrosis and endothelial cell reaction, but this is rarely seen in a mucosal biopsy [13]. The mucosal biopsy may show only mild perivascular inflammation with reactive endothelial cells in the submucosal vessels (Fig. 34.18B) and usually milder enteritis (Fig. 34.18C) or colitis. Subsequent to the vascular injury, the colon may show ulceration or an ischemic pattern of injury in severe cases. Diagnosis of Behçet's disease is based on a synthesis of all clinical, endoscopic, and pathologic information.

In addition to the aforementioned entities, infectious colitis, diverticular disease—associated colitis, and immunodeficiency-associated colitis (covered previously under UC mimics) may also occasionally mimic CD.

Summary

In summary, IBD has many differential diagnoses—clinical, endoscopic, or histologic. An accurate diagnosis of IBD requires the synthesis of all clinical, endoscopic, and pathologic information using a multidisciplinary approach.

Disclosure

The authors declared no financial conflict of interest.

References

[1] Villanacci V, Antonelli E, Geboes K, Casella G, Bassotti G. Histological healing in inflammatory bowel disease: a still unfulfilled promise. World J Gastroenterol 2013;19(7):968−78.

[2] Gordon IO, Agrawal N, Willis E, et al. Fibrosis in ulcerative colitis is directly linked to severity and chronicity of mucosal inflammation. Aliment Pharmacol Ther 2018;47:922−39.

[3] Patil DT, Kissiedu J, Rodriguez ER, et al. Mesenteric arteriovenous dysplasia/vasculopathy is distinct from fibromuscular dysplasia. Am J Surg Pathol 2016;40:1316−25.

[4] Odze RD, Goldblum JR. Odze and Goldblum surgical pathology of the GI tract, liver, biliary tract, and pancreas. Philadelphia, PA: Saunders/Elsevier; 2015. Print.

[5] Laine L, Kaltenbach T, Barkun A, et al. SCENIC international consensus statement on surveillance and management of dysplasia in inflammatory bowel disease. Gastroenterology 2015;148(3):639−51.

[6] Lee H, Westerhoff M, Shen B, Liu X. Clinical aspects of idiopathic inflammatory bowel disease: a review for pathologists. Arch Pathol Lab Med 2016;140:413−28.

[7] Cohen-Mekelburg S, Schneider Y, Gold S, Scherl E, Steinlauf A. Advances in the diagnosis and management of colonic dysplasia in patients with inflammatory bowel disease. Gastroenterol Hepatol 2017;13(6):357−62.

[8] Akram S, et al. Adult autoimmune enteropathy: Mayo Clinic Rochester experience. Clin Gastroenterol Hepatol 2007;5:1282−90.

[9] Masia R, et al. Gastrointestinal biopsy findings of autoimmune enteropathy: a review of 25 cases. Am J Surg Pathol 2014;38:1319−29.

[10] Zhou W, Huang Y, Lai J, et al. Anti-inflammatory biologics and anti-tumoral immune therapies-associated colitis: a focused review of literature. Gastroenterol Res 2018;11(3):174−88.

[11] Weidner AS, Panarelli NC, Geyer JT, Bhavsar EB, Furman RR, Leonard JP, et al. Idelalisib-associated colitis. Histologic findings in 14 patients. Am J Surg Pathol 2015;39:1661−7.

[12] Limsrivilai J, Shreiner AB, Pongpaibul A, Laohapand C, Boonanuwat R, Pausawasdi N, et al. Meta-analytic Bayesian model for differentiating intestinal tuberculosis from Crohn's disease. Am J Gastroenterol 2017;112(3):415−27.

[13] Koklü S, Yüksel O, Onur I, et al. Ileocolonic involvement in Behçet's disease: endoscopic and histological evaluation. Digestion 2010;81:214e7.

Chapter 35

Transmural endoscopic imaging in inflammatory bowel disease

Francesca N. Raffa and David A. Schwartz

Inflammatory Bowel Disease Center, Division of Gastroenterology, Hepatology, and Nutrition, Department of Medicine, Vanderbilt University Medical Center, Nashville, TN, United States

Chapter Outline

Abbreviations

CD	Crohn's disease
IBD	inflammatory bowel disease
EUA	examination under anesthesia
EUS	endoscopic ultrasound
MRI	magnetic resonance imaging
OCT	optical coherence tomography

Introduction

Endoluminal imaging plays a crucial role in the primary diagnosis and management of patients with inflammatory bowel disease (IBD). Advances in the field of ultrasonography and fiber optics have enhanced the ability to assess the gastrointestinal tract transmurally. This chapter provides an overview of current transmural imaging modalities, specifically rectal endoscopic ultrasonography and optical coherence tomography (OCT), and their clinical utility in the endoscopic assessment of IBD patients.

Rectal endoscopic ultrasound

Perianal fistulizing disease is a common phenotype in aggressive Crohn's disease (CD), occurring in 10% of patients at the time of diagnosis and approximately 25% of patients 20 years after onset [1]. Physical examination, including examination under anesthesia (EUA) in the operating room, is often unreliable in fistula detection due to significant perianal scarring and inflammation. Accurate identification of perianal anatomy and assessment of disease activity have crucial implications for treatment, as misclassification of fistulae and failure to identify occult tracts can result in recurrent disease, abscesses, or progression from a simple tract to a complex fistulizing process [2]. The imaging modalities of choice for the initial assessment of perianal fistulae include pelvic magnetic resonance imaging (MRI) and rectal endoscopic ultrasound (EUS) [3]. Studies comparing the accuracy of these modalities have shown that both are highly sensitive in the initial assessment of CD perianal fistulae (87% in MRI vs 91% in rectal EUS) [4]. The choice of which modality to utilize should depend on the local expertise.

Atlas of Endoscopy Imaging in Inflammatory Bowel Disease. DOI: https://doi.org/10.1016/B978-0-12-814811-2.00035-9
© 2020 Elsevier Inc. All rights reserved.

Rectal EUS involves the insertion of an endoscopic transducer probe into the anal canal and distal rectum, while the patient is positioned on their left lateral side (Figs. 35.1 and 35.2). Active fistulae typically appear hypoechoic on ultrasound corresponding to inflammation within the fistula tract but may be internally hyperechoic if they contain gas or air either due to suppurative inflammation or patency of the track itself (Fig. 35.3). An abscess is identified sonographically as an anechoic or hypoechoic mass in the perianal tissues (Fig. 35.4). Hydrogen peroxide is sometimes used to create hyperechoic bubbles on EUS and may be injected into a cutaneous fistula site to more clearly delineate the fistulous tract. A similar effect may be achieved by applying gentle pressure to visualize the motion of air bubbles through the tract itself [2]. Compared to MRI, rectal EUS is rapid, inexpensive, and has the advantage that it can be performed during a diagnostic colonoscopy or flexible sigmoidoscopy. However, rectal EUS is limited in which it cannot be utilized in patients with severe anorectal stenosis [5].

In addition to the initial assessment of perianal fistulae, rectal EUS can also be used to monitor fistula healing and to guide both the choice and timing of medical and surgical therapies. Prospective studies have shown that EUS-guided medical and surgical treatment improves short-term fistula resolution and reduces recurrence rates. In a small randomized study by Spradlin et al., rectal EUS guidance improved clinical outcomes in patients with perianal CD. One of five (20%) in the control group (no EUS guidance) and four of five (80%) in the EUS group had complete cessation in fistula drainage [6]. Furthermore, EUS may also help direct endoscopic therapies for perianal fistulae, including placement of setons or drains (Fig. 35.5) [7]. During EUA, transanal ultrasound may also be used to assess the anatomy of anal sphincters, fistulas, and abscess (Fig. 35.6).

Endoscopic optical coherence tomography

Endoscopic OCT is another endoscopic gastrointestinal imaging modality that can be utilized to assess disease activity in IBD patients. The technique is analogous to ultrasonography, employing light (between 700 and 1500 nm of wavelength) rather than sound waves to generate an image of mucosal structures. An OCT probe is attached to the endoscope and placed in contact with the surface of interest, revealing an image with $7-10\,\mu m$ resolution and $2-3\,mm$ depth depending on the wavelength of light [8]. In addition to disease activity, OCT has found utility in detecting transmural inflammation, a hallmark of Crohn's disease. Shen et al. have shown in both ex vivo and in vivo studies that disruption of the layered structure of the colon wall on OCT is an accurate indicator of transmural inflammation. Their retrospective study of 70 patients found that 36 of 40 CD patients and 5 of 30 ulcerative colitis patients had the disrupted layered structure on OCT. The OCT image pattern of disrupted layered structure had a sensitivity and specificity of 90% and

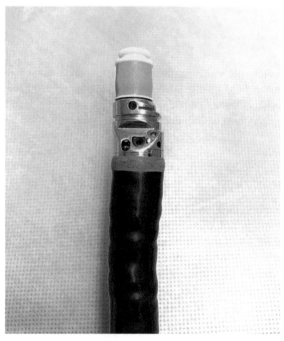

FIGURE 35.1 Radial endoscopic ultrasound probe.

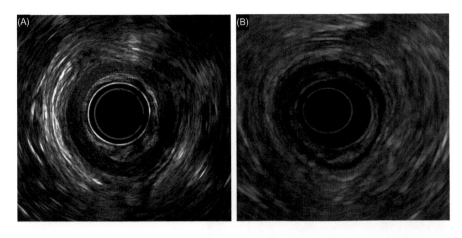

FIGURE 35.2 Rectal endoscopic ultrasound in healthy and diseased: (A) appearance of the normal rectal wall under endoscopic ultrasound and (B) endoscopic ultrasound image showing the alternation of hyperechoic and hypoechoic bands in the normal rectal wall: interface between probe and mucosa (hyperechoic), mucosa and muscularis mucosa (hypoechoic), submucosa (hyperechoic), muscularis propria (hypoechoic), and interface between muscularis propria and serosa (hyperechoic).

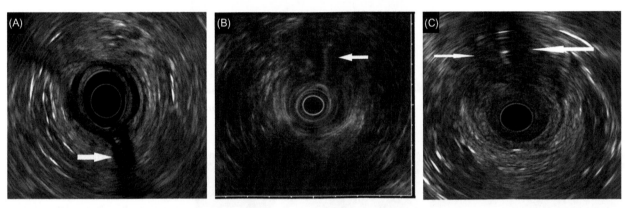

FIGURE 35.3   Several examples of perianal fistulae detected by rectal endoscopic ultrasound. Fistulous tracts are identified as hypoechoic round or oval structures: (A) transsphincteric course of a perianal fistula (*arrow*) and (B) large anterior transsphincteric fistula (*arrow*); (C) rectovaginal fistula (*arrow*).

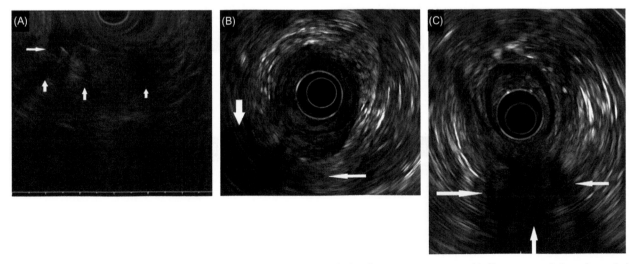

FIGURE 35.4   Perianal abscesses identified by rectal endoscopic ultrasound. An abscess appears sonographically as an anechoic or hypoechoic mass, noted by arrows in (A), (B), and (C) (*arrows*).

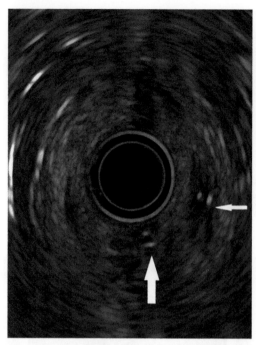

FIGURE 35.5 Rectal endoscopic ultrasound showing perianal fistula with seton thread in place (*arrows*).

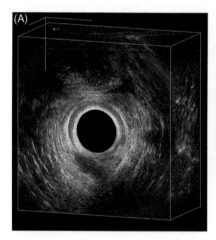

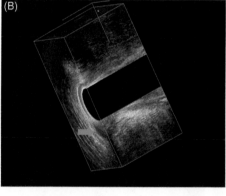

FIGURE 35.6 Transanal ultrasound for the evaluation of perianal abscess in Crohn's disease: (A) 2-D ultrasound and (B)3-D ultrasound (*green arrows*). *Courtesy Bo Shen, MD, Columbia University Irving Medical Center-New York Presbyterian Hospital.*

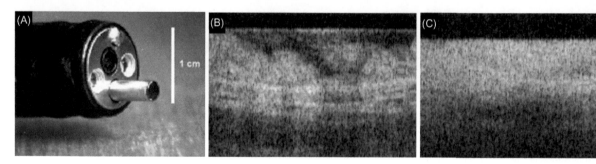

FIGURE 35.7 Distinction between ulcerative colitis and Crohn's disease on optical coherence tomography: (A) probe-based optical coherence tomography, (B) preserved layered colon wall structure in ulcerative colitis, (C) disrupted layered colon wall structure in Crohn's colitis, indicating the presence of transmural disease. *Courtesy Bo Shen, MD, Columbia University Irving Medical Center-New York Presbyterian Hospital.*

83%, respectively, for CD. OCT allows for a minimally invasive "optical biopsy" and is thus a safe and valuable tool in differentiating CD from ulcerative colitis (Fig. 35.7) [9]. The applications of OCT in IBD are discussed in Chapter 20, Confocal laser endomicroscopy and other image-enhanced endoscopies in inflammatory bowel disease.

Ultrasound elastography

Transabdominal ultrasound elastography has been investigated for the distinction between inflammatory and fibrotic strictures in CD. We expect that one day, the probe-based endoscopic elastography will be available for clinical use (Chapter 20: Confocal laser microscopy and other image-enhanced endoscopies in inflammatory bowel disease).

Summary and recommendations

Rectal EUS is highly accurate in the assessment of perianal fistulizing CD in experienced hands. The imaging modality can be used in combination with MRI and EUA. Besides, rectal EUS has been used to monitor fistula healing and to guide the choice and timing of medical and surgical therapies. Furthermore, rectal EUS may also be helpful in directing endoscopic therapies for perianal fistulae, such as endoscopic stricture dilation, endoscopic stricturotomy, endoscopic incision and drainage of abscess, and endoscopic fistulotomy. The EUS approach may be extended to its application in the ileal pouch-anal anastomosis. In contrast the clinical use of OCT for the assessment of transmural disease process is limited due to the shallower depth of penetration.

Disclosure

The authors declared no financial conflict of interest.

References

[1] Kucharzik T, Kannengiesser K, Petersen F. The use of ultrasound in inflammatory bowel disease. Ann Gastroenterol 2017;30(2):135−44.

[2] Wise PE, Schwartz DA. The evaluation and treatment of Crohn perianal fistulae: EUA, EUS, MRI, and other imaging modalities. Gastroenterol Clin North Am 2012;41(2):379−91.

[3] Paine E, Shen B. Endoscopic therapy in inflammatory bowel diseases. Gastrointest Endosc 2013;78(6):819−35.

[4] Schwartz DA, Wiersema MJ, Dudiak KM, Fletcher JG, Clain JE, Tremaine WJ, et al. A comparison of endoscopic ultrasound, magnetic resonance imaging, and exam under anesthesia for evaluation of Crohn's perianal fistulas. Gastroenterology 2001;121(5):1064−72.

[5] Botti F, Losco A, Viganò C, Oreggia B, Prati M, Contessini Avesani E. Imaging techniques and combined medical and surgical treatment of perianal Crohn's disease. J Ultrasound 2015;18(1):19−35.

[6] Spradlin NM, Wise PE, Herline AJ, Muldoon RL, Rosen M, Schwartz DA. A randomized prospective trial of endoscopic ultrasound to guide combination medical and surgical treatment for Crohn's perianal fistulas. Am J Gastroenterol 2008;103(10):2527−35.

[7] Shen B. Exploring endoscopic therapy for the treatment of Crohn's disease-related fistula and abscess. Gastrointest Endosc 2017;85(6):1133−43.

[8] Kirtane TS, Wagh MS. Endoscopic optical coherence tomography (OCT): advances in gastrointestinal imaging. Gastroenterol Res Pract 2014;2014 376367(1−7).

[9] Shen B, Zuccaro Jr G, Gramlich TL, Gladkova N, Trolli P, Kareta M, et al. In vivo colonoscopic optical coherence tomography for transmural inflammation in inflammatory bowel disease. Clin Gastroenterol Hepatol 2004;2(12):1080−7.

Chapter 36

Inflammatory bowel disease—associated bleeding

Bo Shen

Center for Inflammatory Bowel Diseases, Columbia University Irving Medical Center-New York Presbyterian Hospital, New York, NY, United States

Chapter Outline

Abbreviations

CD	Crohn's disease
GI	gastrointestinal
IBD	inflammatory bowel disease
UC	ulcerative colitis

Introduction

Bloody bowel movement is a common presentation of active ulcerative colitis (UC). In contrast, brisk bleeding in Crohn's disease (CD) is uncommon. Rather, patients with CD often present with iron-deficiency anemia or anemia of chronic disease. The patterns of gastrointestinal (GI) hemorrhage in inflammatory bowel disease (IBD) ranges from obscure bleeding to massive, life-threatening bleeding [1,2]. Reported frequency of bleeding in UC was 17% of patients with UC [3], while the reported frequency of severe GI bleeding as a main presenting symptom in CD ranged from 0.6% to 4% [3–8]. Exacerbation factors for GI bleeding include the use of nonsteroidal antiinflammatory drugs or thrombocytopenia, deficiency in coagulating factors, and portal hypertension in conditions such as primary sclerosing cholangitis. Selective cyclooxygenase-2 inhibitors, such as rofecoxib, may also be associated with GI bleeding in CD [4]. Active localized bleeding can be managed endoscopically with modalities, such as endoclips and topical spray or injection.

Bleeding conditions in IBD is summarized and categorized in Table 36.1.

Bleeding from underlying diseases

Bleeding in CD (Figs. 36.1 and 36.2), UC (Fig. 36.3), or ileal pouches (Fig. 36.4) often results from severe mucosal inflammation, ulcer lesion eroding blood vessels or inflammatory, vascular polyps. Patients with IBD, especially those with UC, are prone to the development of inflammatory polyps (Fig. 36.5). It has been believed that the presence of inflammatory pseudopolyps is associated with an increased risk for the development of colitis-associated neoplasia.

TABLE 36.1 Classification of bleeding in inflammatory bowel disease.

	Category	Subcategory/definition	Example
Location	Esophagus and stomach		
	Small bowel		
	Colon and rectum		
	Anal area		
	Anastomosis		
Etiology	Inflammation	Ulcerative colitis	
		Crohn's disease	
	Ulcer	Disease associated	UC or CD
		Anastomosis related	
		Infectious	Cytomegalovirus
	Polyps	Inflammatory/pseudopolyps	
		Adenomatous or dysplastic polyps	
	Malignancy	Cancer	
		Lymphoma	
	Anal-area disease	Hemorrhoids	
		Anal fissure	
	Concurrent systemic disease	Primary sclerosing cholangitis	Variceal bleeding
		Bleeding disorders	Amyloidosis, idiopathic thrombocytopenic purpura
		Transplanted gut	
	Iatrogenic	Diagnostic endoscopy	Mucosal biopsy
		Therapeutic endoscopy	Balloon dilation, polypectomy, and endoscopic stricturotomy
Severity	Mild	Nontransfusion required	
	Moderate	Transfusion-required	
	Severe	Transfusion-required hemodynamically unstable	
Speed	Acute		
	Chronic		

CD, Crohn's disease; *UC*, ulcerative colitis.

However, a recent study showed that the presence of inflammatory polyps did not appear to increase the risk for colitis-associated neoplasia [5]. Therefore routine surveillance biopsy or removal of inflammatory pseudopolyps is controversial, mainly due to the risk of procedure-associated bleeding (Fig. 36.5).

Bleeding from surgical sites

Bowel resection and anastomosis, strictureplasty, construction of ostomies, and restorative proctocolectomy with ileal pouch−anal anastomosis are common surgical treatment modalities for patients with CD or UC. Ulcers at the surgical anastomosis, suture lines, or staple lines are common, which can result in brisk or obscure bleeding. For example,

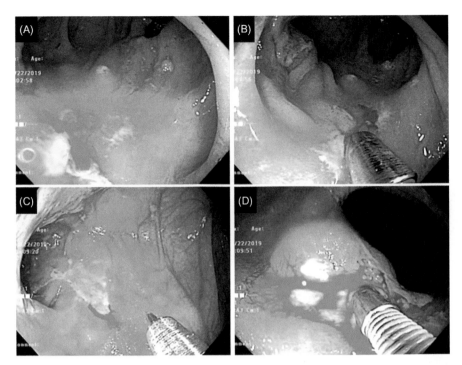

FIGURE 36.1 Crohn's disease with bleeding ulcers. (A and B) Superficial bleeding ulcer in the descending colon, treated with a topical injection of 50% glucose; (C and D) bleeding from an ulcer in the hepatic flexure controlled by a topical injection of 50% glucose in a separate patient.

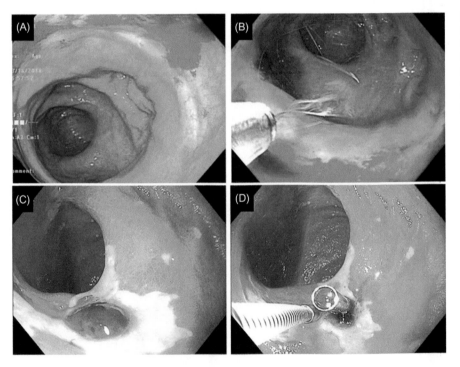

FIGURE 36.2 Crohn's disease bleeding. (A and B) Friable colon mucosa with spontaneous bleeding, which was controlled by the spray of 50% glucose; (C and D) large ulcerated area in the ileocecal valve with a bleeding, visible bleeding vessel controlled with endoclips and intralesional injection of epinephrine.

bleeding can occur at the ileocolonic or ileorectal anastomosis site (Figs. 36.6−36.8A and B), [6] vertical staple lines of the ileal pouch (Fig. 36.8C and D) [7,8], surgical strictureplasty site [9], or stoma (Fig. 36.9).

Bleeding from fecal diversion

Fecal diversion with ileostomy, colostomy, or jejunostomy is an effective surgical treatment modality for the management refractory IBD in the distal bowel or perianal area. Due to the lack of nutrients for intestinal epithelia, particularly

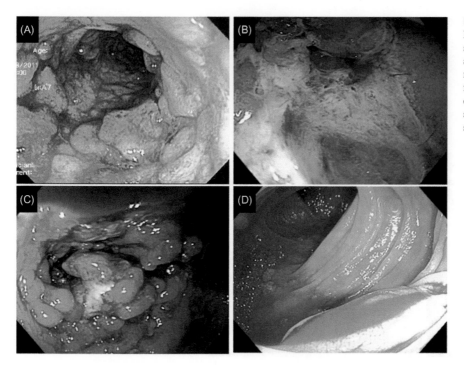

FIGURE 36.3 UC with bleeding. (A) Erythematous and edematous mucosa with spontaneous bleeding; (B) ulcer bleeding at the splenic flexure; (C) bleeding inflammatory polyps; (D) excessive bleeding from a mucosal biopsy in a patient with concurrent UC (in remission) and primary sclerosing cholangitis with portal hypertension. *UC*, Ulcerative colitis.

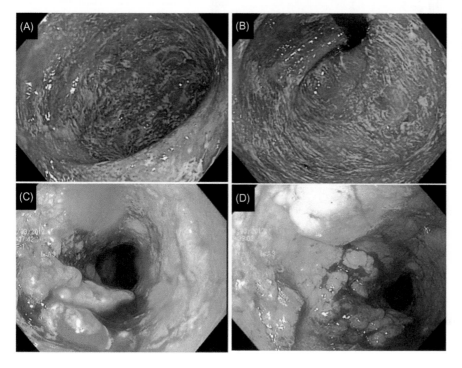

FIGURE 36.4 Hemorrhagic pouchitis and cuffitis. (A and B) Diffuse hemorrhagic pouchitis and (C and D) cuffitis with bleeding inflammatory polyps.

deficiency in luminal short-chain fatty acids, patients with diverted colon, rectum, or ileal pouch are prone to the development of diversion-associated inflammation of the bowel with extremely friable mucosa, with spontaneous bleeding or bleeding after gentle insufflation of air or carbon dioxide during endoscopy (Fig. 36.10) [10]. Long-standing diversion may also cause strictures in the distal bowel (Fig. 36.10C). Endoscopic biopsy or therapy may carry additional risk for bleeding.

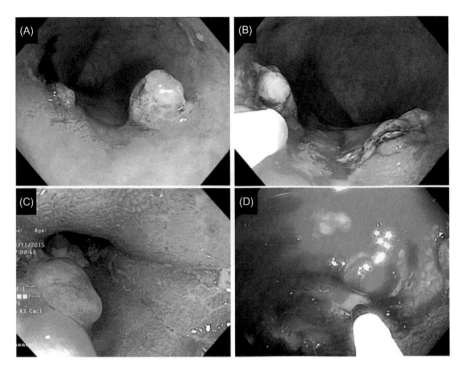

FIGURE 36.5 Polypectomy in UC. (A and B) Inflammatory polyps in UC can cause spontaneous bleeding. Their malignant potential is controversial. Polypectomy is occasionally performed; (C and D) polypectomy of vascular inflammatory lesions carries a higher risk for bleeding than the removal of spontaneous adenomas. The postpolypectomy bleeding was controlled by the spray of 50% glucose. *UC*, Ulcerative colitis.

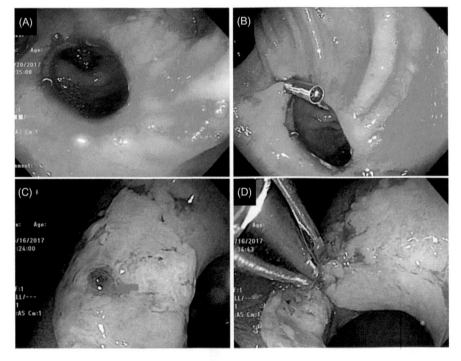

FIGURE 36.6 Anastomosis bleeding in patients with Crohn's disease. (A and B) Stapled side-to-side ulcerated and friable ileocolonic anastomosis with active bleeding, treated with endoclips; (C and D) ulcerated handsewn end-to-side ileocolonic anastomosis with a visible vessel (*green arrow*), treated with endoclips.

Bleeding from endoscopic intervention

Interventional IBD or endoscopic therapy for IBD plays a growing role in the management of IBD- or IBD surgery—associated complications, such as strictures, fistulae, and anastomotic leaks [11]. The endoscopic treatment modalities include polypectomy, balloon dilation or electroincision of strictures, fistulotomy, sinusotomy, endoscopic mucosal resection, and endoscopic submucosal dissection. Those invasive procedures are associated with a risk of bleeding

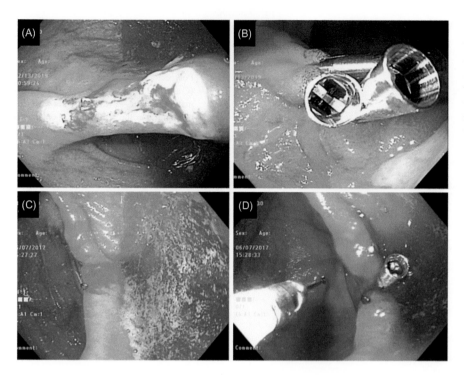

FIGURE 36.7 Anastomosis bleeding in patients with Crohn's disease. (A and B) Linear ulcer on stapled side-to-side ileocolonic anastomosis with active bleeding, treated with endoclips and (C and D) bleeding at the staple line without ulceration in a separate patient with ileocolonic resection and anastomosis, which was treated with an endoclips.

(Figs. 36.11 and 36.12) [12]. Endoscopic evaluation plays a key role in the evaluation and management of bleeding.

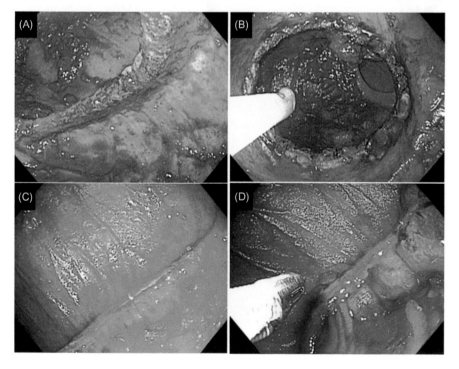

FIGURE 36.8 Bleeding at staple line ulcers that caused iron-deficiency anemia. (A and B) Bleeding from ileorectal anastomosis ulcers, which was treated with argon plasma coagulation and (C and D) bleeding from vertical staple line ulcers in the ileal pouch, which was treated with topical injection of epinephrine.

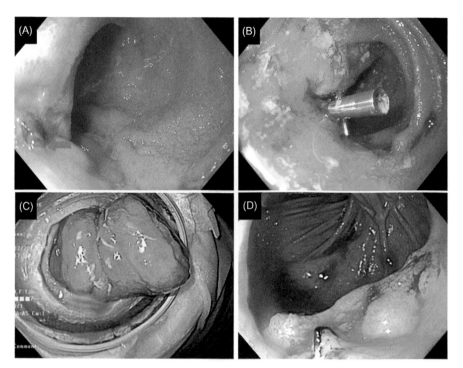

FIGURE 36.9 Crohn's disease bleeding after bowel surgery. (A and B) Visible vessel with bleeding 10 cm from stoma in a patient with ileostomy (*green arrow*), which was treated with endoscopic clips; (C) ulcerated stoma, which had resulted bleeding; and (D) staple line ulcer with bleeding in a patient with strictureplasty of the small intestine.

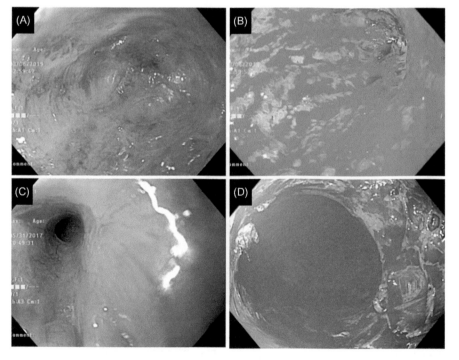

FIGURE 36.10 Diversion-associated proctitis and pouchitis with bleeding. (A and B) Diverted rectum with severe bleeding after gentle carbon dioxide insufflation during endoscopy and (C and D) diverted ileal pouch with a distal stricture and severe diversion-associated pouchitis.

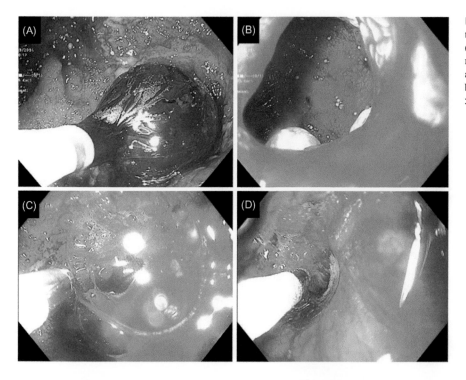

FIGURE 36.11 Endoscopic balloon dilation and procedure-associated bleeding. (A) Balloon dilation of ileocolonic anastomosis stricture; (B) excessive bleeding after balloon dilation; and (C and D) bleeding was controlled by the spray of 50% glucose.

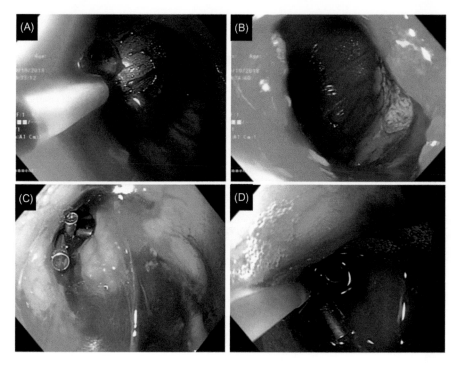

FIGURE 36.12 Endoscopic balloon dilation and procedure-associated bleeding. (A) Balloon dilation of ileocolonic anastomosis stricture; (B) excessive bleeding after balloon dilation; and (C and D) bleeding was controlled by endoclips and the spray of 50% glucose.

The common rescuing endoscopic treatment modalities include the deployment of endoclips, spray, or injection of hypertonic glucose or epinephrine and application of argon plasma coagulation (Figs. 36.11 and 36.12).

Summary and conclusion

GI bleeding is a common presentation in patients with UC, CD, or ileal pouches, which results from severe mucosal inflammation, bleeding polyps, disease- or surgery-associated ulcers. In addition, endoscopic therapy for IBD- or IBD surggery—associated complications may be associated with a risk for procedure (such as balloon dilation and endoscopic stricturotomy) associated bleeding. Endoscopic plays a key role for the diagnosis and management.

References

[1] Egawa T, Kuroda T, Ogawa H, Takeda A, Kanazawa S, Harada H, et al. A case of Crohn's disease with recurrent massive life-threatening hemorrhage from terminal ileum. Hepatogastroenterology 1999;46:1695—8.

[2] Veroux M, Angriman I, Ruffolo C, Barollo M, Buffone A, Madia C, et al. Severe gastrointestinal bleeding in Crohn's disease. Ann Ital Chir 2003;74:213—15 discussion 216.

[3] Farmer RG, Easley KA, Rankin GB. Clinical patterns, natural history, and progression of ulcerative colitis. A long-term follow-up of 1116 patients. Dig Dis Sci 1993;38:1137—46.

[4] Gornet JM, Hassani Z, Modiglian R, Lémann M. Exacerbation of Crohn's colitis with severe colonic hemorrhage in a patient on rofecoxib. Am J Gastroenterol 2002;97:3209—10.

[5] Mahmoud R, Shah SC, Ten Hove JR, Torres J, Mooiweer E, Castaneda D, , et al.Dutch Initiative on Crohn and Colitis No association between pseudopolyps and colorectal neoplasia in patients with inflammatory bowel diseases. Gastroenterology 2019;156:1333—44.

[6] Riss S, Bittermann C, Zandl S, Kristo I, Stift A, Papay P, et al. Short-term complications of wide-lumen stapled anastomosis after ileocolic resection for Crohn's disease: who is at risk? Colorectal Dis 2010;12:e298—303.

[7] Nishikawa T, Hata K, Yoshida S, Murono K, Yasuda K, Otani K, et al. Successful endoscopic treatment of stapled J-pouch ileoanal canal anastomotic hemorrhage by argon plasma coagulation: a case report. J Med Case Rep 2016;10:309.

[8] Fazio VW, Kiran RP, Remzi FH, Coffey JC, Heneghan HM, Kirat HT, et al. Ileal pouch anal anastomosis: analysis of outcome and quality of life in 3707 patients. Ann Surg 2013;257:679—85.

[9] Gardiner KR, Kettlewell MG, Mortensen NJ. Intestinal haemorrhage after strictureplasty for Crohn's disease. Int J Colorectal Dis 1996;11:180—2.

[10] Wu XR, Liu XL, Katz S, Shen B. Pathogenesis, diagnosis, and management of ulcerative proctitis, chronic radiation proctopathy, and diversion proctitis. Inflamm Bowel Dis 2015;21:703—15.

[11] Shen B, Kochhar G, Navaneethan U, Liu X, Farraye FA, Gonzalez-Lama Y, et al. Role of interventional inflammatory bowel disease in the era of biologic therapy: a position statement from the Global Interventional IBD Group. Gastrointest Endosc 2019;89:215—37.

[12] Lan N, Stocchi L, Delaney CP, Hull TL, Shen B. Endoscopic stricturotomy versus ileocolonic resection in the treatment of ileocolonic anastomotic strictures in Crohn's disease. Gastrointest Endosc 2019;90:259—68.

Chapter 37

Endoscopic evaluation of bezoars and foreign bodies in inflammatory bowel diseases

Bo Shen

Center for Inflammatory Bowel Diseases, Columbia University Irving Medical Center-New York Presbyterian Hospital, New York, NY, United States

Chapter Outline

Abbreviations

CD	Crohn's disease
IBD	inflammatory bowel disease
IPAA	ileal pouch—anal anastomosis
UC	ulcerative colitis

Introduction

Bezoars in the gastrointestinal tract are accumulated indigestible foreign materials, which can produce concretions. According to their compositions, Bezoars have been classified into following types: (1) phytobezoars (vegetable or fruit fibers), (2) lactobezoars (milk or dairy proteins), (3) trichobezoars (hairs), (4) lithobezoars (calcified, stone-like materials), and (5) pharmacobezoars (medications) [1]. Besides, the retention of video capsule endoscope in patients with inflammatory bowel disease (IBD) can occur, especially in those with strictures or bowel-altering surgeries. Capsule retention has been a well-recognized adverse sequala in clinical practice [2].

Approximately 70%−80% of patients with Crohn's disease (CD) would eventually require surgery for medically refractory disease, particularly those with strictures, fistulas, or abscesses. Commonly performed surgical modalities in CD include bowel resection and anastomosis, strictureplasty, and ileostomy. Patients with strictureplasty are prone to the development of retention of foods or bezoars. In contrast, it is estimated that 20%−30% of patients with ulcerative colitis (UC) would require colectomy for medically refractory disease, poor tolerance of medications, or colitis-associated neoplasia. Restorative proctocolectomy with ileal pouch—anal anastomosis (IPAA) with "J" or "S" pouches is most commonly performed in these patients. The construction of continent ileostomies, such as Kock pouch or Barnett Continent Intestinal Reservoir, is an alternative option, for selected patients with poor anal sphincter function who are not candidates for or failed pelvic pouches. The presence of the nipple valve in the Kock pouch or Barnett pouch makes the patient vulnerable to the development or retention of bezoars or foreign bodies [3,4].

Staple techniques have been extensively used in UC or CD surgeries. Misfired or dislodged staples during or after surgery can result in pain, bleeding, or anemia.

Atlas of Endoscopy Imaging in Inflammatory Bowel Disease. DOI: https://doi.org/10.1016/B978-0-12-814811-2.00037-2
© 2020 Elsevier Inc. All rights reserved.

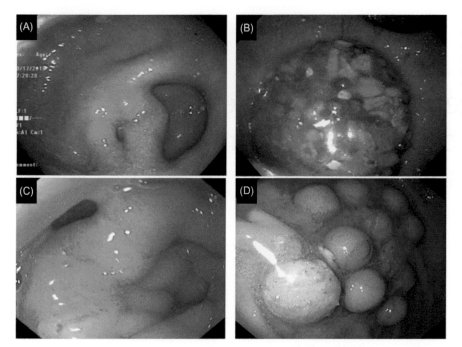

FIGURE 37.1 Food retention in the stomach due to Crohn's disease—associated pyloric stricture. (A—D) Severe strictures at the pylorus in two patients resulting from Crohn's disease, with retained foods in the stomach. *(C and D) Courtesy Dr. Yubei Gu of Shanghai Jiaotong University Ruijin Hospital.*

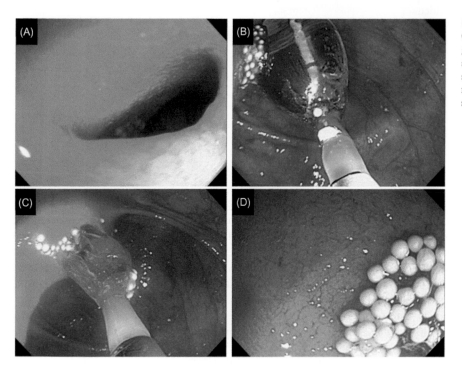

FIGURE 37.2 Ileocecal valve stricture of Crohn's disease and retained mesalamine granules. (A—C) A tight ileocecal valve stricture, which was treated with endoscopic balloon dilations and (D) retained medicine granules resulting from the stricture.

Bezoars and foreign bodies associated with strictures

Patients with narrowed bowel lumen from intrinsic or extrinsic causes are prone to the development of retention of foods, bezoars, and foreign body. Stricturing CD is a common disease phenotype. Also, anastomotic strictures in both patients with UC or CD who undergo various surgical operations can often occur. Primary strictures in CD can develop at the pylorus (Fig. 37.1) or ileocecal valve (Fig. 37.2), resulting in the retention of bezoars in the stomach or the distal ileum. In addition, the retention of bezoars or foreign bodies can also develop in the strictures at the ileocolonic

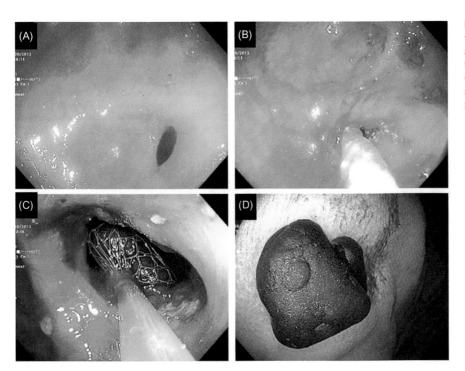

FIGURE 37.3 Lithobezoar in a patient with a tight stricture at the ileocolonic anastomosis. (A) The severe stricture at the end-to-side ileocolonic anastomosis; (B) endoscopic balloon dilation of the stricture; and (C and D) endoscopic removal of the large bezoar with a net.

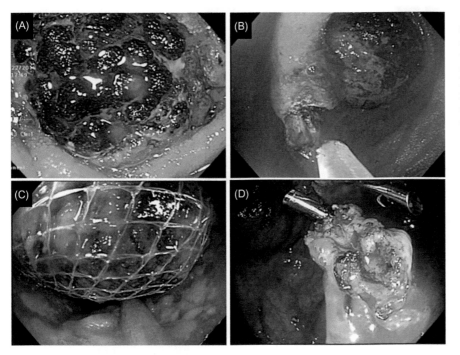

FIGURE 37.4 Retained lithobezoar resulting from stricture at the ileocolonic anastomosis in Crohn's disease. (A) A trapped large lithobezoar at the stricture; (B) endoscopic stricturotomy with needle knife to treat the stricture; (C) endoscopic retrieval of the bezoar with a net; and (D) deployment of endoclips at the stricturotomy site to prevent the reformation of the stricture.

anastomosis (Figs. 37.3 and 37.4), ileostomy or colostomy (Fig. 37.5), or strictureplasty site (Fig. 37.6). Fecal bezoars can occasionally be present in the diverticulum of the colon, resulting in symptoms (Fig. 37.7). Besides, symptoms related to partial bowel obstruction [5], bezoars may cause ulcers or bowel perforation [6]. In the majority of patients the strictures and associated retention of foods, bezoars, or foreign bodies can be successfully managed with endoscopic therapy, such as balloon dilation and stricturotomy (for strictures), and retrieval (for bezoars).

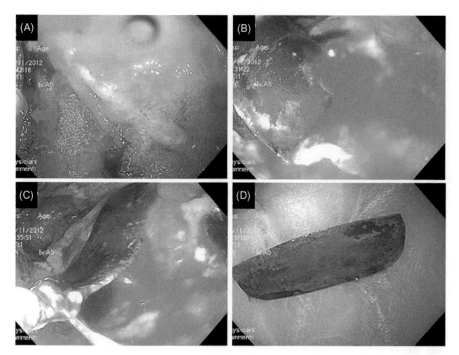

FIGURE 37.5 Phytobezoar in the strictured neo-distal ileum in a patient with an ileostomy for refractory Crohn's disease. (A) A tight, angulated stricture 15 cm from the stoma, which was gently dilated with an endoscopic balloon (*green arrow*) and (B–D) retained phytobezoar in the ileum proximal to the stricture, which was removed with endoscopic biopsy forceps.

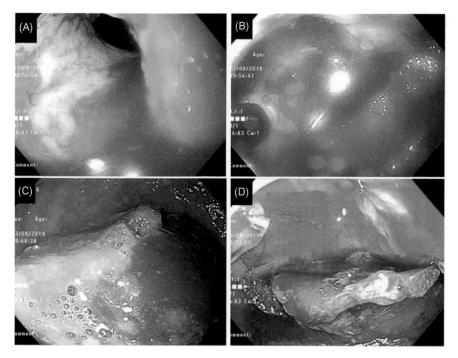

FIGURE 37.6 Phytobezoars in the strictureplasty site in Crohn's disease. (A) An ulcerated stricture at the outlet of strictureplasty site in the distal ileum; (B) endoscopic stricturotomy with an insulated-tip knife of the stricture; and (C and D) retained phytobezoars in the lumen of strictureplasty site.

Bezoars and foreign bodies in ileal pouches and diverted bowels

Structural and functional disorders, such as anastomotic strictures, pelvic floor dysfunction with dyssynergic defecation, can occur in patients with ileal pouches. Severe structural or functional outlet obstructions can lead to symptomatic fecal bezoars (Fig. 37.8). Bezoars or foreign bodies can cause pressure ulcers or even fistulas (Figs. 37.8 and 37.9). Endoscopic therapy of the strictures or pelvic biofeedback has been helpful for the treatment and prevention.

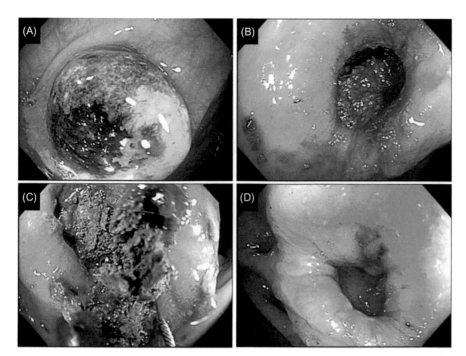

FIGURE 37.7 Trapped fecal bezoar in diverticulum in CD. Diverticulosis is not common in patients with CD. (A and B) A large fecal bezoar caught in the tight-necked diverticulum and (C and D) endoscopic removal of the fecal bezoar, during and after. *CD*, Crohn's disease.

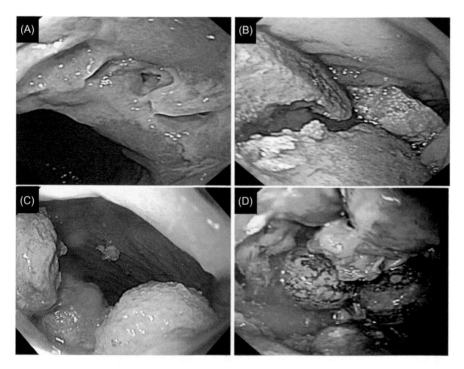

FIGURE 37.8 Dyssynergic defecation in a patient with the ileal pouch. (A) Severe cuffitis resulting from excessive straining and (B−D) retained solid fecal materials in the pouch body, due to the functional outlet obstruction.

The anatomy of continent ileostomies has made the patient vulnerable to the development and retention of bezoars or foreign bodies (Fig. 37.10). Small bezoars or foreign bodies may be removed endoscopically, while large ones often require surgical intervention with ultrasound or laser lithotripsy (in collaboration with urology) and retrieval or incision and retrieval (in collaboration with colorectal surgery or general surgery).

Fecal diversion with ileostomy or colostomy is performed for the treatment of downstream refractory disease. Bezoars can develop in patients with long-standing diverted bowel (Fig. 37.11), from which some of the patients may be symptomatic.

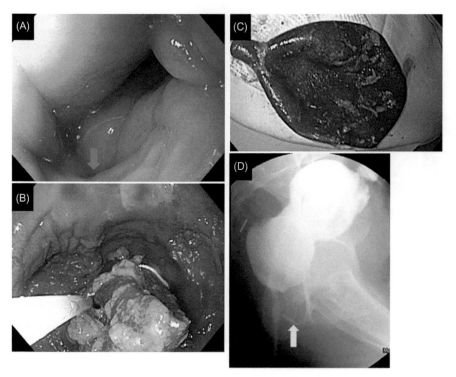

FIGURE 37.9 PVF resulting from a retained foreign body in a patient with ileal pouch−anal anastomosis. (A) The internal opening of PVF at the anal canal (*green arrow*); (B and C) a rubber foreign body in the pouch lumen; and (D) PVF was highlighted from Gastrografin enema via pouch (*yellow arrow*). *PVF*, Pouch-vaginal fistula.

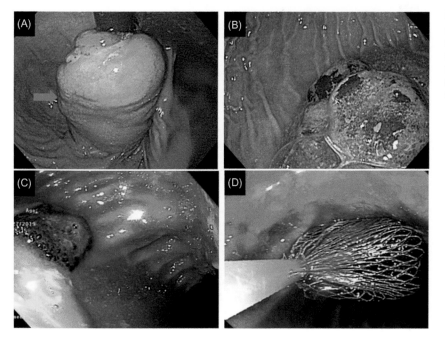

FIGURE 37.10 Fecal bezoars in the Kock pouch. (A−C) The presence of the surgically constructed nipple valve structure (*green arrow*) makes the patients prone to the retention of solid materials in the pouch body and (D) smaller fecal bezoars may be removed endoscopically with tools, such as the net.

Dislodged or misfired staples

Stapling techniques are routinely used for IBD surgeries. Dislodged staples may lead to various conditions, including bleeding, anemia, and pain (Figs. 37.12 and 37.13). In patients with IPAA, the dislodged or misfired staples may cause vaginal fistula or cuffitis (Fig. 37.14).

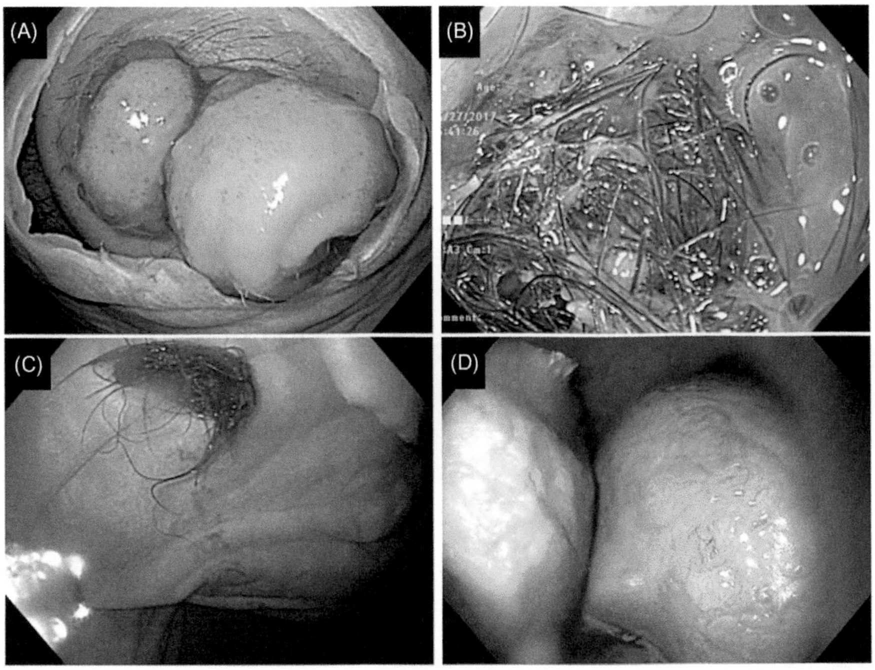

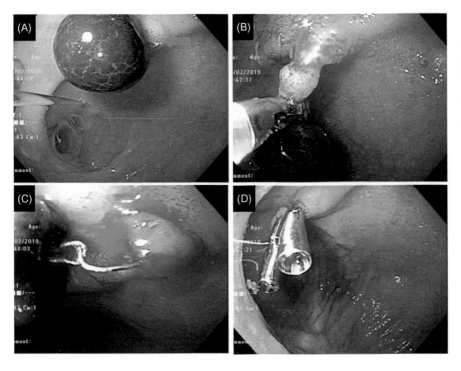

FIGURE 37.11 Bezoars in the diverted bowel. (A—C) A loop ileostomy (A) was created for a patient with Crohn's disease of the ileal pouch. Trichobezoars were found in and removed from the diverted pouch; (D) fecal bezoars found in long-standing diverted colon in a patient with refractory perianal Crohn's disease.

FIGURE 37.12 Staple induced bleeding and chronic iron-deficiency anemia. The 32-year-old male patient with an ileal pouch presented with intermittent hematochezia with unknown etiology. (A—D) Endoscopic removal of a dislodged staple along the mid-pouch staple line resulted in the resolution of her bleeding and iron-deficiency anemia.

Summary and recommendations

Patients with IBD-associated structural complications or surgeries may develop bezoars or the retention of foreign bodies. Patients with strictures, anastomoses, strictureplasties, or continent ileostomies may have an increased risk. Dislodged or misfired staples may lead to bleeding, anemia, pain, and even bowel inflammation. Endoscopic evaluation and management are recommended, particularly in symptomatic patients.

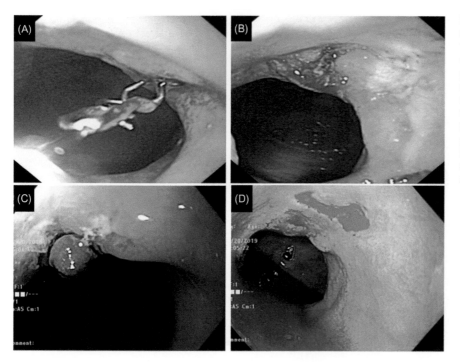

FIGURE 37.13 Dislodged staples in the ileal pouch–anal anastomosis. (A and B) The 35-year-old male patient presented with dyschezia and intermittent sharp anal pain. He had a short cuff with dislodged staples at the anterior wall of the anastomosis. Endoscopic removal of the dislodged staples resulted in the resolution of his symptoms; (C and D) the 50-year-old patient with an ileal pouch presented with intermittent bleeding, which was resolved after endoscopic removal of a dislodged staple at the anastomosis.

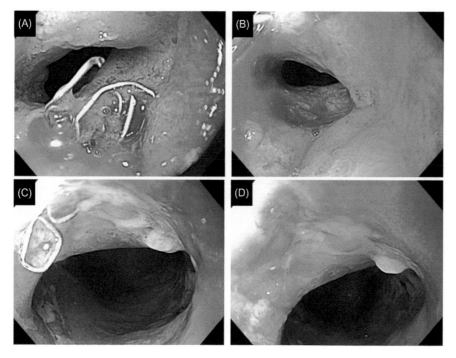

FIGURE 37.14 Misfired staples at the ileal pouch–anal anastomosis. (A and B) The 36-year-old female patient presented with an intermittent vaginal discharge of liquid fecal materials and gas. The symptoms resolved after endoscopic removal of misfired staples at the anterior wall of anastomosis; (C and D) cuffitis associated with a misfired staple and after the endoscopic removal of the staple.

Grant support

None.

Disclosure

Dr. Bo Shen is a consultant for Abbvie, Jansen, and Takeda.

References

[1] Andrus CH, Ponsky JL. Bezoars: classification, pathophysiology, and treatment. Am J Gastroenterol 1988;83:476−8.

[2] Fernandez-Urien I, Carretero C, Gonzalez B, Pons V, Caunedo A, Valle J, et al. Incidence, clinical outcomes, and therapeutic approaches of capsule endoscopy-related adverse events in a large study population. Rev Esp Enferm Dig 2015;107:745−52.

[3] Lian L, Fazio V, Shen B. Endoscopic treatment for pill bezoars after continent ileostomy. Dig Liver Dis 2009;41:e26−8.

[4] Wu XR, Mukewar S, Kiran RP, Hammel JP, Remzi FH, Shen B. The presence of primary sclerosing cholangitis is protective for ileal pouch from Crohn's disease. Inflamm Bowel Dis 2013;19:1483−9.

[5] Harrington S, Mohamed S, Bloch R. Small bowel obstruction by a primary phytobezoar in Crohn's disease. Am Surg. 2009;75:93−4.

[6] Nabeel-Zafar S, Traverso P, Asare M, Romero N, Changoor NR, Hughes K, et al. Small bowel perforation subsequent to mushroom bezoar as a presentation of Crohn's disease. Am Surg. 2013;79:E278−80.

Chapter 38

Macroscopic examination of surgically resected specimens in inflammatory bowel disease

Xianrui Wu and Ping Lan

Department of Colorectal Surgery, The Sixth Affiliated Hospital of Sun Yat-Sen University, Guangzhou, P.R. China

Chapter Outline

Abbreviations

CD Crohn's disease
GI gastrointestinal
IBD inflammatory bowel disease
IPAA ileal pouch—anal anastomosis
ITB intestinal tuberculosis
UC ulcerative colitis

Introduction

Inflammatory bowel disease (IBD) exemplified by Crohn's disease (CD) and ulcerative colitis (UC) describes chronic inflammatory conditions of the gastrointestinal (GI) tract characterized by intermittent relapse and remission. Although the precise etiology remains unclear, it is believed to arise in genetically susceptible individuals exposed to environmental exposures, resulting in an abnormal immune response to the intestinal microbiome. Historically, IBD was considered as diseases of residents from highly affluent Western countries. These notions have been shattered since recent epidemiological studies have demonstrated that the development of IBD is not limited by geographical borders, ethnicity, or socioeconomic status [1].

Patients with IBD may present with bloody diarrhea, abdominal pain, urgency, tenesmus, fatigue, weight loss, and fever. Since the clinical manifestations are variable and nonspecific, the diagnosis of IBD can be difficult in some cases. The extensive differential diagnosis of IBD varies with the site of involvement and the chronicity of the clinical presentation. It includes but is not limited to ischemic colitis, lymphoma, intestinal tuberculosis (ITB), medication-associated colitis, and infectious colitis.

Crohn's disease

CD is a progressive disease with a destructive and disabling course over time. The cycle of inflammation and tissue repair can cause fibrosis and consequently obstructive intestinal strictures Depending on severity, transmural lesions

Atlas of Endoscopy Imaging in Inflammatory Bowel Disease. DOI: https://doi.org/10.1016/B978-0-12-814811-2.00038-4
© 2020 Elsevier Inc. All rights reserved.

can become fistulas and abscesses. It may involve the entire GI tract from mouth to the perianal area. The aim of treatment strategies is to induce and maintain remission with long-term goals of preventing complications and halting the progressive course of disease [2]. Since surgery is not curative, medical therapy remains the mainstay of treatment for most patients with CD. The selection of medications is guided by disease severity and response to previous therapies. Endoscopic therapy such as stricturotomy is emerging to become a main treatment modality for mechanical complications of IBD.

Approximately one in two patients with CD is expected to have at least one surgical procedure once diagnosed. Patients who have persistent symptoms despite best medical management or who develop complications (e.g., perforation, abscess, fistula, hemorrhage, stricture, fistula, or neoplasm) and children who suffer from growth retardation are candidates for operative management. Occasionally, severe colonic disease, in combination with perianal sepsis, requires bowel diversion for symptom control before medications can be used safely. Common procedures used to treat CD patients include bowel resection and strictureplasty with or without the construction of a diverting stoma [2,3]. The choice among procedures depends upon the indication for operative intervention and the locations of the disease. The decision to operate should be discussed within a multidisciplinary team and should include appropriate preoperative imaging, patient counseling, optimization of nutritional status, and prophylaxis for thromboembolic events.

The most common surgical modalities for small-bowel and/or large-bowel CD are ileocolonic resection, small bowel resection, and partial colectomy, with ileoileal, ileocolonic, colo-colonic, or colorectal anastomoses. Those procedures may be replaced with or performed in combination with stricturoplasty or diverting ostomy. Bowel resection may be performed for a medically refractory inflammatory phenotype of CD (Fig. 38.1). More often, bowel resection and anastomosis are performed in patients with structural complications, such as strictures (Figs. 38.2−38.4) and fistulae (Figs. 38.5 and 38.6). The progression of disease phenotypes often starts with an inflammatory disorder and ends with fibrostenotic and/or fistulizing entities. Concurrent inflammation is even present in late-stage disease. Surgically resected bowel specimens often show creeping fat around the mesenteric edge (Figs. 38.1−38.4). Wider resection of mesentery with the fat may provide a better surgical outcome of ileocolonic resection for CD [4].

Ulcerative colitis

UC is characterized by recurring episodes of inflammation limited to the mucosal layer of the colon, which commonly starts in the rectum and can extend to proximal segments of the colon in a continuous fashion. UC is primarily treated with medicines, with the most widely used drugs being 5-aminosalicylates, corticosteroids, immunomodulators, and antitumor necrosis factor, antiintegrin, antiinterleukin, or anti−Janus kinase agents. The choice of medication depends on disease location, severity and extent. Current therapeutic strategies aim for inducing and maintaining remission with the goal to prevent disability, colectomy, and colorectal cancer [5].

It is estimated that up to 30% of patients with UC will eventually require a surgical intervention. Colectomy is indicated in UC patients with medically refractory disease or colitis-associated neoplasia or who develop life-threatening

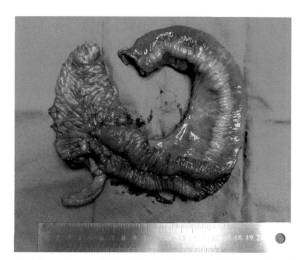

FIGURE 38.1 Inflammatory phenotype of Crohn's disease with right hemicolectomy. Multiple fissuring ulcers, polypoid nodules located at the terminal ileum, and creeping fat along the mesentery edge (*green arrow*).

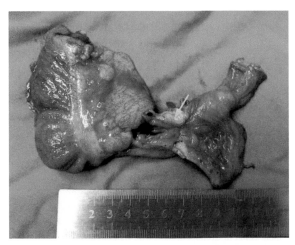

FIGURE 38.2 Fibrostenotic phenotype of CD of the terminal ileum with ileocecal resection. The stricture was 3 cm in length at the terminal ileum, 5 cm from the ileocecal valve (*yellow arrow*). *CD*, Crohn's disease.

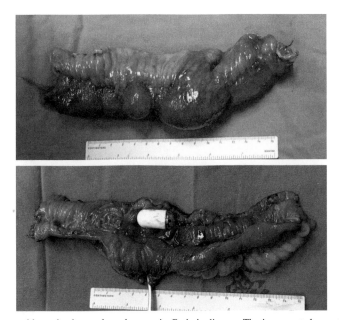

FIGURE 38.3 Small bowel stricture with retained capsule endoscope in Crohn's disease. The incarcerated capsule was found at the site of stricture located at 150 cm from the ileocecal valve proximately. The patient underwent partial small bowel resection after failed endoscopic retrieval.

complications such as colonic perforation, uncontrolled hemorrhage, and toxic megacolon. Restorative proctocolectomy with ileal pouch—anal anastomosis (IPAA) has become the surgical treatment of choice for most patients with UC. The whole procedure can be completed in a one-stage, two-stage, or three-stage fashion, which should be decided on a case-by-case basis. When surgery is elective, it is typically done in two stages starting with a total proctocolectomy with the creation of an IPAA as well as a temporary protective loop ileostomy (the first stage). The protective ileostomy is subsequently taken down in the second operation (the second stage). However, for emergency or urgent indications for severe or fulminant, a subtotal colectomy and Hartmann procedure along with an end ileostomy is typically recommended as the first stage to decrease the risk of immediate postoperative complications such as an anastomotic leak or pelvic sepsis. The IPAA is then created and anastomosed to the anal canal or a short rectal stump with a diverting ileostomy (the second stage), which is eventually reversed to restore intestinal continuity (the third stage) [6].

Macroscopic as well as microscopic examination of surgically resected specimens are important for intra-, immediate post-, and postoperative management (Figs. 38.7—38.9). UC uniformly involves the rectum, which may extend proximally in some patients. Patients with long-standing UC with prior exposure to topical therapy, such as mesalamine

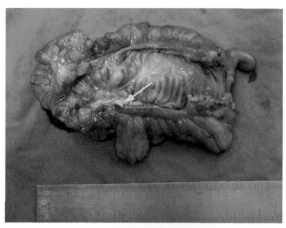

FIGURE 38.4 Primary stricture at the sigmoid colon in Crohn's disease with sigmoid colectomy.

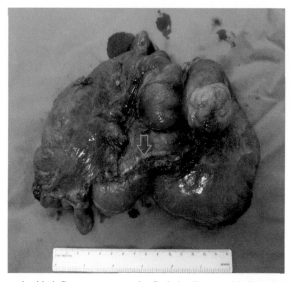

FIGURE 38.5 Ileocolonic fistula presented with inflammatory mass in Crohn's disease with ileocolonic resection (*green arrow*). The disease involved 50 cm of the terminal ileum as well as the ileocecal valve.

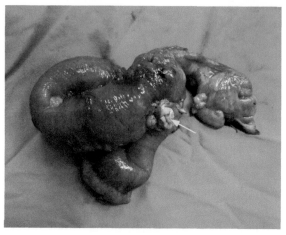

FIGURE 38.6 Entero-cutaneous fistula in Crohn's disease with small bowel resection (*yellow arrow*).

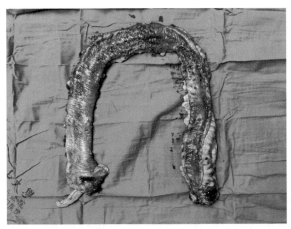

FIGURE 38.7 Subtotal colectomy with a temporary ileostomy for severe ulcerative colitis. The patient had massive hemorrhage refractory to medical treatment.

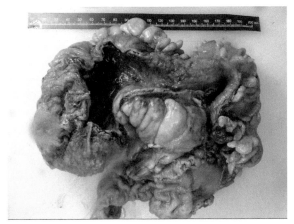

FIGURE 38.8 The patient was diagnosed with high-grade dysplasia on the background of ulcerative colitis. Restorative proctocolectomy with ileal pouch−anal anastomosis and a diverting ileostomy was performed. Flat high-grade dysplasia was also detected by microscopic examination on the surgically resected specimen.

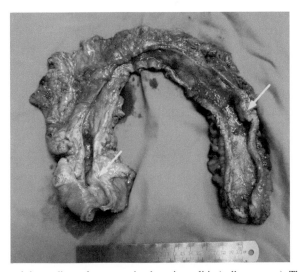

FIGURE 38.9 Simultaneous ascending and descending colon cancer in ulcerative colitis (*yellow arrows*). The patient underwent subtotal colectomy as the first stage of restorative proctocolectomy and ileal pouch−anal anastomosis procedure.

or corticosteroid suppositories, enemas, and foams, may show a relative sparing of the rectum (Fig. 38.10). This should be distinguished from segmental involvement of the large bowel with true rectal sparing in Crohn's colitis. Furthermore, features of incidental CD, such as mesenteric fat creeping and fistula, may be detected in the macroscopic examination in patients with a preoperative diagnosis of UC (Fig. 38.11). The patients with diagnosis of incidental CD are at risk for the development of CD of the ileal pouch (Fig. 38.12), even pouch failure [7]. Finally, careful macroscopic and microscopic examination of colectomy specimens may identify incidental dysplastic or neoplastic lesions, which might have been missed in routine preoperative surveillance colonoscopy. UC patients with a pre- or intraoperative diagnosis of colitis-associated neoplasia have an increased risk for pouch neoplasia [8,9]. The majority of patients with pouch neoplasia have the lesions detected in the anal transition zone or cuff. Some of the patients may not have endoscopically visible lesions as they have laterally spreading neoplastic lesions (Fig. 38.13).

Inflammatory bowel disease—like conditions

The various inflammatory, infectious, vascular, and neoplastic lesions can mimic clinical presentations of IBD, especially CD. The differential diagnosis is ideally made before surgery. Despite advances in endoscopy and imaging

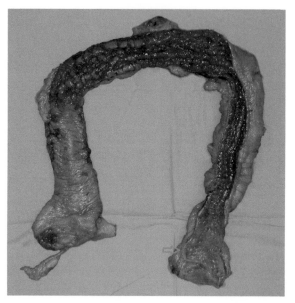

FIGURE 38.10   Proctocolectomy for refractory ulcerative colitis. Severe disease was mainly located at the proximal descending colon, whole transverse colon, and distal ascending colon, with a mild inflammation the rectum (*green arrow*), which was related to a long-term use of topical antiinflammatory agents.

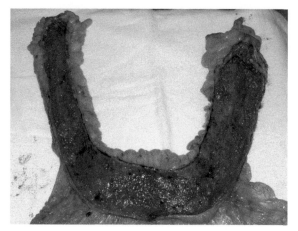

FIGURE 38.11   Colectomy for refractory UC. There can be fat creeping in severe UC, which is not limited to the mesenteric edge, in contrast to Crohn's disease. *UC*, Ulcerative colitis.

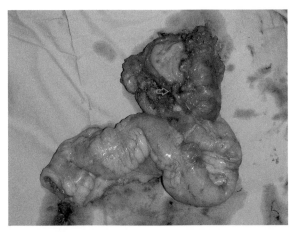

FIGURE 38.12 Excised ileal pouch for CD with severe stricture and transmural disease (*green arrow*). The patient underwent restorative proctoco-lectomy and ileal pouch−anal anastomosis for refractory ulcerative colitis. Postoperatively, he was incidentally found to have CD with mesenteric fat creeping and transmural granulomas in the frozen section. Subsequently, he developed CD of the pouch with ulcers and strictures refractory to medical and endoscopic therapy and had pouch excision and permanent diversion. Pouch failure is defined as a condition in which the pouch is excised or per-manently diverted with creation of an ileostomy. *CD*, Crohn's disease.

FIGURE 38.13 Excised ileal pouch for adenocarcinoma. Notice that the tumor at the anal transition zone was a lateral spreading type, with no visi-ble lesion on the surface of the anal transition zone (*green arrow*). The patient had had colitis-associated cancer before restorative proctocolectomy.

technology, surgical intervention may still be needed for diagnosis, differential diagnosis, and treatment. For example, intestinal Behçet's disease can present with ulcers, strictures, fistula, and even perforation [10] (Figs. 38.14 and 38.15). Ischemic bowel may present with stricture, necrosis, or perforation (Fig. 38.16).

ITB has posed a great challenge on the differential diagnosis of CD, particularly in an endemic area (Fig. 38.17). Surgical intervention may be needed to treat disease-associated complications, such as obstruction, fistula, abscess, as well as the differential diagnosis.

Neoplastic lesions of the gastrointestinal tract can also mimic IBD such as polyposis (Fig. 38.18), intestinal lym-phoma (Figs. 38.19 and 38.20), and sarcoma (Fig. 38.21). Of note, long-standing CD or UC is associated with a risk for the development of adenocarcinoma (Fig. 38.22).

Summary and recommendations

IBD, including CD and UC, describes chronic inflammatory conditions of the GI tract. Medical therapy is the mainstay of treatment for most patients with IBD. However, approximately 50% of CD patients and 20%−30% of UC patients will eventually require surgery. IBD patients, who are not amenable to medical treatment, or who develop complica-tions or colitis-associated neoplasia, are candidates for surgery. Common procedures applied to CD patients are bowel

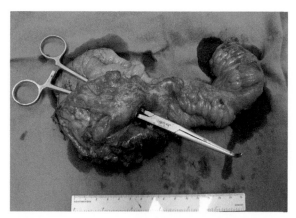

FIGURE 38.14 Behçet's disease involving the terminal ileum and cecum with ileocecal resection. Chronic perforation to the posterior peritoneum was identified.

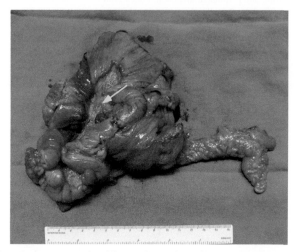

FIGURE 38.15 Behçet's disease at ileocolonic anastomosis with resection and neo-ileocolonic anastomosis. A large, deep, well-demarcated, and clean-based ulcer was developed at the previous ileocolic anastomosis (*yellow arrow*).

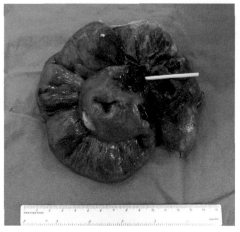

FIGURE 38.16 Perforated ischemic small bowel with ileal resection. An inflammatory and necrotic mass was formed involving 30 cm of the distal ileum at a distance of 10 cm from the ileocecal valve. Ischemic bowel can mimic presentations of Crohn's disease.

FIGURE 38.17 Intestinal tuberculosis at the ileocecum, as well as the ascending colon, with right hemicolectomy. The disease was presented with inflammatory mass.

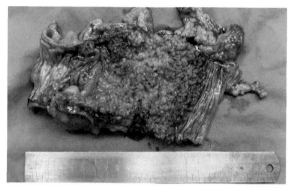

FIGURE 38.18 Polyposis of colon presented with a mass and partial bowel obstruction. The lesion can mimic pseudopolyps from chronic colitis. The patient underwent right hemicolectomy and surgical pathology showed hyperplastic polyps.

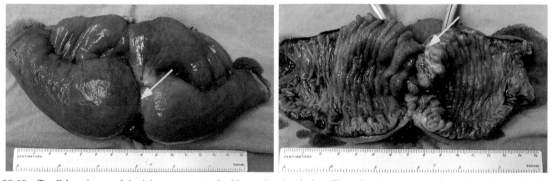

FIGURE 38.19 T-cell lymphoma of the jejunum presented with a stricturing lesion. The stricturing mass was located at 100 cm distal to the ligament of Treitz. The patient underwent partial jejunectomy and surgical pathology showed monomorphic epithelial intestinal T-cell lymphoma.

resection and strictureplasty with or without the construction of a diverting stoma. Restorative proctocolectomy with IPAA has become the surgical treatment of choice for most patients with UC. Macroscopic along with microscopic examination of surgically resected specimens are important for the diagnosis, differential diagnosis, prediction of outcome, and postoperative management of IBD. In patients with inconclusive preoperative diagnosis and those with non-classic features on macroscopic examination, on-site frozen section of surgical specimens is recommended. For endoscopists, solid knowledge on the macroscopic features of surgical specimens may help improve their skills in endoscopic diagnosis, dysplasia surveillance, and therapeutic intervention.

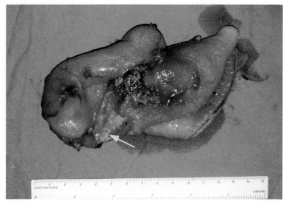

FIGURE 38.20 B-cell lymphoma of the ileum. The obstructing mass was located at 40 cm proximal to the ileocecal valve (*yellow arrow*). The patient underwent partial small bowel resection; and surgical pathology showed diffuse large B-cell lymphoma.

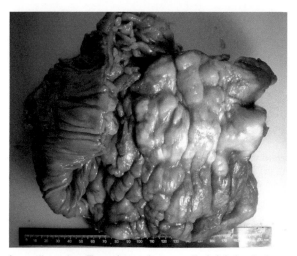

FIGURE 38.21 Granulocytic sarcoma of ascending colon. The patient underwent radical right hemicolectomy.

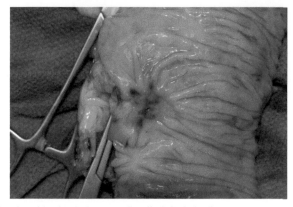

FIGURE 38.22 Adenocarcinoma of the sigmoid colon in long-standing ileal Crohn's disease. Isolated, irregular, deep, and malignant ulcer (*green arrow*) was found in the absence of mucosal inflammation in the adjacent area.

Disclosure

The authors declared no financial conflict of interest.

References

[1] Kaplan GG, Ng SC. Globalisation of inflammatory bowel disease: perspectives from the evolution of inflammatory bowel disease in the UK and China. Lancet Gastroenterol Hepatol 2016;1:307–16.

[2] Torres J, Mehandru S, Colombel JF, Peyrin-Biroulet L. Crohn's disease. Lancet 2017;389:1741–55.

[3] Gionchetti P, Dignass A, Danese S, Magro Dias FJ, Rogler G, Lakatos PL, et al. 3rd European evidence-based consensus on the diagnosis and management of Crohn's disease 2016: Part 2: surgical management and special situations. J Crohns Colitis 2017;11:135–49.

[4] Coffey CJ, Kiernan MG, Sahebally SM, Jarrar A, Burke JP, Kiely PA, et al. Inclusion of the mesentery in ileocolic resection for Crohn's disease is associated with reduced surgical recurrence. J Crohns Colitis 2018;12:1139–50.

[5] Ungaro R, Mehandru S, Allen PB, Peyrin-Biroulet L, Colombel JF. Ulcerative colitis. Lancet 2017;389:1756–70.

[6] Magro F, Gionchetti P, Eliakim R, Ardizzone S, Armuzzi A, Barreiro-de Acosta M, et al. Third European evidence-based consensus on diagnosis and management of ulcerative colitis. Part 1: Definitions, diagnosis, extra-intestinal manifestations, pregnancy, cancer surveillance, surgery, and ileo-anal pouch disorders. J Crohns Colitis 2017;11:649–70.

[7] Melton GB, Fazio VW, Kiran RP, He J, Lavery IC, Shen B, et al. Long-term outcomes with ileal pouch-anal anastomosis and Crohn's disease: pouch retention and implications of delayed diagnosis. Ann Surg 2008;248:608–16.

[8] Kariv R, Remzi FH, Lian L, Bennett AE, Kiran RP, Kariv Y, et al. Preoperative colorectal neoplasia increases risk for pouch neoplasia in patients with restorative proctocolectomy. Gastroenterology 2010;139:806–12.

[9] Derikx LA, Kievit W, Drenth JP, de Jong DJ, Ponsioen CY, Oldenburg B, , et al.Dutch Initiative on Crohn and Colitis Prior colorectal neoplasia is associated with increased risk of ileoanal pouch neoplasia in patients with inflammatory bowel disease. Gastroenterology 2014;146:119–28.

[10] Chou SJ, Chen VT, Jan HC, Lou MA, Liu YM. Intestinal perforations in Behçet's disease. J Gastrointest Surg 2007;11:508–14.

Index

9780128148112